AF323823

# STABILITY AND TIME-OPTIMAL CONTROL OF HEREDITARY SYSTEMS

## WITH APPLICATION TO THE ECONOMIC DYNAMICS OF THE US

Series on Advances in Mathematics for Applied Sciences – Vol. 60

# STABILITY AND TIME-OPTIMAL CONTROL OF HEREDITARY SYSTEMS

## WITH APPLICATION TO THE ECONOMIC DYNAMICS OF THE US

### 2nd Edition

E. N. Chukwu

North Carolina State University

World Scientific

*New Jersey • London • Singapore • Hong Kong*

*Published by*

World Scientific Publishing Co. Pte. Ltd.

P O Box 128, Farrer Road, Singapore 912805

*USA office:* Suite 1B, 1060 Main Street, River Edge, NJ 07661

*UK office:* 57 Shelton Street, Covent Garden, London WC2H 9HE

**British Library Cataloguing-in-Publication Data**
A catalogue record for this book is available from the British Library.

**STABILITY AND TIME-OPTIMAL CONTROL OF HEREDITARY SYSTEMS**
**With Application to the Economic Dynamics of the US (Second Edition)**

Copyright © 2001 by World Scientific Publishing Co. Pte. Ltd.

*All rights reserved. This book, or parts thereof, may not be reproduced in any form or by any means, electronic or mechanical, including photocopying, recording or any information storage and retrieval system now known or to be invented, without written permission from the Publisher.*

For photocopying of material in this volume, please pay a copying fee through the Copyright Clearance Center, Inc., 222 Rosewood Drive, Danvers, MA 01923, USA. In this case permission to photocopy is not required from the publisher.

ISBN  981-02-4674-9

Printed in Singapore by Uto-Print

# Dedication

It is with gratitude and great joy that I dedicate this book to the following good and courageous men and women whose faith in my family sustained us through our recent difficulties. My entire family, including my wife Regina, and our children Ezeigwe, Emeka, Uchenna, Obioma, Ndubusi, and Chika are most humbled by the friendship from so many. Our trials were very harrowing, but with the support and belief of all these fine women and men, our faith in God and belief in the principles of truth and justice were reaffirmed.

Dr. A. Schafer (Chair, Committee of American Mathematical Society, Human Rights of Mathematicians, American Mathematical Society), Dr. William Browder (Princeton University) President, American Mathematical Society), Dr. M. Brin, Professor O. Hájek, Professor G. Leitmann, Professor R. Martin, Professor A. Fauntleroy, Professor E. Burniston, Dr. G. Christensen, Mr. R. O'Connor, Bishop F. Joseph Gossman, Father O. Howard, Father G. Wilson, Mr. H. Ray Daley, Jr., Dr. J. Nathan, Mr. J. Downs, Mr. G. Berens and Mrs. Marie Berens, Professor T. Hallam, Dr. H. Liston, Dr. S. Lenhart, Dr. S. Middleton, Professor H. Petrea, Professor and Mrs. C. Wolf, Dr. G. Briggs, Dr. C. Jefferson, Reverend Fr. Doherty, Reverend Fr. J. Scannel, Mr. A. MacNair, Mr. J. Barry, Reverend M. P. Shugrue, Dr. B. Westbrook, Dr. D. M. Lotz, Ms. S. McKenna, Ms. J. Hunt, Rev. G. Lewis, Rev. M. J. Carter, Mr. and Mrs. B. E. Spencer.

My family and I are eternally grateful to the following political leaders who provided support and assistance during a very difficult time that occurred while I was in the process of developing and writing this book. Without their support and assistance I could not have completed this work, and our difficulties would have overwhelmed us.

The Honorable Terry Sanford, U.S. Senator for North Carolina.
The Honorable Jesse Helms, U.S. Senator for North Carolina.
The Honorable Albert Gore, U.S. Senator for Tennessee.
The Honorable Clairborne Pell, U.S. Senator for Rhode Island.
The Honorable David Price, U.S. Congressman for the 4th District, for North Carolina.

vi                                         *Dedication*

The Honorable Rufus L. Edmisten, North Carolina Secretary of State.

My family and I rejoice in your generosity, and thank you with all our hearts and souls. It is with humility that I pay tribute to my undergraduate teacher Professor Harold Ward and my Ph.D. supervisor and friend Professor O. Hájek.

Ethelbert Nwakuche Chukwu<br>
January, 1991<br>
Raleigh, NC, USA

# Acknowledgment
# (First Edition)

The author is indebted to Professors T. Angell and A. Kirsch for their kind permission to use their recent published paper. The author has consulted the work of Professors J. Hale, Haddock and Terjéki, T. Banks, M. Jacobs and Langenhop, A. Manitius, Drs. G. Kent and D. Salamon, to all whom he is greatly indebted. He acknowledges the enduring inspiration of Professor O. Hájek, L. Cesari, H. Hermes and J. P. LaSalle, and R. Bellman. It was Bellman who invited the author to write this book. He is particularly appreciative of the joyous efforts of Mrs. Vicki Grantham in typing the manuscript. The author acknowledges with immense gratitude the support of NSF for the summer of 1991, which enabled him to complete some of the original work reported here. It was a pleasure to work with the professional staff of Academic Press, especially Charles B. Glaser and Camille Pecoul, and the author offers his gratitude to them.

Ethelbert Nwakuche Chukwu
January 1991
Raleigh, NC. USA

# Acknowledgment
# (Second Edition)

Words of gratitude and thanks cannot suffice. Nevertheless, I thank Editors N. Bellomo, Anthony Doyle and Ye Qiang for doing such a good and wonderful job of publishing this second edition. Of course, my eternal gratitude rests with Professor Otomar Hájek on whose work on differential game my economic theory on the Wealth of Nations also rests. A big thanks to secreterial technologist Mrs Joyce Sorensen for preparing this manuscript.

# Contents

## Chapter 2    General Linear Equations     49



## Chapter 3    Lyapunov–Razumikhin Methods of Stability in Delay Equations     79



## Chapter 4    Global Stability of Functional Differential Equations of Neutral Type     97



## Chapter 5    Synthesis of Time-Optimal and Minimum-Effort Control of Linear Ordinary Systems     113



## Chapter 6    Control of Linear Delay Systems     187

# Preface

This monograph introduces the theory of stability and of time-optimal control of hereditary systems. It is well known that the future state of realistic models in the natural sciences, economics and engineering depends not only on the present but on the past state and the derivative of the past state. Such models that contain past information are called hereditary systems. There are simple examples from biology (predator-prey, Lotka–Volterra, spread of epidemics, e.g., AIDS epidemic), from economics (dynamics of capital growth of global economy), and from engineering (mechanical and aerospace: aircraft stabilization and automatic steering using minimum fuel and effort, control of a high-speed closed air circuit wind tunnel; computer and electrical engineering: fluctuations of current in linear and nonlinear circuits, flip-flop circuits, lossless transmission line). Such examples are used to study the stability and the time-optimal control of hereditary systems. Within this theory the problem of minimum energy and effort control of systems are also explored. Conditions for stability for both linear and nonlinear systems are given. Questions of controllability are posed and answered. Their application to the growth of the global economy yield broad, obvious but startling policy prescription. For the problem of finding some control to use in order to reach, in minimum time, an equilibrium point from any initial point and any initial state, existence theorems are given. Some effort is made to construct an optimal control strategy for the simple linear systems cited in the examples. For finite dimensional linear systems, time-optimal, and minimum effort feedback controls are constructed.

It is assumed that the reader is familiar with advanced calculus and has some knowledge of ordinary differential equations. Some familiarity with real and functional analysis is helpful. The contents of this book were offered as a year-long course at North Carolina State University in the 1989–1991 sessions. The students were mainly from engineering, economics, mathematical biology, and Applied Mathematics. Though in many aspects the book is self-contained, theorems often are stated and the appropriate references cited. The references should help a further exploration of the subject.

There are thirteen chapters in this book. The first chapter is devoted to examples from applications in which delays are important in time-optimal, minimum effort, and stability problems. The second chapter studies linear systems. The fundamental matrices of linear delay and linear neutral matrices are calculated and used to obtain the variation of parameter formulae for nonhomogeneous systems. The stability theory of linear and perturbed linear systems is considered.

Chapter 3 presents the basic Lyapunov and Razumikhin theories and their extensions by Hale and LaSalle and recently by Haddock and Terjéki. On the basis of these, a simple quadratic form is used to deduce a necessary and sufficient condition for stability of linear delay equations. The fourth chapter treats the global stability of functional differential equals of neutral type. The second part of the book begins in Chapter 5. It deals with the construction of minimum time and minimum effort optimal feedback control of linear ordinary differential systems. It is both an introduction and a model for such studies of linear hereditary systems. Most of the theory presented here is only available in research articles and theses. It will be most interesting to present parallel results for hereditary systems. Chapter 6 presents the theory for the underlying controllability assumptions required for the existence of optimal control of delay equations. The geometric theory of time-optimal control is then presented in Chapter 7, where numerous examples are considered. Here also the maximum principle both in Euclidean and function space is formulated. The fundamental work of Angell and Kirsch is stated: It gives a maximum principle in a setting that guarantees the existence of non-trivial regular multipliers for nonlinear problems involving point-wise control constraints. It presents an introduction to the synthesis of optimal feedback control of linear hereditary systems. The synthesis of optimal control of simple examples is attempted. In Chapter 8 controllability of nonlinear systems is treated. Interconnected systems are studied in Chapter 9. Their universal principles for the control of interconnected organizations are formulated. They have fundamental economic policy implications of immense practical interest. Chapters 10 through 12 consider the corresponding controllability and time-optimal control of linear and nonlinear neutral equations. Chapter 13 presents the stability theory of large-scale hereditary systems in the spirit of Michel and Miller.

## The World

"Should we not question the very economic models often adopted by states which also as a result of international pressures and forms of conditioning cause the aggravate the situations of injustice and violence in which the life of whole people is degraded and trampled upon?"

Pope John Paul II, Evagelium Vitae p26, Liberia Editrice 1995.

## Asia

"The promotion of sustained economic development is the 'single most important target' for the United Nations". International peace and stability will be enhanced only when all countries enjoy a minimum of standard of economic self-sufficiency and well-being."

Indian Prime Minister Inder Kumar Gujral, Annual General assembly debate in 1997.

## G-8 Summit Communiqué

"Our challenge is to build on and sustain the process of globalization and to ensure that its benefits are wide spread more widely to improve the quality of life of people everywhere. Of the major challenges facing the world on the threshold of the 21st century this summit has focused on three: Achieving sustainable economic growth and development throughout the world in a way which will enable developing countries to grow faster and reduce poverty, restore growth to emerging Asian economies and sustain the liberalization of trade in goods and services and of investments in a stable international economy.

Building lasting growth in our economies in which all can participate, creating jobs and combating social exclusion:

In an interdependent world, we must work to build sustainable economic growth in all countries."

This book provides the tools needed to understand and do research in the many of the applied areas cited. It also provides a basis for the computer aided numerical work which will help in forecasting and prediction. For example the theoretical implication in Section 10.4, Theorem 10.4.1, Theorem 10.4.3 for economic policy is now explicitly evident from the models: the solidarity function q=[T,g,e,d,M1,M1,fo], where q is defined in (1.10.75). The possibilities of solidarity, Q, the solidarity set, should be dominated by P the possibilities of private initiative, p=[Co,Io,Xo,Mo,n,w,x0,yo,po] defined (1.10.76). The controllability theorems may now be understood: they give conditions which will ensure that the initial economic state of gross national product, interest rate, employment, value of capital stock, prices (and cumulative balance of payment can be controlled to another state simultaneously subject to scarcity. The MATLAB program needed is detailed and can easily be mastered through the U.S.A. example.

# Preface to the Second Edition

The excellent review and favorable reception of the first edition and the recent G-8 affirmation of the great importance of economic growth of nations, (an application of the theory treated in this book) have inspired the author to bring out this new edition. Some minor errors contained the first edition are corrected. More detailed economic model of the growth of wealth of nations is presented as an application of hereditary systems theory. This second edition contains section 1:10 which details a careful derivation of a neutral differential dynamics as the growth of wealth of nations. It contains a computer program, an appendix that confronts the economic data of the U.S.A. and the model dynamics by identifying its coefficients. This inclusion is motivated by the recent universal opinion of the world leaders on the need for and the importance of better national economic models.

# Chapter 1

# Examples of Control Systems Described by Functional Differential Equations

## 1.1  Ship Stabilization and Automatic Steering

A ship is rolling in the waves. The differential equation satisfied by the angle of tilt $x$ from the normal upright position is given by

$$m\frac{d^2x(t)}{dt^2} + b\frac{dx(t)}{dt} + kx(t) = 0 \tag{1.1.1}$$

where $m > 0$ is the mass, $b > 0$ is the damping constant, and $k > 0$ is a constant. The damping constant $b$ determines how fast the ship returns to its upright position. It is known that if $b^2 < 4mk$, the ship is "underdamped" and the ship oscillates as it returns to its upright position. If $b^2 > 4mk$, it is "overdamped", there is no oscillation, but it returns to its upright position more slowly the larger $b$ becomes. When $b^2 = 4mk$, it is "critically damped" and upright position is achieved more rapidly. This "roll-quenching" was a very important problem tackled by engineers for ships and destroyers of the second world war. In one such research by Minorsky [5, 6] ballast tanks, partially filled with water, are introduced in each side of the ship in order to obtain the best value of $b$. A servomechanism is introduced and is designed to reduce the angle of tilt $x$, and its velocity, $\frac{dx}{dt}$, to 0 as fast as possible. What the contrivance does is to introduce an input to the natural damping of the rolling ship, a term proportional to the velocity at an earlier instant $t-h : q\dot{x}(t-h)$. The delay $h$ is present because the servomechanism cannot respond instantaneously. It takes this time delay $h$ to respond. Also introduced by the servomechanism is a control with components $(u_1, u_2)$, which yields the following equations:

$$\dot{x}(t) = y(t) + u_1(t)\,,$$

$$\dot{y}(t) = -\frac{b}{m}y(t) - \frac{q}{m}y(t-h) - -\frac{k}{m}x(t) + u_2(t)\,. \tag{1.1.2}$$

1

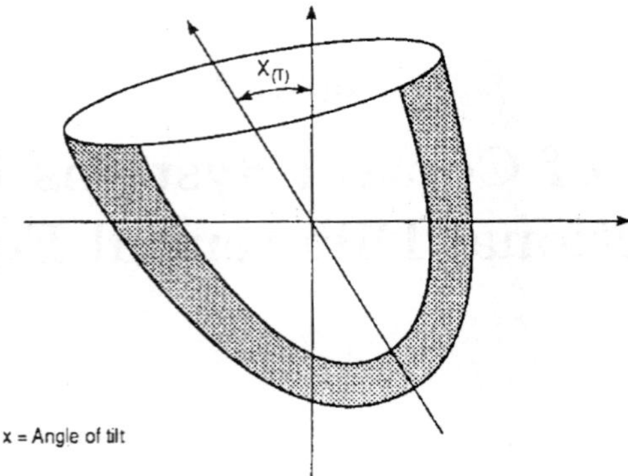


Fig. 1.1.1

We note that (1.1.1) is equivalent to

$$\dot{x}(t) = y(t)$$
$$\dot{y}(t) = -\frac{b}{m}y(t) - \frac{k}{m}x(t) .$$

(1.1.3)

Equation (1.1.2) can be written in matrix form

$$\dot{\underline{x}}(t) = A_0\underline{x}(t) + A_1\underline{x}(t-h) + Bu(t) ,$$

(1.1.4)

where

$$A_0 = \begin{bmatrix} 0, & 1 \\ -\dfrac{k}{m}, & -\dfrac{b}{m} \end{bmatrix} , \quad A_1 = \begin{bmatrix} 0 & 0 \\ 0 & -\dfrac{q}{m} \end{bmatrix} , \quad B = \begin{bmatrix} 1 & 0 \\ 0 & 1 \end{bmatrix} , \quad \underline{x} = \begin{bmatrix} x \\ y \end{bmatrix} .$$

Using matrix notation, (1.1.3) assumes the form

$$\dot{\underline{x}}(t) = A_0\underline{x}(t) + A_1\underline{x}(t-h) .$$

(1.1.5)

The point $(0, 0)$ can be regarded as an equilibrium position and the aim of the gadget is to steer the angle of tilt and its velocity to the equilibrium position. There are three problems. The first is the problem of stability: Determine necessary and sufficient conditions on $A_0$, $A_1$ such that the solution of (1.1.5) satisfies

$$\underline{x}(t) \to 0 \quad \text{as} \quad t \to \infty .$$

(1.1.6)

But it is undesirable to wait forever for the system to attain the upright position. The second problem is: Is it possible that a control $u$, introduced by the servomechanism, can drive the system to the equilibrium in finite time? Is the system controllable? The third problem is: Find a control strategy $u$ that will drive the angle of tilt and its velocity to zero in minimum time. Such a control is called a time-optimal control. Our ultimate goal in optimal control theory is to get an optimal control as a function of the appropriate state space, i.e., to obtain the "feedback" or "closed loop" control. The major advantage of such a control, as opposed to an "open loop" one with $u$ as a function of $t$, is that the system in question becomes self-correcting and automatic.

## 1.2  Predator-Prey Interactions

Let $x(t)$ be the population at time $t$ of a species of fish called prey, and let $y(t)$ be the population of another species called the predator, which lives off the prey. Under the assumption that without the predator present the prey will increase at a rate proportional to $x(t)$, and that the feeding action of the predator reduces the growth rate of the prey by an amount proportional to the product $x(t)y(t)$, we have

$$\dot{x}(t) = a_1 x(t) - b_1 x(t)y(t) \,.$$

If the predator eats the prey and breeds at a rate proportional to its number and the amount of food available, then

$$\dot{y}(t) = -a_2 y(t) + b_2 x(t)y(t) \,,$$

where $a_1, a_2, b_1, b_2$ are positive constants. The system of two equations

$$\dot{x}(t) = a_1 x(t) - b_1 x(t)y(t) \,,$$
$$\dot{y}(t) = -a_2 y(t) + b_2 x(t)y(t) \,, \tag{1.2.1}$$

is rather naive. A more realistic model assumes that the birthrate of the prey will diminish as $x(t)$ grows because of overcrowding and shortage of available food. In this model it is assumed that there is a time delay of period $h$ for the predator to respond to changes in the sizes of $x$ and $y$. Thus

$$\dot{x}(t) = a_1 \left[ 1 - \frac{x(t)}{p} \right] x(t) - b_1 x(t)y(t) \,,$$
$$\dot{y}(t) = -a_2 y(t) + b_2 x(t-h)y(t-h) \,, \tag{1.2.2}$$

where $p$ is a positive constant.

Volterra [9] studied (1.2.1) and (1.2.2) and various generalizations of (1.2.2). One such system is given by

$$\dot{x}(t) = \left[ a_1 - c_1 x(t) - b_1 y(t) - \int_{-h}^{0} g_1(s)y(t+s)ds \right] x(t),$$
$$\dot{y}(t) = \left[ -a_2 - c_2 y(t) + b_2 x(t) + \int_{-h}^{0} g_2(s)x(t+s)ds \right] y(t) \,, \tag{1.2.3}$$

where $g_1, g_2$ are continuous, nonnegative functions, and $c_1, c_2$ are constants.

In the interaction between the prey and the predator it is important to ask whether there are equilibrium states that may be reached for the systems (1.2.3).

If these states are not zero, then neither the predator nor the prey is extinct and the following is true:

$$a_1 - c_1 x(t) - b_1 y(t) - \int_{-h}^{0} g_1(s)y(t+s)ds = 0\,,$$

$$a_2 - c_2 y(t) + b_2 x(t) + \int_{-h}^{0} g_2(s)x(t+s)ds = 0\,.$$

The function $z^* = (x^*, y^*)$, which solves the above functional equation, is the equilibrium; it may well be the saturation levels of the given species. In this case it may be desirable that every solution $(x(t), y(t)) = z(t)$ of (1.2.3) satisfies $z(t) \to z^*$ as $t \to \infty$. This is an asymptotic stability property. The state $z^*$ is not attained in finite time. A more desirable objective is to "manage" the interaction and to have the population as near as possible to its equilibrium position, and to prevent near periodic outbreaks of predator population $y$ beyond its equilibrium. We aim at coexistence of the two species at their equilibrium states, which are also the saturation levels of the two populations.

One management strategy is fishing. Man is interested in harvesting one or both species at some rates $e_i$, $i = 1, 2$. For this situation the dynamics of the interaction are

$$\dot{x}(t) = \left[a_1 - c_1 x(t) - b_1 y(t) - \int_{-h}^{0} g_1(s)y(t+s)ds\right] x(t) - e_1(t)x(t)\,,$$

$$\dot{y}(t) = \left[-a_2 - e_2 y(t) + b_2 x(t) + \int_{-h}^{0} g_2(s)x(t+s)ds\right] y(t) - e_2(t)y(t)\,. \tag{1.2.4}$$

If $u_1(t) = e_1 x(t)$, $u_2(t) = e_2 y(t)$, then $u_i(t)$ is the harvest rate that is proportional to the population density. The effort level $e_i(t)$ is a positive function with dimension 1/time, which measures the total effort at time $t$ made to harvest a given species, and $u_i$ is a piecewise continuous function $[0, T] \to [0, b]$, $T > 0$, $b > 0$. Thus $e_i(t)$ can be considered a control function. The function $u = (u_1, u_2)$ represents the rate the fisherman

   (i) selectively kills the prey $x$,
  (ii) selectively kills the predator $y$, or
 (iii) kills both $x$ and $y$.

There are two problems. By harvesting, what conditions on the systems' parameters ensure that the two species can be driven to the equilibrium $z^*$ in finite time. This is a problem of controllability. If $e_i(t)$ is negative in (1.2.4), $u$ can be said to represent the rate at which laboratory-reared

   (i) fishes $x$, or
  (ii) predators $y$

are released into the system.

One can use a combination of release and harvesting strategies to drive $x, y$ to the saturation level $z^*$ in finite time.

The second question is optimality: What is the best harvesting strategy that will, as quickly as possible, drive the system from any initial population to this equilibrium? Good fish management is interested in the time-optimal control problem for the system (1.2.4).

A more general version of this system is given by

$$\dot{z}(t) = L(t, z_t)z(t) + B(z(t), t)e(t) \,,$$

where $L(t, \phi)$ is a $2 \times 2$ matrix function, which is linear in $z(t)$; and $B(z(t), t)$ is the matrix

$$\begin{bmatrix} \beta_1(x(t), t) & 0 \\ 0 & \beta_2(y(t), t) \end{bmatrix} \,.$$

A more general $n$-dimensional version of this system is given by

$$\dot{x}(t) = L(t, x_t)x(t) + B(x(t), t)u(t) \,, \tag{1.2.5}$$

where $B$ is an $n \times m$ matrix function, $u$ is in $E^m$, and

$$L(t, x_t) = \sum_{k=0}^{\infty} A_k(t)x(t - w_k) + \int_{-h}^{0} A(t, s)x(t + s) \,, \qquad 0 \le w_k \le h \,.$$

Here $A_k$ is an $n \times n$ continuous matrix function, $A(t, s)$ is integrable in $s$ for each $t$, and there is a function $a(t)$ that is integrable such that

$$\left| \int_{-h}^{0} A(t, s)\phi(s)ds \right| \le a(t)\|\phi\| \,.$$

The three questions that we posed for (1.2.4) can also be asked for (1.2.5) or for the general nonlinear equation

$$\dot{x}(t) = f(t, x_t) + B(t, x_t)u(t) \,, \tag{1.2.6}$$

where $f : E \times C \to E^n$, $B : E \times C \to E^{n \times m}$ is an $n \times m$ matrix valued function, and $u$ is a $m$-vector valued function. Here $E = (-\infty, \infty)$, $E^n$ is the $n$-dimensional Euclidean space, and $C = C([-h, 0], E^n)$ is the space of continuous functions from $[-h, 0] \to E^n$. The symbol $x_t \in C$ is a function defined $x_t(s) = x(t + s)$ $s \in [-h, 0]$.

In the discussion above we required all populations to be driven to the equilibrium, which can be taken as a target. A more general target can be assumed to be

$$T = \{\phi \in C : H\phi = \rho\} \,. \tag{1.2.7}$$

Here $H$ is a linear operator and $\rho \in C$. The target represents the final configuration at which it is desirable for the species to lie after harvesting.

## 1.3   Fluctuations of Current

Let us consider an electric circuit in which two resistances, a capacitance and inductance, are connected in series. Assume that current is flowing through the loop, and its value at time $t$ is $x(t)$ amperes. We use also the following units: volts for the voltage, ohms for the resistance $R$, henry for the inductance $L$, farads for the capacitance $c$, coulombs for the charge on the capacitance, and seconds for the time $t$. It is well known that with these systems of units, the voltage drop across the inductance is $L\frac{dx(t)}{dt}$, and that across the resistances it is $(R+R_1)x(t)$. The voltage drop across the capacitance is $q/c$ where $q$ is the charge on the capacitance. It is also known that $x(t) = \frac{dq}{dt}$. A fundamental law of Kirchhoff states that the sum of the voltage drops around the loop must be equal to the applied voltage:

$$L\frac{dx}{dt} + (R + R_1)x(t) + \frac{1}{c}q = 0\,.$$

On differentiating with respect to $t$ we deduce

$$L\frac{d^2x(t)}{dt^2} + (R + R_1)\frac{dx}{dt} + \frac{1}{c}x(t) = 0\,. \tag{1.3.1}$$

In Fig. 1.3.2, the voltage across $R_1$ is applied to a nonlinear amplifier $A$. The output is provided with a special phase-shifting network $P$. This introduces a constant time lag between the input and output $P$. The voltage drop across $R$ in series with the output $P$ is

$$e(t) = qg(\dot{x}(t - h))\,;$$

$q$ is the gain of the amplifier to $R$ measured through the network. The equation becomes

$$L\frac{d^2x(t)}{dt^2} + R\dot{x}(t) + qg(\dot{x}(t - h)) + \frac{1}{c}x(t) = 0\,.$$

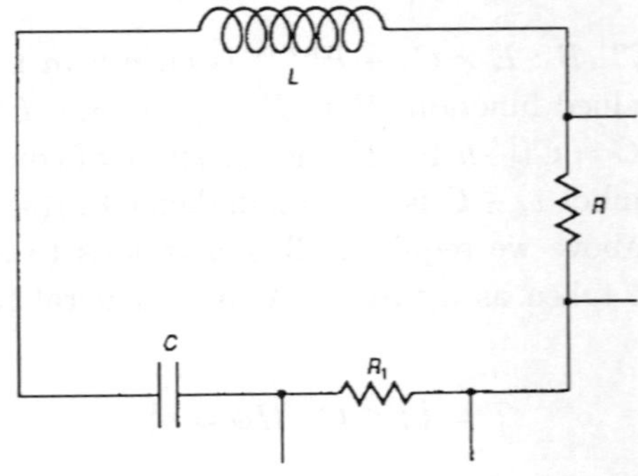


Fig. 1.3.1

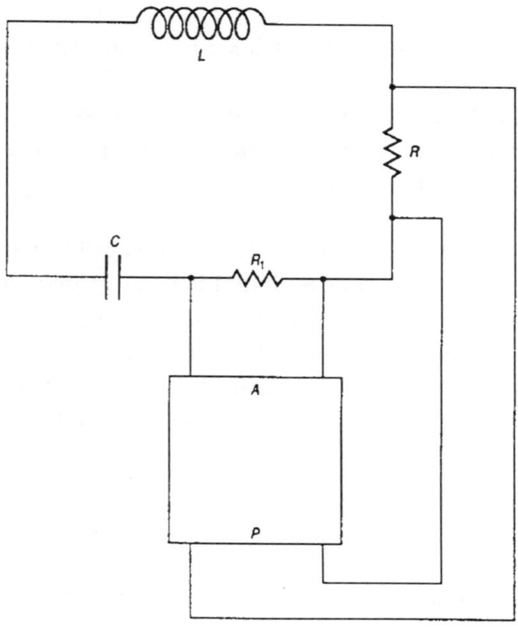


Fig. 1.3.2

Finally, a control device is introduced to help stabilize the fluctuations of the current. If $\dot{x}(t) = y(t)$, the "controlled" system may be described by

$$\dot{x}(t) = y(t) + u_1(t)\,,$$

$$\dot{y}(t) = -\frac{R}{L}y(t) - \frac{q}{L}g(y(t-h)) - \frac{1}{cL}x(t) + u_2(t)\,. \tag{1.3.2}$$

The control $u = (u_1, u_2)$ is "created" and introduced by the stabilizer.

The three basic questions are now natural: With $u = 0$, what are the properties of the systems parameter which will ensure that $x^2(t) + y^2(t) \to 0$ as $t \to \infty$? Will "admissible controls" $u$ (say $|u_j| \leq 1$, $j = 1, 2$) bring any wild fluctuations of current (any initial position) to a normal equilibrium in finite time? Can this be done in minimum time?

## 1.4  Control of Epidemics

### 1.4.1  *First model*

In this subsection we formulate a theory of epidemics, a problem of time-optimal control theory. It is a slight modification of Bank's presentation of the work of Gupta and Rink [2]. We assume as basic that there are four types of individuals in our population:

(i)   susceptibles: $X_1(t)$,
(ii)  exposed and infected but not yet infectious (infective) individuals: $X_2(t)$,
(iii) infectives: $X_3(t)$, and
(iv)  removed individuals: $X_4$. We include in (iv) those who have recovered and gained immunity and those who have been removed from the population because of observable symptoms.

We assume a constant rate $A$ of inflow of new susceptible members. The latent period $h_1$ is the period from the time an individual is exposed and becomes infected to the time he becomes infective. The infectious period $h_2$ is the period the individual is in the infectious (infective) class. The incubation period will denote the time from exposure and infectedness to the time of removal (i.e., $h_1 + h_2 = \tau$). After normalization by setting $x_i = X_i/N$, where $N$ is the population, the dynamics of interaction is

$$
\begin{aligned}
\dot{x}_1(t) &= -\beta x_1(t)x_3(t) + A\,, \\
\dot{x}_2(t) &= \beta x_1(t)x_3(t) - \beta x_1(t - h_1)x_3(t - h_1)\,, \\
\dot{x}_3(t) &= \beta x_1(t - h_1)x_3(t - h_1) - \beta x_1(t - h_1 - h_2)x_3(t - h_1 - h_2)\,, \\
\dot{x}_4(t) &= \beta x_1(t - h_1 - h_2)x_3(t - h_1 - h_2) - A\,.
\end{aligned}
\tag{1.4.1}
$$

The constant $\beta$ is the average number of individuals per unit time that any individual will encounter, sufficient to cause infection. To control the epidemic we allow two strategies:

(i)   The removal at some rates $u_i$ $i = 2, 3$ of both the infectives and exposed and infected individuals. The removal rate $u_i(t) = e_i(t)b_i(x_i(t))$ is proportional to a function $b_i$ of the population density and the effort level $e_i(t)$.
(ii)  Active immunization, the injection of dead or live but attenuated disease microorganisms into members of the population, resulting in antibodies in the population of vaccinated individuals. If $E_1(t)$ represents the reliable rate per day at which members of the population are being actively vaccinated at time $t$, the normalized rate is

$$
e_1(t) = E_1(t)/N\,.
$$

We assume that the effective immunization rate is $b_1(x_1(t))e_1(t) = u_1(t)$. If we assume that there is a delay $h$ ($h_i \le h$, $i = 1, 2$) days before the antibodies become effective, then the rate is $b_1(x_1(t))e_1(t - h) = u_1(t)$. Thus the dynamical system is

$$
\begin{aligned}
\dot{x}_1(t) &= -\beta x_1(t)x_3(t) - b_1(x_1(t))e_1(t - h) + A\,, \\[4pt]
\dot{x}_2(t) &= \beta x_1(t)x_3(t) - \beta x_1(t - h_1)x_3(t - h_1) - u_2(t)\,, \\[4pt]
\dot{x}_3(t) &= \beta x_1(t - h_1)x_3(t - h_1) - \beta x_1(t - h_1 - h_2)x_3(t - h_1 - h_2) - u_3(t)\,.
\end{aligned}
\tag{1.4.2}
$$

We use control strategies (i) and (ii) to reduce the total number of infected to an "acceptable" size in finite time. This health policy may be used to "prevent an epidemic" in minimum time, where preventing an epidemic means that the solution of (1.4.2) satisfies

(i) $|x_2(t)| + |x_3(t)| \leq A$ for all $t \geq 0$.
(ii) $\max_{s \in [0,T]} |x_3(s)| \leq B_1$ where $A_1 B_1 A$, $B$ are prescribed constants.

The solution of (1.4.2) may be viewed as lying in the space $C([-h, 0], E^3)$ the space of continuous functions mapping $[-h, 0]$ into $E^3$ with the sup norm. Or it may be viewed as lying in $E^3$. We define a subset $T \subset E^3$ by

$$T = \{x \in E^3 : |x_2| + |x_3| \leq A_1 ,\ |x_3| \leq B\} .$$

The best health policy may be that solutions of (1.4.2) hit the target, $T$, in minimum time. Also desirable is to have solutions of (1.4.2) to reach and remain in $T$, and to do so as fast as possible.

The time-optimal control problem formulated in relation to the epidemic models above can also be formulated for the more general nonlinear system

$$\dot{x}(t) = f(t, x(t),\ x(t - h_1) \cdots x(t - h_n),\ u(t),\ u(t - h)), \qquad (1.4.3)$$

with target set

$$T = \{x \in E^n : Hx = r,\ r \in E^n\} ,$$

and $H$ an $n \times n$ matrix. Equation (1.4.3) is a special case of the delay system

$$\dot{x}(t) = f(t, x_t, u_t) , \qquad x_\sigma = \phi , \qquad (1.4.4)$$

which includes the ordinary differential equation

$$\dot{x}(t) = k(t, x(t), u(t)) . \qquad (1.4.5)$$

In (1.4.4) $x_t$ is a function defined by $x_t(s) = x(t + s)\ s \in [-h, 0]$.
The symbol $u_t$ is defined similarly.

### 1.4.2  *Control of epidemics: AIDS*

Just as in I, we formulate a theory of acquired immunodeficiency syndrome (AIDS) epidemic as a problem of time-optimal control theory. It is a modification of the recent works of Castillo-Chavez, Cooke, Huang, and Levin [14]. We assume as basic that there are five classes (of sexually active male homosexuals with multiple partners):

$x_1$: susceptible individuals,
$x_2$: those infectious individuals who will go on to develop full-blown AIDS,
$x_3$: infectious individuals who will not develop full-blown AIDS,

$x_4$: those former $x_3$ who are no longer sexually active, and

$x_5$: those former $x_2$ who have developed full-blown AIDS.

Note that if an individual enters $x_5$ or $x_4$, he no longer enters into the dynamics of the disease. In contrast to our earlier treatment of the theory of epidemics, a latent class is excluded, i.e., those exposed individuals who are not yet infectious, since only a very short time is spent in that class. It is our assumption that an individual with full-blown AIDS has no sexual contacts and is therefore not infectious. Also, once infected, an individual is immediately infectious, and sexual inactivity or acquisition of AIDS are at the constant rates of $a_3$ and $a_2$, respectively, per unit time. We let $\Lambda$ denote the total recruitment rate into the susceptible class, defined to be those individuals who are sexually active. Let $\mu$ be the natural mortality rate, and $d$ the disease-induced mortality due to AIDS. Suppose $p$ is the fraction of the susceptible individuals that after becoming infectious will enter the AIDS class, and $(1-p)$ the fraction that become infectious but will not develop full-blown AIDS. We use the following diagram:

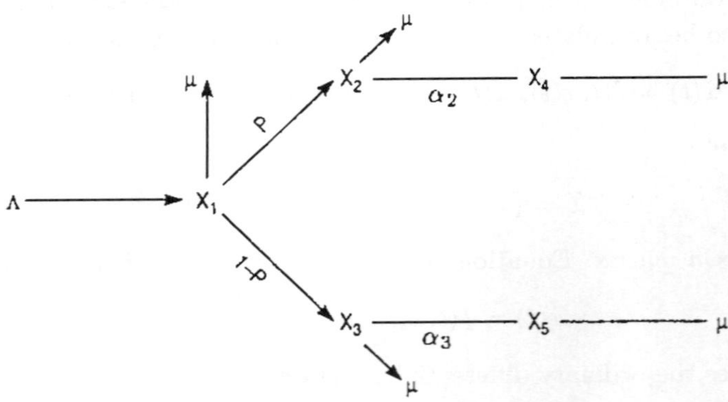


Fig. 1.4.1

to determine the epidemiological dynamics:

$$\frac{dx_1(t)}{dt} = \Lambda - \lambda C(T)(t)x_1(t)\frac{W(t)}{T(t)} - \mu x_1(t), \qquad (1.4.6)$$

$$\frac{dx_2(t)}{dt} = \lambda_p C(T)(t)x_1(t)\frac{W(t)}{T(t)} - (\alpha_2 + \mu)x_2(t), \qquad (1.4.7)$$

$$\frac{dx_3(t)}{dt} = \lambda(1-p)C(T)(t)x_1(t)\frac{W(t)}{T(t)} - (\alpha_3 + \mu)x_3(t), \qquad (1.4.8)$$

$$\frac{dx_4(t)}{dt} = \alpha_2 x_2(t) - (d+\mu)x_4(t), \qquad (1.4.9)$$

$$\frac{dx_5(t)}{dt} = \alpha_3 x_3(t) - \mu x_5(t), \qquad (1.4.10)$$

where

$$W = x_2 + x_3 \quad \text{and} \quad T = W + x_1 \,. \tag{1.4.11}$$

In the above formulation, $C(T)$ represents the mean number of sexual partners an average individual has per unit time when the total population is $T$, and the constant $\lambda$ denotes the average sexual risk per partner. Thus the expression $\lambda C(T)SW/T$ denotes the number of newly infected individuals per unit time. Very often the Michaelis–Menten type contact law is accepted for $C(T)$:

$$C(T) = \frac{\beta T}{1 + kT} \,, \tag{1.4.12}$$

where $\beta, k$ is constant. Systems (1.4.6–1.4.10) are ordinary differential equations. Further to the assumption above where individuals become immediately infectious and the infectious period is equal to the incubation period, we designate $h_2$ to be the fixed period of infection for $x_2$, and $h_3$ to be that of $x_3$. We set $x_{20}(t)$, $x_{30}(t)$, to be infectious $x_2(t)$ and $x_3(t)$ at time $t = 0$, and $x_{40}(t), x_{50}(t)$ living $x_4(t), x_5(t)$ at time $t = 0$. Clearly $x_{40}, x_{50}$ have compact support, i.e., $x_{40}(t), x_{50}(t)$ vanish as $t \to \infty$. Note that $x_{20}(t) = x_{30}(t) = 0$ for $t > \max(h_1, h_2)$. If $H(t)$ is the unit step or Heaviside function, i.e., $H(t) = \{ {0, \ t<0 \atop 1, \ t>1} $, then the system is modified to be

$$\dot{x}_1(t) = \Lambda - \lambda C(T)(t)x_1(t)\frac{W(t)}{T(t)} - \mu x_1(t) \,, \tag{1.4.13}$$

$$x_2(t) = x_{20}(t) + \lambda p \int_{t-h_2}^{t} C(T)(s)x_1(s)\frac{W(s)}{T(s)} H(s)e^{-\mu(t-s)} ds \,, \tag{1.4.14a}$$

$$x_3(t) = x_{30}(t) + \lambda(1 - p) \int_{t-h_3}^{t} C(T)(s)x_1(s)\frac{W(s)}{T(s)} H(s)e^{-\mu(t-s)} ds \,, \tag{1.4.14b}$$

$$x_4(t) = x_{40}(t) + \lambda p \int_{0}^{t-h_2} C(T)(s)x_1(s)\frac{W(s)}{T(s)} H(s)e^{-\mu(t-s)}$$

$$- d(t - s - h_2)ds \,, \tag{1.4.15a}$$

$$x_5(t) = x_{50}(t) + \lambda(1 - p) \int_{0}^{t-h_3} C(T)(s)\frac{x_1(s)W(s)}{T(s)} H(s)e^{-\mu(t-s)} ds \,, \tag{1.4.15b}$$

with initial conditions,

$$x_1(t) = x_{10}(t) \,, \qquad t \in [-\max(h_1, h_2), 0]$$

$$x_2(t) = x_{20}(t) \,, \tag{1.4.16}$$

$$x_3(t) = x_{30}(t) \,.$$

This is a delay differential equation whose existence and uniqueness of solutions are easily proved. We now restrict our analysis to the case $p = 1$ (i.e., we model AIDS as a progressive disease). The study of the steady states reduces to the following set of equations:

$$\dot{x}_1(t) = \Lambda - \lambda CT(t)x_1(t)\frac{x_2(t)}{T(t)} - \mu x_1(t)\,,$$

$$\dot{x}_2(t) = \lambda|C(T)(t)x_1(t)\frac{x_2(t)}{T(t)} - CT(t-h)x_1(t-h)\frac{x_2(t-h)}{T(t-h)}e^{-\mu h} - \mu x_2(t)\,,$$

with infection free state $(\frac{\Delta}{\mu}, 0)$ as equilibrium. It is possible to allow $h_2, h_3$ to be distributed. For this we follow the generalizations of [14] and define the survivorship functions $P_2(s), P_3(s)$, which are the proportions of those individuals who become $x_2$ or $x_3$ infective at time $t$ and if alive are still infectious at time $t + s$. These are nonnegative nonincreasing with

$$P_2(0) = P_3(0) = 1\,,$$

$$\int_0^\infty P_2(s)ds < \infty\,, \qquad \int_0^\infty P_3(s)ds < \infty\,. \tag{1.4.17}$$

Under these assumptions, $-\dot{P}_2(s)$ and $-\dot{P}_3(s)$ are the removal rates of individuals from $x_2$ and $x_3$ into $x_4$, and $x_5$ classes $s$ time units after infection. We therefore have

$$\dot{x}_1(t) = \Lambda - \lambda C(T)(t)x_1(t)\frac{W(t)}{T(t)} - \mu x_1(t)\,, \tag{1.4.18}$$

$$x_2(t) = x_0(t) + p\int_0^t \lambda C(T)(s)x_1(s)\frac{W(s)}{T(s)}e^{-\mu(t-s)}P_2(t-s)ds\,, \tag{1.4.19}$$

$$x_3(t) = x_{30}(t) + (1-p)\int_0^t \lambda C(T)(s)x_1(s)\frac{W(s)}{T(s)}e^{-\mu(t-s)}P_3(t-s)ds\,, \tag{1.4.20}$$

$$x_4(t) = p\int_0^t \left\{ \int_0^\tau \lambda C(T)(s)x_1(s)\frac{W(s)}{T(s)}e^{-\mu(\tau-s)}[-\dot{P}_2(\tau-s)e^{-(\mu+d)(t-\tau)}ds]\right\}d\tau$$

$$+ x_{40}e^{-(\mu+d)t} + x_{40}(t)\,, \tag{1.4.21}$$

$$x_5(t) = (1-p)\int_0^t \left\{ \int_0^\tau \lambda C(T(s))x_1(s)\frac{W(s)}{T(s)}e^{-\mu(\tau-s)}[-\dot{P}_3(\tau-s)\right.$$

$$\left. \times e^{-\mu(t-\tau)}]ds\right\}d\tau + x_{50}e^{-\mu t} + x_{50}(t)\,. \tag{1.4.22}$$

Now denote by $B$ the rate of infection — the number of new cases of infection per unit of time:

$$B(t) = C(T(t))x_1(t)\frac{W(t)}{T(t)}. \tag{1.4.23}$$

Since $\mu$ is the mortality rate, we denote the rate of attrition by $A(t)$: In (1.4.6) this is given by $A(t) = -\mu x_1(t)$. Thus, (1.4.6) becomes

$$\dot{x}_1(t) = \Lambda - B(t) - A(t). \tag{1.4.24a}$$

This can be generalized. If $\Lambda$ is time varying, then $\Lambda(t)$ is defined as the total rate of recruitment into the susceptible class at time $t$. Assuming a certain fraction of the total dies in each future time period after recruitment, then $m(\tau)d\tau$ is the natural mortality density, i.e., the fraction that dies in any small interval of length $d\tau$ around the time point $\tau$. Then $\Lambda(t)m(t)d\tau$ will die in a small interval about $t + \tau$, and if we replace $t$ by $t - \tau$ then $\Lambda(t - \tau)m(\tau)d\tau$ of the total recruitment made in a small interval about $t - \tau$ will die. The total natural death at time $t$ of all previous total recruitments is $\int_0^\infty \Lambda(t - \tau)m(\tau)d\tau$. Equation (1.4.24a) becomes

$$\dot{x}_1(t) = \Lambda(t) - B(t) - \int_0^\infty \Lambda(t - \tau)m(\tau)d\tau. \tag{1.4.24b}$$

This can be considered an integral equation of renewal theory, where $\dot{x}_1(t)$ and $B(t)$ are known and $\Lambda(t)$ is unknown. If we define the $n$th convolution, $m^n(\tau)$ of $m(\tau)$ recursively by the relations

$$m^{(1)}(\tau) = m(\tau), \quad m^{n+1}(\tau) = \int_0^\tau m^n(\sigma)m(\tau - \sigma)d\sigma, \tag{1.4.25a}$$

and the replacement density, $r(\tau) = \sum_{n=1}^\infty m^{(n)}(\tau)$, then the unique solution of (1.4.24b) is

$$\Lambda(t) = \dot{x}_1(t) + B(t) + \int_0^\infty \dot{x}_1(t - \tau)r(\tau)d\tau. \tag{1.4.25b}$$

If the mortality density is exponential, $m(\tau) = \mu e^{-\mu\tau}$ then

$$m^n(\tau) = \frac{\mu^n \tau^{n-1}}{(n-1)!}e^{-\mu\tau} \tag{1.4.26}$$

so that $r(\tau) = \mu$. With this (1.4.25b) becomes

$$\Lambda(t) = \dot{x}_1(t) + \int_0^\infty \dot{x}_1(t - \tau)r(\tau)d\tau + B(t)$$

or

$$\Lambda(t) = \dot{x}_1(t) + \mu x_1(t) + B(t).$$

From this analysis one has

$$\dot{x}_1(t) + \int_0^\infty \dot{x}_1(t - \tau)r(\tau)d\tau = \Lambda(t) - B(t)\,. \qquad (1.4.27)$$

Since $r(\tau) \geq 0$ for all $\tau$, there is a number $0 \leq h(t) = h < \infty$ such that

$$\int_0^\infty \dot{x}_1(t - \tau)r(\tau)d\tau = \dot{x}_1(t - h)\int_0^\infty r(\tau)d\tau\,,$$

and we can then postulate that for some finite number $a_{-1}$,

$$\dot{x}_1(t - h(t))a_{-1} = \dot{x}_1(t - h(t))\int_0^\infty r(\tau)d\tau\,. \qquad (1.4.28)$$

We have derived the neutral equation

$$\dot{x}_1(t) + a_{-1}\dot{x}_1(t - h(t)) = \Lambda(t) - B(t)\,, \qquad (1.4.29)$$

which now replaces (1.4.13). Because of the delays in (1.4.14) and (1.4.15) there is a possibility of oscillation and endemic equilibrium. One can now study the stability of the disease as well as its controllability by using any of the control strategies of Section 1.4.1, i.e., allowing removal and active immunization. The results are available in Chukwu [15].

## 1.5   Wind Tunnel Model; Mach Number Control

Delay equations of type (1.4.4) are also encountered in the design of an aircraft control facility. For example, in the control of a high-speed closed-air unit wind tunnel known as the National Transonic Facility (NTF), which is at present under development at NASA Langley Research Center, a linearized model of the Mach number dynamics is the system of three state equations with a delay, given by

$$\dot{x}_1(t) = -ax_1(t) + akx_2(t - h) + k_1u_1(t)\,,$$

$$\dot{x}_2(t) = x_3(t) + k_2u_2(t)\,, \qquad (1.5.1)$$

$$\dot{x}_3(t) = -w^2x_2(t) - 2\xi wx_3(t) + w^2u_3(t)\,,$$

with $a = \frac{1}{1.964}$, $k = 0.117$, $w = 6$, $\xi = 1.6$, and $h = 0.33s$. $|u_i| \leq K$ with $K$ a positive constant. The state variables $x_1, x_2, x_3$ represent deviations from a chosen operating point (equilibrium point) of the following quantities: $x_1 = $ Mach number, $x_2 = $ actuator position guide vane angle in a driving fan), and $x_3 = $ actuator rate. The delay represents the time of the transport between the fan and the test section. Though in many cases only $k_1 = k_2 = 0$ are considered ([1, 4]), we assume here that there is a control device that creates the control variable with three components that constitute an input to the rate of the Mach number, the actuator velocity, and the actuator servomechanism. The parameter values correspond to one particular operating point. Equation (1.5.1) can be written as

$$\dot{x}(t) = A_0 x(t) + A_1 x(t-h) + Bu(t) , \qquad (1.5.2)$$

where

$$A_0 = \begin{bmatrix} -a & 0 & 0 \\ 0 & 0 & 1 \\ 0 & -w^2 & -2\xi w \end{bmatrix} , \qquad A_1 = \begin{bmatrix} 0 & ak & 0 \\ 0 & 0 & 0 \\ 0 & 0 & 0 \end{bmatrix} , \qquad B = \begin{bmatrix} k_1 & 0 & 0 \\ 0 & k_2 & 0 \\ 0 & 0 & w_2 \end{bmatrix} .$$

The following problem can be proposed (Manitius and Tran [4]): Find an optimal control subject to its constraints such that the solution of (1.5.2), with this control ($k_1 = k_2 = 0$) and with an initial configuration

$$(x_1(0) = 0.1 , \ x_2(\theta) = x_1(\theta)/k , \ \theta \in [-h, 0] , \ x_3(\theta) = 0) ,$$

will hit a target

$$G = \{(x_1, x_2, x_3) : |x_1(t)| \le 0.002 \ |x_2(t)| \le 30 \ |x_3(t)| \le 30\}$$

in minimum time $T$ and remain there forever after.

**Remark 1.5.1** In (1.3.2) and (1.5.1) the controls $(u_1, u_2)$ and $(u_1, u_2, u_3)$ introduced are such that additional variables have been added in the equations describing the relations between velocity and position. One can question the physical validity of such an effort since velocity is the derivative of position, and no extra control may be allowed in the relation. Such an objection is shallow. Just as a bird flying in the wind with $x_1$ as its position may have its resultant velocity $\dot{x}_1 = y + u_1$, where $u_1$ is the velocity of the wind (control), so may the control device bring in an additional $u_1$ to bear in (1.3.2) and an additional $(u_1, u_2)$ to bear in (1.5.1). These are possibilities. If these components are zero, the Euclidean controllability of these equations can still be studied.

## 1.6   The Flip-Flop Circuit

In the general system (1.4.4) whose special cases are given in the previous examples, we deduced the equilibrium position by setting $f(t, x_t, u_t) = 0$. The state $x_t^*$ for which this is true is the equilibrium state. The problem of stability is to deduce conditions on $f$ for which every solution $x$ of (1.4.4) satisfies $x_t \to x_t^*$ as $t \to \infty$. It is possible that some dynamical systems possess multiple equilibria and are therefore suited to be used as a memory storage device in the design of a digital computer. The flip-flop circuit has such dynamics [8]. It is the basic element in a digital computer, and a standard model is given in Figs. 1.6.1 and 1.6.2 in [8].

In this model the portion between 0 and 1 is a lossless transmission line with inductance $L$ and capacitance $C$. The current $i$ flowing through the line and the voltage $v$ across it are both functions of $x$ and $t$. In [8] the function $g(v)$ is a nonlinear function of $v$ and gives the current in the indicated box in the direction shown. The lossless transmission line can be described by the hyperbolic partial differential equations

$$\frac{\partial v}{\partial x} = -L\frac{\partial i}{\partial t}, \quad \frac{\partial i}{\partial x} = -C\frac{\partial v}{\partial t}, \quad 0 < x < 1, \quad t > 0 \tag{1.6.1}$$

with boundary conditions

$$E - v(0,t) - Ri(0,t) = 0,$$

$$C\frac{\partial v(1,t)}{\partial t} = i(1,t) - g(v(1,t)). \tag{1.6.2}$$

If we let $s = (LC)^{-\frac{1}{2}}$, $z = (L/C)^{\frac{1}{2}}$, $k = (z - R)/(z + R)$, and $\alpha = 2E/(z + R)$, we can convert (1.6.1) with boundary conditions (1.6.2) into a neutral functional differential equation, i.e., a system with delay in the derivative of the state as well as in the state itself. Indeed, the general solution of (1.6.1) is

$$v(x,t) = \phi(x - st) + \psi(x + st),$$

$$i(x,t) = \frac{1}{z}[\phi(x - st) - \psi(x + st)] \tag{1.6.3}$$

which is equivalent to

$$2\phi(x - st) = v(x,t) + zi(x,t),$$

$$2\psi(x + st) = v(x,t) - zi(x,t).$$

This implies that

$$2\phi(-st) = v\left(1, t + \frac{1}{s}\right) + zi\left(1, t + \frac{1}{s}\right),$$

$$2\psi(st) = v\left(1, t - \frac{1}{s}\right) - zi\left(1, t - \frac{1}{s}\right).$$

We use these expressions in the general solution (1.6.3) and in the first expression of (1.5.2) at the moment $(t - (\frac{1}{s}))$ to deduce

$$i(1,t) - ki\left(1, t - \frac{2}{s}\right) = \alpha - \frac{1}{z}v(1,t) - \frac{k}{z}v\left(1, t - \frac{2}{s}\right).$$

On using the second boundary condition of (1.6.2) and setting $u(t) = v(1,t)$, we obtain the equation

$$\dot{u}(t) - k\dot{u}\left(t - \frac{2}{s}\right) = f\left(u(t), u\left(t - \frac{2}{s}\right)\right), \tag{1.6.4}$$

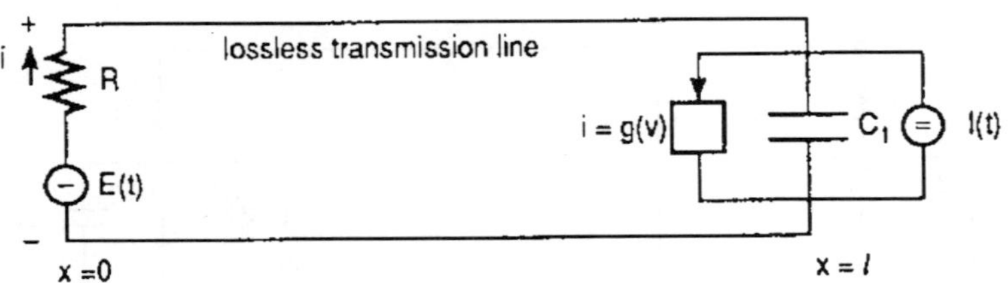


Fig. 1.6.1

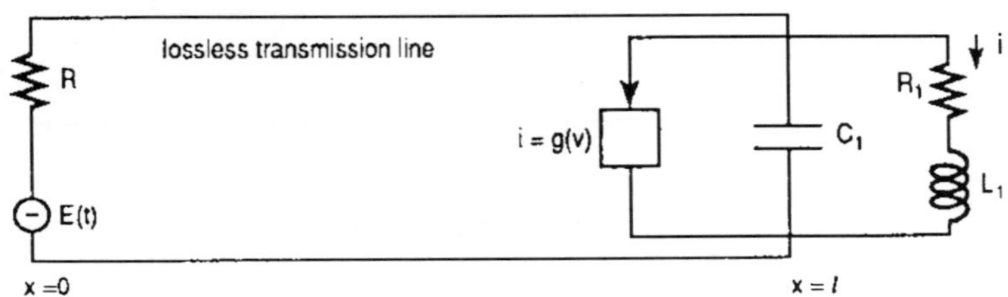


Fig. 1.6.2

where

$$c_1 f(u(t), u(t-h)) = \alpha - \frac{1}{2}u(t) - \frac{k}{z}u(t-h) - g(u(t)) + kgu(t-h)).$$

For the flip-flop circuit to operate as a memory storage device, one equilibria it may possess is assumed to be globally asymptotically stable: Criteria on the circuit parameters for a single equilibria point to be globally, asymptotically stable are deduced by a Lyapunov stability theory for a more general system than (1.6.4), namely

$$\frac{d}{dt}[x(t) - A_{-1}x(t-h)] = f(t, x_t), \tag{1.6.5}$$

where $A_{-1}$ is an $n \times n$ matrix and $f : E \times C \to E^n$ is a continuous function.

Equation (1.6.5) can be obtained from a lossless transmission line in which conditions (1.6.2) are more complicated. See Ref. [7]. In Fig. 1.6.3, for example, (1.6.1) is satisfied and the following boundary conditions hold:

$$v(t, 0) = u_0(t) - R_0 i(t, 0) + E_0(t),$$

$$v(t, 1) = u_1(t) + R_1 i(t, 1), \tag{1.6.6}$$

$$u_0(t) = -L_0 i_0(t), \qquad u_1(t) = L_1 i_1(t), \tag{1.6.7}$$

$$i(t, 0) - i_0(t) = -c_0 \dot{u}_0(t), \qquad i(t, 1) - i_1(t) = c_1 \dot{u}_1(t). \tag{1.6.8}$$

If we integrate (1.6.1) along its characteristics, then we have

$$x_1(t) = \sqrt{c}v(t, 0) + \sqrt{L}i(t, 0) = \sqrt{c}v(t+h, 1) + \sqrt{L}i(t+h, 1),$$

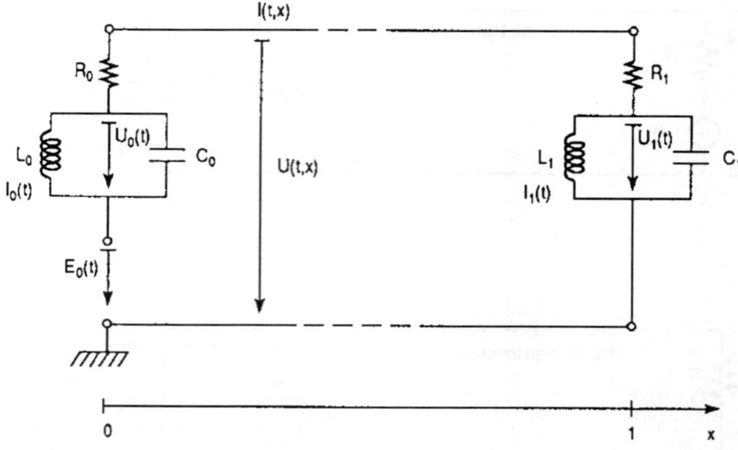


Fig. 1.6.3

$$x_2(t) = \sqrt{c}\,v(t,1) - \sqrt{L}\,i(t,1) = \sqrt{c}\,v(t+h,0) - \sqrt{L}\,i(t+h,0)\,,$$

where $h = \sqrt{cL}$. Setting $x_3(t) = 2\sqrt{L}\,i_0(t)$, $x_4(t) = 2\sqrt{L}\,i_1(t)$, and $u(t) = 2\sqrt{c}\,E_0(t)$, and using the boundary conditions (1.6.6)–(1.6.8), we deduce an equation of the form

$$\frac{d}{dt}[x(t) - A_{-1}x(t-h) - B_{-0}u(t)] = A_0x(t) + A_1x(t-h) + B_0u(t)\,, \qquad (1.6.9)$$

where

$$A_0 = \begin{bmatrix} -\alpha_0 & 0 & \alpha_0 & 0 \\ 0 & -\alpha_1 & 0 & -\alpha_1 \\ -\alpha_2 & 0 & 0 & 0 \\ 0 & \alpha_3 & 0 & 0 \end{bmatrix}, \quad A_1 = \begin{bmatrix} 0 & \alpha_0 & 0 & 0 \\ \alpha_1 & 0 & 0 & 0 \\ 0 & \alpha_2\alpha_4 & 0 & 0 \\ -\alpha_3\alpha_5 & 0 & 0 & 0 \end{bmatrix},$$

$$B_0 = \begin{bmatrix} 0 \\ 0 \\ \alpha_2\beta_0 \\ 0 \end{bmatrix}, \quad A_1 = \begin{bmatrix} 0 & \alpha_4 & 0 & 0 \\ \alpha_5 & 0 & 0 & 0 \\ 0 & 0 & 0 & 0 \\ 0 & 0 & 0 & 0 \end{bmatrix}, \quad B_{-0} = \begin{bmatrix} \beta_0 \\ 0 \\ 0 \\ 0 \end{bmatrix},$$

and where

$$\alpha_0 = \frac{1}{c_0}\frac{\sqrt{c}}{R_0\sqrt{c}+\sqrt{L}}\,, \qquad \alpha_2 = \frac{1}{L_0}\frac{R_0\sqrt{c}+\sqrt{L}}{\sqrt{c}}\,,$$

$$\alpha_1 = \frac{1}{c_1}\frac{\sqrt{c}}{R_1\sqrt{c}+\sqrt{L}}\,, \qquad \alpha_3 = \frac{1}{L_1}\frac{R_1\sqrt{c_1}+\sqrt{L}}{\sqrt{c}}\,,$$

$$\alpha_4 = \frac{R_0\sqrt{c}-\sqrt{L}}{R_0\sqrt{c}+\sqrt{L}}\,, \qquad \alpha_5 = \frac{R_1\sqrt{c}-\sqrt{L}}{R_1\sqrt{c}+\sqrt{L}}\,.$$

Equation (1.6.9) is a neutral functional differential equation with control $u$ that is related to the initial power source $E_0(t)$.

Equations (1.6.4), (1.6.5), and (1.6.9) are special cases of the system

$$\frac{d}{dt}[D(t,x_t)] = f(t,x_t,u(t))\,, \tag{1.6.10}$$

where $D : E \times C \to E^n$, $f : E \times C \times E^m \to E^n$ are continuous functions with

$$D(t,x_t) = x(t) - g(t,x_t)\,, \tag{1.6.11}$$

and $g : E \times C \to E^n$ is continuous.

## 1.7 Hyperbolic Partial Differential Equations with Boundary Controls

A final example involves the derivation of (1.6.10) from linear hyperbolic partial differential equations with boundary control. The derivation is treated in [3]. Let the wave equation for $w(t,x)$ be given by

$$w_{tt} - c^2 w_{xx} = 0\,, \qquad t \in [0,T]\,, \qquad x \in [0,1]\,, \tag{1.7.1}$$

with boundary conditions

$$\begin{aligned}
A_0(t)w_{tt}(t,0) + B_0(t)w_{tx}(t,0) &= G_0(t, w(t,0)\,, w_t(t,0)\,, w_x(t,0))\,, \\
A_1(t)w_{tt}(t,1) + B_1(t)w_{tx}(t,1) &= G_1(t, w(t,1)\,, w_t(w,1)\,, w_x(t,1))\,,
\end{aligned} \tag{1.7.2}$$

and initial conditions

$$w(0,x) = f_0(x)\,, \qquad w_t(0,x) = f_1(x) \quad \text{on} \quad [0,1]\,, \tag{1.7.3}$$

where the initial functions satisfy the boundary conditions at $t = 0$. Here the subscript $t$ or $x$ denotes partial derivative. The prime in the sequel denotes total derivative. The terms $G_0$, $G_1$ contain the controls if

$$G_0(t,\rho,\sigma,\tau) = F_0(t,u(t),\rho,\sigma,\tau)\,,$$

$$G_1(t,\rho,\sigma,\tau) = F_1(t,v(t),\rho,\sigma,\tau)\,,$$

and $\binom{u}{v}$ is the control to be chosen from a prescribed class. We assume that $A_i$, $B_i$ are continuously differentiable, $G_i$ are absolutely continuous in $t$, continuously differentiable in other arguments with $G_{it}$ dominated by a square integrable function. We assume further that

$$A_0(t) - B_0(t)/c \neq 0, \qquad A_1(t) + B_1(t)/c \neq 0, \qquad (1.7.4)$$

for $t \in [0, T]$.

We assume a D'Alamberto solution of the form

$$w(t, x) = \phi(t + x/c) + \psi(t - x/c), \qquad (1.7.5)$$

and substitute this in (1.7.2). On setting

$$\alpha_i(s) = \left[A_i(s) + \frac{1}{c}B_i(s)\right], \qquad \beta_i(s) = \left[A_i(s) - \frac{1}{c}B_i(s)\right], \qquad i = 0, 1,$$

the substitution yields

$$\alpha_0(t)\phi''(t) + \beta_0(t)\psi''(t) = G_0\left(t, f_0(0) + \int_0^t \phi'(s)ds + \int_0^t \psi'(s)ds,\right.$$

$$\times \left. \phi'(t) + \psi'(t)\frac{1}{c}\phi'(t) - \frac{1}{c}\psi'(t)\right)$$

$$\equiv \tilde{G}_0(\phi'(\cdot), \psi'(\cdot), t), \ t \geq 0,$$

and

$$\alpha_1\left(t - \frac{1}{c}\right)\phi''(t) + \beta_1\left(t - \frac{1}{c}\right)\psi''\left(t - \frac{2}{c}\right)$$

$$= G_1\left(t - \frac{1}{c}, f_0(0) + \int_0^t \phi'(s)ds + \int_0^{t-\frac{2}{c}} \psi'(s)ds, \phi'(t)\right.$$

$$\left. + \psi'\left(t - \frac{2}{c}\right), \ \frac{1}{c}\phi'(t) - \frac{1}{c}\psi'\left(t - \frac{2}{c}\right)\right)$$

$$\equiv \tilde{G}_1(\phi'(\cdot), \psi'(\cdot), t), \quad t \geq \frac{1}{c}.$$

It follows from condition (1.7.4) that we can multiply the last pair of equations by the matrix

$$\begin{bmatrix} 0 & \dfrac{1}{\alpha_1(t - \frac{1}{c})} \\[3ex] \dfrac{1}{\beta_0(t)} & \dfrac{\alpha_0(t)}{\alpha_1(t - \frac{1}{c})\beta_0(t)} \end{bmatrix}$$

to deduce the equations

$$\phi''(t) + \frac{\beta_1(t - \frac{1}{c})\psi''(t - \frac{2}{c})}{a_1(t - \frac{1}{c})} = \frac{1}{\alpha_1(t - \frac{1}{c})}\tilde{G}_1(\phi'(\cdot), \psi'(\cdot), t),$$

$$\psi''(t) = \frac{\alpha_0(t)\beta_1(t - \frac{1}{c})}{\alpha_1(t - \frac{1}{c})\beta_0(t)}\psi''\left(t - \frac{2}{c}\right)$$

$$= \frac{1}{\beta_0(t)}\tilde{G}_0(\phi'(\cdot), \psi'(\cdot), t) - \frac{\alpha_0(t)G_1(\phi'(\cdot), \psi'(\cdot), t)}{\alpha_1(t - \frac{1}{c})\beta_0(t)},$$

$$\frac{1}{c} \leq t.$$

Denoting $(\phi', \psi')$ by $(y, z)$, this equation is a neutral equation of the form

$$\dot{y}(t) + h_1(t)\dot{z}\left(t - \frac{2}{c}\right) = H_1(t, y(\cdot), z(\cdot)),$$

$$\dot{z}(t) + h_2(t)\dot{z}\left(t - \frac{2}{c}\right) = H_2(t, y(\cdot), z(\cdot)).$$

$$(1.7.6)$$

The data of (1.7.3) can now be used to produce initial and terminal functions of $(y, z)$ for (1.7.6). Since there are controls on $G_i$ and therefore on $H_i$, the system (1.7.6) is a control system for $t \in [\frac{1}{c}, T]$ with initial and terminal values for $y$ given on $[0, \frac{1}{c}]$ and at $t = T$, and the corresponding values of $z$ given on $[-\frac{1}{c}, \frac{1}{c}]$ and $[T - \frac{2}{c}, T]$. Because of the smoothness conditions on $G_i$ and therefore on $H_i$, (1.7.6) may be argued from the fundamental existence theorem below to have a solution $(y, z) = (\phi', \psi')$, which is absolutely continuous with square integrable derivatives. One argues that this $(\phi, \psi)$ utilized in (1.7.5) yields a solution of (1.7.1) in the generalized sense or in the sense of almost everywhere, i.e., $w(t, x) = \phi(t + \frac{x}{c}) + \psi(t - \frac{x}{c})$ is continuously differentiable with $w_t$, $w_x$ being absolutely continuous and possessing square integrable partials that satisfy (1.7.1) a.e.

It is interesting to note that the boundary conditions of (1.7.2) cover the usual ones for transverse vibrations of a string or longitudinal vibrations in an elastic rod with elastically supported ends. Just as in earlier examples, it is there at the boundary that controls are introduced. It sometimes requires some ingenuity to introduce in practice the type of control that is required for the system $\frac{d}{dt}[D(t)x_t] = f(t, x_t) + B(t)u(t)$ (or of (1.6.10)) to have a desired effect on $B(t)$. One can then relate this to the control device. An appropriate time-optimal control problem can now be formulated for such systems.

## 1.8  Control of Global Economic Growth

Let $x(t)$ denote the value of capital stock at time $t$. Let $u(t)$, $0 \leq u(t) \leq 1$ be the fraction of the stock that is allocated to investment. This investment is used to increase productive capacity. We can assume the value of this investment $I(t)$ is given by

$$I(t) = ku(t)x(t)$$

where $k$ is the constant of proportionality. There is depreciation, $D(t)$, of capital stock, and it is proportional to capital stock

$$D(t) = -\delta x(t).$$

The net investment is

$$N(t) = I(t) - \delta x(t),$$

and it is used to produce new capital. Thus net capital formation $\dot{x}(t)$ is given by

$$\dot{x}(t) = ku(t)x(t) - \delta x(t). \tag{1.8.1}$$

If $-1 \le u(t) \le 0$, we can interpret $u(t)$ to be the fraction of the value of stock that is used for payment of taxes, etc. Thus, in general, $u(t)$ satisfies $-1 \le u(t) \le 1$. Such models are very naive and unrealistic. As pointed out by Takayama [10, p. 705], this implies that adjustment of desired stock of capital is instantaneous and frictionless, an implication that has no valid empirical foundation. There is a time lag that needs to be incorporated to the control procedure of the firm. Investment in new capital equipment does not yield new productive capacity until the equipment is delivered, installed, and tested. There is a time delay $h > 0$. It is more realistic to express $I(t)$ as a function of the present and past values of $u(t)$ and $x(t)$ and of time:

$$I(t) = g(t, x(t), u(t), x(t-h), u(t-h)).$$

In general,

$$I(t) = B(t, x_t, u_t)u_t \equiv \int_{-h}^{0} d_s H(t, x(t+s), u(t+s))u(t+s).$$

Here the past history of $x$ and $u$ over an entire interval of length $h$ enters equations through the Riemann–Stieltjes integral. Realism dictates that depreciation be incorporated into the model. We assume that the value of the capital stock that has depreciated is not just proportional to $x(t)$ as in (1.8.1), but is a function $g$ of $(x(t), u(t))$ and is given by $g(x(t), u(t))x(t)$ at time $t$. Suppose this value decreases by a factor $P(a)$ at time $a$; ($P(0) = 1$, $P(L) = 0$ where $L$ is the lifetime of the asset). If $p(a) = \frac{dP(a)}{da}$, the depreciation density or mortality density, $p(\tau)d\tau$, represents the fraction lost to productive use in any small time interval of length $d\tau$ around the time $\tau$. Therefore, $g(x(\tau), u(\tau))x(\tau)p(\tau)d\tau$ will disappear in a small interval about $t + \tau$. If we replace $t$ by $t - \tau$, we can say that at time $t$ the amount

$$g(x(\tau-h), u(\tau-h))x(\tau-h)(\tau)p(\tau)d\tau$$

in a small time interval about $t - \tau$ will disappear. The total evaporation at time $t$ is

$$-\int_0^L g(x(\tau - h), u(\tau, -h))x(\tau - h)p(\tau)d\tau\,,$$

or

$$-\int_0^L g(x(\tau - h), u(\tau - h))x(\tau - h)dp(\tau)$$

if we use Riemann–Stieltjes integral. A general representative expression is

$$L(t, x_t, u_t)x_t \equiv \int_{-h}^0 d_s\eta(t, s, x(t + s), u(t + s)), x(t + s)$$

where we use Riemann–Stieltjes integral. We can therefore state that net capital formation $\dot{x}(t)$ is given by

$$\dot{x}(t) = L(t, x_t, u_t)x_t + B(t, x_t, u_t)u_t\,. \tag{1.8.2}$$

We postulate initial capital endowment function as $x_s = \phi$, and initial fraction of stock allocated to investment as $u_\sigma = v$. The key elements in (1.8.2) are the time patterns of capital $x(t)$ and the capital policy described by the function $u(t)$. The problem of optimal capital policy is to describe $u(t)$ subject to its constraints $-1 \leq u(t) \leq 1$, such that the system with initial endowment function $\phi$ will hit a prescribed target $x_1$ while minimizing the firms' power or effort defined by some function $E(u)$, and perhaps to do it in minimum time. Possible definitions of effort or power are given by

$$E_1(u(t)) = \int_\sigma^{t_1} |u(t)|^2 dt\,, \tag{1.8.3a}$$

$$E_2(u(t)) = \sup_{0 \leq t \leq t_t} |u(t)|\,, \tag{1.8.3b}$$

where

$$u \in U = \{u \in E,\ u \text{ piecewise continuous } |u(t)| \leq 1\ 0 \leq t \leq t_1\}\,.$$

This represents maximum investment thrust available to the firms.

$$E_3(u(t)) = \left(\int_0^{t_1} |u(t)|^p dt\right)^{\frac{1}{p}}\,, \tag{1.8.3c}$$

where

$$u \in U = \{u \in E,\ u \text{ piecewise continuous } |u(t)| \leq 1\ 0 \leq t \leq t_1\}\,.$$

If $p = 2$, $E(u(t))$ represents the investment energy or power that must be minimized.

$$E_4(u(t)) = \int_0^{t_1} |u(t)| dt \equiv \|u\|_{L_1}\,, \tag{1.8.3d}$$

where $u \in U = \{u : \|u\|_{L_1} \leq 1\}$. The expression $E_4(u(t))$ can be called investment; the problem is to minimize it when we achieve our growth target from the initial endowment $\phi$ to our prescribed target $x_1$. We may wish to maximize some attainable value of the social welfare criterion

$$J = \int_0^T E(u) \exp(-\gamma t) dt,$$

where $E$ is a specified "well-behaved" utility function and $\gamma$ is a fixed discount rate. $E$ is usually a strictly concave monotone increasing in $u$ with second derivative defined everywhere and such that

$$\lim_{u \to 0} E(u) = \infty.$$

The fixed time $T$ is the term of years in which the above objectives will be fulfilled. To solve this problem one must first solve the problem of controllability. The controllability question is investigated in Chaps. 8 and 12. Problems of optimality are reported in Chap. 7. Our research here deals with $-1 \leq u(t) \leq 1$, the situation $0 \leq u(t) \leq 1$ will be explored.

In the previous discussion $x(t)$ is a real number, the value of one stock. We can denote $x(t) = (x_1(t) \cdots x_n(t))$ to be the value of $n$ capital stocks at time $t$, with investment and tax strategy $u = (u_1 \cdots u_n)$, where $-1 \leq u_j(t) \leq 1$. We therefore consider (1.8.2) as the equation of the net capital function for $n$ stocks in a region that is isolated. These stocks are linked to $\ell$ other such systems in the globe, and the "interconnection" or "solidarity function" is given by

$$g_i(x_{1t}, \ldots, x_{\ell t}, \ u_{1t}, \ldots, u_{\ell t}).$$

This function describes the effects of other subsystems on the $i$th subsystem as measured locally at the $i$th location. Thus,

$$\dot{x}_i(t) = L_i(t, x_{it}, u_{it}) x_{it} + B_i(t, x_{it}, u_{it}) u_{it} + g_i(t, x_{1t}, \ldots, x_{\ell t}, u_{1t} \cdots u_{\ell t}), \quad i = 1, \ldots, \ell$$

$$(1.8.4)$$

is the decomposed interconnected system whose free subsystem is

$$\dot{x}_i(t) = L_i(t, x_{it}, u_{it}) x_{it} + B_i(t, x_{it}, u_{it}) u_{it}. \qquad (S_i)$$

We now introduce the following notation: Let $\sum_{i=1}^{\ell} n_i n, \ \sum_{i=1}^{\ell} m_i = m$.

$$\phi = [\phi_1, \ldots, \phi_\ell] \in E^n, \quad x = [x_1, \ldots, x_\ell] \in E^n, \quad u = [u_1, \ldots, u_\ell] \in E^n,$$

$$L(t, x_t, u_t) = [L_1(t, x_{1t}, u_1) \cdots L_\ell(t, x_{\ell t}, u_{\ell t})],$$

$$g(t, x_t, u_t) = [g_1(t, x_t, u_t) \cdots g_\ell(t, x_t, u_t)],$$

$$B(t, x_t, u_t) = \text{diag}[B_1(t, x_{1t}, u_{1t}) \cdots B_\ell(t, x_{\ell t} u_{(t)}].$$

Then (1.8.3) is given as

$$\dot{x}(t) = L(t, x_t, u_t) + B(t, x_t, u_t)u_t + g(t, x_t, u_t)\,. \tag{S}$$

To clarify our terminology we introduce the following notations and definitions. Let $E = (-\infty, \infty)$, and $E^r$ be the $r$-dimensional Euclidean space with norm $|\cdot|$. The symbol $C$ denotes the space of continuous functions mapping the interval $[-h, 0]$, $h > 0$, $h \in E$ into $E^n$ with the sup norm $\|\cdot\|$ defined by $\|\phi\| = \sup_{-h \leq s \leq 0} |\phi(s)|$, $\phi \in C$. If $t \in [\sigma, t_1]$, we let $x_t \in C$ be defined by $x_t(s) = x(t+s)$, $s \in [-h, 0]$. The symbol $L_\infty([\sigma, t_1], E^m) = L_\infty$ denotes the space of essentially bounded measurable functions on $[\sigma, t_1]$ with the norm $\max_{1 \leq j \leq m} \sup_{t \in [\sigma, t_1]} |u_j(t)| = \|u\|_\infty$.

**Definition 1.8.1**    The system (S) is Euclidean controllable on $[\sigma, t_1]$ if for any function $\phi \in C$ and any vector $x_1 \in E$, there is a control $u \in L_\infty([\sigma, t_1], E^m)$ such that the solution $x(t) = x(t, \sigma, \phi, u)$ of (S) satisfies $x_\sigma(\cdot, \sigma, \phi, u) = \phi$, $x(t_1, \sigma, \phi, u) = x_1$. It is Euclidean null controllable on $[\sigma, t_1]$ if $x_1 = 0$ in the above definition.

The system (S) is controllable on $[\sigma, t_1]$, $t_1 > \sigma + h$, if for each $\phi, \psi \in C$ there is a control function $u \in L_\infty([\sigma, t_1], E^m)$ such that the solution $x(t) = x(t, \sigma, \phi, u)$ satisfies $x_\sigma(\cdot) = \phi$, $x_{t_1} = \psi$. It is null controllable on $[\sigma, t_1]$ if $\psi \equiv 0$ in the preceding definition. In $C$ we drop the qualifying phrase "on the interval $[\sigma, t_1]$", if controllability obtains on every interval $[\sigma, t_1]$, with $t_1 > \sigma + h$. In $E^n$ we drop "on the interval $[\sigma, t_1]$" if Euclidean controllability holds on every interval $[\sigma, t_1]$, $t_1 \geq \sigma$.

The problem of controllability of (S) will be explored in Chaps. 8 and 9. Conditions are stated for the controllability of the isolated system $S_i$. Assuming that the subsystem $(S_i)$ is controllable we shall deduce conditions for (S) to be controllable when the solidarity function is "nice".

In (1.8.2) we postulated that capital growth is described by a functional differential equation with delay. The crucial assumption is that the net capital formation $\dot{x}(t)$ is given by the nonlinear function, $I(t)$, the gross investment, on the right. A more general situation can be obtained. Arrow [11, p. 184] showed that indeed it is realistic to have

$$I(t) = \dot{x}(t) + \int \dot{x}(t-s)r(s)ds\,,$$

where $r(t) \geq 0$ is the replacement density and $\dot{x}(t)$ the net capital formation. Since $I(t)$ is finite valued and $r(t) \geq 0$, the mean value theorem for integrals allows one to write

$$\int_0^\infty \dot{x}(t-s)r(s)ds = \dot{x}(t-h(t)) \int_0^\infty r(s)ds\,,$$

where $0 \leq h \leq \infty$. It is therefore reasonable to postulate that

$$I(t) = \dot{x}(t) + \dot{x}(t-h(t))a\,,$$

where $0 \le h < \infty$ and $a$ is determined from the replacement density. As a consequence of this hypothesis we can reasonably postulate that

$$\dot{x}(t) + a\dot{x}(t-h) = L(t, x_t, u_t)x_t + B(t, xt, u_t)u_t, \qquad (1.8.4)$$

in place of (1.8.2). This is a functional differential equation of neutral type. Analogous systems (1.8.4) (Si) and (S) can be formulated.

## 1.9    The General Time-Optimal Control Problem and the Stability Problem

The general form of equation studied is

$$\frac{d}{dt}[D(t)x_t] = f(t, x_t, u(t)), \qquad x_\sigma = \varphi, \qquad (1.9.1)$$

where

$$D(t)x_t = x(t) - g(t, x_t),$$

$g(t, \varphi)$ is linear in $\varphi$, and $f(t, \varphi, u)$ may be nonlinear. It includes the delay system

$$\dot{x}(t) = f(t, x_t, u(t)), \qquad x_\sigma = \varphi, \qquad (1.9.2)$$

and the ordinary differential system

$$\dot{x}(t) = k(t, x(t), u(t)), \qquad x(0) = x_0. \qquad (1.9.3)$$

The initial state $\varphi$ can appropriately be given as the space $C$ of continuous functions from $[-h, 0]$ into $E^n$ with the uniform norm. In this case, finding a solution of (1.9.2) is equivalent to finding a solution of the integral equation

$$x(t) = \varphi(0) + \int_\sigma^t f(s, x_s, u(s))ds, \qquad t \ge \sigma,$$
$$x(\sigma + \theta) = \varphi(\theta), \qquad -h \le \theta \le 0.$$

Other spaces with this property of equivalence of solution may be taken, and will be considered in subsequent discussions. If spaces other than $C$ are used, the final point and the solution lie in $C$ for $t \ge \sigma + h$. Often the final point we reach may be taken to be in $E^n$, or in $C$. The space $W_p^1$, consisting of all absolutely continuous functions from $[-h, 0] \to E^n$ that have $p$-integrable derivatives where $1 \le p \le \infty$, will play a significant role in our investigations. Another space of initial functions that is found to be very useful in application is the space $L_p([-a, 0], E^n)$, i.e., the space of $p$-integrable functions with the usual norm. If we appropriate this as the state of initial conditions, the solutions of (1.9.1) can be considered in the product space

$$M^p = E^n \times L_p([-h, 0], E^n) \times L_p([0, T], E^m),$$

$1 \leq p \leq \infty$, endowed with the norm

$$\|\varphi\| = [\|\|\varphi^0\|^p + \|\varphi^1\|_p^p + \|\varphi^2\|^p]^{\frac{1}{p}}\,, \qquad \varphi = (\varphi^0, \varphi^1, \varphi^2) \in M^p\,.$$

With the state space selected, the target is either a point or a subset of the appropriately selected state space.

The admissible controls are measurable vector valued functions with values constrained to lie in a compact convex set $\mathcal{U}$ of $E^m$. Often $\mathcal{U}$ is specialized to be the unit cube,

$$C^m = \{u \in E^m : |u_j| \leq 1,\ h = 1, 2, \ldots, m\}.$$

The time-optimal control problem is now formulated as follows: Determine an admissible control $u^*$ such that the solution $x(\varphi, \sigma, u^*)$ of (1.9.1) hits a continuously moving target point or set in the appropriate space, in minimum time, $t^* \geq 0$. Such controls $u^*$ are called time-optimal controls. Our ultimate goal is to get an optimal control $u^*$ as a function of the appropriate state space, i.e., to obtain the "feedback" or "closed loop" control. The major advantage of such a feedback-optimal control as against an "open-loop" one with $u$ as a function of $t$ is that the system in question becomes self-correcting and automatic.

Thus if $C$ is the state space, we hunt for a measurable function $m : C \to E^m$ such that

(i) $m(x_t) \equiv u(t),\ u(t) \in \mathcal{U}$.

(ii) $m$ is optimal feedback control for (1.9.1) in the following sense. In addition to (1.9.1), consider the differential equation

$$\frac{d}{dt}[D(t)y_t] = f(t, y_t, m(y_t))\,. \tag{1.9.4}$$

Then each optimal control of (1.9.1) is a solution of (1.9.4), and conversely each solution of (1.9.4) is an optimal control solution of (1.9.1). Once found, the time-optimal control problem for the system (1.9.1) is completely solved.

In (1.2.4) for instance, the optimal feedback control is the fishing strategy that would drive the system to the target (equilibrium) as quickly as possible.

Sometimes the theory and techniques of the solution of the time-optimal problem can be appropriated to tackle the problem of minimizing a cost function which, for example, is given in (1.8.3). The general situation will be treated in Chaps. 7 and 9.

### 1.9.1  *The stability problem*

The complete solution of the time-optimal problem is dependent on the solution of the corresponding stability problem, which is formulated as follows: Let $x^*$ be a solution of

$$f(t, x_t, 0) \equiv g(t, x_t) = 0. \tag{1.9.5}$$

Find necessary and sufficient conditions on $D$ and $g$ such that every solution $x(\phi)$ of

$$\frac{d}{dt}[D(t, x_t)] = g(t, x_t), \quad t \geq \sigma, \quad x_\sigma = \phi, \tag{1.9.6}$$

is uniformly, globally, asymptotically stable in the following sense: For each $\phi$, the solution $x(\phi)$ of (1.9.6) satisfies

$$x_t(\phi) \to x^* \quad \text{as} \quad t \to \infty. \tag{1.9.7}$$

These two problems of stability and optimal control are the primary questions we attempt to address in this monograph.

**Remark 1.9.1**    In the time-optimal or minimum-effort problem treated here, the controls of essential interest are small and bounded in components: $|u_j(t)| \leq 1$, $j = 1, \ldots, m$. Because of this constraint, the global asymptotic stability of the system without control are needed and will be studied to ensure global constrained controllability of our dynamics, on which rest the existence of an optimal control. This contrasts with the situation in which the constraint $|u_j(t)| \leq 1, j = 1, \ldots, m$ is removed, and we require admissible controls to be square integrable. For such "big" controls one could study feedback stabilization and optimal control with quadratic cost functional. Though such a study is important, we shall not pursue it in depth because the main applications we have in mind for this introductory text have essentially bounded controls.

## 1.10   Economic Models with Delay

In the construction of dynamic economic models, the kind of "time", "continuous" or "discrete", dictates whether a differential equation or a difference equation is the model. It has been argued strongly and persuasively by G. Gandolfo that mixed differential-difference equations, functional differential equations are much more suitable than differential equations alone or difference equations alone for an adequate treatment of dynamic economic phenomena [20]. Functional differential equations were used by Kalecki to model capital stock. More recent studies include Gandolfo [20], Chukwu [19, 21] and [26]. Kalecki [27] argued that the growth of capital stock $x(t)$ of a single firm is given by

$$\dot{x}(t) = I(t) = a_0 x(t) + a_1 x(t - h), \tag{1.10.1}$$

where $a_i$ are constants and the delay $h$ represents the time lag between the decision to invest and the deliveries of capital equipment. The crucial assumption for (1.10.1) is that the net capital formation $\dot{x}(t)$ is given by $I(t)$. To obtain (1.10.1), Kalecki assumes that the decision to invest $B$ is given by

$$B(t) = a(1 - c)y(t) - kx(t) + \varepsilon,$$

$a$, $c$, $k$ are constants, $\varepsilon$ may be time varying, $y$ is income (output), $x(t)$ denotes the stock capital assets at time $t$. Later as an exercise, Kalecki suggests that the decision to invest should be

$$B(t) = a(1 - c)y(t) - kx(t) + \varepsilon + v\frac{dy}{dt}\,.$$

The outcome of this analysis is the functional differential equation of neutral type,

$$\frac{d}{dt}(x(t) - a_{-1}x(t - h)) = a_0 x(t) - a_1 x(t - h) + a_2\,.$$

This is a functional differential equation of neutral type which describes the growth of capital stock of a single firm. We can introduce $b(t)u(t)$ at the right-hand side of this system to obtain

$$\dot{x}(t) + a_{-1}\dot{x}(t - h) = a_0 x(t) + a_1 x(t - h) + b(t)u(t)\,.$$

We interpret $b(t)u(t)$ as follows. If $0 \leq u(t) \leq 1$, then $u(t)$ is the fraction of "available capital assets, $b(t)$" at time $t$ that is allocated to investment. If $-1 \leq u(t) \leq 0$, then $u(t)$ is the fraction allocated to consumption or for payment of taxes. Thus $-1 \leq u(t) \leq 1$ and $u(t)$ is an investment consumption strategy which is appropriated as a control to drive an initial capital endowment $\phi$ to a target while minimizing a cost function.

There are at least two ways that time delays emerge in the dynamics of economic variables: there is some time lag between the time economic decisions are made and the time the decisions bear fruit. See [35] and Chukwu [28, 29]. There is a "hidden" way, the way rational expectation (see Ray C. Fair [30], Luigi Amoroso [31], J. B. Taylor [32, Chap. 3]).

For dynamic economic systems we appropriate the argument of Fair that the "rational expectations hypothesis" can better be approximated by assuming that the expected values of the economic model are a function of the current and past values. Indeed we assume that aggregate demand $z$ is a sum of investment $I$, consumption $C$, net export $X$, and government outlay $G$:

$$z = I + C + X + G\,, \tag{1.10.2}$$

where

$$C = C_0 + C_1(y(t) - T(t)) + C_2(y(t - h) - T(t - h)) + C_3(\dot{y}(t) - \dot{T}(t))$$
$$+ C_4(\dot{y}(t - h)$$
$$- \dot{T}(t - h)) + C_5 R(t) + C_6 R(t - h) + C_7(ML)\,. \tag{1.10.3}$$

Here,

$$
\begin{aligned}
C &= \text{private consumption}\,, & M &= \text{money supply}\,, \\
\tilde{L} &= \text{money demand} = Md\,, & ML &= \tilde{L} - M\,,
\end{aligned}
\tag{1.10.4}
$$

$$\tilde{L} = M_0 + M_1 y(t) + M_1 y(t-h) + M_3 R(t) + M_4 R(t-h)$$

$$+ M_5 \dot{R}(t-h) + M_6 p(t) , \tag{1.10.5}$$

$$I = I_0 + I_1 y(t) + I_2 y(t-h) - I_3 \dot{y}(t) + I_4 \dot{y}(t-h) + I_5 R(t) + I_6 R(t-h)$$

$$+ I_8 L(t) + I_9 L(t-h) - I_{11} K(t) - I_{13}(M - \tilde{L}) , \tag{1.10.6}$$

where

$$K = \text{value of capital stock}, \quad L = \text{employment},$$

$$M = \text{money supply} = M1, \quad \tilde{L} = \text{Liquidity} = \text{money demand} = Md .$$

Also

$$G = g_0 + g_1 y(t) + g_2 y(t-h) + g_4 \dot{y}(t-h) + g_5 R(t) + g_8 L(t) , \tag{1.10.7}$$

$$(g_0 = \text{federal budget net expenditure}) \tag{1.10.8}$$

where

$$X = x - m = \text{Export} - \text{Import}$$

$$x = \text{export function} = x(p, e, d, \tau, y, L)$$

where $p$ = price, $y$ = income, $e$ = exchange rate, $\tau$ = tariffs, $L$ = employment. We know that $\partial x / \partial e < 0, \partial x / \partial p < 0$, if the elasticity of substitution among importables (ESM) is greater than 1 and if the elasticity of transformation between exportables (ETE) is greater than ETDF, the elasticity of transformation between production for domestic market and for foreign markets. We can therefore assume a linear model

$$x = x_0 + x_1 y(t) + x_2 y(t-h) + x_3 \dot{y}(t) + x_4 \dot{y}(t-h) + x_5 R(t) + x_6 L(t)$$

$$+ x_{10} \dot{L}(t-h) + x_{16} \tau(t) + x_{15} e(t) - x_{12} p(t) + z_{17} d(t) , \quad (x_{15} > 0, x_{12} > 0) . \tag{1.10.9}$$

(It can be assumed that

$$M = M_0 + M_1 y + M_2 p + M_3 e + M_4 L + M_5 \tau \tag{1.10.10}$$

where obviously, $M_1 > 0$, $M_2 > 0$, $M_3 > 0$.) Since income $y$ has a predictable positive effect on trade flows and increase in imports we have $M_1 > 0$. Thus net export

$$X = X_0 + X_1 y(t) + X_2 y(t-h) + X_3 \dot{y}(t) + X_4 \dot{y}(t-h) + X_5 R(t) + X_8 L(t)$$

$$+ X_{10} \dot{L}(t-h) + X_{12} p(t) + X_{16} \tau(t) + X_{15} e(t) + X_{17} d(t) , \tag{1.10.11}$$

where

$$X_0 = X_0 - M_0, \qquad X_1 = X_1 - M_1, \text{ etc}.$$

Therefore, on gathering results we have that aggregate demand

$$z = C + I + X + G$$
$$= z_0 + z_1 y(t) + z_2 y(t-h) + z_3 \dot{y}(t) + z_4 \dot{y}(t-h) + z_5 R(t) + z_6 R(t-h)$$
$$+ z_8 L(t) + z_9 L(t-h) + z_{10} \dot{L}(t-h) + z_{11} K(t) + z_{13} \dot{R}(t-h) + z_{18} p(t)$$
$$- z_{14} T(t) - z_{15} e(t) + z_{16} \tau(t) + z_{17} d(t) - z_{19} T(t-h)$$
$$- z_{20} \dot{T}(t) - z_{21} \dot{T}(t-h), \tag{1.10.12}$$

where

$$z_0 = g_0 + I_0 - M_0(C_7 + I_{13}) + C_0 + X_0,$$
$$z_1 = g_1 + I_1 - M_1(I_{13} + C_7) + C_1 + X_1,$$
$$z_2 = g_2 + I_2 + C_2 + X_2 + M_2(I_{13} + C_7),$$
$$z_3 = g_3 - I_3 + C_3 + x_3,$$
$$z_4 = g_4 + I_4 + C_4 + X_4 + M_4(C_7 + I_{13}),$$
$$z_5 = g_5 + I_5 + C_5 + X_5 - C_7 M_7,$$
$$z_6 = I_6 + C_6 - C_7 M_6,$$
$$z_8 = g_8 + I_8 + X_8,$$
$$z_9 = I_9,$$
$$z_{10} = X_{10},$$
$$z_{11} = -I_{11},$$
$$z_{13} = (I_{13} + C_7)M_6,$$
$$z_{14} = -C_1,$$
$$z_{15} = X_{15},$$
$$z_{16} = X_{16},$$
$$z_{17} = X_{17},$$
$$z_{18} = (C_7 + z_{13})M_6,$$
$$z_{19} = C_2,$$

$$z_{20} = C_3 \,,$$

$$z_{12} = C_4 \,.$$

By the market principle of supply and demand,

$$\frac{dy(t)}{dt} = \lambda_1 (z(t) - y(t)) \,, \tag{1.10.14}$$

where $\lambda_1$ is the speed of response of supply to demand, the speed of adjustment. The reciprocal of the speed of adjustment $(1/\lambda_1)$ is the mean time lag, i.e. the time necessary for about 63% of the discrepancy between $y$ and $z$ (or between the actual and desired value of the variable) to be eliminated [33, p. 94]. From (1.10.14) the following equation emerges:

$$\frac{dy}{dt} - a_{-11}\dot{y}(t-h) - a_{-13}\dot{L}(t-h) = \lambda_1 \sigma_1^{-1}[(z_1 - 1 - z_{13}M_1)y(t)$$

$$+ z_2 y(t-h) + (z_5 - z_{13}M_2)R(t)$$

$$+ z_6 R(t-h) + z_8 L(t)$$

$$+ z_9 L(t-h) + z_{11}k(t) - z_{13}M_3 p(t)$$

$$- z_{14}T(t) - z_{15}e(t)$$

$$+ z_{16}\tau(t) + z_{19}T(t-h) - z_{20}\dot{T}(t)$$

$$+ z_{21}\dot{T}(t-h) + z_{17}d(t)] + z_0 \lambda_1 \sigma_1^{-1} \,,$$

where

$$\sigma_1 = 1 - \lambda_1 z_3 \,, \qquad a_{-11} = \lambda_1 \sigma_1^{-1} z_4 \,, \qquad a_{-13} = \lambda_1 \sigma_1^{-1} z_{10} \,.$$

Set

$$a_{01} = \lambda_1 \sigma_1^{-1}(z_1 - 1 - z_{13}M_1) \,,$$

$$a_{11} = z_2 \lambda_1 \sigma_1^{1} \,,$$

$$a_{12} = (z_5 - z_{13}M_2)\lambda_1 \sigma_1^{-1} \,,$$

$$a_{13} = z_6 \lambda_1 \sigma_1^{-1} \,,$$

$$a_{14} = z_8 \lambda_1 \sigma_1^{-1} \,,$$

$$a_{15} = z_9 \lambda_1 \sigma_1^{-1} \,,$$

$$a_{16} = z_{11} \lambda_1 \sigma_1^{-1} \,,$$

$$a_{17} = z_{19} \lambda_1 \sigma_1^{-1} \,,$$

$$a_{18} = \lambda_1 \sigma_1^{-1} z_{13} M_3 \,,$$

and set

$$q_1(t) = \lambda_1 \sigma_1^{-1}[g_0 - z_{14}T(t) + z_{19}T(t-h) - z_{20}\dot{T}(t)$$

$$- z_{21}\dot{T}(t-h) - z_{15}e(t) + z_{16}\tau(t) + z_{17}d(t)], \qquad (1.10.17)$$

$$r_1(t) = \lambda_1 \sigma_1^{-1}[(C_0 + I_0 + X_0) - M_0(I_{13} + C_7)]. \qquad (1.10.18)$$

Then the dynamics of GNP in our economic system is

$$\frac{dy}{dt} - a_{-11}\dot{y}(t-h) - a_{13}\dot{L}(t-h) = a_{01}y(t) + a_{11}y(t-h) + a_{12}R(t) + a_{13}R(t-h)$$

$$+ a_{14}L(t) + a_{15}L(t-h)$$

$$+ a_{16}k(t) - a_{18}p(t) + q_1(t) + r_1(t). \qquad (1.10.19)$$

The rate of interest is determined by the typically Kenesian dynamics,

$$\frac{dR}{dt} = \lambda_2(L - M). \qquad (1.10.20)$$

This yields

$$\dot{R}(t) = \lambda_2 M_1 y(t) + \lambda_2 M_2 y(t-h) + \lambda_2 M_3 R(t) + \lambda_2 M_4 R(t-h)$$

$$+ \lambda M_5 \dot{R}(t-h) + \lambda_2 M_6 p(t) + \lambda M_0 - \lambda_2 M1. \qquad (1.10.21)$$

If we set

$$\begin{aligned}
q_2(t) &\equiv -\lambda_2 M1, & \sigma_2(t) &= -\lambda_2 M_0, \\
a_{-22} &= \lambda_2 M_5, & a_{21} &= \lambda_2 M_1, \\
a_{22} &= \lambda_2 M_2, & a_{23} &= \lambda_2 M_3, \\
a_{25} &= \lambda_2 M_6, & a_{24} &= \lambda_2 M_4,
\end{aligned} \qquad (1.10.22)$$

then

$$\dot{R}(t) - a_{-22}\dot{R}(t-h) = a_{21}y(t) + a_{22}y(t-h) + a_{33}R(t)$$

$$+ a_{24}R(t-h) + a_{25}p(t) - \sigma_2(t) + q_2(t). \qquad (1.10.23)$$

### 1.10.1 *Prices*

To obtain the dynamics of domestic prices we appropriate as our background the treatment by Gandolfo and Padoan [33]. Price of output is determined as

$$D\log p = \alpha_{12}\log(\hat{p}/p) + \alpha_{13}DM + \alpha_{14}\log(M/M_d) \qquad (1.10.24)$$

where

$$\hat{p} = \gamma_7 (p^f \cdot e)^{\beta_{12}} w^{\beta_{13}} \text{ prod} - \beta_{14}\,, \tag{1.10.25}$$

$$DM = \text{proportional rate of change of } M1$$
$$\text{or money supply} \tag{1.10.26}$$

$$M1 = \text{Nominal stock of money}$$

$$p^f = \text{import price level (in foreign currency)}$$

$$e = \text{exchange rate (country--dollar spot exchange rate)}$$

$$M_d = \tilde{L} = \text{money demand}$$

$$W = \text{money wage rate}$$

$$= \text{wage/labor}$$

$$\text{prod} = \text{labor productivity} = n\,. \tag{1.10.27}$$

We observe that

$$D \log p = \frac{\dot{p}(t)}{p(t)}\,, \tag{1.10.28}$$

and if we linearize we obtain

$$\frac{\dot{p}}{p(t)} = p_0 + p_1 p^f(t) \cdot e(t) + p_2 W(t) - p_3 n(t)$$
$$- p_4 p(t) + p_5 \dot{M}1 + p_6 (M1 - L)\,, \tag{1.10.29}$$

where $L$ is the liquidity function,

$$P_0 = \log \gamma_7 = \beta_{12} \gamma_7 (a_1 p_0^f e_0 + a_0) + \beta_{13}(C_0 \omega_0 - C_1 \omega_0)$$
$$+ \beta_{14}(p_0 - p_1 n_0)\,. \tag{1.10.30}$$

Thus

$$\dot{p}(t) = p(t)[p_0 + p_1 p^f(t) \cdot e(t) + p_2 w(t) - p_3 n(t)$$
$$- p_4 p(t) + p_5 \dot{M}_1 + p_6 (M - L)]\,. \tag{1.10.31}$$

Observe that

$$p_6(M - L) = p_6(M - M_0 - M_1 y(t) - M_2 y(t - h) - M_3 R(t)$$
$$- M_4 R(t - h) - M_6 p(t) - M_7 \dot{R}(t - h))\,, \tag{1.10.32}$$

where

$$L = M_0 + M_1 y(t) + M_2 y(t-h) + M_3 R(t)$$
$$+ M_4 R(t-h) + M_6 p(t) + M_5 \dot{R}(t-h)\,. \tag{1.10.33}$$

Set

$$q_5(t) = p_1 p^f(t) \cdot e(t) + p_5 \dot{M}_1(t) + p_6 M_1(t)\,,$$
$$\sigma_5(t) = p_0 - p_3 n(t) + p_2 w(t) - p_6 M_0(t)\,. \tag{1.10.34}$$

Then

$$\dot{p}(t) = p(t)[(M_6 + p_4)p(t)] - p_6 M_1 y(t)$$
$$- p_6 y(t-h) - p_6 M_3 R(t) - M_4 p_6 R(t-h)$$
$$- p_6 M_7 \dot{R}(t-h) + [q_5(t) + \sigma_5(t)]p(t)\,. \tag{1.10.35}$$

### 1.10.2  *Balance of payment*

The Balance of Payment is given by $B = X - F - T$ where $X$ is defined in (1.10.11) and

$$F + T = f_0 + f_1 R(t) + f_2 R(t-h) + f_4 \dot{R}(t-h)$$
$$+ f_5 y(t) + f_6 y(t-h) + f_8 L(t) + f_9 L(t-h)$$
$$+ f_{11} \dot{L}(t-h) + f_{12} E(t-h) + f_{13} \dot{E}(t-h)\,. \tag{1.10.36}$$

Hence

$$B = b_0 + b_1 y(t) + b_2 y(t-h) + b_3 p(t)$$
$$+ b_4 \dot{y}(t-h) + b_5 R(t) + b_6 R(t-h)$$
$$+ b_7 e(t) + b_8 \dot{R}(t-h) + b_9 L(t)$$
$$+ b_{10} L(t-h) + b_{12} \dot{L}(t-h)$$
$$+ b_{15} d(t) + b_{13} \tau(t) + b_{17} B(t-h)\,, \tag{1.10.37}$$

where

$$p, e, d = \text{have the usual meaning, and where } E$$
$$\text{is the cumulative balance of payment}\,,$$

$$E(t) = \int_0^t B(s)ds\,. \tag{1.10.38}$$

Thus the differential equation for the cumulative balance of payment is

$$\frac{dE}{dt} = B(t) \,; \tag{1.10.39}$$

or

$$\frac{dE(t)}{dt} = b_0 + b_1 y(t) + b_2 y(t-h) + b_4 \dot{y}(t-h) + b_5 R(t)$$
$$+ b_6 R(t-h) + b_8 \dot{R}(t-h) + b_9 L(t) + b_{10} L(t-h)$$
$$+ b_{12} \dot{L}(t-h) + b_3 p(t) + b_7 e(t) + b_{15} d(t)$$
$$+ b_{13} \tau(t) + b_{17} B(t-h) \,. \tag{1.10.40}$$

On setting

$$X_0 = x_0 - M_0 \,, \qquad b_0 = X_0 - f_0 \,,$$

and

$$-r_6(t) = x_0 - m_0 = X_0 \,. \tag{1.10.41}$$

Import quotas and taxes (tariff, which is privately firm generated and partially controlled); and

$$q_6(t) = b_7 e(t) + b_8 \tau(t) + b_{15} d(t) - f_0 \,, \tag{1.10.42}$$

government control instruments: exchange rate, $e$, tariffs, foreign credit, interest equalization tax, $f_0$, preferential arrangement (which reduce trade barriers and enhance trade flows between nations) $d$, transportation and distance between partners. The dynamics of cumulative balance of payment becomes

$$\dot{E}(t) = b_1 y(t) + b_2 y(t-h) + b_4 \dot{y}(t-h) + b_5 R(t)$$
$$+ b_6 R(t-h) + b_8 \dot{R}(t-h) + b_9 L(t) + b_{10} L(t-h)$$
$$+ b_{12} \dot{L}(t-h) + b_{17} B(t-h) - r_6(t) + q_6(t) \,. \tag{1.10.43}$$

### 1.10.3 *Employment and capital stock*

We now derive the equation of Capital Stock and Employment. The definition of national income from the expenditure side is

$$y = \tilde{C} + \tilde{I} + \tilde{X} + \tilde{G} \,, \tag{1.10.44}$$

where

$$\tilde{C} = y_{10} + y_{11} y(t) + y_{12} \dot{y}(t) + y_{13} R(t) + y_{15}(M - \tilde{L}) + y_{18}(y - T) \,, \tag{1.10.45}$$
$$\tilde{I} = I_0 + I_1 \,,$$

and where

$$I_1 = \frac{1}{h} \int_{t-h}^{t} D(s)ds\,, \tag{1.10.46}$$

and

$$\frac{dk(t)}{dt} = D(t-h)\,, \tag{1.10.47}$$

which is the rate of deliveries of new equipment. We postulate that

$$D(t) = a(1-c)y(t) - k_0 k(t) + k_{13}\dot{k}(t) + L_5 L(t) + L_6 p(t) + v\dot{y}(t)\,. \tag{1.10.48}$$

Now,

$$I_1(t) = \frac{1}{h} \int_{t-h}^{t} \frac{dk(s+h)ds}{ds} = \frac{1}{h} \int_{t}^{t+h} \frac{dk(\tau)d\tau}{d\tau}$$

$$= \frac{1}{h} k(t) \Big|_{t}^{t+h} = \frac{1}{h}[k(t+h) - k(t)]\,. \tag{1.10.49}$$

Also

$$\tilde{X} = x_0 + x_1 y(t) + x_2 y(t-h) + x_5 R(t) + x_8 L(t)$$

$$+ X_{10}\dot{L}(t) + x_{11}e(t) + x_{12}\tau(t) + x_{13}d(t)\,, \tag{1.10.50}$$

$$\tilde{G} = g_{s0} + g_{s1} y(t) + g_{s4}\dot{y}(t) + g_{s5} R(t) + g_{s8} L(t)\,.$$

Thus

$$y(t) = z_{s0} + z_{s1} y(t) + z_{s2} y(t-h) + z_{s4}\dot{y}(t) + z_{s5} R(t)$$

$$+ z_{s8} L(t) + z_{s10}\dot{L}(t) + z_{s13} M_1 - z_{s14} T(t) + z_{s15}e(t)$$

$$+ z_{s16}\tau(t) + z_{s17}d(t) + \frac{1}{h}[k(t+h) - k(t)]\,. \tag{1.10.51}$$

It follows that

$$y(t) = (1 - z_{s1})^{-1}[z_{s0} + z_{s2} y(t-h) + z_{s4}\dot{y}(t)$$

$$+ z_{s5} R(t) + z_{s8} L(t) + z_{s10}\dot{L}(t) + z_{s13} M_1$$

$$- z_{s14} T(t) + z_{s15}e(t) + z_{s16}\tau(t) + z_{s17}d(t)]$$

$$+ \frac{1}{(1 - z_{s1})h}[k(t+h) - k(t)]\,. \tag{1.10.52}$$

Substitute $D(t) = \dot{k}(t+h)$ and (1.10.52) into (1.10.48) then

$$\dot{k}(t+h) = \frac{a(1-c)}{(1 - z_{s1})h}[k(t+h) - k(t)]$$

$$+ \frac{a(1-c)}{1-z_{s1}}[z_{s0} + z_{s13}M1 - z_{s14}T(t) + z_{s15}e(t)$$

$$+ z_{s16}\tau(t) + z_{s17}d(t)]$$

$$+ \frac{a(1-c)}{1-z_{s1}}[z_{s1}y(t) + z_{s4}\dot{y}(t) + z_{s5}R(t) + z_{s8}L(t)$$

$$+ z_{s10}\dot{L}(t) + z_{s11}k(t) + k_{13}\dot{k}(t)]$$

$$+ L_4 R(t) + L_5 L(t) + L_6 p(t). \tag{1.10.53}$$

Now to switch $t$ to $t - h$ to deduce that

$$\dot{k}(t) + a_{-1}\dot{k}(t-h) = a_0 k(t) + a_1 k(t-h) + a_2 y(t-h)$$

$$+ a_3 \dot{y}(t-h) + a_4 R(t-h) + a_5 L(t-h)$$

$$+ a_6 \dot{L}(t-h) + a_8 p(t-h) + b_0 q_0 + b_1 p_1, \tag{1.10.54}$$

where

$$a_{-1} = -k_{13}\left(\frac{a(1-c)}{1-z_{s1}}\right),$$

$$a_0 = \frac{a(1-c)}{(1-z_{s1})h},$$

$$a_1 = \frac{-a(1-c)}{1-z_{s1}}(1/h)z_{s11},$$

$$a_2 = \frac{a(1-c)}{(1-z_{s1})}(z_{s1}),$$

$$a_3 = \frac{a(1-c)}{1-z_{s1}}z_{s4},$$

$$a_4 = \frac{a(1-c)}{(1-z_{s1})}z_{s5} + L_4, \tag{1.10.55}$$

$$a_5 = \frac{a(1-c)}{(1-z_{s1})} \cdot z_{s8} + L_5,$$

$$a_6 = \frac{a(1-c)}{(1-z_{s1})} \cdot z_{s10},$$

$$a_8 = L_6,$$

$$b_0 q_0 + b_1 p_1 = \frac{a(1-c)}{(1-z_{s1})}[z_{s0} + z_{s13}M1 - z_{s14}T(t) + z_{s15}e(t)$$

$$+ z_{s16}\tau(t) + z_{s17}d(t)],$$

$$z_{s0} = g_0 + x_0 + y_{10} + I_0.$$

Let

$$q_4 = g_0 + z_{s13}M1 - z_{s14}T(t) + z_{s15}e(t) + z_{s16}\tau(t) + z_{s17}d(t);  \tag{1.10.56}$$

and

$$\sigma_4(t) = x_0 + y_{10} + I_0;  \tag{1.10.57}$$

then the dynamics of capital stock emerges as

$$\dot{k}(t) + a_{-1}\dot{k}(t-h) - a_3\dot{y}(t-h) - a_6\dot{L}(t-h)$$

$$= a_0 k(t) - a_1 k(t-h) + a_2 y(t-h) + a_4 R(t-h)$$

$$+ a_5 L(t-h) + a_8 p(t) + \sigma_4(t) + q_4(t).  \tag{1.10.58}$$

**Remark**  We note that the coefficients of the dynamics in (1.10.54) and therefore of (1.10.58) are identified in (1.10.55). These are obtained by a simple MATLAB linear regression using the arx command on (1.10.47), (1.10.48), (1.10.49), (1.10.50), (1.10.51) and (1.10.52). To obtain the functional differential equation satisfied by employment we recall the Cobb–Douglas equation

$$y = f(k, L) = k^\alpha L^{1-\alpha};  \tag{1.10.59}$$

and the relation

$$L = m(\omega)k, \qquad \dot{L}(t) = m(\omega)\dot{k}(t),  \tag{1.10.60}$$

where

$$m(\omega) = \left[(1-\alpha)\frac{1}{\omega}\right]^{1/\alpha}.  \tag{1.10.61}$$

Because

$$\dot{k}(t) = \frac{\dot{L}(t)}{m(\omega)}.  \tag{1.10.62}$$

The capital stock equation becomes

$$\frac{\dot{L}(t)}{m(\omega)} + \left(\frac{a_{-1}}{m(\omega)} - a_6\right)\dot{L}(t-h) - a_3\dot{y}(t-h) = \frac{a_0}{m(\omega)}L(t) - \frac{a_1}{m(\omega)}L(t-h)$$

$$+ a_2 y(t-h) + a_4 R(t-h)$$

$$+ a_5 L(t-h) + a_8 p(t)$$

$$+ \sigma_4(t) + q_4(t).  \tag{1.10.63}$$

Multiply both sides of this equation by $m(\omega)$:

$$\dot{L}(t) + (a_1 - m(\omega)a_6)\dot{L}(t - h) - a_3 m(\omega)\dot{y}(t - h)$$

$$= a_0 L(t) - a_1 L(t - h) + m(\omega)a_2 y(t - h)$$

$$+ a_4 m(\omega)R(t - h) + a_5 m(\omega)L(t - h)$$

$$+ m(\omega)a_8 p(t) + m(\omega)\sigma_4(t) + m(\omega)q_4(t). \qquad (1.10.64)$$

Recall that profit $P = y - \omega L - rK$ where $\omega$ is the wage of labor per unit time, $r$ is the rent per unit time of the use of capital. We built into our model the maximization of profit so that

$$\frac{\partial P}{\partial L} = 0, \qquad \frac{\partial P}{\partial k} = 0. \qquad (1.10.65)$$

As a consequence $L = m(\omega)k$ where

$$m(\omega) = \left[(1 - \alpha)\frac{1}{\omega}\right]^{1/\alpha}.$$

Let

$$
\begin{aligned}
a_{-1} - m(\omega)a_6 &= l_{-01}, & a_3 m(\omega) &= l_{-03}, \\
a_0 &= l_0, & -a_1 &= l_1, \\
m(\omega)a_2 &= l_2, & m(\omega)a_4 &= l_4, & (1.10.66) \\
m(\omega)a_5 &= l_5, & m(\omega)a_8 &= l_8, \\
m(\omega)\sigma_4(t) &= -\sigma_3(t), & m(\omega)q_4(t) &= q_3.
\end{aligned}
$$

Then the dynamics which we are seeking is

$$\dot{L}(t) - l_{-01}\dot{L}(t - h) - l_{-03}\dot{y}(t - h) = l_0 L(t) - l_1 L(t - h) + l_2 y(t - h) + l_4 R(t - h)$$

$$+ l_5 L(t - h) + \sigma_3(t) + q_3(t). \qquad (1.10.67)$$

The equations we have displayed can be put in matrix form as follows

$$\dot{x}(t) - A_{-1}\dot{x}(t - h) = A_0 x(t) + A_1 x(t - h) - \sigma_{(t)} + q(t), \qquad (1.10.68)$$

where

$$x = \begin{bmatrix} y \\ R \\ L \\ k \\ p \\ E \end{bmatrix}, \qquad (1.10.69)$$

$$A_{-1} = \begin{bmatrix} a_{-11} & 0 & a_{-13} & 0 & 0 & 0 \\ 0 & a_{-22} & 0 & 0 & 0 & 0 \\ I_{-03} & 0 & -l_{01} & 0 & 0 & 0 \\ a_3 & 0 & a_6 & -a_{-1} & 0 & 0 \\ 0 & -M_7 p_6 p(t) & 0 & 0 & 0 & 0 \\ b_4 & b_8 & b_{12} & 0 & 0 & 0 \end{bmatrix}, \tag{1.10.70}$$

$$A_0 = \begin{bmatrix} a_{01} & a_{12} & a_{14} & a_{16} & -a_{18} & 0 \\ a_{21} & a_{23} & 0 & 0 & q_{25} & 0 \\ 0 & 0 & a_{33} & 0 & a_{35} & 0 \\ 0 & 0 & 0 & a_{44} & a_{45} & 0 \\ a_{51} & a_{52} & 0 & 0 & a_{55} & 0 \\ a_{61} & a_{62} & a_{63} & 0 & 0 & 0 \end{bmatrix}, \quad \text{or} \tag{1.10.71}$$

$$A_0 = \begin{bmatrix} a_{01} & a_{12} & a_{14} & a_{16} & -a_{18} & 0 \\ a_{21} & a_{23} & 0 & 0 & a_{25} & 0 \\ 0 & 0 & l_0 & 0 & M a_8 & 0 \\ 0 & 0 & 0 & a_0 & a_8 & 0 \\ -p_6 M_1 p & -p_6 M_3 p & 0 & 0 & (M_6 + p_4)p & 0 \\ b_1 & b_5 & b_9 & 0 & 0 & 0 \end{bmatrix}.$$

$$a_{01} = \gamma_1 \sigma_1^{-1}(z_1 - 1 - z_{13} M_1),$$

$$a_{12} = (z_5 - z_{13} M_2)\gamma_1 \sigma^{-1},$$

$$a_{14} = \gamma_1 \sigma_1^{-1} z_8 = \gamma_1 \sigma_1^{-1}(g_8 + I_8 + x_8),$$

$$a_{16} = \gamma_1 \sigma_1^{-1} z_{11} = \gamma_1 \sigma_1^{-1}(-I_{11}),$$

$$a_{18} = \gamma_1 \sigma_1^{-1} z_{13} M_3 = -\gamma_1 \sigma_1^{-1} M_3 M_5 (I_{13} + C_7),$$

$$a_{21} = \gamma_2 M_1,$$

$$a_{23} = \gamma_2 M_3,$$

$$a_{25} = \gamma_2 M_6,$$

$$a_{33} = a_0 = l_0 = \frac{a(1-c)}{(1-z_{s1})h},$$

$$a_{35} = l_8 = m(\omega)a_8 = \left[(l-\alpha)\frac{1}{\omega}\right]^{1/\alpha} L_6, \tag{1.10.72}$$

$$a_{44} = a_0 = \frac{a(1-c)}{(1-z_{s1})h}\,,$$

$$a_{45} = a_8 = L_6\,,$$

$$a_{51} = -p_6 M_1 p(t)\,,$$

$$a_{52} = -p_6 M_3 p(t)\,,$$

$$a_{55} = -(M_6 + p_4)p(t)\,,$$

$$a_{61} = b_1\,,$$

$$a_{62} = b_5\,,$$

$$a_{63} = b_9\,.$$

$$A_1 = \begin{bmatrix} a_{111} & a_{112} & a_{113} & 0 & 0 & 0 \\ a_{121} & a_{122} & 0 & 0 & 0 & 0 \\ a_{131} & a_{132} & a_{133} & 0 & 0 & 0 \\ a_{141} & a_{142} & a_{143} & -\tilde{a}_1 & 0 & 0 \\ a_{151} & a_{152} & 0 & 0 & 0 & 0 \\ a_{161} & a_{162} & a_{163} & a_{164} & a_{165} & a_{166} \end{bmatrix} \qquad (1.10.73)$$

$$\text{or} \qquad A_1 = \begin{bmatrix} a_{11} & a_{13} & 1_{15} & 0 & 0 & 0 \\ a_{22} & a_{24} & 0 & 0 & 0 & 0 \\ l_2 & l_4 & l_5 - l_1 & 0 & 0 & 0 \\ a_2 & a_4 & a_5 & -a_1 & 0 & 0 \\ -p_6 p(t) & -M_4 p_6 p(t) & 0 & 0 & 0 & 0 \\ b_2 & b_6 & b_{10} & 0 & 0 & b_{17} \end{bmatrix}$$

where

$$a_{111} = a_{11} = z_2 \gamma_1 \sigma_1^{-1} = \gamma_1 \sigma_1^{-1}[g_2 + I_2 + C_2 + x_2 + M_2(I_{13} + C_7)]\,,$$

$$a_{112} = a_{13} = \gamma_1 \sigma_1^{-1}[I_6 + C_6 - C_7 M_6]\,,$$

$$a_{113} = a_{15} = \gamma_1 \sigma_1^{-1} z_9 = \gamma_1 \sigma_1^{-1} I_9\,,$$

$$a_{114} = 0\,,$$

$$a_{115} = 0\,,$$

$$a_{116} = 0\,,$$

$$a_{121} = a_{22} = \gamma_2 M_2\,,$$

$$a_{122} = a_{24} = \gamma_2 M_4 \,,$$

$$a_{123} = 0 \,,$$

$$a_{124} = 0 \,,$$

$$a_{125} = 0 \,,$$

$$a_{126} = 0 \,,$$

$$a_{131} = l_2 = m(\omega)\frac{a(1-c)}{(1-z_{s1})}z_{s2} \,,$$

$$= \frac{m(\omega)a(1-c)x_2}{1-(x_1+g_{s1}+y_{11})} \,,$$

$$a_{132} = l_4 = m(\omega)a_4 = \frac{a(1-c)}{(1-z_{s1})}z_{s5} + L_4$$

$$a_{133} = l_5 - l_1 = m(\omega)a_5 - l_1 = \frac{m(\omega)a(1-c)z_{s8}}{1-z_{s1}} + L_5 - l_1 \,,$$

$$a_{134} = 0 \,,$$

$$a_{135} = 0 \,,$$

$$a_{136} = 0 \,,$$

$$a_{141} = a_2 = \frac{a(1-c)}{(1-z_{s1})}z_{s2} \,,$$

$$a_{142} = a_4 = \frac{a(1-c)}{(1-z_{s1})}z_{s5} + L_4 \,,$$

$$a_{143} = a_5 = \frac{a(1-c)}{(1-z_{s1})} \cdot z_{s8} + L_5$$

$$a_{144} = 0 = a_{145} = a_{146} \,,$$

$$a_{151} = -p_6 p(t) \,,$$

$$a_{152} = -M_4 p_6 p(t) \,,$$

$$a_{153} = 0 \,,$$

$$a_{154} = 0 \,,$$

$$a_{155} = 0 \,,$$

$$a_{156} = 0 \,,$$

$$a_{161} = b_2 \,,$$

$$a_{162} = b_6 \,,$$

$$a_{163} = b_{10} \,,$$

$$a_{164} = 0\,,$$

$$a_{165} = 0\,,$$

$$a_{166} = b_{17}\,. \tag{1.10.74}$$

Let

$$q = \begin{bmatrix} T_1 \\ g_0 \\ e \\ \tau \\ d \\ M_1 \\ \dot{M}_1 \\ f_0 \end{bmatrix}, \tag{1.10.75}$$

where $T_1 = -z_{14}T(t) + z_{19}T(t-h) - z_{20}\dot{T}(t) - z_{21}\dot{T}(t-h)$.

$$\sigma = \begin{bmatrix} C_0 \\ I_0 \\ X_0 \\ M_0 \\ n \\ w \\ x_0 \\ y_{10} \\ p_0 \end{bmatrix}, \tag{1.10.76}$$

$$\xi_1 = \gamma_1 \sigma_1^{-1}\,,$$

$$q_1 = \gamma_1 \sigma_1^{-1}[g_0 + z_{13}M1 + T_1 - z_{15}e(t) + z_{16}\tau(t) + z_{17}d(t)]\,,$$

$$\sigma_1(t) = \gamma_1 \sigma_1^{-1}[C_0 + I_0 + X_0 - z_{13}M_0]\,,$$

$$q_2(t) = -\gamma_2 M1\,,$$

$$-\sigma_2(t) = \gamma_2 M_0\,,$$

$$q_5(t) = p_1 p^f(t) \cdot e(t) + p_5 \dot{M}1 + p_6 M1\,, \tag{1.10.77}$$

$$\sigma_5(t) = P_0 - p_3 n(t) + p_2 w(t) - p_6 M_0\,,$$

$$q_4(t) = g_0 + z_{s13}M - z_{s14}T(t) + z_{s15}e(t) + z_{s16}\tau(t) + z_{s17}d(t)\,,$$

$$\sigma_4(t) = x_0 + y_{10} + I_0 \,,$$

$$q_3(t) = m(\omega)q_4(t) \,,$$

$$\sigma_3(t) = m(\omega)\sigma_4(t) \,.$$

$$q_6 = b_7 e(t) + b_8 \tau(t) + b_{15} d(t) - f_0 \,,$$

$$-\sigma_6(t) = x_0 - M_0 = X_0 \,. \tag{1.10.78}$$

$$B_1 = \begin{bmatrix}
-\xi_1 & \xi_1 & -z_{15}\xi_1 & -\xi_1 z_{16} & -\xi_1 z_{17} & \xi_1 z_{18} & 0 & 0 \\
0 & 0 & 0 & 0 & 0 & -\gamma_2 & 0 & 0 \\
-m(\omega)z_{s14} & m(\omega) & m(\omega)z_{s15} & mz_{s16} & mz_{s17} & mz_{s13} & 0 & 0 \\
-z_{s14} & 1 & z_{s15} & z_{s16} & z_{s17} & z_{s13} & 0 & 0 \\
0 & 0 & p_1 p f(t) & 0 & 0 & p_6 & p_5 & 0 \\
0 & 0 & b_7 & b_8 & b_{15} & 0 & 0 & -1
\end{bmatrix} \,, \tag{1.10.79}$$

$$B_2 = \begin{bmatrix}
-\xi_1 & \xi_1 & \xi_1 & \xi_1(I_{13} + C_7) & 0 & 0 & 0 & 0 & 0 \\
0 & 0 & 0 & -\gamma_2 & 0 & 0 & 0 & 0 & 0 \\
0 & m(\omega) & 0 & 0 & 0 & 0 & m & m & 0 \\
0 & 1 & 0 & 0 & 0 & 0 & 1 & 1 & 0 \\
0 & 0 & 0 & -p_6 & -p_3 & p_2 & 0 & 0 & 1 \\
0 & 0 & 0 & -1 & 0 & 0 & 1 & 0 & 0
\end{bmatrix} \,. \tag{1.10.80}$$

Then

$$\dot{x}(t) + A_{-1}\dot{x}(t - h) = A_0 x(t) + A_1 x(t - h) + B_1 q + B_2 \sigma \,. \tag{1.10.81}$$

Compared with (1.10.81):

$$D(t, x_t) = x(t) + A_{-1}x(t - h) \,,$$

$$\frac{d}{dt}(D(t, x_t)) = \frac{d}{dt}(x(t) - A_{-1}x(t - h)) \,,$$

$$f(t, x_t, u(t)) = A_0 x(t) + A_1 x(t - h) + u(t) \,,$$

where

$$u(t) = B_1 q(t) + B_2 \sigma(t) \,.$$

We consider the general nonlinear functional differential equation

$$\dot{x}(t) - A_{-1}\dot{x}(t - h) = f(t, x_t, u(t)) + B(t, x_t)u(t) \,, \tag{1.10.82}$$

where $f : E \times C^0 \times E^m \to E^n$ is a nonlinear function. Here $E^r$ is the $r$-dimensional Euclidean space with norm $|\cdot|$. The symbol $C^0$ denotes the space on continuous functions mapping the interval $[-h, 0]$, $h > 0$, into $E^n$, with the sup norm $\|\cdot\|$, defined by $\|\Phi\| = \sup_{-h \le s \le 0} |\Phi(s)|$, $\Phi \in C$. The control matrix function $B :$ $E \times C^0 \to E^{n \times m}$ is possibly nonlinear. The controls are square integrable functions $u \in L_2([\sigma, t_1], E^m)$, $\sigma, t_1 \in E$, $t_1 > \sigma$, and $L_2$ is the space of measurable functions $u$ defined on intervals $[\sigma, t_1]$ for which $|u|^2$ is summable. If $t \in [\sigma, t_1]$, we let $x_t \in C^0$ be defined by $x_t(s) = x(t + s)$, $-h \le s \le 0$. With $L_2$ as the set of admissible controls the state space is either $E^n$, or $W_2^{(1)}$, the Sobolev space of absolutely continuous functions $x : [-h, 0]$ with the property that $t \to \dot{x}(t) \in L_2([-h, 0], E^n)$. The targets are points in $E^n$ or function in $W_2^{(1)}$.

The concept of function space controllability is very appropriate. Once a target is hit, it remains (hopefully) on a certain growing function for some time $T$. For example, once employment is brought to a desired level, it is important to keep it there at that level for some $T$.

The control variable of our economic system are of two kinds and of the form

$$u = B_1 p - B_2 g \,;$$

where $g$ is the control instrument of government (taxes, money, supply, public consumption, exchange rate, subsidy, preferential trade arrangement, tariff), and where $p$ is the control instrument of private initiative (autonomous consumption, investment, net export, money holding wages productivity) $B_i$, $i = 1, 2$ are the respective control matrices. It can be proved that any function target $x_\tau = (y_\tau, R_\tau, L_\tau, K_\tau, p_\tau, E_\tau)$, can be reached from any position if and only if the number of effective control instruments is equal to the number $(S)$ of target functions. This is a resurgence of "Tinbegen's static controllability condition". There exist a set of policy instruments which is capable of moving the initial function state into some other desired position in a finite time. As a result levels of national income, interest rate, employment, prices, value of capital stock and cumulative balance of payment can be controlled simultaneously. As a result, inflation and employment can be controlled at the same time, provided all the control instruments are in force. This seems to be a basic insight and argument of Robert Eisner in his recent book [34, Chap. 8]. See the Theorem reported in Chukwu [29, pp. 81–89].

## References


1. E. S. Armstrong and J. S. Tripp, "An Application of Multivariate Design Techniques to the Control of the National Transonic Facility," NASA Technical paper 1887, 1981.
2. H. T. Banks, "Modelling and Control in Biomedical Sciences," *Lecture Notes in Biomathetics* **6**, Springer-Verlag (1975).
3. G. A. Kent, "Optimal Control of Functional Differential Equations of Neutral Type," Ph.D. Thesis, Brown University (1971).

4. A. Manitius and H. Tran, "Numerical Simulation of a Nonlinear Feedback Controller for a Wind Tunnel Model Involving a Time Delay," *Optimal Control Application and Methods* **7** (1986) 19–39.

5. N. Minorsky, *Nonlinear Oscillations*, D. Van Nostrand Co., Inc., Princeton (1962).

6. N. Minorsky, "Self-Existed Oscillations in Dynamical Systems Possessing Retarded Actions", *J. Appl. Mech.* **9** (1942) 65–71.

7. D. Salamon, *Control and Observation of Neutral Systems*, Pitman Advanced Publishing Program, Boston (1984).

8. M. Slemrod, "The Flip-Flop Circuit as a Neutral Equation" in *Delay and Functional Differential Equations and Their Applications*, 387–392, Academic Press, New York (1972).

9. V. Volterra, "Sur la théorie mathématique des phénomènes héréditaires," *J. Math. Pures Appl.* **7** (1928) 249–298.

10. A. Takayama, *Mathematical Economics*, 2nd Edition, Cambridge University Press (1984).

11. K. J. Arrow, *Production and Capital*, Collected Papers of Kenneth J. Arrow, The Belknap Press of Harvard University Press, Cambridge, Massachusetts (1985).

12. K. L. Cooke and J. A. Yorke, "Equation Modelling Population Growth, Economic Growth and Gonorrhea Epidemiology," in *Ordinary Differential Equations*, edited by L. Weiss, Academic Press, New York, 1972.

13. F. Brauer, "Epidemic Models in Population of Varying Size," in "Mathematical Approaches to Problems in Resource Management and Epidemiology," C. Castillo-Chavez, S. A. Levin, and C. A. Shoemaker (eds.), *Lecture Notes in Biomathematics* **81**, Springer-Verlag, 1989.

14. C. Castillo-Chavez, K. Cooke, W. Huang, and S. A. Levin, "The Role of Long Periods of Infectiousness in the Dynamics of Acquired Immunodeficiency Syndrome (AIDS)," in "Mathematical Approaches to Problems in Resource Management and Epidemiology," C. Castillo-Chavez, S. A. Levin, and C. A. Shoemaker (eds.), *Lecture Notes in Biomathematics* **81**, Springer-Verlag, 1989.

15. E. N. Chukwu, "Mathematical Control of AIDS Epidemic" Preprint.

16. J. Hale, *Theory of Functional Differential Equations*, Springer-Verlag, New York, 1977.

17. O. Lopes, "Forced Oscillations in Nonlinear Neutral Differential Equations, *SIAM J. Appl. Math.* **29** (1975) 196–207.

18. D. Salamon, *Control and Observation of Neutral Systems*, Pitman Advanced Publishing Program, Boston (1984).

19. E. N. Chukwu, "Stability and Time Optimal Control of Hereditary Systems," Academic Press, New York (1992).

20. G. Gandolfo, *Economic Dynamics: Methods and Models*, North-Holland, Amsterdam, 1980, p. 519.

21. E. N. Chukwu, "Mathematical Controllability Theory of Capital Growth of Nations," *Applied Mathematics and Computation* **52** (1992) 317–344.

22. J. Hale, "Forward and Backward Continuation for Neutral Function Differential Equations," *J. Diff. Eq.* **9** (1971) 168–181.

23. W. R. Melvin, "A Class of Neutral Functional Differential Equations," *J. Diff. Eq.* **12** (1972) 524–534.

24. W. R. Melvin, "Topologies for Neutral Functional Differential Equation," *J. Diff. Eq.* **13** (1973) 24–31.

25. R. A. Adams, *Sobolev Spaces*, Academic Press, New York, 1975.

26. E. N. Chukwu, "Optimal Control of the Growth of Income of Nations," *Applied Mathematics and Computation* **62** (1994) 279–309.

27. M. A. Kalecki, "Macrodynamic Theory of Business Cycles," *Econometrica* **3** (1935) 327–344.

28. E. N. Chukwu, "Control in $W_2^{(1)}$ of Nonlinear Interconnected Systems of Neutral Type," *J. Austr. Math. Soc. Series B* **62** (1994) 286–312.

29. E. N. Chukwu, "Control of Interconnected Nonlinear Delay Differential Equations in $W_2^{(1)}$," *Proc. Indian Acad. Science* **105** (1995) 73–98.

30. R. C. Fair, *Specification, Estimation, and Analysis of Macroeconometric Models*, Harvard University Press, Cambridge, MA, 1984.

31. L. Amoroso, *Economia di mercato Bologna Zuggi*, 1949.

32. J. B. Taylor, *Macroeconomic Policy in a World Economy*, W. W. Norton, New York, 1993.

33. G. Gandolfo and P. C. Padoan, *The Italian Continuous Time Model Theory and Empirical Results, Economic Modeling*, Butterworth (1990) 91–132.

34. R. Eisner, *The Misunderstood Economy: What Counts and How to Count It*. Harvard Business School Press, Cambridge, MA, 1984.

35. R. G. D. Allen, *Mathematical Economics*, Macmillan, London, 1973.

36. E. N. Chukwu, "On the Controllability of Nonlinear Economic Systems with delay, The Italian Example Applied Math and Computation," 95(2 and 3) September 15, 1998.

37. E. N. Chukwu, "Differential Models and Neutral Systems for Controlling the Wealth of Nations, World Scientific, Singapore, 2001. [For graphs].

# Chapter 2

# General Linear Equations

## 2.1 The Fundamental Matrix of Retarded Equations

Let $E = (-\infty, \infty)$, and $E^n$ be a real $n$-dimensional Euclidean vector space. Let $h > 0$. Suppose $C = C([-h, 0], E^n)$, the space of continuous functions from the interval $[-h, 0]$ into $E^n$ with the sup norm $\|\cdot\|$ defined by $\|\phi\| = \sup_{s \in [-h,0]} |\phi(s)|$. Consider the linear retarded differential equation

$$\dot{x}(t) = A_0 x(t) + \sum_{i=1}^{N} A_i x(t - i\tau) + f(t) , \tag{2.1.1}$$

$$x(0) = x_0 \in E^n , \quad x(s) = \phi(s) , \quad s \in [-\tau N, 0) , \quad \tau > 0 ,$$

where $A_i$, $i = 0, \ldots, N$ are constant $n \times n$ matrices, $f$ is continuous on $E$, and $x$ is an $n$ vector. If $\phi(0) = x_0$, and $\tau N = h$ and $\phi \in C$, then a unique solution $x(\phi, f)$ of (2.1.1) exists on $[-h, \infty)$ and coincides with $\phi$ on $[-h, 0]$. The proof of this is contained in Hale [3, pp. 14, 142]. We now represent this solution in terms of the fundamental matrix of the system

$$\dot{x}(t) = A_0 x(t) , \qquad x(0) = x_0 , \tag{2.1.2}$$

obtained from (2.1.1) by setting $f = 0$ and $A_i = 0$, $i = 1, \ldots, N$.

Let $X$ be an $n \times n$ matrix function whose column vectors $x_1, \ldots, x_n$ constitute a fundamental set of solutions of (2.1.2) with the property that $X(0) = I$, the identity matrix. Then $X(t) = e^{A_0 t}$ and $x(t, x_0) = e^{A_0 t} x_0$ is the unique solution of (2.1.2). Using this we know that the solution of the nonhomogeneous differential equation

$$\dot{x}(t) = A_0 x(t) + f(t) , \qquad x(0) = x_0 \tag{2.1.3}$$

is by the method of variation of parameter given as

$$x(t, f, x_0) = e^{A_0 t} x_0 + \int_0^t e^{A_0(t-s)} f(s) ds . \tag{2.1.4}$$

49

If $x = x(\phi, f)$ is a solution of (2.1.1), which coincides with $\phi$ on $[-h, 0]$, then by (2.1.4), we have

$$x(t) = \phi(t), \quad t \in [-h, 0], \tag{2.1.5a}$$

$$x(t) = e^{A_0 t}\phi(0) + \int_0^t e^{A_0(t-s)}\left[\sum_{i=1}^{N} A_i x(s - \tau i) + f(s)\right] ds. \tag{2.1.5b}$$

Clearly (2.1.5) satisfies (2.1.1). And (2.1.5) is unique follows from its explicit calculation by the so-called method of steps: On the interval $[0, h]$ the function $x$ is uniquely represented by

$$x(t) = e^{A_0 t}\phi(0) + \int_0^t e^{A_0(t-s)}\left[\sum_{i=1}^{N} A_i \phi(s - \tau i) + f(s)\right] ds.$$

With $x$ defined and continuous on the interval $[0, h]$, we use (2.1.5b) to obtain $x$ on the interval $[h, 2h]$. The process is continued.

We now use the method outlined above to calculate the fundamental matrix solution $U$ of the system

$$\dot{x}(t) = A_0 x(t) + \sum_{i=1}^{N} A_i x(t - i\tau), \tag{2.1.6}$$

$$x(0) = x_0, \quad \phi \equiv 0 \quad \text{on} \quad [-h, 0],$$

in terms of $e^{A_0 t}$. $U$ is a matrix solution of (2.1.6), which also satisfies the initial condition

$$U(t) = \begin{cases} 0, & t < 0, \\ I, & t = 0, \, I \, \text{identity matrix}. \end{cases}$$

Consider a simple version of (2.1.6). The next result is valid:

**Proposition 2.1.1**   *Consider*

$$\dot{x}(t) = A_0 x(t) + A_1 x(t - h),$$

$$x(0) = x_0, \quad x(s) \equiv 0, \quad s \in [-h, 0). \tag{2.1.7}$$

*Then*

$$U(0) = I,$$

$$U(t) = e^{A_0 t}, \quad t \in [0, h]$$

$$U(t) = e^{A_0 t} + \sum_{i=1}^{k} \int_{ih}^{t} e^{A_0(t-s_i)} A_1 ds_i \cdots \tag{2.1.8}$$

$$\int_h^{s_2} e^{A_0(s_2 - s_1)} A_1 e^{A_0(s-h)} ds_1, \quad t \in (kh, (k+1)h],$$

*where we understand that $\sum_{i=1}^{0} = 0$. Thus*

$$U(t) = e^{A_0 t} + \int_h^t e^{A_0(t-s_1)} A_1 e^{A_0(s_1-h)} ds_1 + \cdots$$

$$+ \int_{kh}^t e^{A_0(t-s_k)} A_1 ds_k \cdots \int_{2h}^{s_1} e^{A_0(s_2-s_1)} A_1 \cdot e^{A_0(s_1-h)} ds_1 .$$

*We next give an expression of $U(t)$ for the general $N$-delay terms of (2.1.6).*

Let $x_0 \in E^n$ and $U_k(t) = U(t + (k-1)h)$, $t \in [0,h]k = 1, 2, \ldots$ and define the $k$-system as follows:

$$\frac{d}{dt} \begin{bmatrix} U_1(t)x_0 \\ U_2(t)x_0 \\ \vdots \\ U_N(t)x_0 \\ \vdots \\ U_k(t)x_0 \end{bmatrix} = \begin{bmatrix} A_0 & & & & & \\ A_1 & A_0 & & 0 & & \\ & & \ddots & & & \\ A_N & A_{N-1} & \cdots & A_0 & & \\ 0 & 0 & \cdots & A_N & A_{N-1} & A_1 A_0 \end{bmatrix} \cdot \begin{bmatrix} U_1(t)x_0 \\ U_2(t)x_0 \\ \vdots \\ U_N(t)x_0 \\ \vdots \\ U_k(t)x_0 \end{bmatrix} ,$$

where $0$ is the zero $n \times n$ matrix. Because $U_k(t)x_0$ is the function translated by $(k-1)h$ of the solution $x(t; x_0)$ of (2.1.6), the $k$th system is equivalent to step-by-step integration of (2.1.6). By induction,

$$U_\ell(t)x_0 = e^{A_0 t} U_\ell(0)x_0 + \sum_{i=1}^{\min(\ell-1,N)} \int_0^t e^{A_0(t-s)} A_i U_{\ell-i}(s)ds , \qquad (2.1.9a)$$

where $U_\ell(0)x_0 = U_{\ell-1}(h)x_0$. From this one obtains the following recursive formula for $U_k(t) : U_1(t) = e^{A_0 t}$:

$$U_2(t) = e^{A_0(t+h)} + \int_0^t e^{A_0(t-s)} A_1 e^{A_0 s} ds ,$$

$$U_k(t) = e^{A_0 t} U_{k-1}(h) + \sum_{i=1}^{\min(k-1,N)} \int_0^t e^{A_0(t-s)} A_i U_{k-1} ds, \quad k \geq 2 .$$

This is a simple induction formula. It is desirable to write $U_k(t)$, $k \geq 1$ in terms of $e^{A_0 t}$ and $A_i$ only. For this we let $x_\ell = U_\ell(0)x_0$ $\ell = 1, \ldots, k$. Then from (2.1.9a),

$$U_\ell(t)x_0 = V_\ell(t)x_\ell + V_{\ell-1}x_2 + \cdots + V_2(t)x_{\ell-1} + V_1(t)x_1 .$$

The matrices $V_1(t) \cdots V_k(t)$ are defined by induction as

$$V_1(t) = e^{A_0 t} ,$$

$$V_k(t) = \sum_{i=1}^{\min(k-1,N)} \int_0^t e^{A_0(t-s)} A_i V_{k-1}(s)\,ds\,, \quad k = 2, 3, \ldots.$$

These formulas for $V_k(t)$ will enable us to give a simpler expression of $U_k(t)$ than (2.1.9a), and then for $U(t)$.

We now consider (2.1.6). The explicit form of $U(t)$ is complicated. We introduce some notation. Define the index $\Lambda(j,k)$ for all $j = 1, 2, \ldots$ and $k = 1, 2$ by

$$\Lambda(j,k) = \{(i_1, \ldots, i_j) : 1 \le i_1, \ldots, i_j \le N \quad \text{and} \quad i_1 + \cdots i_j = k\}.$$

Note that $\Lambda(j,k) = \phi$ for $j > k$. The following finite dimensional analogue of a result of Nakagiri [6] gives an explicit representation of $U(t)$ in terms of $e^{A_0 t}$ and $A_i\; i = 1, \ldots, N$.

**Proposition 2.1.2**    *Define the matrix operator $V_k$ ($k = 1, 2 \cdots$) as follows:*

$$V_1(t) = e^{A_0 t}\,,$$

$$V_k(t) = \sum_{j=1}^{k-1} \sum_{\Lambda(j,k-1)} \int_0^t e^{A_0(t-s_{j-1})} A_{i_1} \cdots \tag{2.1.9b}$$

$$\times \int_0^{s_1} e^{A_0(s_1-s)} A_{i_j} e^{A_0 s}\,ds\,ds_1 \cdots ds_{j-1}, \quad k \ge 2\,.$$

*The fundamental matrix $U(t)t \ge 0$ of (2.1.6) is given by*

$$U(t) = \sum_{i=1}^{k} V_i(t - (i-1)\tau)\,, \quad t \in [(k-1)\tau, k\tau]\,. \tag{2.1.10a}$$

*Writing out some expressions of $V_k$, we have*

$$V_1(t) = e^{A_0 t}\,,$$

$$V_2(t) = \int_0^t e^{A_0(t-s)} A_1 e^{A_0 s}\,ds\,,$$

$$V_3(t) = \int_0^t e^{A_0(t-s)} A_2 e^{A_0 s}\,ds \tag{2.1.10b}$$

$$+ \int_0^t e^{A_0(t-s_1)} A_1 \int_0^{s_1} e^{A_0(s_1-s)} A_1 e^{A_0(s)}\,ds\,ds_1\,.$$

*The expression (2.1.10) becomes simpler if $A_i$ ($i = 1 \cdots N$) commute with $e^{A_0 t}$. We have:*

**Corollary 2.1.1**    *Assume that for each $i$ and for all $t \ge 0$, $A_i$ commutes with $e^{A_0 t}$, i.e., $A_i e^{A_0 t} = e^{A_0 t} A_i$, $i = 1, \ldots, N$. Then*

$$U(t) = e^{A_0 t} + \sum_{i=2}^{k} \sum_{j=1}^{i-1} \sum_{\Lambda(j,i-1)} \frac{1}{j!}(t - (i-1)\tau)^j$$

$$\times\, e^{A_0(t-(i-1)\tau)} A_{i_1} \cdots A_{i_j}, \quad t \in [(k-1)\tau, k\tau].$$
(2.1.11)

For a special case of (2.1.7), namely

$$\dot{x}(t) = A_0 x(t) + A_1 x(t-1),$$
(2.1.12)

we give an alternative description [7] of the fundamental matrix $U$. We note that by definition

$$\frac{\partial}{\partial t}U(t-s) = A_0 U(t-s) + A_1 U(t-s-1),$$
(2.1.13)

for $(t,s) \in [s,T] \times [0,T]$, where

$$U(t-s) = I(n \times n \text{ identity matrix for } t = s)$$
$$= 0,\, (t,s) \in [-1,s) \times [0,T].$$
(2.1.14)

Let $U_k(\tau) = U(\tau + k)$ for $\tau \in [0,1]$ $k = 0,1,\ldots$, and assume $s = 0$ in (2.1.13). Then by substituting in (2.1.13) we have

$$\frac{d}{d\tau}U_0(\tau) = A_0 U_0(\tau), \quad U_0(0) = I,$$

$$\frac{d}{d\tau}U_1(\tau) = A_1 U_0(\tau) + A_0 U_1(\tau), \quad U_1(0) = U_0(1),$$
(2.1.15)

$$\vdots$$

$$\frac{d}{d\tau}U_k(\tau) = A_1 U_{k-1}(\tau) + A_0 U_k(\tau), \quad U_k(0) = U_{k-1}(1).$$

It follows that the solution of (2.1.13) and (2.1.14) over the interval $t \in [k, k+1]$ is given by $U(t) = U_k(t-k)$. Set $Z_k(\tau) = [U_0^T(\tau), \ldots, U_k^T(\tau)]^T$ to convert (2.1.15) into the system

$$\frac{d}{dt}Z_k(\tau) = B_k Z_k(\tau) \quad \text{for} \quad \tau \in [0,1],$$
(2.1.16)

$$U_k(\tau) = E_k Z_k(\tau),$$
(2.1.17)

where $Z_k(\tau)$ is an $n(k+1) \times n$ matrix, and $B_k$ and $E_k$ are respectively $n(k+1) \times n(k+1)$ and $n \times n(k+1)$ matrices given by

$$
B_k = \begin{bmatrix} A_0 & 0 & 0 & \cdots & 0 & 0 \\ A_1 & A_0 & 0 & \cdots & 0 & 0 \\ 0 & A_1 & A_0 & \cdots & 0 & 0 \\ \vdots & \vdots & \vdots & & \vdots & \vdots \\ 0 & 0 & 0 & & A_1 & A_0 \end{bmatrix}, \quad E_k = [0,\ldots,0,I].
$$

Because the unique solution of (2.1.16) is

$$
Z_k(\tau) = e^{B_k \tau} Z_k(0), \tag{2.1.18}
$$

we have

$$
U_k(\tau) = E_k e^{B_k \tau} Z_k(0). \tag{2.1.19}
$$

We note that

$$
Z_0(0) = I,
$$

$$
Z_k(0) = \begin{bmatrix} I \\ \cdots\cdots \\ e^{B_{k-1}} Z_{k-1}^{(0)} \end{bmatrix}, \quad k = 1, 2, \ldots. \tag{2.1.20}
$$

From the definition $U_k(\tau)$ and from (2.1.19), we deduce that

$$
U(t) = U_k(t - k) = E_k e^{B_k(t-k)} Z_k(0) \tag{2.1.21}
$$

for $t \in [k, k+1]$. The essential difference between the fundamental matrix solution of the ordinary differential equation (2.1.2) and that of the delay system is that $U(t)$ may be singular for some values of $t \in [0, \infty)$, whereas $e^{A_0 t}$ is always nonsingular. Indeed, consider the system

$$
\dot{x}(t) = A_0 x(t) - e^{A_0} x(t-1).
$$

We have

$$
U(t) = e^{A_0 t}, \quad t \in [0\,1].
$$

$$
U(t) = e^{A_0 t} - e^{A_0 t}(t-1), \quad t \in [1, 2].
$$

It is clear that $U(2) = 0$, so that the fundamental matrix is singular at $t = 2$.

**Problem 2.1.1**    Find fundamental matrix solution $U(t)$ using both (2.1.10) and (2.1.8). Check the validity of solutions $\dot{x}(t) = -x(t) + x(t-1)$ on $(0, 3]$.

### 2.1.1  *Using (2.1.10a) and (2.1.10b)*

$$U(t) = \sum_{i=1}^{k} V_i(t - (i-1)\tau), \quad t \in [(k-1)\tau, k\tau].$$

On $[0,1]$,

$$V_1(t) = e^{A_0(t)} = e^{-t}.$$

On $[1,2]$,

$$U(t) = V_1(t) + V_2(t-1),$$

$$U(t) = e^{-t} + \int_0^{t-1} e^{A_0(t-1-s)} A_1 e^{A_0 s} ds = e^{-t} + \int_0^{t-1} e^{-(t-1)} ds,$$

$$U(t) = e^{-t} + (t-1)e^{-(t-1)}.$$

On $[2,3]$,

$$U(t) = V_1(t) + V_2(t-1) + V_3(t-2).$$

$$U(t) = e^{-t} + (t-1)e^{-(t-1)} + \int_0^{t-2} e^{A_0(t-2-s_1)}$$

$$\times A_1 \int_0^{s_1} e^{A_0(s_1-s)} A_1 e^{A_0(s)} ds ds_1,$$

$$U(t) = e^{-t} + (t-1)e^{-(t-1)} + \int_0^{t-2} e^{-(t-2-s_1)}(1) \int_0^{s_1} e^{-s_1} ds ds_1,$$

$$U(t) = e^{-t} + (t-1)e^{-(t-1)} + \int_0^{t-2} e^{-(t-2-s_1)} s_1 e_1^{-s_1} ds_1,$$

$$U(t) = e^{-t} + (t-1)e^{-(t-1)} + \int_0^{t-2} s_1 e^{-(t-2)} ds_1 = e^{-t} + (t-1)e^{-(t-1)}$$

$$+ \left. \frac{1}{2} s_1^2 e^{(t-2)} \right|_0^{t-2},$$

$$U(t) = e^{-t} + (t-1)e^{-(t-1)} + \frac{1}{2}(t-2)^2 e^{-(t-2)} = e^{-t} + (t-1)e^{-(t-1)}$$

$$+ \frac{1}{2}(t^2 - 4t + 4)e^{-(t-2)},$$

$$U(t) = e^{-t}[1 + e(t-1) + \frac{1}{2}e^2(t^2 - 4t + 4)].$$

Fundamental Matrix Solution:

$$U(t) = e^{-t}, \quad t \in [0,1],$$

$$U(t) = e^{-t}[1 + e(t-1)],$$

$$U(t) = e^{-t}\left[1 + e(t-1) + \frac{1}{2}e^2(t^2 - 4(t+4))\right].$$

Check validity by substitution. Does $U(t)$ satisfy $\dot{x}(t) = -x(t) + x(t-1)$; $\dot{x}(t) + x(t) - x(t-1) = 0$ on $[0,1]$ $U(t) = e^{-t}$, $\dot{U}(t) = -e^{-t}$, $U(t-1) = 0$. Substitute

$$-e^{-t} + e^{-t} + 0 = 0 \quad \therefore \text{ solution valid on } [0,1].$$

On $[1, 2]$,

$$U(t) = e^{-t}[1 + e(t-1)],$$

$$\dot{U}(t) = -e^{-t}[1 + e(t-1)] + e^{-(t-1)},$$

$$U(t-1) = e^{-(t-1)}.$$

Substituting into the equation, we have

$$-e^{-t}[1 + e(t-1)] + e^{-(t-1)} + e^{-t}[1 + e(t-1)] - e^{-(t-1)} = 0.$$

The solution is valid on $[1, 2]$.

    On $[2, 3]$,

$$U(t) = e^{-t}\left[1 + e(t-1) + \frac{1}{2}e^2(t^2 - 2t + 4)\right],$$

$$\dot{U}(t) = -e^{-t}\left[1 + e(t-1) + \frac{1}{2}e^2(t^2 - 4t + 4)\right] e^{-t}[e + e^2 t - 2e^2],$$

$$U(t-1) = e^{-(t-1)}[1 + e(t - 1 - 1)].$$

Substituting

$$-e^{-t}\left[1 + e(t-1) + \frac{1}{2}e^2(t^2 - 2t + 4)\right] + e^{-t}[e + e^2 t - 2e^2]$$

$$+ e^{-t}\left[1 + e(t-1) + \frac{q}{2}e^2(t^2 - 2t + 4)\right]$$

$$- e^{-(t-1)}[1 + e(t-2)]$$

$$= e^{-t}[e + e^2 t - 2e^2] - e^{-(t-1)}[1 + e(t-2)]$$

$$= e^{-t}[e + e^2 t - 2e^2 - e - e^2(t-2)]$$

$$= e^{-t}[e^2(t-2) - e^2(t-2)] = 0.$$

Therefore the solution is valid on $[2, 3]$.

## 2.1.2  *Using (2.1.8)*

On $[0, 1]$,

$$U(t) = e^{A_0 t} = e^{-t}.$$

On $[1, 2]$,

$$U(t) = e^{A_0 t} + \int_h^t e^{A_0(t-s_1)} A_1 e^{A_0(s_1-h)} ds_1,$$

$$U(t) = e^{-t} + \int_1^t e^{-(t-s_1)}(1)e^{-(s-h)} ds_1 = e^{-t} + \int_1^t e^{-(t-1)} ds_1,$$

$$U(t) = e^{-t} + (t-1)e^{-(t-1)} = e^{-t}(1 + e(t-1)),$$

$$U(t) = e^{-t}, \quad t \in [0, 1],$$

$$U(t) = e^{-t}(1 + e(t-1)), \quad t \in [1, 2],$$

$$U(t) = e^{-t}\left[1 + e(t-1) + \frac{1}{2}e^2(t^2 - 1 + 4)\right], \quad t \in [2, 3].$$

We now check the validity of this by substitution. Does $U(t)$ satisfy

$$\dot{x}(t) + x(t) - x(t-1) = 0?$$

On interval $[0, 1]$,

$$U(t) = e^{-t}, \quad \dot{U}(t) = -e^{-t}, \quad U(t-1) = 0.$$

Substituting in the equation, we have

$$e^{-t} - e^{-t} - 0 = 0. \quad \text{The solution is valid on } t \in [0, 1].$$

On interval $[1, 2]$,

$$U(t) = e^{-t}(1 + e(t-1)),$$

$$\dot{U}(t) = e^{-t}e - e^{-t}(1 + e(t-1)),$$

$$U(t-1) = e^{-(t-1)} = e^{-t}e.$$

Substituting above formulas in the equation, we have

$$e^{-t}e - e^{-t}(1 + e(t-1)) + e^{-t}(1 + e(t-1)) - e^{-t}e = 0.$$

The solution is valid on $t \in [1, 2]$.

On interval $[2h, 3h]$,

$$U(t) = e^{A_0 t} + \int_h^t e^{A_0(t-s_1)} A_1 e^{A_0(s_1-h)} ds_1$$

$$+ \int_{2h}^{t} e^{A_0(t-s_2)} A_1 \left( \int_{2h}^{s_2} e^{A_0(s_2-s_1)} A_1 e^{A_0(s_1-2h)} ds \right),$$

$$U(t) = e^{-t} + (t-1)e^{-(t-1)}$$

$$+ \int_{2}^{t} e^{-(t-s_2)}(1) \left( \int_{2}^{s_2} e^{-(s_2-s_1)}(1)e^{-(s_1-2)} \right) ds_1 ds_2,$$

$$U(t) = e^{-t} + (t-1)e^{-(t-1)} + \int_{2}^{t} e^{-(t-s_2)}(s_2-2)e^{-(2-s_2)} ds_2,$$

$$U(t) = e^{-t} + (t-1)e^{-(t-1)} + \int_{2}^{t} (s_2-2)e^{(2-t)} ds_2,$$

$$U(t) = e^{-t} + (t-1)e^{(1-t)} + e^{(2-t)} \left( \frac{1}{2}(t^2-4) - 2(t-2) \right),$$

$$U(t) = e^{-t} + (t-1)e^{(1-t)} + e^{(2-t)} \left( \frac{1}{2}t^2 - 2t + 2 \right),$$

$$U(t) = e^{-t} \left[ 1 + e(t-1) + \frac{1}{2}e^2(t^2-4t+4) \right].$$

This is the same solution obtained for $t \in [2,3]$ using (2.1.10a) and (2.1.10b).

## 2.2   The Variation of Constant Formula of Retarded Equations

Using the fundamental solution $U(t)$ of Sec. (2.1), we represent the solution $x(\phi, f)$ of the nonhomogeneous equation (2.1.1). We will show that this is given by

$$x(\phi, f)(t) = x(\phi, 0)(t) + \int_{0}^{t} U(t-s)f(s)ds, \quad t \geq 0. \tag{2.2.1}$$

In particular, if $N = 1$, then for $t \geq 0$,

$$x(\phi, 0)(t) = U(t)\phi(0) + A_1 \int_{-\tau}^{0} U(t-\theta-\tau)\phi(\theta)d\theta. \tag{2.2.2}$$

The variation of constant formula is presented for a very general linear retarded functional differential equation,

$$\dot{x}(t) = L(t, x_t) + f(t), \quad t \geq \sigma,$$
$$x_\sigma = \phi, \tag{2.2.3}$$

where $f$ is locally integrable on $E$. In this case by a solution we mean a function $x$ that satisfies (2.2.3) almost everywhere. We assume $f$ is a function mapping $[\sigma, \infty) \rightarrow E^n$; $L : [0, \infty) \times C \rightarrow E^n$ is continuous and linear in $x_t$, i.e., for any numbers $\alpha, \beta$ and functions $\phi, \psi \in C$ we have the relation $L(t, \alpha\phi+\beta\psi) = \alpha L(t, \phi) + \beta L(t, \psi)$. It follows from the Riesz Representation Theorem [4, p. 143] that

$$L(t, \phi) = \int_{-h}^{0} [d_\theta \eta(t, \theta)] \phi(\theta) \,, \qquad (2.2.4)$$

where $\eta(t, \phi)$ is an $n \times n$ matrix function that is measurable in $(t, \phi) \in E \times E$, such that

$$\eta(t, \theta) = 0 \quad \text{for} \quad \theta \geq 0, \quad \eta(t, \theta) = \eta(t, -h) \quad \text{for} \quad \theta \leq -h \,.$$

Also $\eta(t, \theta)$ is continuous from the left in $\theta$ on $(-h, 0)$, of bounded variation in $\theta$ on $[-h, 0]$ for each $t$. Also there is a locally integrable function $m : (-\infty, \infty) \to E$ such that

$$|L(t, \phi)| \leq m(t)\|\phi\|, \quad \forall t \in (-\infty, \infty), \quad \phi \in C \,.$$

Included in (2.2.3) is the system

$$\dot{x}(t) = \sum_{k=1}^{N} A_k(t) x(t - w_k) + \int_{-h}^{0} A(t, \theta) x(t + \theta) d\theta + f(t) \,, \qquad (2.2.5)$$

where $A_k(t)$ are $n \times n$ matrices. $A(t, \theta)$ is an $n \times n$ matrix function that is integrable in $\theta$ for each $t$, and there is a locally integrable function $a(t)$ such that

$$\left| \int_{-h}^{0} A(t, \theta) \phi(\theta) d\theta \right| \leq a(t)\|\phi\|, \quad \forall t \in E, \quad \phi \in C \,.$$

The fundamental matrix of

$$\dot{x}(t) = L(t, x_t) \,, \quad x_\sigma = \phi \qquad (2.2.6)$$

is the $n \times n$ matrix $U(t, s)$, which is a solution of the equation

$$\frac{\partial U(t, s)}{\partial t} = L(t, U_t(\cdot, s)), \quad t \geq s \quad \text{a.e. in } s \text{ and } t \,,$$

$$U(t, s) = \begin{cases} 0, & s - h \leq t < s, \\ I, & \text{identity map } t = s, \end{cases} \qquad (2.2.7)$$

where $U_t(\cdot, s)(\theta) = U(t + \theta, s)$, $-h \leq \theta \leq 0$. If $L(t, \phi) \equiv L(\phi)$ is independent of $t$, then

$$U(t, s) = U(t - s, 0) = U(t - s) \,. \qquad (2.2.8)$$

For the general system, under the conditions on $L, f$ we have:

**Proposition 2.2.1** *For any $\sigma \in E$, $\phi \in C$ there exists a unique solution $x(\sigma, \phi)$ that is defined and continuous on $[\sigma - h, \infty)$ and satisfies (2.2.3) a.e. on $(\sigma, \infty)$. This solution is given as*

$$x(\sigma, \phi, f)(t) = x(\sigma, \phi, 0)(t) + \int_\sigma^t U(t,s)f(s)ds, \quad t \geq \sigma, \tag{2.2.9}$$

$$x_\sigma = \phi,$$

*where $x(\sigma, \phi, 0)$ is the solution of the homogeneous equation (2.2.6) through $(\sigma, \phi)$, and $U(t,s)$ is the fundamental matrix, which is continuous for $t \geq s$ and measurable in $t, s$. Equation (2.2.9) gives $x(\sigma, \phi, f)(t)$ as a point of $E^n$. To obtain a solution in $C$, we write*

$$x_t(\sigma, \phi, f) = x_t(\sigma, \phi, 0) + \int_\sigma^t U_t(\cdot, s)f(s)ds, \quad t \geq \sigma, \tag{2.2.10}$$

*where it is understood that the integral equation is an integral in $E^n$. If we define $x_t$ by*

$$x_t(\sigma, \phi, 0) \stackrel{\text{def}}{=} T(t, \sigma)\phi \tag{2.2.11}$$

*and*

$$U_t(\cdot, s) = T(t, s)X_0, \tag{2.2.12}$$

*where*

$$X_0(\theta) = \begin{cases} 0, & -h \leq \theta < 0, \\ I, & \theta = 0, \end{cases} \tag{2.2.13}$$

*then $T(t, \sigma)$ is a continuous linear map and (2.2.10) becomes*

$$x_t(\sigma, \phi, f) = T(t, \sigma)\phi + \int_\sigma^t T(t, s)X_0 f(s)ds, \quad t \geq \sigma. \tag{2.2.14}$$


## 2.3 The Fundamental Solution of Linear Neutral Functional Differential Equations

In this section we consider linear neutral functional differential equations that have the form

$$\frac{d}{dt}\left[x(t) - \sum_{j=1}^M A_{-1j}x(t - h_j)\right] = A_0 x(t) + \sum_{j=1}^M A_{1j}x(t - h_j) + f(t), \quad \text{for } t \geq 0,$$

$$x(t) = \phi(t), \quad \phi \in C, \quad t \in [-h, 0), \quad x(0) = x_0. \tag{2.3.1}$$

Here $0 \leq h_1 \leq h_2 < \cdots \leq h_M = h$, and $A_{-1j}, A_0, A_{1j}, j = 1, \ldots, M$, are constant matrices. Equation (2.3.1) has a simpler form in

$$\frac{d}{dt}[x(t) - A_{-1}x(t - h)] = A_0 x(t) + A_1 x(t - h) + f(t) \tag{2.3.2}$$

with the same initial conditions. A more general form of (2.3.1) is the system

$$\frac{d}{dt}[Dx_t] = Lx_t + f(t), \quad t \geq 0,$$
$$x_\sigma = \phi \in C, \quad \phi(0) = x_0 \in E^n,$$
(2.3.3)

where

$$Lx_t = \int_{-h}^0 [d\eta(\theta)]x(t+\theta)$$
(2.3.4)

and

$$Dx_t = x(t) - \int_{-h}^0 d\mu(\theta)x(t+\theta).$$
(2.3.5)

In (2.3.4) and (2.3.5) the functions $\eta, \mu$ are $n \times n$ matrices on $[-h, 0]$ of bounded variation that are understood to have been extended on $E$ by requiring, for example, that $\eta(\theta) = \eta(-h)$, for $\theta \leq -h$, and $\eta(\theta) = \eta(0)$, for $\theta \geq 0$. The integration is given by the Riesz Representation Theorem in the Lebesgue–Stieltjes sense. It is assumed in (2.3.5) that $D$ is atomic at zero in the following sense:

**Definition 2.3.1**  The map $D : E \times C \to E^n$, which is linear in the second variable and which is given by

$$D(t, \phi) = \int_{-h}^0 d_\theta \eta(t, \phi)\phi(\theta)$$
(2.3.6)

is atomic at $\alpha$ on $E \times C$ if, whenever

$$A(t, \alpha) = \eta(t, \alpha^+) - \eta(t, \alpha^-),$$

we have $\det(A(t, \alpha)) \neq 0, \forall t \in E$. In (2.3.2), $D(t, \phi) = \phi(0) - A_{-1}\phi(-h)$ and $D$ is atomic at zero. In (2.3.5) we assume that the variation of $\mu$, $\text{Var}_{[-s,0]}\mu \to 0$ as $s \to 0$.

**Definition 2.3.2**  The fundamental matrix solution of (2.3.2) is the solution $U(t)$ of

$$\dot{U}(t) - A_{-1}\dot{U}(t-h) = A_0 U(t) + A_1 U(t-h),$$
(2.3.7)

for $t \neq kh\ h = 0, 1, 2, \ldots$ and $U(t) = 0, t < 0, U(0) = I$ identity. For the system

$$\frac{d}{dt}[Dx_t] = Lx_t,$$
(2.3.8)

the fundamental matrix is the $n \times n$ matrix solution (of bounded variation on compact intervals that is continuous from the right) of the equation

$$D(U_t) = I + \int_0^t L(U_s)ds, \quad t \geq 0,$$

$$U_0(\theta) = \begin{cases} 0, & -h \leq \theta < 0, \\ I, & (\text{identity}) \, \theta = 0. \end{cases}$$

(2.3.9)

It is important to get an expression of the fundamental matrix solution of (2.3.8). Because it is complicated for the general situation, we consider

$$\frac{d}{dt}\left[x(t) - \sum_{j=1}^{M} A_{-1j}x(t - h_j)\right] = A_0 x(t) + \sum_{j=1}^{M} A_{1j}x(t - h_j),$$

$$x(t) = \phi(t), \quad \phi \in C, \quad x(0) = x_0 \in E^n.$$

(2.3.10)

Our aim is to represent $U$ explicitly in terms of $e^{A_0 t}$, $A_{-1j}$ and $A_{1j}$ $(j = 1, \ldots, M)$. The expression we give is a result of a much wider statement in Banach space setting by Datko [2].

**Proposition 2.3.1**  *Consider (2.3.10).  The fundamental matrix $U$ of (2.3.10) satisfies the identity*

$$U(t) = e^{A_0 t} + \sum_{j=1}^{M} \mathcal{X}(t - h_j)A_{-1j}U(t - h_j) + \int_0^t e^{A_0(t-s)}$$

$$\times \left[\sum_{j=1}^{M} \mathcal{X}(s - h_j)(A_{1j} + A_0 A_{-1j})U(s - h_j)\right] ds,$$

(2.3.11)

*where $\mathcal{X}(s) = 0$ if $s < 0$, and $\mathcal{X}(s) = 1$ if $s \geq 0$.*

**Proof**  Let $e^{A_0 t}$ be the fundamental solution of

$$\dot{x}(t) = A_0 x(t).$$

Then the integrated form of (2.3.10) is the equation

$$x(t, 0, \phi) = \sum_{j=1}^{M} A_{-1j}x(t - h_j, 0, \phi) + e^{A_0 t}\left[\phi(0) - \sum_{j=1}^{M} A_{-1j}\phi(-h_j)\right]$$

$$+ \int_0^t e^{A_0(t-s)}\left[\sum_{j=1}^{M}(A_{1j} + A_0 A_{-1j})x(s - h_j, 0, \phi)\right] ds,$$

if $t \geq 0$ and $x(t, 0, \phi) = \phi(t)$, $\phi \in C$ if $t \in [-h, 0]$. Let the Laplace transform of this solution $x(t, 0, 0, \phi)$ be denoted by $\bar{x}(s, \phi)$. If we observe that the Laplace transform $\mathcal{L}\{e^{A_0 t}\}(s) = (sI - A_0)^{-1}$, then

$$\overline{x}(s,\phi) = \left(\sum_{j=1}^{M} A_{-1j}e^{-sh_j}\right)\overline{x}(s,\phi)$$

$$+ \sum_{j=1}^{M} A_{-1j}e^{-shj}\int_{-hj}^{0} e^{-s\sigma}\phi(\sigma)d\sigma$$

$$+ (sI - A_0)^{-1}\left[\phi(0) - \sum_{j=1}^{M} A_{-1j}\phi(-hj)\right] \tag{L}$$

$$+ (sI - A_0)^{-1}\left(\sum_{j=1}^{M}(A_{1j} + A_0 A_{-1j})e^{-shj}\overline{x}(s,\phi)\right)$$

$$+ (sI - A_0)^{-1}\sum_{j=1}^{M}(A_{1j} + A_0 A_{-1j})e^{-shj}\int_{-hj}^{0} e^{-s\sigma}ds.$$

This can be expressed as follows:

$$\overline{x}(s,\phi) = \left[\left(I - \sum_{j=1}^{M} A_{-1j}e^{-shj}\right) - (sI - A_0)^{-1}\right.$$

$$\left.\times \sum_{j=1}^{M}(A_{1j} + A_0 A_{-ij})e^{-shj}\right]\overline{x}(s,\phi),$$

$$= (sI - A_0)^{-1}\left[\sum_{j=1}^{M} A_{1j}e^{-shj}\int_{-hj}^{0} e^{-s\sigma}\phi(\sigma)d\sigma + \phi(0)\right] \tag{L}$$

$$+ \sum_{j=1}^{M}\left[(I - (sI - A_0)^{-1}A_0)A_{1j}e^{-shj}\right.$$

$$\left.\times \int_{-hj}^{0} e^{-s\sigma}\phi(\sigma)d\sigma - (sI - A_0)^{-1}A_{-1j}\phi(ds)\right].$$

Consider now the left-hand side of (L) and the equation in $\overline{S}$ given by

$$\left[\left(I - \sum_{j=1}^{M} A_{-1j}e^{-shj}\right) - (sI - A_0)^{-1}\cdot\sum_{j=1}^{M}(A_j + A_0 A_{-1j})e^{-shj}\right]\overline{S}(s)$$

$$= (sI - A_0)^{-1}.$$

We now invert this expression and transfer all terms except $U(t)$, (where $U(t) = \mathcal{L}^{-1}(\overline{S}(s))(t)$, the fundamental matrix of (2.3.10)), to the right-hand side. This yields (2.3.11). Notice that we have assumed $|U(t)| = 0$ if $t < 0$. $\qquad\square$

Assuming $e^{A_0 t}$ is known, (2.3.11) is used to compute $U(t)$ by the method of steps over intervals of the form $[kh_1, (k+1)h_1]$. For example, we consider (2.3.2).

**Proposition 2.3.2**   *The fundamental matrix solution of*

$$\frac{d}{dt}[x(t) - A_{-1}x(t-h)] = A_0 x(t) + A_1 x(t-h) \tag{2.3.12}$$

*is given as follows:*

$$U(t) = e^{A_0 t}, \quad t \in [0, h],$$

$$U(t) = e^{A_0 t} + A_{-1}e^{A_0(t-h)}$$

$$+ \int_h^t e^{A_0(t-s)}[A_1 + A_0 A_{-1}]e^{A_0(s-h)}ds, \quad t \in [h, 2h],$$

$$U(t) = e^{A_0 t} + \sum_{k=1}^{2} A_{-1}^k e^{A_0(t-(j-1)h)} \tag{2.3.13}$$

$$+ A_{-1}\int_h^t e^{A_0(t-s_1)}(A_1 + A_0 A_{-1})e^{A_0(s-h)}ds_1$$

$$+ \int_{2h}^t e^{A_0(t-s_2)}(A_1 + A_0 A_{-1})(e^{A_0 s_2} + A_{-1}e^{A_0(s_2-h)})$$

$$\cdot \int_h^{s_2} e^{A_0(s_2-s_1)}(A_1 + A_0 A_{-1})e^{A_0(s_2-h)}ds_1 \bigg] ds_2,$$

*for* $t \in [2h, 3h]$.

One can express the solution of the homogeneous system (2.3.12) with initial function $\phi \in C$ in terms of $U$, and furthermore obtain a solution of (2.3.2) in terms of the homogeneous solution and another term. It is contained in the following statement whose proof is given in Hale [4].

**Proposition 2.3.3**   *If $U$ is the fundamental solution of (2.3.12), the solution $x(\phi)$ of (2.3.12) with $x_0(\phi) = \phi$ is given by*

$$x(\phi, 0)(t) = U(t)[\phi(0) - A_{-1}\phi(-h)] + A_1 \int_{-h}^0 U(t - \theta - h)\phi(\theta)d\theta$$

$$- A_{-1}\int_{-h}^0 [dU(t - \theta - h)]\phi(\theta). \tag{2.3.14}$$

*The solution $x(\phi, f)$ of (2.3.2) with $x_0(\phi) = \phi$ is given by the variation of the constant formula*

$$x(\phi, f)(t) = x(\phi, 0)(t) + \int_0^t U(t - s)f(s)ds. \qquad (2.3.15)$$

*Also, if (2.3.8) is considered, and $U$ is the fundamental matrix as defined by (2.3.9) while $x(\phi, 0)$ is the solution of (2.3.8), then the unique solution of (2.3.3) is the function $x(\phi, f)$ with $x_0(\phi, f) = \phi$ and*

$$x(\phi, f)(t) = x(\phi, 0)(t) + \int_0^t U(t - s)f(s)ds. \qquad (2.3.16)$$

## 2.4  Linear Systems Stability [4]

In this section we tackle at various levels the problem of global uniform asymptotic stability of linear hereditary systems that was posed in Sec. 1.9. First we characterize this desirable property by examining the properties of the characteristic equations of the linear systems. Next in Sec. 3.2 we shall use a Lyapunov–Razumikhin type theorem based on work by Haddock and Terjéki to derive a computable sufficient condition, in terms of the system's coefficients, of asymptotic stability of the linear system

$$\dot{x}(t) = L(t, x_t), \qquad (2.4.1)$$

where $L$ is some variant of (2.2.4). We now specify our notion of stability. Note that $L(t, 0) = 0, \forall t \in E$.

**Definition 2.4.1**  The solution $x = 0$ of (2.4.1) is called uniformly stable if for any $\sigma \in E$, $\epsilon > 0$ there is a $\delta = \delta(\epsilon)$ such that if $\|\phi\| \leq \delta$, then $\|x_t(\sigma, \phi)\| \leq \epsilon$, $\forall t \geq \sigma$. The solution $x = 0$ of (2.4.1) is uniformly asymptotically stable if it is uniformly stable and there is a positive constant $\lambda > 0$, such that for every $\eta > 0$ there is a $T(\eta)$ such that if $\|\phi\| < \lambda$, then $\|x_t(\sigma, \phi)\| < \eta$, $\forall t \geq \sigma + T(\eta)$ for each $\sigma \in E$. If $y(t)$ is any solution of (2.4.1), then $y$ is said to be uniformly stable if the solution $z = 0$ of

$$\dot{z}(t) = L(t, z_t + y_t) - L(t, y_t)$$

is uniformly stable. The concept of uniform asymptotic stability is defined similarly. In the definition given above, $\delta$ is independent of $\sigma$ and therefore stability is uniform. If $\delta = \delta(\epsilon, \sigma)$, we will have mere stability. However, the two concepts coincide if $L(t, \phi)$ is periodic in $t$ and $L(t + w, \phi) = L(t, \phi)$, or if $L$ is time invariant.

The concept of uniform boundedness is useful. It is defined as follows: The solution $x(\sigma, \phi)$ of (2.4.1) is uniformly bounded if for any $\alpha > 0$ there is a $\beta(\alpha) > 0$ such that for all $\sigma \in E$, $\phi \in C$ and $\|\phi\| \leq \alpha$ we have $\|x(\sigma, \phi)(t)\| \leq \beta$

for all $t \geq \sigma$. Recall that $L$ satisfies for some locally integrable $M : (-\infty, \infty) \to E$ the inequality

$$|L(t, \phi)| \leq M(t)\|\phi\|, \quad \forall t \in E, \quad \phi \in C. \tag{2.4.2}$$

Stated below is a characterization of the two concepts of stability in terms of the fundamental matrix solution $U$ of (2.4.1).

**Theorem 2.4.1** *Consider (2.4.1) where $M$ in (2.4.2) satisfies, for some constant $M_1$, the inequality*

$$\int_t^{t+h} M(s)ds \leq M_1, \quad \forall t \in E. \tag{2.4.3}$$

*Then the solution $x = 0$ of (2.4.1) is uniformly stable if and only if there is a constant $k$ such that*

$$|U(t, s)| \leq k, \quad \forall t \geq s, \quad t, s \in E. \tag{2.4.4}$$

*Also the trivial solution $x = 0$ of (2.4.1) is uniformly asymptotically stable if and only if there are constants $k > 0$ and $\alpha > 0$ such that for all $s \in E$,*

$$|U(t, s)| \leq ke^{-\alpha(t-s)}, \quad t \geq s. \tag{2.4.5}$$

*If the solution $x(\sigma, \phi)$ of (2.4.1) defines the operator $T$ as follows:*

$$T(t, \sigma)\phi = x_t(\sigma, \phi), \quad t \geq \sigma,$$

*then we can replace (2.4.4) and (2.4.5) respectively by*

$$|T(t, \sigma)| \leq k, \quad t \geq \sigma; \tag{2.4.6}$$

*and*

$$|T(t, \sigma)| \leq ke^{-\alpha(t-\sigma)}, \quad t \geq \sigma. \tag{2.4.7}$$

Asymptotic stability can also be deduced from some properties of the characteristic equation of a homogeneous linear differential difference equation with constant coefficients (2.1.7), namely

$$\Delta(\lambda) = I\lambda - A_0 - A_1 e^{-\lambda h} = 0, \tag{2.4.8}$$

where $\lambda$ is a complex number. Equation (2.4.8) is obtained from (2.1.7) by substituting $x = e^{\lambda t}\xi$ in (2.1.7) where $\lambda$ is a constant and $\xi$ is a nontrivial vector. We have:

**Proposition 2.4.1** *Let $U(t)$ be the fundamental matrix solution of (2.1.7). Let $\alpha = \max\{\mathrm{Re}\,\lambda : \Delta(\lambda) = 0\}$. If $\alpha < 0$, then we have*

$$|U(t-s)| \leq ke^{\alpha(t-s)}, \quad t \geq s, \tag{2.4.9}$$

*and (2.1.7) is asymptotically stable. In this case if $x(\phi)$ is a solution of (2.1.7), which coincides with $\phi$ on $[-h, 0]$, then*

$$\|x_t(\phi)\| \le ke^{\alpha t}\|\phi\|, \quad t \ge 0, \tag{2.4.10}$$

*and all solutions approach zero exponentially as $t \to \infty$.*

Consider the more general situation

$$\dot{x}(t) = L(x_t), \tag{2.4.11}$$

where $L$ is a continuous linear function mapping $C$ into $E^n$ and described by

$$L(\phi) = \int_{-h}^{0} [d\eta(\theta)]\phi(\theta), \quad \phi \in C, \tag{2.4.12}$$

where $\eta$ is an $n \times n$ matrix $\eta(\theta), -h \le \theta \le 0$, whose elements are of bounded variation. Let

$$\Delta(\lambda) = I\lambda - \int_{-h}^{0} e^{\lambda\theta}d\eta(\theta). \tag{2.4.13a}$$

The characteristic equation for (2.4.11) is

$$\det \Delta(\lambda) = 0. \tag{2.4.13b}$$

The following Theorem is valid.

**Theorem 2.4.2** *Let all the roots of the characteristic equation (2.4.13) of (2.4.11) have negative real parts. Then there are positive constants $\alpha > 0$, $k > 0$ such that if $x(\phi)$ is a solution of (2.4.11) with $x_0(\phi) = \phi$, then*

$$\|x_t(\phi)\| \le k\|\phi\|e^{-\alpha t}, \quad t \ge 0, \tag{2.4.14}$$

*and*

$$\|U(t)\| \le ke^{-\alpha t}, \quad t \ge 0. \tag{2.4.15}$$

*The definition of uniform stability and uniform asymptotic stability can be applied to the neutral linear system (2.3.8), i.e.,*

$$\frac{d}{dt}[Dx_t] = Lx_t, \tag{2.4.16}$$

*where*

$$Lx_t = \int_{-h}^{0} [d\eta(\theta)]x(t+\theta), \tag{2.4.17}$$

*and*

$$Dx_t = x(t) - \int_{-h}^{0} [d\mu(\theta)]x(t+\theta), \tag{2.4.18}$$

*and the variation* $\mathrm{Var}_{[-s,0]}\mu$ *of* $\mu$ *satisfies* $\mathrm{Var}_{[-s,0]}\ \mu \to 0$ *as* $s \to 0$ *and* $\mu$ *has no singular part. This means that*

$$D\phi = D_0\phi + \int_{-h}^{0} A(\theta)\phi(\theta)d\theta\,,$$

*where*

$$D_0\phi = \phi(0) + \sum_{k=1}^{\infty} A_k\phi(-w_k)\,,$$

*and*

$$0 < w_k \le h\,, \quad \sum_{k=1}^{\infty}|A_k| + \int_{-h}^{0}|A(s)|ds < \infty\,.$$

We now give conditions in terms of the characteristic equation

$$\det \Delta(\lambda) = 0\,, \quad \Delta(\lambda) = \lambda D(e^{\lambda}I) - L(e^{\lambda}I) \tag{2.4.19}$$

for uniform asymptotic stability.

**Theorem 2.4.3** *Suppose in* (2.4.16) *where* $L$ *and* $D$ *are given as in* (2.4.17) *and* (2.4.18), *we assume that* $\Delta(\lambda)$ *in* (2.4.19) *satisfies*

$$\sup\{\mathrm{Re}\,\lambda : \det \Delta(\lambda) = 0\} < 0\,.$$

*Then the zero solution of* (2.4.16) *is uniformly asymptotically stable. Furthermore, if* $x(\phi)$ *is the solution of* (2.4.16) *with* $x_0 = \phi$, *and if* $U$ *is the fundamental matrix solution of* (2.4.16), *then there exist constants* $k > 0$, $\alpha > 0$ *such that*

$$\|x_t(\phi)\| \le k\|\phi\|e^{-\alpha t}\,, \qquad t \ge 0\,,$$
$$\|U(t)\| \le ke^{-\alpha t}\,, \qquad\quad t \ge 0\,,$$
$$\mathrm{Var}_{[t-1,t]}U \le ke^{-\alpha t}\,, \qquad t \ge 0\,.$$

For the autonomous system (2.4.16) stability implies uniform stability, but asymptotic stability does not imply uniform asymptotic stability. It is know in Brumley [1] that we can have the eigenvalues $\lambda$ with $\mathrm{Re}\,\lambda < 0$ and have some solutions unbounded. To have the same situation as in retarded equations where for autonomous systems or periodic systems uniform asymptotic stability is equivalent to asymptotic stability, we require that $D$ be stable in the following sense:

**Definition 2.4.2** Suppose $D : C \to E^n$ is linear, continuous, and atomic at 0. Then $D$ is called stable if the zero solution of the homogeneous difference equation,

$$Dy_t = 0\,, \quad t \ge 0\,,$$
$$y_0 = \psi\,, \quad D\psi = 0\,, \tag{2.4.20}$$

is uniformly asymptotically stable.

Let the characteristic equation of $D$ be

$$\Delta_D(\lambda) = I - \sum_{k=1}^{\infty} A_k e^{-\lambda w_k} - \int_{-h}^{0} e^{\lambda\theta} A(\theta)\,d\theta\,.$$

Stable operators are characterized by the following theorem:

**Theorem 2.4.4** *The following assertions are equivalent:*

(i) *$D$ is stable.*
(ii) $\alpha_D = \sup\{\operatorname{Re}\lambda : \det\Delta_D(\lambda) = 0\} < 0.$
(iii) *Any solution of the nonhomogeneous equation*

$$Dy_t = h(t), \quad t \geq 0\,,$$

*where $h : [0, \infty) \to E^n$ is continuous, satisfies*

$$\|y_t\| \leq b e^{-\alpha t}\|y_0\| + b \sup_{0\leq s\leq t} |h(s)|, \quad t \geq 0\,, \qquad (2.4.21)$$

*where $\alpha > 0$, $b > 0$ are some constants.*
(iv) *Let $D_\phi = \phi(0) - \int_{-h}^{0}[d\mu(\theta)]\phi(\theta)$, $\operatorname{Var}_{[-s,0]}\mu \to 0$ as $s \to 0$, and $\mu$ has no singular part. Then all solutions of the characteristic equation*

$$\det\left[I - \int_{-h}^{0} e^{\lambda\theta} d\mu(\theta)\right] = 0$$

*satisfy $\operatorname{Re}\lambda \leq -\delta$ for some $\delta > 0$.*

There are simple examples of stable operators:

1. Let $D\phi = \phi(0)$. This corresponds to the retarded functional differential equations.
2. $D\phi = \phi(0) - A\phi(-h)$, where the eigenvalues of $A$ have moduli $< 1$. $D$ is stable because the eigenvalues of $\det[I - Ae^{-\lambda h}] = 0$ have $\operatorname{Re}\lambda \leq -\delta < 0$. It is illuminating to state a corollary of Theorem 2.4.3.

**Corollary 2.4.3** *Consider the system*

$$\frac{d}{dt}[x(t) - A_{-1}x(t - h)] = A_0 x(t) + A_1 x(t - h)\,. \qquad (2.4.22)$$

*Suppose*

$$\alpha_0 = \sup\{\operatorname{Re}\lambda : \lambda(I - A_{-1}e^{-\lambda h}) = A_0 + A_1 e^{-\lambda h}\}, \quad \alpha_0 < 0$$

*and $x(\phi)$ is a solution of (2.4.22), which coincides with $\phi$ on $[-h, 0]$.*
   *Then $|x(\phi)(t)| \leq k e^{-\alpha t}\|\phi\|$, $t \geq 0$ for some $k \geq 0$ and $\alpha > 0$.*

## 2.4  Examples

**Example 2.4.1**  [cf. (9)] Consider the delay system

$$\ddot{x}(t) + b\dot{x}(t) + q\dot{x}(t - h) + kx(t) = 0 \,, \tag{2.4.23}$$

where $a, q, k$, and $h$ are positive constants. If we assume

$$b > q \,, \tag{2.4.24}$$

then (2.4.23) is uniformly asymptotically stable. We shall prove this by showing that every root of the characteristic equation

$$\lambda^2 + b\lambda + q\lambda e^{-\lambda h} + k = 0 \tag{2.4.25}$$

has negative real part. Indeed, suppose that $\lambda = \alpha + i\beta$ is a root of (2.4.25) with $\alpha \geq 0$, and aim at a contradiction. Suppose $\beta = 0$. Then the left-hand of (2.4.25) will be positive yielding a contradiction. Hence $\beta \neq 0$. Note that

$$\mathrm{Im}[\lambda^2 + b\lambda + q\lambda e^{-\lambda h} + k)/\beta]$$

$$\geq 2\alpha + b + qe^{-\alpha h}\left(\cos \beta h - \frac{\alpha h \sin \beta h}{\beta h}\right)$$

$$> b - qe^{-\alpha h}(1 + \alpha h) > b - q > 0 \,.$$

Since the imaginary part of (2.4.25) is zero we have deduced a contradiction. Hence every $\lambda$ has negative real part.

**Example 2.4.2**  Consider (2.4.23). The following is another sufficient condition that depends on the magnitude of $h$ for the uniform asymptotic stability of (2.4.23):

$$b + q > 0, \quad (b + q + k^{\frac{1}{2}})h < \pi/2 \,. \tag{2.4.26}$$

To see this, assume that $\lambda = \alpha + i\beta$ with $\alpha \geq 0$. Then

$$0 = |\lambda^2 + b\lambda + q\lambda e^{-\lambda h} + k| \geq |\lambda|^2 - (b + q)|\lambda| - k \,.$$

From this we deduce that

$$|\lambda| \leq \frac{b + q + [(b + q)^2 + 4k]^{\frac{1}{2}}}{2} \leq b + q + k^{\frac{1}{2}} \,.$$

Hence

$$|\lambda h| < \pi/2 \,.$$

As before, $\lambda = \alpha + i\beta$, $\alpha \geq 0$, $\beta = 0$ is impossible for (2.4.23). Hence $|\beta h| < \pi/2$. With this,

$$\mathrm{Im}(\lambda^2 + b\lambda + q\lambda e^{-\lambda h} + k)/\beta$$

$$= 2\alpha + b + qe^{-\alpha h}\left(\cos\beta h - \frac{\alpha}{\beta}\sin\beta h\right)$$

$$> 2\alpha - qe^{-\alpha h}\alpha h\frac{\sin\beta h}{\beta h}$$

$$= 2\alpha - q\alpha h \geq 0\,.$$

This contradiction proves the required result.

**Example 2.4.3**  The same methods yield the following sufficient conditions for uniform asymptotic stability of

$$\dot{x}(t) = a_0 x(t) + a_1 x(t-h) : h > 0\,, \quad a_0 < 0 + a_1 < 0, \quad a_0 + |a_1| \leq 0\,. \quad (2.4.27)$$

$$\dot{x}(t) = -a_1\dot{x}(t-h) : a_1 > 0, \quad 0 \leq a_1 h < \pi/2\,. \tag{2.4.28}$$

$$\ddot{x}(t) + b\dot{x}(t) + q\dot{x}(t-h_1) + kx(t) + px(t-h_2) = 0 : b, q, k, p, h_1, h_2 > 0\,,$$

$$b > ph_2\,, \quad [b + q + (k+p)^{\frac{1}{2}}]h_1 < \frac{\pi}{2}\,.$$

## 2.5  Perturbed Linear Systems Stability

In this section we consider conditions that will ensure that the stability properties of the homogeneous linear equation

$$\dot{x}(t) = L(t, x_t) \tag{2.5.1}$$

will be shared by the system

$$\dot{x}(t) = L(t, x_t) + f(t, x_t)\,, \tag{2.5.2}$$

where $f(t, 0) = 0$.

In (2.5.1) there is a locally integrable function $m : (-\infty, \infty)$ such that

$$|L(t, \phi)| \leq m(t)\|\phi\|, \quad \forall t \in (-\infty, \infty)\,. \tag{2.5.3}$$

Also the function $f : E \times C \to E^n$ is continuous.

The following lemma is well known:

**Lemma 2.5.1**  **(Bellman's Inequality):**  *Let $u$ and $\alpha$ be real valued continuous functions on an interval $[a, b]$, and assume that $\beta$ is an integrable function on $[a, b]$ with $\beta \geq 0$ and*

$$u(t) \leq \alpha(t) + \int_a^t \beta(s)u(s)ds, \quad a \leq t \leq b\,. \tag{2.5.4}$$

*Then*

$$u(t) \le \alpha(t) + \int_a^t \beta(s)\alpha(s)\left[\exp\int_s^t \beta(\tau)d\tau\right] ds, \quad a \le t \le b. \tag{2.5.5}$$

*If we assume further that $\alpha$ is nondecreasing, then*

$$u(t) \le \alpha(t)\exp\left(\int_a^t \beta(s)ds\right), \quad a \le t \le b. \tag{2.5.6}$$

**Theorem 2.5.1** *Suppose the system (2.5.1) is uniformly stable and there is a locally integrable function $\gamma : [0,\infty) \to E$ such that*

$$|f(t,\phi)| \le \gamma(t)\|\phi\|, \quad t \in [0,\infty),$$

*then (2.5.2) is uniformly stable on $[0,\infty)$. Suppose (2.5.1) is*

(i) *uniformly asymptotically stable, so that there exist some constants $k \ge 1$, $\alpha > 0$ such that every solution of (2.5.1) satisfies*

$$\|x_t(\cdot,\sigma,\phi)\| \le k\|\phi\|e^{-\alpha(t-\sigma)}, \quad t \ge \sigma;$$

(ii) $f(t,\phi) = f_1(t,\phi) + f_2(t,\phi),$

*where*

$$|f_1(t,\phi)| \le \pi(t)\|\phi\|, \quad |f_2(t,\phi)| \le \epsilon\|\phi\|, \quad \forall t \ge \sigma, \phi \in C,$$

*where $\epsilon > 0$ is the constant $\epsilon = \alpha/2k$ and $\pi$ is an integrable function with*

$$\Pi = \int_\sigma^\infty \pi(t)dt < \infty.$$

*Then every solution $x(\phi,\sigma)$ of (2.5.2) with $x_\sigma(\sigma,\phi) = \phi$ satisfies*

$$\|x_t(\cdot,\sigma,\phi)\| \le M\|\phi\|e^{-\alpha/2(t-\sigma)}, \quad t \ge \sigma.$$

**Remark 2.5.1** Condition (ii) of the second part of Theorem 2.5.1 is not too severe. See [5, p. 86]. They are natural generalizations of conditions stated in the famous theorem of Lyapunov on stability with respect to the first approximation.

**Proof** From Theorem 2.4.1 and Condition (2.4.6), $|T(t,\sigma)| \le k$, $t \ge \sigma$ for some $k > 0$, because (2.5.1) is uniformly stable. Also $|T(t,\sigma)X_0| \le k$, $\forall t \ge \sigma$ where $X_0$ is given in (2.2.13). The variation of constant formula (2.2.14) applied to (2.5.2) implies

$$\|x_t(\phi,\sigma)\| \le \|T(t,\sigma)\phi)\| + \int_\sigma^t \|T(t,s)X_0f(s,x_s)ds\|$$

$$\le k\|\phi\| + k\int_\sigma^t \gamma(s)\|x_s\|ds, \quad t \ge \sigma.$$

Apply Bellman's Inequality (2.5.6) to this to deduce that

$$\|x_t(\phi, \sigma)\| \leq k\|\phi\| \exp\left[k \int_\sigma^t \gamma(s) ds\right]$$

$$\leq k\|\phi\| \exp\left[k \int_\sigma^\infty \gamma(s) ds\right],$$

for all $\phi \in C$. This proves that (2.5.2) is uniformly stable on $[0, \infty)$.

For the second statement, recall that by Theorem 2.4.1

$$\|T(t, \sigma)\| \leq ke^{-\alpha(t-\sigma)}, \quad \|T(t, \sigma)X_0\| \leq ke^{-\alpha(t-\sigma)} \quad \text{for} \quad t \geq \sigma.$$

It follows from the variation of constant formula applied to (2.5.2) that

$$\|x_t(\sigma, \phi)\| \leq k\|\phi\|e^{-\alpha(t-\sigma)} + k \int_\sigma^t e^{-\alpha(t-s)}[\epsilon\|x_s\| + \gamma(s)]ds.$$

Thus if $z_t = e^{\alpha t}x_t$, then

$$\|z_t\| \leq k\|\phi\|e^{\alpha\sigma} + k \int_\sigma^t [\epsilon + \gamma(s)]\|z_s\|ds.$$

Using Bellman's Inequality (2.5.6) once again we deduce from this estimate that

$$\|z_t\| \leq k\|\phi\|e^{\alpha\sigma} \exp\left(\int_\sigma^t k\epsilon + k\gamma(s))ds\right)$$

$$\leq k\|\phi\|e^{\alpha\sigma} \exp\left(k\epsilon(t - \sigma) \exp \int_\sigma^\infty k\gamma(s)ds\right).$$

This yields

$$\|x_t\| \leq \Pi k\|\phi\|e^{-\alpha(t-\sigma)}e^{k\epsilon(t-\sigma)}$$

$$\leq M\|\phi\|e^{-\frac{\alpha}{2}(t-\sigma)}, \quad t \geq \sigma,$$

where $\epsilon = \alpha/2k$, $M = k\Pi$. The proof is complete. $\qquad\square$

Condition (ii) of Theorem 2.5.1 can be relaxed by using a recent sharper generalization of Bellman's Inequality due to En-Hao-Yang [3].

**Lemma 2.5.2** *Let $C(J, E^+)$ be the Banach space of continuous functions taking $J$ into $E^+ = [0, \infty)$. Let the following conditions be satisfied:*

(i) *$a(t) \in C(J, E^+)$ is a nondecreasing function with $a(t) \geq 1$, $\forall t \in J$.*
(ii) *The functions $u(t)$, $f_i(t) \in C(J, E^+)$.*
(iii) *The inequality*

$$u(t) \leq a(t) + \sum_{i=1}^n \int_0^t f_i(s)(u(s))^{r_i} ds, \quad t \in [0, \tau] \equiv J \subseteq E^+$$

*holds where $r_i \in (0, 1]$.*

*Then we have the inequality*

$$u(t) \le a(t) \prod_{i=1}^{n} G_i(t) \,, \quad t \in J \,,$$

*where*

$$G_i(t) = \begin{cases} \left[ 1 + (1 - r_i) \left( \prod_{k=1}^{i-1} G_k(t) \right) \int_0^t f_i(s) ds \right]^{\frac{1}{2 - r_i}} \,, & 0 < r_i < 1 \,, \\[2em] \exp\left\{ \left( \prod_{k=1}^{i-1} G_k(t) \right) \int_0^t f_i(s) ds \right\} \,, & r_i = 1 \,, \quad i = 1, \dots, n \,. \end{cases}$$

*Here we let $\prod_{i=1}^{0} G_k(t) = 1$, $t \in J$.*

**Theorem 2.5.2**    *In (2.5.2) assume that*

(i)   $f(t, 0, 0) = 0$.

(ii)   *The system (2.5.1) is uniformly asymptotically stable.*

(iii)   *For each $\phi \in C$, $t \ge \sigma \ge 0$ we have*

$$|f(t, \phi)| \le \sum_{i=1}^{n} c_i(t) \|\phi\|^{r_i} + c_{n+1}(t) \,,$$

*where $r_i \in (0, 1]$ are constants, $c_j(t) \in C(E^+, E^+)$ are known functions, $j = 1, 2, \dots, n + 1$.*

(iv)   $\int_0^\infty c_j(s) e^{\alpha s} ds < \infty$, $j = 1, \dots, n + 1$.

*Then all solutions of (2.5.2) satisfy*

$$\|x_t(\sigma, \phi)\| \le M(\sigma, \phi) e^{-\alpha(t)}$$

*for some constant $M = M(\sigma, \phi)$ and $\alpha > 0$.*

**Proof**    Just as before, using uniform asymptotic stability of (2.5.1) and the variation of constant formula, we have

$$\|x_t(\sigma, \phi)\| \le k \|\phi\| e^{-\alpha(t - \sigma)} + k \int_\sigma^t e^{-\alpha(t - s)} \left[ \sum_{i=1}^{n} c_i(s) \|x_s\|^{r_i} + c_{n+1}(s) \right] ds \,,$$

$$\|x_t(\sigma, \phi)\| e^{\alpha t} \le k \|\phi\| e^{\alpha \sigma} + k \int_\sigma^t e^{\alpha s} c_{n+1}(s) ds + k \int_\sigma^t e^{\alpha s} \left[ \sum_{i=1}^{n} c_i(s) \|x_s\|^{r_i} \right] ds \,.$$

Now apply the lemma to deduce the inequality

$$\|x_t(\sigma, \phi)\| e^{\alpha t} \le D(t) \prod_{i=1}^{n} R_i(t) \,,$$

where

$$D(t) = 1 + k\|\phi\|e^{\alpha\sigma} + k \int_{\sigma}^{t} e^{\alpha s} c_{n+1}(s)ds\,,$$

and

$$R_i(t) = \begin{cases} \left[ 1 + (1 - r_i)\left(\prod_{k=1}^{i-1} R_k(t)\right) \int_{\sigma}^{t} c_i(s)e^{\alpha s}ds \right]^{\frac{1}{2-r_i}}, & 0 < r_i < 1\,, \\ \exp\left\{ \left(\prod_{k=1}^{i-1} R_k(t)\right) \int_{\sigma}^{t} c_i(s)e^{\alpha s}ds \right\}, & r_i = 1\,, \quad i = 1,\ldots,n\,, \end{cases}$$

here in $\prod_{k=1}^{0} R_k(t) = 1$, $t \geq \sigma$. Consequently,

$$\|x_t(\sigma,\phi)\| \leq e^{-\alpha t}D(\sigma,\phi)\,,$$

where

$$\lim_{t\to\infty} D(t)\prod_{i=1}^{n} R_i(t) = D(\sigma,\phi) < \infty\,,$$

by hypothesis (iv). Hence

$$\|x_t(\sigma,\phi)\| \leq D(\sigma,\phi)e^{-\alpha t}\,,$$

completing the proof. $\square$

**Example 2.5.1** [14] We have seen that under certain conditions on $f$ the stability properties of (2.5.1) are shared by those of (2.5.2), (at least locally). Indeed, if $f$ is smooth enough, a nonlinear system

$$\dot{x}(t) = f(t, x_t)$$

can be written (locally) in the form (2.5.2), where $L(t, x_t) = D_2 f(t, x_\infty)x_t$ and $D_2 f(t, x_\infty)$ is the Fréchet derivative with respect to $x_t$ about the equilibrium $x_\infty$. This insight enables us to determine stability about equilibrium of nonlinear systems. The next example from Brauer illustrates this. Let $x(t)$ be the number of a population susceptible to an infectious disease. We assume that births and deaths other than those caused by the disease are dependent on the number of susceptible members. If $g(x)$ represents excess of births over deaths in a unit time when the susceptible population is $x$, then $x(t)$ satisfies the following:

$$\dot{x}(t) = g(x(t)) - \beta x(t)\left[ x(t - h_1 - h_2) - x(t - h_1) + \int_{t-h_1-h_2}^{t-h_1} g(x(s))ds \right]. \quad (2.5.7)$$

**Proposition 2.5.1** *Suppose in* (2.5.7)

    (i) $g(0) = g(k) = 0$, $g(x) > 0$ *for* $0 < x < k$ *where $k$ is some positive constant (the carrying capacity in the absence of disease).*

(ii) $g(x) > xg'(x)$,    $0 < x < k$.
       *Then*

        (a) *The equilibrium $x_\infty = 0$ is unstable.*
        (b) *The equilibrium $x_\infty = k$ is stable if*

(iii) $\beta h_2 k < 1$, *and unstable if* $\beta h_2 k > 1$.
       *Then we also have*

        (c) *The equilibrium $x_\infty = \frac{1}{\beta h_2}$ is stable if $\beta h_2 k > 1$ or if $g'(\frac{1}{\beta h_2}) > 0$ and $a + ch_2 > 0$ or $\beta h_2 g(\frac{1}{\beta h_2}) - 2q'(\frac{1}{\beta h_2}) > 0$ with $a = \beta h_2 g(x_\infty) - g'(x_\infty)$, $c = -\beta x_\infty g'(x_\infty)$.*

**Proof**    The equilibria for (2.5.7) are solutions $x_\infty$ of

$$g(x_\infty) - \beta h_2 x_\infty g(x_\infty) = 0.$$

If $g(x_\infty) = 0$, which implies that $x_\infty = 0$ (and we have extinction), or $x_\infty = k$ (disappearance of the disease), or $x_\infty = \frac{1}{\beta h_2}$ (endemic equilibrium), then we now linearize about the equilibrium $x_\infty$:

$$\dot{z}(t) = [g'(x_\infty) - \beta h_2 g(x_\infty)]z(t) - \beta x_\infty z(t - h_1 - h_2) - \beta x_\infty g'(x_\infty)$$

$$\times \int_{t-h_1-h_2}^{t-h_1} z(s)ds, \tag{2.7.8}$$

and construct the characteristic equation — the condition on $\lambda$ such that $z(t) = ce^{\lambda t}$ is a solution of (2.7.8):

$$\lambda + a = be^{-\lambda h_1}(1 - e^{-\lambda h_2}) + c \int_{-h_2-h_1}^{-h_1} e^{\lambda s}ds$$

$$= (\lambda b + c)e^{-\lambda h_1}\left[\frac{1 - e^{-\lambda h_2}}{\lambda}\right]. \tag{2.7.9}$$

At $x_\infty = 0$, (2.7.8) is

$$\dot{z}(t) = g'(0)z(t).$$

This is unstable by hypothesis (ii). For the equilibrium $x_\infty = k$, we observe that $a = -g'(k) > 0$, $b = \beta k$, $c = -\beta k g'(k) > 0$, since $g(k) = 0$, $g'(k) < 0$. But the roots of (2.7.9) have negative real part (by [8]) if $\beta h_2 k < 1$ and unstable if $\beta h_2 k > 1$, since $g(k) = 0$, $g'(k) < 0$ and all the roots of (2.7.9) satisfy real $\lambda < 0$ if $h_1 \geq 0$, $h_2 \geq 0$, $0 < |c|h_2 < a$, $|b|h_2 \leq 1$, and we have

$$a = -g'(k) > 0, \quad b = \beta k, \quad c = -\beta k g'(k) > 0.$$

At the equilibrium $x_\infty = \frac{1}{\beta h_2}$ $a = \frac{1}{\beta h_2}g(\frac{1}{\beta \tau}) - g'(\frac{1}{\beta h_2}) > 0$, by (ii) $b = \frac{1}{h_2}$, $c = -g'(\frac{1}{\beta h_2})/h_2$. Thus $a - c\tau > 0$ if $\beta h_2 k > 0$, so that $g(\frac{1}{\beta h_2}) > 0$. Thus if $g'(\frac{1}{\beta h_2}) < 0$

so that $c > 0$, the equilibrium is stable. If $g'(\frac{1}{\beta h_2}) > 0$ so that $c < 0$, we have stability if $a + c\tau > 0$ or $\beta h_2 g(\frac{1}{\beta h_2}) - 2g'(\frac{1}{\beta h_2}) > 0$. We sum up: If $\beta h_2 k < 1$, the only stable equilibrium is $x_\infty = k$, and the disease disappears. If $\beta h_2 k > 1$, there is an endemic equilibrium $x_\infty = \frac{1}{\beta h_2}$ and this is stable if $\beta h_2 k$ is sufficiently close to 1 and unstable (with oscillation about $x_\infty$) if $\beta h_2 k$ is sufficiently large. $\qquad\square$

## References


1. W. E. Brumley, "On the Asymptotic Behavior of Solutions of Differential Difference Equations of Neutral Type," *J. Differential Equations* **7** (1970) 175–188.
2. R. Datko, "Linear Autonomous Neutral Differential Equations in a Banach Space," *J. Differential Equations* **25** (1977) 258–274.
3. En-Hao Yang, "Perturbations of Nonlinear Systems of Ordinary Differential Equations," *J. Math. Anal. Appl.* **103** (1984) 1–15.
4. J. Hale, *Theory of Functional Differential Equations*, Springer-Verlag, New York, 1977.
5. J. Hale, *Ordinary Differential Equations*, Wiley, New York, 1969.
6. S. Nakagiri, "On the Fundamental Solution of Delay-Differential Equations in Banach Space," *J. Differential Equations* **41** (1981) 349–368.
7. R. B. Zmood and N. H. McClamrock, "On the Point-Wise Completeness of Differential-Difference Equations," *J. Differential Equations* **12** (1972) 474–486.
8. F. Brauer, "Epidemic Models in Population of Varying Size," in "Mathematical Approaches to Problems in Resource Management and Epidemiology," edited by C. Castillo-Chavez, S. A. Levin, C. A. Shoemaker, *Lecture Notes in Biomathematics* **81**, Springer-Verlag, 1989.
9. R. D. Driver, *Ordinary and Delay Differential Equations*, Springer-Verlag, New York, 1977.
10. E. N. Chukwu, "Null Controllability of Nonlinear Delay Systems with Restrained Controls," *J. Math Analysis and Applications* **76** (1980) 283–296.

# Chapter 3

# Lyapunov–Razumikhin Methods of Stability in Delay Equations

## 3.1 Lyapunov Stability Theory

For linear and quasilinear systems the methods of the previous chapter are adequate. For inherently nonlinear systems the ideas of Lyapunov as extended by LaSalle and Hale are powerful tools. For hereditary systems these methods require the construction of Lyapunov functionals. The main theory is now presented.

Consider the nonlinear system

$$\dot{x}(t) = f(t, x_t), \tag{3.1.1}$$

where

$$f(t, 0) = 0. \tag{3.1.2}$$

Suppose $V : E \times C \to E$ is continuous, and $x(\sigma, \phi)$ is a solution of (3.1.1). We define

$$\dot{V}(t, \phi) = \lim_{r \to 0^+} \sup \frac{1}{r}[V(t + r, x_{t+r}(t, \phi)) - V(t, \phi)].$$

The following result is valid [3, p. 105].

**Theorem 3.1.1** *Suppose $f : E \times C \to E^n$ in (3.1.1) takes $E \times$ (bounded sets of $C$) into bounded sets of $E^n$, and the functions $u, v, w : E^+ \to E^+$ are continuous, nondecreasing functions with the property that $u(s), v(s)$ are positive for $s > 0$, $u(0) = v(0) = 0$. Suppose there exists a continuous function $V : E \times C \to E$ such that*

 (i)  *$u(|\phi(0)|) \leq V(t, \phi) \leq v(\|\phi\|)$, and*
 (ii)  *$\dot{V}(t, \phi) \leq -w(|\phi(0)|)$.*

*Then $x = 0$ is uniformly stable.*
*If $u(s) \to \infty$ as $s \to \infty$, then the equation is uniformly bounded.*
*If $w(s) > 0$ for $s > 0$, then the solution $x = 0$ of (3.1.1) is uniformly asymptotically stable.*

79

**Example 3.1.1**   Consider the system

$$\dot{x}(t) = -ax(t) - bx(t - h) \,. \tag{3.1.3}$$

We now show that if $a > |b|$, then (3.1.3) is uniformly asymptotically stable. We use the Lyapunov functional

$$V(\phi) = \frac{1}{2}\phi^2(0) + \mu \int_{-h}^{0} \phi^2(s)ds \,,$$

where $\mu > 0$. Note that (3.1.3) is the same as

$$\dot{\phi}(0) = -a\phi(0) - b\phi(-h) \,,$$

so that

$$\dot{V} = \phi(0)[-a\phi(0) - b\phi(-h)] + \mu(\phi^2(0) - \phi^2(-h))$$

$$= (a - \mu)\phi^2(0) - b\phi(0)\phi(-h) - \mu\phi^2(-h) \,.$$

This is a negative definite quadratic form in $\phi(0)$ and $\phi(-h)$ if

$$a > \mu > 0, \quad 4(a - \mu)\mu > b^2 \,.$$

Note that if $\mu = \frac{a}{2}$, $|b|$ is as large as possible. As a consequence, $|b| < a$. It is now very easy to see that conditions (i) and (ii) are satisfied with $u(|\phi(0)|) = \frac{1}{2}\phi^2(0)$ and $v(\|\phi\|) = V(\phi)$, and $w(|\phi(0)|) \equiv \sigma\phi^2(0)$ for some $\sigma > 0$.

We now apply Theorem 3.1.1 to an inherently nonlinear system.

**Example 3.1.2**   Consider

$$\dot{x}(t) = a(t)x^3(t) + b(t)x^3(t - h) \,. \tag{3.1.4}$$

Assume that there are some constants

$$\delta > 0 \quad \text{and} \quad q, \quad \text{such that}$$

$$a(t) \leq -\delta < 0, \quad 0 < q < 1, \tag{3.1.5}$$

with

$$|b(t)| < q\delta \,. \tag{3.1.6}$$

Then (3.1.4) is uniformly asymptotically stable. Indeed, consider the functional

$$V(\phi) = \frac{\phi^4(0)}{4} + \frac{\delta}{2} \int_{-h}^{0} \phi^6(s)ds \,,$$

which obviously satisfies condition (i) of Theorem 3.1.1. If we differentiate, we obtain

$$\dot{V} = \phi^3(0)\dot{\phi}(0) + \delta/2[\phi^6(0) - \phi^6(-h)]$$

$$= \phi^3(0)[a(t)\phi^3(0) + b(t)\phi^3(-h)] + \delta/2[\phi^6(0) - \phi^6(-h)]$$

$$= \phi^6(0)[a(t) + \delta/2] - \delta/2\phi^6(-h) + b(t)\phi^3(0)\phi^3(-h)$$

$$\leq -\delta/2\phi^6(0) - \delta/2\phi^6(-h) + q\delta\phi^3(0)|\,|\phi^3(-h)| < 0$$

by (3.1.5) and (3.1.6). Thus conditions (ii) of uniform asymptotic stability of Theorem 3.1.1 are verified.

For autonomous systems an interesting theory is available for investigating the asymptotic behavior of solutions. Consider

$$\dot{x}(t) = f(x_t),$$
$$x_0 = \phi, \tag{3.1.7}$$

where $f : C \to E^n$ is continuous on some open set

$$S = \{\phi \in C : \|\phi\| < H\}.$$

$H$ is constant. We denote the solution of (3.1.7) by $x(\phi)$ and define a path or motion in $C$ through $\phi$ as the set $\bigcup_{0 \leq t < \alpha} x_t(\phi)$ where the solution $x(\phi)$ is defined on $[0, \alpha)$.

**Definition 3.1.1**  A point $\psi \in C$ is said to be in $\Gamma^+(\phi)$, the positive limiting set of $\phi$, if the solution $x(\phi)$ is defined on $[-h, \infty)$ and there exists a sequence of nonnegative real numbers $\{t_k\}$ such that $t_k \to \infty$ and $k \to \infty$ and $\|x_{t_k}(\phi) - \psi\| \to 0$ as $k \to \infty$. A subset $M$ of $C$ is said to be an invariant set if for any $\phi \in M$, the solution $x(\phi)$ of (3.1.7) is defined on $(-\infty, \infty)$, and $x_t(\phi) \in M$ for all $t \in (-\infty, \infty)$. The next two statements are technical but important. They help validate the so-called invariance principle of LaSalle and Hale for differential equations with delay.

**Lemma 3.1.1**  *If $x(\phi)$ is a solution of (3.1.7), which is defined on $[-h, \infty)$ and satisfies $\|x_t(\phi)\| \leq \alpha_1 < \alpha$ for all $t \in [0, \infty)$, then the set of functions $\{x_t(\phi) : t \geq 0\}$ belongs to a compact subset of $C$.*

**Proof**  Recall that a metric space $X$ is compact if we can select from any sequence $\{x_n\}$ of points on $X$ a subsequence $\{x_{nk}\}$ convergent to the space $X$. In the space of functions, the following Arzela-Ascoli Theorem is used to prove that a set is compact: $\qquad\qquad\square$

**Arzela-Ascoli Theorem**  *A necessary and sufficient condition for the space $S$ of continuous functions $x : [a, b] \to E^n$ with the sup norm to be compact is as follows:*

(i) *There exists a constant $k$ such that*

$$|x(t)| \leq k, \qquad (x \in S,\, t \in [a, b]).$$

(ii) *$S$ is equicontinuous, i.e., for every $\epsilon > 0$ there is $\delta$ such that if*

$$|t - t'| < \delta, \qquad then \quad |x(t) - x(t')| < \epsilon \quad for\ all \quad x \in S, \quad t, t' \in [a, b].$$

(iii) *Condition (ii) is satisfied whenever*

$$|\dot{x}(t)| \leq M \quad for\,all \quad t \in [a,b], \quad x \in S\,,$$

*where $M > 0$ is a constant.*

To conclude the proof of the lemma, we observe that if $x(\phi)$ is a solution of (3.1.1), which is bounded, then $\dot{x}(\phi)(t) = f(x_t(\phi))$ so that $|f(x_t(\phi))| \leq L$ for some $L$ since $\|x_t(\phi)\| \leq \alpha$, and $f$ maps bounded sets into bounded sets. Since $\|\dot{x}(\phi)(t)\| \leq L$ for all $t \geq 0$, the Arzela-Ascoli Theorem guarantees that $\{x_t(\phi) : t \geq 0\}$ belongs to a compact set.

**Lemma 3.1.2**   *Let $x(\phi)$ be a solution of (3.1.1) which is defined on $[-h,\infty)$ and $\|x_t(\phi)\| \leq \alpha_1 < \alpha$ for all $t \in [0,\infty)$. Then the $\Gamma^+(\phi)$ the positive limiting set of $\phi$ is nonempty, compact, connected, invariant, and distance $(x_t(\phi), \Gamma^+(\phi)) \to 0$ as $t \to \infty$.*

**Definition 3.1.2**   A map $V : C \to E^+$ is a Lyapunov functional on a set $G$ in $C$ relative to the equation

$$\dot{x}(t) = f(x_t)\,. \tag{3.1.7}$$

If $V$ is continuous on $\overline{G}$, the closure of $G$ and

$$\dot{V}(\phi) = \dot{V}(\phi) = \limsup_{r \to 0^+} \frac{1}{r}[V(x_r(\phi)) - V(\phi)] \leq 0$$

on $G$.

Let

$$S = \{\phi \in \overline{G} : \dot{V}(\phi) = 0\}\,,$$

and let the set $M$ be the largest set in $S$ invariant with respect to (3.1.7).

**Theorem 3.1.2**   **[3, p. 119]** *Suppose $V : C \to E^+$ is a Lyapunov functional on $G$ and $x_t(\phi)$ is a bounded solution of (1.1.7), which remains in $G$ for each $t$. Then $x_t(\phi) \to M$ as $t \to \infty$.*

**Theorem 3.1.3**   **[3, p. 119]** *Let $U_\ell = \{\phi \in C : V(\phi) < \ell\}$, where $V : C \to E^+$ is a Lyapunov functional, and $\ell$ is a constant. Suppose there is a constant $k = k(\ell)$ such that if $\phi \in U_\ell$, then $|\phi(0)| < k$, for some constant $k$. Then any solution $x_t(\phi)$ of (3.1.7) with $\phi \in U_\ell$ approaches $M$ as $t \to \infty$.*

**Corollary 3.1.3**   *Let $V : C \to E^+$ be continuous. Suppose there are nonnegative functions $u(r)$, $v(r)$ such that $u(r) \to \infty$ as $r \to \infty$ and $u(|\phi(0)|) \leq V(\phi)$, $\dot{V}(\phi) \leq -v(|\phi(0)|)$. Then every solution of (3.1.7) is bounded. If we assume further that $v(r)$ is positive definite, so that $v(0) = 0, v(r) > 0, r > 0$, then every solution approaches zero as $t \to \infty$.*

**Example 3.1.3**   Consider the system of two masses and three linear springs. Let $x_i$ denote the displacement of the mass $m_i$ from its equilibrium position, with positive displacements measured to the right. The masses are subject to external forces $F_i(t)$. This coupled system is described by the equations

$$M_1\ddot{x}_1 = -k_1 x_1 + k_2(x_2 - x_1) + F_1\,,$$

$$M_2\ddot{x}_2 = -k_2(x_2 - x_1) - k_3 x_2 + F_2\,.$$

In matrix form this becomes

$$M\ddot{x} + Bx = F\,,$$

where

$$M = \begin{bmatrix} M_1 & 0 \\ 0 & M_2 \end{bmatrix}\,, \quad x = \begin{bmatrix} x_1 \\ x_2 \end{bmatrix}\,, \quad B = \begin{bmatrix} k_1 + k_2 & -k_2 \\ -k_2 & k_2 + k_3 \end{bmatrix}\,, \quad F = \begin{bmatrix} F_1, & 0 \\ 0, & F_2 \end{bmatrix}\,.$$

Obviously $M$ is symmetric and positive definite, $B$ is symmetric and positive semidefinite. The external force $F$ can be taken to be a state feedback that has delay and is given by

$$H(x_t) = \int_0^h F(s)x(t - s)ds\,.$$

Thus the system considered is

$$M\ddot{x}(t) + Bx(t) = \int_0^h F(s)x(t - s)ds\,; \tag{3.1.8}$$

$M, B$ are $2 \times 2$ symmetric matrices, and $F$ is a $2 \times 2$ matrix that is continuously differentiable. We now consider a coupled system of $n$ such masses. Then $M_i$, $i - 1, \ldots, n$. $A, B, F$ are $n \times n$ symmetric matrices. If we set

$$M = B - \int_0^h F(s)ds\,,$$

(3.1.8) can be rewritten as

$$\dot{x}(t) = y(t)\,,$$

$$A\dot{y}(t) = -Mx(t) + \int_0^h F(s)[x(t - s) - x(t)]ds\,. \tag{3.1.9}$$

**Theorem 3.1.4** *In (3.1.9) assume that $A$ and $M$ are positive definite and $F(s)$ is positive semidefinite for $s \in [0, h]$, and assume that $F$ has the property that there exists a $c \in [0, h]$ such that $\dot{F}(c) < 0$. Then every solution of (3.1.9) satisfies $x(t) \to 0$ as $t \to \infty$.*

**Proof** The proof of this result is due to Hale [3, p. 124]. It uses the following Lyapunov functional:

$$V(\phi, \psi) = \frac{1}{2}\phi^T(0)M\phi(0) + \frac{1}{2}\psi^T(0)A\psi(0)$$

$$+ \frac{1}{2} \int_0^h [\phi(-s) - \phi(0)]^T F(s)[\phi(s) - \phi(0)]ds \,,$$

where $T$ denotes vector transpose and $\phi, \psi$ are the initial values of $x$ and $y$ of (3.1.9). One verifies easily that

$$\dot{V}(\phi, \psi) = -\frac{1}{2}[\phi(-h) - \phi(0)]^T F(h)[\phi(-h) - \phi(0)]$$

$$+ \frac{1}{2} \int_0^h [\phi(-s) - \phi(0)]\dot{F}(s)[\phi(-s) - \phi(0)]ds \leq 0 \,, \tag{3.1.10}$$

by conditions of the theorem. From the hypothesis there exists an interval $I_c$ containing $c$ such that $\dot{F}(s) < 0$ for all $s \in I_c$. It follows from (3.1.10) that $\dot{V}(\phi, \psi) = 0$ implies $\phi(-s) = \phi(0)$ for $s \in I_c$. For a solution $(x, y)$ to belong to the largest invariant set where $\dot{V}(\phi, \psi) = 0$ we must have $x(t - s) = x(t)$, $\forall t \in (-\infty, \infty)$, $s \in I_c$. This implies that $x(t) = b$ is a constant. From (3.1.9), $y = 0$ and thus $Mb = 0$. Because $M > 0$, $b = 0$. Therefore the largest set where $\dot{V}(\phi, \psi) = 0$ is $(0,0)$, the origin. But $V$ satisfies the conditions of Theorem 3.1.3 and its Corollary 3.1.1. The conclusion is valid. Every solution of (3.1.9) approaches $(0,0)$ as $t \to \infty$.     $\square$

## 3.2   Razumikhin-Type Theorems

There is an extensive theory in the application of the Lyapunov stability method for ordinary differential equations. Often the Lyapunov functions constructed for ordinary systems can be fruitfully exploited to investigate the stability of analogous functional differential equations. The theorems of Razumikhin are the bridge. These results are based on the properties of a continuous function $V : E^n \to E$ whose derivative $\dot{V}(t, \phi(0))$ along solutions $x(t, \phi)$ of the system

$$\dot{x}(t) = f(t, x_t) \,,$$

$$x_\sigma = \phi \,, \tag{3.2.1}$$

is defined to be

$$\dot{V}(t, \phi(0)) = \varlimsup_{r \to 0^+} [V(t + r, x(t, \phi)(t + r) - V(t, \phi)] \,.$$

The following results are contained in Hale [3, p. 127].

**Theorem 3.2.1**   *Let $f : E \times C \to E^n$ take $E^+$ (bounded sets of $C$) into bounded sets of $E^n$. Suppose $u, v, w : E^+ \to E^+$ are continuous, nondecreasing functions $u(s), v(s)$ positive for $s > 0$, $u(0) = v(0) = 0$. Let $V : E \times E^n \to E$ be a continuous function such that*

$$u(|x|) \leq V(t, x) \leq v(|x|), \quad \forall \, (t, x) \in E \times E^n \,,$$

*and*

$$\dot{V}(t,\phi(0)) \leq -\omega(|\phi(0)|) \quad if \quad V(t+\theta,\phi(\theta)) \leq V(t,\phi(0)), \quad \theta \in [-h,0].$$

*Then the solution $x = 0$ of (3.2.1) is uniformly stable.*

*If $w(s) > 0$ for $s > 0$ and there is a continuous nondecreasing function $p(s) > s$ for $s > 0$ such that $\dot{V}(t,\phi(0)) \leq -w(|\phi(0)|)$ if $V(t+\theta,\phi(\theta)) < p(V(t,\phi(0)))$, $\theta \in [-h,0]$. Then the solution $x = 0$ of (3.2.1) is uniformly asymptotically stable. If $u(s) \to \infty$ as $s \to \infty$, then the solution $x = 0$ is globally uniformly asymptotically stable.*

**Example 3.2.1**  Consider the scalar system

$$\dot{x}(t) = -ax(t) - bx(t-h). \tag{3.2.2}$$

**Proposition 3.2.1**  *Assume that $a,b$ are constants such that $|b| \leq a$. Then the solution $x = 0$ of (3.2.2) is uniformly stable. If in addition there exists some $q > 1$ such that $|b|q < a$, then (3.2.2) is uniformly asymptotically stable.*

**Proof**  Consider $V = x^2/2$.

$$\dot{V}(x(t)) = -ax^2(t) - bx(t)x(t-h)$$

$$\leq -[a - |b|]x^2(t)$$

if $|x(t-h)| < |x(t)|$. Thus $\dot{V}(x(t)) \leq 0$ if $V(x(t-h)) \leq V(x(t))$. It follows from Theorem 3.2.1 that $x = 0$ is uniformly stable. For the second part, for some $q > 1$, choose $p(s) = q^2 s$. If $p(V(x(t) > V(x(t-h))$, i.e., $|x(t-h)| < q|x(t)|$, then $\dot{V}(x(t)) \leq -(a - |b|q)x^2(t)$. Since $a - |b|q > 0$, the second part of the theorem guarantees that the solution $x = 0$ of (3.2.2) is uniformly asymptotically stable. $\square$

**Example 3.2.2**  Consider

$$\dot{x}(t) = y(t),$$

$$\dot{y}(t) = -a(t,y(t)) - f(x(t)) + \int_{-h}^{0} g(x(t+\theta))y(t+\theta)d\theta. \tag{3.2.3}$$

**Proposition 3.2.2**  *Assume in (3.2.3) that:*

   (i) $\frac{a(t,y(t))}{y} > \alpha > 0$, $y \neq 0$, *for some constant $\alpha$;*
   (ii) $f(x)\,\mathrm{sgn}\,x \to \infty$ *as* $|x| \to \infty$, $f(x)\,\mathrm{sgn}\,x > 0$; *and*
   (iii) $|g(x)| \leq L$, $\quad \forall x \in E$, *where $Lhq < \alpha$.*

*Then (3.2.3) is uniformly asymptotically stable.*

**Proof**   Consider

$$V(x,y) = \int_0^{x(t)} f(s)ds + y^2/2 \,,$$

$$\dot{V}(x(t),y(t)) = y(t)\dot{y}(t) + f(x(t))y(t)$$

$$= y(t)\left[-a(t,y(t)) - f(x,(t)) + \int_{-h}^0 g(x(t+\theta))y(t+\theta)d\theta\right]$$
$$+ f(x)y(t)$$

$$= -y(t)a(t,y(t)) + y(t)\int_{-h}^0 g(x(t+\theta))y(t+\theta)d\theta$$

$$\leq -\alpha y^2(t) + |y(t)|^2 Lqh \,,$$

if $|y(t+\theta)| < q|y(t)|$, then $q > 1$. Thus

$$\dot{V}(x(t),y(t)) \leq -(\alpha - Lhq)y^2(t) < 0$$

if $\alpha - Lhq > 0$. The result follows.  $\square$

## 3.3   Lyapunov–Razumikhin Methods of Stability in Delay Equations

The method of Lyapunov and its extensions by LaSalle and Hale require the construction of a Lyapunov functional, which can be very difficult to practice. These functionals are nonincreasing along solutions. Lyapunov functions, which are simpler and which are not nonincreasing along solutions, can be appropriated. This is the fundamental idea of Haddock and Terjéki [2]: the development of an invariance principle using Razumikhin functions.

In what follows, the power of the Lyapunov method is married to the simplicity of Razumikhin functions in a theory proposed by Haddock and Terjéki [5]. A quadratic Lyapunov function is shown to yield good stability criteria.

Consider an autonomous retarded functional differential equation of the form

$$\dot{x}(t) = f(x_t) \,, \quad x_0 = \phi \,, \tag{3.3.1}$$

where $f : C \to E^n$ is continuous and maps closed and bounded sets into bounded sets. We also assume that solutions depend continuously on initial data.

**Definition 3.3.1**   Let $\phi \in C$. An element $\psi \in C$ belongs to an $w$-limit set of $\phi$, $\Omega(\phi)$ if the solution $x(t,0,\phi)$ of (3.1.1) is defined on $[-h,\infty)$ and there is a sequence $\{t_n\}$, $t_n \to \infty$ for $n \to \infty$ such that $\|x_{t_n}(\phi) - \psi\| \to 0$ as $n \to \infty$ and $x_{t_n}(\phi)(s) = x(t_n + s,0,\phi)$. A set $M \subseteq C$ is called an invariant set (with respect to (3.3.1)) if for any $\phi \in M$ there is a solution $x(\phi)$ of (3.3.1) that is defined on $(-\infty, \infty)$ such that $x_0 = \phi$ and $x_t \in M$, $\forall t \in (-\infty, \infty)$. A set $M \subseteq C$ is positively invariant if for each $\phi \in M$, $x_t(\phi) \in M$, $\forall t \in [0, \infty)$. It is well known [8] that if $x(\phi)$ is a solution of (3.3.1) that is defined and bounded on $[-h, \infty)$, then

(i)  the set $\{x_t(\phi); \ t \geq 0\}$ is precompact,
(ii)  $\Omega(\phi)$ is nonempty, compact, connected, and invariant, and
(iii)  $x_t(\phi) \to \Omega(\phi)$ as $t \to \infty$.

Haddock and Terjéki [2] applied a well known Lyapunov function argument to these stated properties of $w$-limit sets of solution to obtain an invariance principle that utilizes the easier to construct Lyapunov function instead of the usual Lyapunov functional of Hale–LaSalle [5]. To state the result we need the following definitions:

**Definition 3.3.2**  A function $V : E^n \to E$ that has derivatives is called a Lyapunov function or a Razumikhin function.

Let $V$ be a Lyapunov function. The upper right hand derivative of $V$ with respect to (3.1.1) is defined by

$$\dot{V}[\phi] = \limsup_{r \to 0^+} \frac{1}{r} \left( V[\phi(0) + r f(\phi)] - V[\phi(0)] \right). \tag{3.3.2}$$

Because $V$ has continuous first partial derivative, we have

$$\dot{V}[\phi] = \sum_{i=1}^{n} \frac{\partial V[\phi(0)]}{\partial x_i} f_i(\phi), \tag{3.3.3}$$

where $f_i$ is the $i$-component of $f$. We note that if $x$ is a solution of (3.3.1),

$$\frac{dV[x(t)]}{dt} = \dot{V}[x_t].$$

**Definition 3.3.3**  Suppose $V$ is a Lyapunov function and $G \subset C$ a set. Let

$$E_V(G) = \left\{ \phi \in G : \max_{-h \leq s \leq 0} V[x_t(\phi)(s)] = \max_{-h \leq s \leq 0} V[\phi(s)], \ \forall t \geq 0 \right\},$$

and denote by $M_V(G)$ the largest subset of $E_V(G)$ that is invariant with respect to (3.3.1).

The following theorem and its consequences are available in [2]

**Theorem 3.3.1**  *Let $V$ be a Lyapunov function and $G$ a closed set that is positively invariant with respect to (3.3.1). Suppose*

$$\dot{V}[\phi] \leq 0 \quad \forall \phi \quad such \ that \quad V[\phi(0)] = \max_{-h \leq s \leq 0} V[\phi(s)].$$

*Then for any $\phi \in G$ such that $x(\phi)(\cdot)$ is defined and bounded on $[-h, \infty)$, $\Omega(\phi) \subseteq M_V(G) \subseteq E_V(G)$ and*

$$x_t(\phi) \to M_V(G) \quad as \quad t \to \infty.$$

**Corollary 3.3.1** *Consider* (3.3.1) *and suppose*

(i) $f(0) = 0$.
  *Suppose in addition there exists a Lyapunov function* $V$ *and a constant* $a > 0$ *such that:*
(ii) $V(0) = 0$ *and* $V[x] > 0$,   $\forall x,\ 0 \neq |x| < a$,
(iii) $\dot{V}[0] = 0$, *and*
(iv) $\dot{V}[\phi] < 0$ *for any* $0 \neq \|\phi\| < a$ *such that*

$$\max_{-h \leq s \leq 0} V[\phi(s)] = V[\phi(0)].$$

*Then the solution* $x = 0$ *of* (3.3.1) *is asymptotically stable.*

We note that if every solution of (3.3.1) is bounded on $[-h, \infty)$ by $a$, say, then Corollary 3.3.1 gives a global asymptotic stability result. We now state a criterion for boundedness in terms of a Lyapunov function $V$.

**Theorem 3.3.2** *Suppose* $u, v, w : E^+ \to E^+$ *are continuous, nondecreasing functions,* $u(s), v(s)$ *positive for* $s > 0$, $u(0) = v(0) = 0$. *Let* $V : E^n \to E$ *be a Lyapunov function such that*

$$u(|x|) \leq V(x) \leq v(|x|), \quad \forall x \in E^n, \tag{3.3.4}$$

*and* $\dot{V}[\phi] \leq -w(|\phi(0)|)$ *for any* $\phi \in C$ *such that*

$$V[\phi(0)] = \max_{-h \leq s \leq 0} V[\phi(s)]. \tag{3.3.5}$$

*If* $u(s) \to \infty$ *as* $s \to \infty$, *then the solutions of* (3.3.1) *are uniformly bounded.*

**Proof** Consider the function $\max_{-h \leq s \leq 0} V[x_t(\phi)(s)] = W[x_t]$. Just as in Haddock and Terjéki [2], $W[x_t]$ is a nonincreasing function of $t$ on $[0, \infty)$. By (3.3.4)

$$u(|x(t)|) \leq W[x_t] \leq v(\|x_t\|), \quad \forall t \in E, \quad \phi \in C.$$

Since $u(s) \to \infty$ as $s \to \infty$, if $\alpha > 0$ is any given constant there is a $\beta > 0$ such that $u(\beta) = v(\alpha)$. As a result, if $\|\phi\| \leq \alpha$, the solution $x(0, \phi)$ with $x(0, \phi) = \phi$ satisfies $|x(\phi)(t)| \leq \beta \ \forall t \geq 0$, which implies uniform boundedness. Indeed,

$$u(|x(\phi)(t)|) \leq V[x(t)] = W(x_t) \leq W(\phi) \leq v(\|\phi\|) \leq v(\alpha) = u(\beta).$$

Therefore $|x(\phi)(t)| \leq \beta\ t \geq 0$ as asserted.

  It now follows from Theorem 3.3.2 and Corollary 3.3.1 that the following is true:

□

**Corollary 3.3.2** *Consider* (3.3.1) *and suppose*

(i) $f(0) = 0$,

(ii) *and there are continuous nondecreasing functions $u(s)$, $v(s)$ positive for $s > 0$, $u(0) = v(0) = 0$; $u(s) \to \infty$ as $s \to \infty$. Let $V$ be a Lyapunov function such that*

(iii) *$V[0] = 0$, $u(|x|) \le V(x) \le v(|x|)$, $\forall\, x \in E^n$,*

(iv) *$\dot{V}[\phi] < 0$ for any $\phi$ such that $\max_{-h \le s \le 0} V[\phi(s)] = V[\phi(0)]$.*

*Then the solution $x = 0$ of (3.3.1) is globally asymptotically stable and every solution $x(\phi)$ of (3.3.1) satisfies $x_t(\phi) \to 0$ as $t \to \infty$.*

We now use Corollary 3.3.2 to give checkable criterion in terms of the system's coefficients for global uniform asymptotic stability and the convergence

$$x_t(\phi) \to 0 \quad \text{as} \quad t \to \infty, \tag{3.3.6}$$

for the nonlinear system (3.1.1) whose linear approximation is

$$\dot{x}(t) = L(x_t), \tag{3.3.7}$$

where $L(x_t) = D_1 f(0) x_t$. $D_1 f(0)$ is the Fréchet derivative of $f(\phi)$ with respect to $\phi$ at 0, and this is the mapping $L(\cdot) : C \to E^n$ given by

$$L(x_t) = A_0 x(t) + \sum_{k=1}^{N} A_k x(t - w_k) + \int_{-h}^{0} A(s) x(t + s)\, ds. \tag{3.3.8}$$

Here $0 \le w_1 \le w_2 \cdots \le w_N = h$, and $A_i$ are $n \times n$ constant matrices. We note that

$$f(\phi) - L(\phi) - f(0) = g(\phi),$$

where $\|g(\phi)\|/\|\phi\| \to 0$ as $\|\phi\| \to 0$, so that, given any $\epsilon > 0$, there is a $\delta(\epsilon) > 0$ such that if $\|\phi\| < \delta$, then $\|g(\phi)\| < \epsilon\|\phi\|$. Thus any $f$ with Fréchet derivative $L$ can be written as

$$f(\phi) = L(\phi) + f(0) + g(\phi), \tag{3.3.9}$$

where for some $\epsilon > 0$

$$|g(\phi)| \le \epsilon\|\phi\| \quad \text{for all} \quad \phi \in \mathcal{O}, \tag{3.3.10}$$

with $\mathcal{O} = \{\phi \in C : \|\phi\| < \delta\}$. We assume that $f$ given in (3.3.1) is of this form where (3.3.10) holds for all $\phi \in C$, since it is a global result that we need. If $f(0) = 0$ and $g(\phi) \equiv 0$, we deduce conditions on the system's coefficients that guarantee global uniform stability. If $f(0) = 0$ but $g(\phi)$ satisfies (3.3.10) for all $\phi \in C$, we deduce results on stability in the first approximation with acting perturbations. We also obtain the global decay (3.3.6) when $f(0) \ne 0$.

**Theorem 3.3.3**　*In (3.3.1) assume that*

(i) *$f(0) = 0$.*

(ii) *There exists a constant, nonsingular, positive definite symmetric matrix $H$ that satisfies $d_1|x(t)|^2 \leq x^T(t)Hx(t) \leq d_2|x(t)|^2$, $\forall t \in E$, $d_i > 0$ $(i = 1, 2)$, and the matrix*

$$B \equiv HA_0 + A_0^T H + 2Iq(\|H\|)\left(\sum_{k=1}^{N}\|A_k\| + a + \epsilon\right) \qquad (3.3.11)$$

*is negative definite; ($I$ is identity matrix), and $q \geq 1$.*
(iii) *$f(\phi) = L(\phi) + g(\phi)$ where $L$ is given in (3.3.8), and*

$$\left|\int_{-h}^{0} A(s)x(t+s)ds\right| \leq a\|x_t\|, \qquad (3.3.12)$$

*a a constant, and*

$$|g(\phi)| \leq \epsilon\|\phi\|, \quad \forall t \in E, \phi \in C, \qquad (3.3.13)$$

*for some $\epsilon > 0$.*

*Then every solution $x(\phi)$ of (3.3.1) with $x_0 = \phi$ satisfies (3.3.6).*

**Proof**    We take inner products in (3.3.9) and use (3.3.12) to obtain

$$(f(x_t), x(t)) = (L(x_t), x(t)) + (f(0), x(t)) + (g(x_t), x(t))$$

$$\leq (L(x_t), x(t)) + (f(0), x(t)) + \epsilon\|x_t\|\,|x(t)|$$

so that

$$(f(x_t), x(t)) \leq (L(x_t), x(t)) + \epsilon\|x_t\|\,|x(t)| + (f(0), x(t)). \qquad (3.3.14)$$

Since $f(0) = 0$, we have the estimate

$$(f(x_t), x(t)) \leq (L(x_t), x(t)) + \epsilon\|x_t\|\,|x(t)|. \qquad (3.3.15)$$

Now consider the function

$$V(x(t)) = (x(t), Hx(t)),$$

where $H$ is given in (ii). Differentiate $V$ along solutions of (3.1.1) and use (3.1.15) to obtain

$$\dot{V}(x(t)) = 2(Hf(x_t), x(t)),$$

$$\dot{V}(x(t)) = 2(Hf(x_t), -Hf(0), x(t)) + 2(Hf(0), x(t)), \qquad (3.3.16)$$

$$\dot{V}(x(t)) \leq 2(HL(x_t), x(t)) + 2\epsilon\|H\|\,\|x_t\|\,|x(t)| + 2(Hf(0), x(t)).$$

We now bring to bear the definition of $L(\phi)$ in (3.3.8) and condition (3.3.12), so that

$$\dot{V} \leq ((HA_0 + A_0^T H)x(t), x(t))$$

$$+ 2\|H\| \sum_{k=1}^{N} \|A_k\| \, |x(t - w_k)| \, |x(t)|$$

$$+ 2a\|H\| \, \|x_t\| \, |x(t)| + 2\epsilon\|H\| \, \|x_t\| \, |x(t)| \, .$$

But if $V[x(t)] = \max_{-h \leq s \leq 0} V[x_t(s)]$, then by Lemma 3.3.2 below we have $\|x_t\| \leq q|x(t)|$, where $q = \alpha_2/\alpha_1$, $\alpha_2$ the greatest and $\alpha_1$ the least eigenvalue of $H$. Thus

$$\dot{V} \leq (Bx(t), x(t)) < 0 \, ,$$

where $B$ is the $n \times n$ matrix defined by

$$B = HA_0 + A_0^T H + 2I_q\|H\| \left( \sum_{k=1}^{N} \|A_k\| + a + \epsilon \right) .$$

This $B$ is negative definite by (ii). Setting $u(s) = d_1 s^2$, $v(s) = d_2 s^2$ in Corollary 3.3.2, we conclude that $x_t(\phi) \to 0$ as $t \to \infty$, and the trivial solution is globally asymptotically stable. $\qquad\square$

**Corollary 3.3.3**  *Consider the linear system*

$$\dot{x}(t) = A_0 x(t) + \sum_{k=1}^{N} A_k x(t - w_k) + \int_{-h}^{0} A(s)x(t + s)ds \, , \qquad (3.3.17)$$

*where $A(s)$ satisfies (3.3.12). Assume there is a positive definite symmetric matrix $H$ such that the matrix*

$$B \equiv HA_0 + A_0^T H + 2Iq\|H\| \left( \sum_{k=1}^{N} \|A_k\| + a \right)$$

*is negative definite.*

*Then (3.3.17) is globally, uniformly, asymptotically stable and every solution $x(\phi)$ of (3.3.17) with $x_0(\phi) = \phi$ satisfies*

$$\|x_t(\phi)\| \leq k\|\phi\|e^{-\alpha t}, \quad t \geq 0 \, ,$$

*for some $k > 0$, $\alpha > 0$.*

**Proof**  This is an immediate consequence of Theorem 3.3.3, since $g(\phi) \equiv 0$ implies $\epsilon$ can be taken as zero, and Theorem 3.3.2 can be applied. $\qquad\square$

The estimate $B < 0$ may well be the best possible. The case

$$\dot{x}(t) = ax(t) + bx(t - h) \, ,$$

considered by Infante and Walker [9], can be used to test this statement. Indeed, assume $H = 1$. Then

$$B = 2a + 2|b| < 0 \quad \text{or} \quad a + |b| < 0 \, .$$

Recall that in [6] uniform asymptotic stability is valid for every $h \in [0, \infty)$ if and only if $a + |b| \leq 0$ and $a + b < 0$. If the system

$$\dot{x}(t) = A_0 x(t) + A_1 x(t - h) \qquad (3.3.18)$$

is considered, and if $H = I$ is taken where the matrices are $n \times n$, then the condition for global asymptotic stability is $B = \frac{1}{2}(A_0 + A_0^T) + 2I(\|A_1\|)$ and is negative definite, which is comparable to a recent result of Mori [7].

The next result can be proved by the methods of this paper and the ideas of observability. The proof is contained in Chukwu [1].

**Proposition 3.3.1**  *Consider the system*

$$\dot{x}(t) = A_0 x(t) + \sum_{i=1}^{N} A_i x(t - i\alpha),$$
$$\alpha = h/N > 0.$$

$$(S)$$

*Assume that*

(i)  $B \equiv HA_0 + A_0^T H + \sum_{k=1}^{N} 2qn M_k H$ *is negative semidefinite, for some positive definite symmetric matrix* $H$ *where* $M_k = \max |(A_k)ij|$.
   *Then* (S) *is asymptotically stable if and only if*

(ii)  $\mathrm{rank}\,[\Delta(\lambda), B] = n$ *for each complex* $\lambda$, *where* $\Delta(\lambda) = \lambda I - A_0 - \sum_{k=1}^{N} A_k e^{-\lambda_k \alpha}$, *and*

(iii)  *for some complex number* $\lambda$,

$$\mathrm{rank} \begin{bmatrix} A_0 - \lambda I & A_1 & \cdots & A_N & B \\ A_1 & & A_N & 0 & 0 \\ A_N & & 0 & 0 & 0 \end{bmatrix} = n + \mathrm{rank} \begin{bmatrix} A_1 & \cdots & A_N \\ A_N & 0 & 0 \end{bmatrix}.$$

**Remark 3.3.1**  Condition (i) of Proposition 3.3.1 ensures that (S) has uniformly bounded solutions.

**Example 3.3.1**  The scalar system

$$\dot{x}(t) = a_0 x(t) + \sum_{k=1}^{N} a_k x(t - wk)$$

has (by Proposition 3.3.1) the following property: If

$$B = 2a_0 + \sum_{k=1}^{N} 2|a_k| \leq 0,$$

the scalar system is uniformly asymptotically stable if and only if for some complex number $\lambda$

$$
\text{rank}
\begin{bmatrix}
a_0 - \lambda & a_1 & \cdots & a_N & B \\
 & a_1 & & a_N & 0 \\
 & & & 0 & \\
 & & & 0 & \\
 & a_N & & 0 & 0
\end{bmatrix}
= 1 + \text{rank}
\begin{bmatrix}
a_1 \cdots a_N \\
a_N \cdots 0
\end{bmatrix} .
$$

This rank condition holds whenever $a_N \neq 0$.

We now state a lemma that was used in the proof of the theorem.

**Lemma 3.3.1** *Let $t \in (-\infty, \infty)$. If $V[x(t)] = \max_{-h \le s \le 0} V[x_t(s)]$. Then $\|x_t\| \le q|x(t)|$ for the same $t$, where $q = \alpha_2/\alpha_1$, $\alpha_2$ the greatest and $\alpha_1$ the least eigenvalue of $H$, and where*

$$
V[x(t)] = (Hx(t), x(t))
$$

*with $H$ a positive definite symmetric matrix, and $(\cdot, \cdot)$ denotes the inner product in $E^n$.*

**Proof**  Because

$$
(Hx(t), x(t)) = \max_{-h \le s \le 0} (Hx(t+s), x(t+s))
$$

and the maximum is attained at some $s^* \in [-h, 0]$, we have

$$
(Hx(t), x(t)) = (Hx(t+s^*), x(t+s^*))
$$

so that

$$
\alpha_1(x(t+s^*)) \le (Hx(t+s^*), x(t+s^*))
$$

$$
= (H(x(t), x(t)) \le \alpha_2(x(t), x(t))
$$

where $\alpha_1$ and $\alpha_2$ are the least and greatest eigenvalues of $H$. Thus

$$
(x(t+s^*), x(t+s^*)) \le \frac{\alpha_2}{\alpha_1}(x(t), x(t))
$$

that is $\|x_t\| \le q|x(t)|$. Note that the argument can be reversed.

A special case of (3.3.1) is the system

$$
\dot{x}(t) = f(x(t)) + \sum_{k=1}^{N} g_k(x(t - w_k)) + G\left(\int_{-h}^{0} A(s)x(t+s)ds\right), \qquad (3.3.19)
$$

where $0 \le w_k < h$, $k = 1, \ldots, N$, $h > 0$ and where $x(s) \in E^n$, $f, g_k : E^n \to E^n$ are continuous. We require that

$$
\left| G\left(\int_{-h}^{0} A(s)x(t+s)ds\right) \right| \le \beta\|x_t\| \qquad (3.3.20)
$$

where $\beta$ is a constant. We assume $f, g_k$ take (bounded sets of $E^n$) into bounded sets of $E^n$ and are continuously differentiable. Let $D_0$ denote the $n \times n$ matrix function $\frac{\partial f_i(x(t))}{\partial x_j}$, and $D_k$ the $n \times n$ matrix function $\frac{\partial g_k^i(x(t-w_k))}{\partial x_j(t-w_k)}$. Let $D_a$, $a = 0, 1, \ldots, N$ be bounded. Let $H$ be a constant $n \times n$ positive definite symmetric matrix. Define

$$J_a = HD_a + D_a^T H, \quad a = 0, 1, \ldots, N. \tag{3.3.21}$$

$\square$

**Theorem 3.3.4**    *In* (3.1.19), *suppose there exists a positive definite constant symmetric matrix $H$ such that the $n \times n$ matrix*

$$B = J_0 + q \left( \sum_{k=1}^{N} \|J_k\| + 2\|H\|\beta \right) I \tag{3.3.22}$$

*is negative definite uniformly on $x(t)$, $x(t - w_k)$. Assume*

$$f(0) = 0 = g_k(0), \quad k = 1, \ldots, N.$$

*Then* (3.3.19) *is globally uniformly asymptotically stable.*

**Proof**    Let $H$ be as stated and consider $V = (Hx(t), x(t))$.  Then there exist positive finite constants $d_1, d_2$ such that

$$d_1|x(t)|^2 \le V \le d_2|x(t)|^2. \tag{3.3.23}$$

Differentiating $V$ along solutions of (3.3.19), and utilizing Lemma 3.1.1 and Lemma 3.1.2 below, we have

$$\dot{V} \le (J_0 x(t), x(t)) + q \left( \sum_{k=1}^{N} \|J_k\| + 2\|H\|\beta \right) |x(t)|^2.$$

By hypothesis,

$$B \equiv J_0 + qI \left( \sum_{k=1}^{N} \|J_k\| + 2\beta\|H\| \right)$$

is negative definite, so that

$$\dot{V} \le (Bx(t), x(t)) < 0. \tag{3.3.24}$$

This inequality (3.3.24) and (3.3.23) suffice for Theorem 3.3.2 to apply. The theorem is proved. To round up we state Lemma 3.3.2.    $\square$

**Lemma 3.3.2**    *Let $\ell : E \times E^n \to E^n$ be a continuously differentiable function, and suppose there is a constant $M > -\infty$ such that*

$$J = \frac{1}{2} \left( \frac{\partial \ell_i}{\partial x_j} + \frac{\partial \ell_j}{\partial x_i} \right) < MI,$$

(*I*, *the identity matrix*) *uniformly in* $x$ *and* $t > \sigma$, $\sigma, t \in E$. *Then the scalar products*

$$L_1 = (\ell(t, x + r) - \ell(t, x), r)\,,$$

$$L_2 = (\ell(t, x + r) - \ell(t, x), y)\,,$$

*satisfy*

$$L_1 \leq M|r|^2\,, \qquad L_2 \leq |M|\,|r|\,|y|\,.$$

**Proof**  The first estimate follows readily from the Mean Value Theorem; and the second follows from the Mean Value Theorem and the Schwartz inequality. Indeed, if $H$ is the $n \times n$ matrix $h_{ij}$ where

$$h_{ij} = \frac{\partial h_i}{\partial x_j}(t; x + \theta_i r)\,,$$

and $\theta_i = \theta_i(x, t)$ satisfies $0 < \theta_i < 1$, the Mean Value Theorem yields the identity

$$L_1 = (Hr, r) = (Jr, r)\,,$$

and because $J$ is symmetric,

$$(Jr, r) < M(r, r) = M|r|^2\,.$$

Also

$$L_2 = (Hr, y)\,,$$

and by the Schwartz Identity,

$$L_2 \leq |(Hr, y)| \leq |Hr|\,|y| \leq \|H\|\,|r|\,|y| = |M|\,|r|\,|y|\,. \qquad \square$$

## References


1. E. N. Chukwu, "Global Behaviour of Linear Retarded Functional Differential Equations," *J. Math. Anal. Appl.* **162** (1991) 277–293.
2. J. R. Haddock and J. Terjéki, "Lyapunov–Razumikhin Functions and an Invariance Principle for Functional Differential Equations," *J. Differential Equations* **48** (1983) 95–122.
3. J. Hale, *Theory of Functional Differential Equations*, Springer-Verlag, New York, 1977.
4. J. Hale, *Ordinary Differential Equations*, Wiley, New York, 1969.
5. J. Hale, "Sufficient Conditions for Stability and Instability of Autonomous Functional Differential Equations," *J. Differential Equations* **1** (1965) 452–482.
6. E. F. Infante and J. A. Walker, "A Lyapunov Functional for a Scalar Differential Difference Equation," *Proc. Royal. Soc.*, Edinburgh **79A** (1977) 307–316.
7. T. Mori, "Criteria for Asymptotic Stability of Linear Time-Delay Systems," *IEEE Transactions Aut. Control* **AC-30** (1985) 158–161.

# Chapter 4

# Global Stability of Functional Differential Equations of Neutral Type

## 4.1  Definitions

Suppose $f$, $g$ are continuous functions taking $[\tau, \infty) \times C \to E^n$. Define the functional difference operator

$$D(\cdot, \cdot) : [\tau, \infty) \times C \to E^n$$

by

$$D(t)\phi = \phi(0) - g(t, \phi) \tag{4.1.1}$$

for $t \in [\tau, \infty)$, $\phi \in C = C([-h, 0], E^n)$. Consider the system

$$\frac{d}{dt} D(t)x_t = f(t, x_t), \qquad x_t(s) = x(t + s), \tag{4.1.2}$$

where $x_t \in C$. We say $x$ is a solution of (4.1.2) with initial value $\phi$ at $\sigma$ if there exists $a \in [\tau, \infty)$, $a > 0$ such that $x : [\sigma - h, \sigma + a] \to E^n$ is continuous $x_\sigma = \phi$, $D(t)x_t$ is continuously differentiable on $(\sigma, \sigma + a)$, and (4.1.2) is satisfied on $(\sigma, \sigma + a)$. In this section we shall consider the following systems: (4.1.2),

$$\frac{d}{dt} D(t)x_t = f(t, x_t) + F(t, x_t), \tag{4.1.3}$$

$$\frac{d}{dt}[D(t)x_t - G(t, x_t)] = f(t, x_t) + F(t, x_t), \tag{4.1.4}$$

under the following, basic assumptions:
$D(t)\phi = \phi(0) - g(t, \phi)$ is defined in (4.1.1) where $g(t, \phi)$ is linear in $\phi$ and is given by

$$g(t, \phi) = \int_{-h}^{0} [d_\theta \mu(t, \theta)]\phi(\theta). \tag{4.1.5a}$$

Here $\mu(t, \theta)$ is an $n \times n$ matrix function of $t \in [\tau, \infty)$, $\theta \in [-h, 0]$, of bounded variation in $\theta$ which satisfies the inequality

97

$$\left| \int_{-s}^{0} d_\theta \mu(t,\theta)\phi(\theta) \right| \le l(s) \sup_{-s \le \theta \le 0} |\phi(\theta)|, \quad \forall\, t \in [\tau,\infty), \quad \phi \in C. \tag{4.1.5b}$$

Whenever (4.1.5) holds for $g$, we say $g$ is uniformly nonatomic at zero. We also assume that $G$ in (4.1.4) is uniformly nonatomic at zero. Furthermore, there is a nonnegative constant $N$ such that

$$|g(t,\phi)| \le N\|\phi\|, \qquad \forall\, (t,\phi) \in [\tau,\infty) \times C. \tag{4.1.6}$$

Basic in our discussions below are the assumptions that $f$, $g$, $D$, $G$, $F$ are smooth enough to ensure that a solution of (4.1.2)–(4.1.4) exists through each $(\sigma,\phi) \in [\tau,\infty) \times C$, is unique and depends continuously upon $(\sigma,\phi)$, and can be continued to the right as long as the trajectory remains in a bounded set in $[\tau,\infty) \times C$. The needed conditions are given in Sec. 7.1.

Our main interest here is the stability theory of (4.1.2), (4.1.3), and (4.1.4). The basic Lyapunov theory on (4.1.2) is first presented. In addition, if in (4.1.2) $f$ has a uniform Lipschitz constant, a converse theorem to the basic Lyapunov theory of Cruz and Hale [2] is formulated. All these are applied in subsequent sections. Some definitions are needed.

**Definition 4.1.1**     The zero solution of (4.1.2) is Uniformly Asymptotically Stable in the Large (UASL) if it is uniformly stable and if the solutions of (4.1.2) are uniformly bounded, and if for any $\alpha > 0$ and any $\epsilon > 0$ and any $\sigma \in [0,\infty)$ there exists a $T(\epsilon,\alpha) > 0$ such that if $\|\phi\| \le \alpha$, then $\|x_t(\sigma,\phi)\| < \epsilon$ for all $t \ge \sigma + T(\epsilon,\alpha)$. The trivial solution (4.1.2) is said to be Weakly Uniformly Asymptotically Stable in the Large (WUASL) if it is uniformly stable and for every $\sigma \in [0,\infty)$ and every $\phi \in C$ we have $\|x_t(\sigma,\phi)\| \to 0$ as $t \to \infty$. The zero solution of (4.1.2) is Exponentially Asymptotically Stable in the Large (EXASL) if there exists $c > 0$ and for any $\alpha > 0$ there exists a $k(\alpha) > 0$ such that if $\|\phi\| \le \alpha$,

$$\|x_t(\sigma,\phi)\| \le k(\alpha)e^{-c(t-\alpha)}\|\phi\|, \qquad \forall\, t \ge \sigma.$$

The zero solution is Uniformly Exponentially Asymptotically Stable in the Large (UEXASL) if there exist $c > 0$, $k > 0$ such that

$$\|x_t(\sigma,\phi)\| \le ke^{-c(t-\sigma)}\|\phi\|, \qquad \forall\, t \ge \sigma. \tag{4.1.7}$$

The trivial solution is Eventually Uniformly Stable (EVUS) if for every $\epsilon > 0$ there is $\alpha = \alpha(\epsilon)$ such that (4.1.2) is uniformly stable for $t \ge \sigma \ge \alpha(\epsilon)$. Other eventual stability concepts can similarly be formulated by insisting that the initial time $\sigma$ satisfy $\sigma \ge \alpha$. For example, the trivial solution of (4.1.2) is called Eventually Weakly Uniformly Asymptotically Stable in the Large (EVWUASL) if it is (EVUS) and if there exists $\alpha_0 > 0$ such that for every $\sigma \ge \alpha_0$ and for every $\phi \in C$, we have $\|x_t(\sigma,\phi)\| \to 0$ as $t \to \infty$.

## 4.2 Lyapunov Stability Theorem

Consider the homogeneous difference equation

$$D(t)x_t = 0\,, \quad t \geq \sigma\,, \quad x_\sigma = \phi\,, \quad D(\sigma)\phi = 0\,, \tag{4.2.1}$$

which was discussed in Sec. 2.4. There it was stated that $D$ is uniformly stable if there are constants $\beta$, $\alpha > 0$ such that for every $\sigma \in [\tau, \infty)$, $\phi \in C$ the solution $x(\sigma, \phi)$ of (4.2.1) satisfies

$$\|x_t(\sigma, \phi)\| \leq \beta e^{-\alpha(t-\sigma)} \|\phi\|\,, \qquad t \geq \sigma\,.$$

Cruz and Hale have given the following two lemmas [2]:

**Lemma 4.2.1** *If $D$ is uniformly stable, then there are positive constants $a, b, c, d$ such that for any $h \in C([\tau, \infty),\ E^n)$, the space of continuous functions from $[\tau, \infty) \to E^n$, and $\sigma \in [\tau, \infty)$ the solution $x(\sigma, \phi, g)$ of*

$$D(t)x_t = M(t)\,, \qquad t \geq \sigma\,, \quad x_\sigma = \phi\,, \tag{4.2.2}$$

*satisfies*

$$\|x_t(\sigma, \phi, g)\| \leq e^{-a(t-\sigma)} \left( b\|\phi\| + C \sup_{\sigma \leq u \leq t} |M(u)| \right) + d \sup_{\sigma \leq u \leq t} |M(u)|\,.$$

*Furthermore, $a, b, c, d$ can be chosen so that for any $s \in [\sigma, \infty)$,*

$$\|x_t(\sigma, \phi, g)\| \leq e^{-a(t-s)} \left( b\|\phi\| + c \sup_{\sigma \leq u \leq t} |M(u)| \right) + d \sup_{s \leq u \leq t} |M(u)|\,,$$

*for $t \geq s + h$.*

**Lemma 4.2.2** *Suppose $D$ is uniformly stable, and $M : [\tau, \infty) \to E^n$ is continuous such that $M(t) \to 0$ as $t \to \infty$. Then the solution $x(\sigma, \phi, M)$ of (4.2.2) approaches zero as $t \to \infty$ uniformly with respect to $\sigma$ in $[\tau, \infty)$ and $\phi$ in closed bounded sets.*

Let $V : [\tau, \infty) \times C \to E$ be a continuous functional and $x(\sigma, \phi)$ be a solution of (4.1.2) through $(\sigma, \phi)$, i.e., a solution of the integral equation

$$D(t)x_t = D(\sigma)\phi + \int_\sigma^t f(s, x_s)ds\,, \qquad t \geq \sigma\,, \quad x_\sigma = \phi\,. \tag{4.2.3}$$

Define $\dot{V}(t, \phi)$ by

$$\dot{V}(t, \phi) = \lim_{r \to 0^+} \frac{1}{r}[V(t + r, x_{t+r}(\sigma, \phi)) - V(t, \phi)]\,.$$

**Theorem 4.2.1** [2] *Let $D$ be uniformly stable, $f : [\tau, \infty) \times C \to E^n$, continuous, and $f$ maps $[\tau, \infty) \times$ (bounded sets of $C$) into bounded sets of $E^n$. Suppose there are continuous, nonnegative, and nondecreasing functions $u(s)$, $v(s)$, $w(s)$ with $u(s)$,*

$v(s) > 0$ *for* $s \neq 0$ *and* $u(0) = v(0) = 0$, *and there is a continuous function* $V : [\tau, \infty) \times C \to E$ *such that*

$$u(|D(t)\phi|) \leq V(t, \phi) \leq v(\|\phi\|),$$
$$\dot{V}(t, \phi) \leq -w(|D(t)\phi|).$$

$\hspace{10cm}$ (4.2.4a)

*Then the solution* $x = 0$ *of* (4.1.2) *is uniformly stable.*

*If* $u(s) \to \infty$ *as* $s \to \infty$, *the solutions of* (4.1.2) *are uniformly bounded.*

*If* $w(s) > 0$ *for* $s > 0$, *then the solution* $x = 0$ *of* (4.1.2) *is uniformly asymptotically stable.*

*The same conclusion holds if the upper bound on* $\dot{V}(t, \phi)$ *is given by* $-w(|\phi(0)|)$.

The following converse theorem to Theorem 4.2.1 was stated and proved in Chukwu [1].

**Theorem 4.2.2**   *In* (4.1.2), *assume*

(i) $f(t, 0) = 0 \; \forall \; t$, *and* $f$ *has a uniform Lipschitz constant, i.e., there exists a constant* $L$ *such that for any* $\phi, \psi \in C$ *and for all* $t \in [\tau, \infty)$

$$|f(t, \phi) - f(t, \psi)| \leq L\|\phi - \psi\|.$$

(ii) *There is a constant* $N > 0$ *such that*

$$|D(t)\phi| \leq N\|\phi\|, \qquad \forall \, t \in [\tau, \infty), \quad \phi \in C.$$

*The zero solution of* (4.1.2) *is* (*UEXASL*), *so that*

$$\|x_t(\sigma, \phi)\| \leq k e^{-c(t-\sigma)}\|\phi\|, \qquad \forall \, t \geq \sigma,$$

*for some* $k > 0$, $c > 0$.

*Then there are positive constants* $M, R$ *and a continuous functional* $V : [\tau, \infty) \times C \to E$ *such that*

$$|D(t)\phi| \leq V(t, \phi) \leq M\|\phi\|, \qquad \dot{V}(t, \phi) \leq -c/2V(t, \phi),$$
$$|V(t, \phi) - V(t, \psi)| \leq R\|\phi - \psi\|, \qquad \forall \, t \in [\tau, \infty), \quad \phi, \psi \in C.$$

$\hspace{10cm}$ (4.2.4b)

We now use Theorem 4.2.1 to study the system

$$\frac{d}{dt}[x(t) - A_{-1}x(t - h)] = f(t, x(t), x(t - h)), \tag{4.2.5}$$

where $A_{-1}$ is an $n \times n$ symmetric constant matrix and $f : [\tau, \infty) \times E^n \times E^n \to E^n$ and $f : [\tau, \infty) \times$ (bounded sets of $E^n$) $\times$ (bounded sets of $E^n$) $\to$ (bounded sets of $E^n$). We also assume that $f$ has continuous partial derivatives.

If we let $D(t)\phi = \phi(0) - A_{-1}\,\phi(-h)$, then (4.2.5) is equivalent to

$$\frac{d}{dt}D(t)\phi = f(t, \phi(0), \phi(-h)). \tag{4.2.6}$$

In (4.2.6) we denote by $D_a$ $(a = 1, 2)$ the $n \times n$ symmetric matrices $\frac{1}{2}(d_{aij} + d_{aji})$, where

$$d_{1ij} = \frac{\partial f_i(t, \phi(0), \phi(-h))}{\partial \phi_j(-h)},$$

$$d_{2ij} = \frac{\partial f_i(t, \phi(0), 0)}{\partial \phi_j(0)}.$$

Let

$$J_a = \frac{1}{2}(AD_a + D_a^T A), \qquad a = 1, 2, \tag{4.2.7}$$

where $A$ is a positive definite constant symmetric $n \times n$ matrix, and $D_a^T$ is the transpose of $D_a$. Our stability study of (4.2.6) will be made under various assumptions on $J_a$.

**Theorem 4.2.3**  *In (4.2.5), we assume that*

(i) *$D\phi = \phi(0) - A_{-1}\phi(-h)$ is uniformly stable,*
(ii) *$f(t, 0, 0) = 0$.*
(iii) *There are some positive definite constant symmetric matrices $A$ and $N$ such that the matrix*

$$J = \begin{pmatrix} \alpha & \beta \\ \beta & \gamma \end{pmatrix}, \tag{4.2.8a}$$

*where*

$$\alpha = 2J_2 + N, \quad \beta = J_1 + J_2 A_{-1} + NA_{-1}, \quad \gamma = A_{-1}NA_{-1} - N \tag{4.2.8b}$$

*and $J_a$ are defined in (4.2.7), is negative definite.*
(iv) *$A_{-1}$ is symmetric.*
   *Then the solution $x \equiv 0$ of (4.2.5) is uniformly asymptotically stable.*

**Remark 4.2.3**  *$J$ in (4.2.8) will be negative definite if*

(i) *the eigenvalues of $\lambda_{2j}$ of $J_2$ in (4.2.7) satisfy $\lambda_{2j} \leq -\lambda_2 < 0$, $(j = 1, 2, \ldots, n)$, $\forall\, \phi(0)$, and $t \geq 0$ where $\lambda_2$ is a constant;*
(ii) *the eigenvalues $\lambda_{3j}$ of the matrix $J_2 + N$ satisfy $\lambda_{3j} \leq -\lambda_3 < 0$, $(j = 1, 2, \ldots, n)$, $\forall\, \phi(0)$, and $t \geq 0$ where $\lambda_3$ is a constant; and*
(iii) *the eigenvalues $\lambda_{4j}$ of $N - A_{-1}NA_{-1}$ satisfy $\lambda_{4j} \geq \lambda_4$, $(j = 1, \ldots, n)$ where $2\lambda_3\lambda_4 > \|J_1 + J_2 A_{-1} + NA_{-1}\|$ and $\|\cdot\|$ denote the matrix norm.*

**Proof**  Let $D\phi = \phi(0) - A_{-1}\phi(-h)$ and define the Lyapunov functional

$$V(t, \phi) = (AD\phi, D\phi) + \int_{-h}^{0} (N\phi(s), \phi(s))ds,$$

where $A$ and $N$ are positive definite symmetric $n \times n$ constant matrices and $(\cdot, \cdot)$ denote inner product in $E^n$. Because $A, N$ are positive definite symmetric matrices there are constants $a_i$, $n_i$, $i = 1, 2$, such that

$$a_i(D\phi)^2 + n_1 \int_{-h}^{0} \phi^2(s)ds \leq V(t, \phi) \leq a_2(D\phi)^2 + n_2 \int_{-h}^{0} \phi^2(s)ds \,.$$

Thus the first inequality of Theorem 4.2.1 is satisfied. That the second inequality of (4.2.4) is also satisfied follows from the following calculations: Because $A$ is symmetric, we have

$$\begin{aligned}
\dot{V}(t, \phi) &= 2(Af(t, \phi(0), \phi(-h)), D\phi) + N(\phi(0), \phi(0)) - N(\phi(-h), \phi(-h)) \\
&\quad + 2(Af(t, \phi(0), 0) - Af(t, 0, 0), D\phi) + 2(Af(t, 0, 0), D\phi) \\
&= 2(Af(t, \phi(0), \phi(-h)) - f(t, \phi(0), 0), D\phi) \\
&\quad + N(\phi(0), \phi(0)) - N(\phi(-h), \phi(-h)) \\
&= 2(J_1\phi(-h), D\phi) + 2(J_2\phi(0), D\phi) + N(\phi(0), \phi(0)) \\
&\quad - N(\phi(-h), \phi(-h)) \,,
\end{aligned}$$

since $f(t, 0, 0) = 0$, and we have used Lemma 3.1.2. Note that $\phi(0) = D\phi + A_{-1}\phi(-h)$, so that

$$\begin{aligned}
\dot{V}(t, \phi) &= 2(J_1\phi(-h), D\phi) + 2(J_2(D\phi + A_{-1}\phi(-h)), D\phi) \\
&\quad + (N(D\phi + A_{-1}\phi(-h)), D\phi + A_{-1}\phi(-h)) - (N\phi(-h), \phi(-h)) \\
&= (J_2 D\phi, D\phi) + (J_2 D\phi, D\phi) + 2(J_1\phi(-h), D\phi) \\
&\quad + 2(J_2 A_{-1}\phi(-h), D\phi) + (ND\phi, D\phi) + 2(NA_{-1}\phi(-h), D\phi) \\
&\quad + (A_{-1}NA_{-1}\phi(-h), \phi(-h)) - (N\phi(-h), \phi(-h)) \\
&= ((2J_2 + N)D\phi, D\phi) + 2((J_1 + A_2 A_{-1} + NA_{-1})\phi(-h), D\phi) \\
&\quad - ((N - A_{-1}NA_{-1})\phi(-h), \phi(-h)) \\
&\equiv (D\phi, \phi(-h))J(D\phi, \phi(-h))^T < 0 \,,
\end{aligned}$$

since $J$ is negative definite. It now follows that (4.2.4) of Theorem 4.2.1 is completely verified. Because $D$ is assumed to be uniformly stable, all the requirements of Theorem 4.2.1 are met: We conclude that the solution $x = 0$ of (4.2.5) is uniformly asymptotically stable. $\qquad\square$

We now consider some examples.

**Example 4.2.1**    Consider the scalar equation

$$\frac{d}{dt}[x(t) + cx(t - h)] + ax(t) = 0 \,,$$

where $a$ and $c$ are constants with $a > 0$ $|c| < 1$ and $a > c/(1 - c^2)\frac{\sqrt{3}}{2} < c < 1$. Lemma 3.1 of Cruz and Hale [2] implies that the operator $D\phi = \phi(0) + c\phi(-h)$ is uniformly stable. We now apply Theorem 4.2.3 with $A_{-1} = -c$, $A = 1$, $J_1 = 0$, $J_2 = -a$, $N = a/2$,

$$\alpha = 2J_2 + N = -2a + \frac{a}{2} = -3a/2$$

$$\beta = J_1 + J_2 A_{-1} + N A_{-1} = ac - \frac{a}{2}c = \frac{ac}{2},$$

$$\gamma = A_{-1}N A_{-1} - N = \frac{a}{2}(c^2 - 1), \qquad \text{and}$$

$$J = \begin{pmatrix} \dfrac{-3a}{2} & \dfrac{ac}{2} \\ \dfrac{ac}{2} & \dfrac{a}{2}(c^2 - 1) \end{pmatrix}$$

is negative definite. We can also use Remark 4.2.3 with $\lambda_2 = a$, $\lambda_3 = a/2$, $\lambda_4 = \frac{a}{2}(1 - c^2)$, so that

$$2\lambda_3\lambda_4 = 2(a/2)(a/2)(1 - c^2) > ac - ac/2,$$

since $a^2(1 - c^2)/2 > ac/2$ if we assume $a(1 - c^2) > c$.

**Example 4.2.2**   Consider the equation

$$\frac{d}{dt}[x(t) + c(t)x(t - h)] + ax(t) + b(t)x(t - h) = 0,$$

where $a > 0$ is a constant, $c(t)$, $b(t)$ are continuous for $t \geq 0$, and there is a $\delta > 0$ such that $c^2(t) \leq 1 - \delta$. Let $D(t)\phi = \phi(0) + c(t)\phi(-h)$, so that by a remark in Cruz and Hale [2, p. 338], $D(t)$ is uniformly stable. Observe that $f(t, x(t), x(t - h)) = -ax(t) - b(t)x(t - h)$. On taking $A = 1$, $A_{-1} = -c(t)$, $N = a/2$, $J_1 = -b(t)$, $J_2 = -a = -\lambda_2$, $J_2 + N = -a/2 = -\lambda_3$, $J_2 A_{-1} = ac$, $N A_{-1} = -ac/2$, $\lambda_4 = N - A_{-1}N A_{-1} = a/2 - ac^2/2$. We observe that

$$2\lambda_3\lambda_4 = a^2/2(1 - c^2(t)) > \left| b(t) + \frac{ac(t)}{2} \right|$$

ensures uniform asymptotic stability. This inequality will hold if $a^2/2\delta > |b(t) + ac(t)/2|$ for all $t \geq 0$.

**Example 4.2.3**   Consider the $n$-dimensional autonomous neutral difference equation

$$\dot{x}(t) + A_{-1}\dot{x}(t - h) = A_0 x(t) + A_1 x(t - h),$$

where $x(t) \in E^n$. $A_{-1}, A_0, A_1$ are constant $n \times n$ matrices $h > 0$. Let $D_1 = \frac{1}{2}(A_1 + A_1^T)$, $D_2 = \frac{1}{2}(A_0 + A_0^T)$, $J_i = AD_i + D_i^T A$, $i = 1, 2$, $A$, a positive definite matrix. Assume $\|A_{-1}\| < 1$ and $A_{-1}$ to be symmetric. For some constant positive definite symmetric matrix $N$, Remark 4.2.3 yields the required condition for uniform asymptotic stability.

## 4.3   Perturbed Systems

The converse Theorem 4.2.2 will now be employed to prove stability results for the system (4.1.3).

**Theorem 4.3.1**    *In System (4.1.3), assume that*

(i) $|g(t,\phi)| \le N\|\phi\|$, $|f(t,\phi) - f(t,\psi)| \le N\|\phi - \psi\|$, $\forall\, \phi, \psi \in C$, $t \in [\tau, \infty)$; *and $D$ is uniformly stable.*

(ii) *Suppose that the zero solution of (4.1.2) is (UEXASL), so that every solution $x(t,\phi)$ satisfies*

$$\|x_t(\sigma,\phi)\| \le k e^{-c(t-\sigma)}\|\phi\|, \qquad \forall\, t \ge \sigma,$$

*for some constants $c > 0$   $k > 0$.*

(iii) *The function $F$ in (4.1.3) satisfies $F = F_1 + F_2$ with $|F_1(t,\phi)| \le v(t)|D(t)\phi|$, $\forall\, t \ge \sigma$, $\phi \in C$, where there is a constant $\xi > 0$ such that for any $p > 0$*

$$\frac{1}{p} \int_t^{t+\rho} v(s)\,ds < \xi, \qquad t \ge \sigma, \tag{4.3.1}$$

*and for some $\epsilon > 0$ we have*

$$|F_2(t,\phi)| \le \epsilon |D(t)\phi|, \qquad \forall\, t \ge \sigma, \quad \phi \in C.$$

*Then every solution $x(\sigma,\phi)$ of (4.1.3) satisfies*

$$\|x_t(\sigma,\phi)\| \to 0 \quad \text{as} \quad t \to \infty$$

*uniformly with respect to $\sigma \in [\tau, \infty)$, and $\phi$ in closed bounded sets. In particular, the null solution of (4.1.3) is WUASL.*

**Proof**    The assumptions we have made in Theorem 4.2.2 guarantee the existence of a Lyapunov function $V$ with the properties of Theorem 4.2.2. Suppose $y = y(\sigma,\phi)$ and $x = x(\sigma,\phi)$ are solutions of (4.1.3) and (4.1.2) respectively with initial values $\phi$ at $\sigma$. If $\dot{V}_{(4.1.3)}$ and $\dot{V}_{(4.1.2)}$ are the derivatives of $V$ along solutions of the systems (4.1.3) and (4.1.2) respectively, then the relations (4.2.4b) yields the inequality

$$\dot{V}_{(4.1.3)}(\sigma,\phi) \le \dot{V}_{(4.1.2)}(\sigma,\phi) + R \lim_{r \to 0^+} \frac{1}{r}[y_{\sigma+r}(\sigma,\phi) - x_{\sigma+r}(\sigma,\phi)]. \tag{4.3.2}$$

But then

$$D(\sigma + r)(y_{\sigma+r} - x_{\sigma+r}) = \int_\sigma^{\sigma+r} \{[f(s,y_s) - f(s,x_s)] + F(s,y_s)\}\,ds. \tag{4.3.3a}$$

Because $g(t, \phi)$ is uniformly nonatomic at zero and (4.1.5) holds, there is an $r_0 > 0$ such that

$$\|y_{\sigma+r} - x_{\sigma+r}\| \leq \frac{1}{1 - c(r_0)} \int_\sigma^{\sigma+r} |\{f(s, y_s) - f(s, x_s) + F(s, y_s)\}| ds$$

for $0 < r < r_0$. With this estimate in (4.3.2) we obtain

$$\dot{V}_{(4.1.3)}(t, \phi) \leq \dot{V}_{(4.1.2)}(t, \phi) + Q\{|F_1(t, \phi)| + |F_2(t, \phi)|\}, \tag{4.3.3b}$$

$\forall\, t \geq \sigma,\ \phi \in C$ where $Q = \frac{R}{1 - c(h_0)}$. It follows from (4.2.4b) that

$$\dot{V}(t, \phi) \leq -c/2 V(t, \phi) + Qv(t)|D(t)\phi| + Q\epsilon|D(t)\phi|$$

$$\leq (-c/2 + Q\epsilon)V(t, \phi) + Qv(t)V(t, \phi). \tag{4.3.4}$$

We now set $\epsilon = -c/4Q$ and choose some $\xi < c/8Q$ so that from (4.3.1)

$$\frac{1}{t - \sigma} \int_\sigma^t v(s)ds < \xi,$$

and

$$\int_\sigma^t v(s)ds < \xi(t - \sigma),$$

and

$$\int_\sigma^t (-c/4 + Qv(s))ds \leq \left(-\frac{c}{4} + Q\xi\right)(t - \sigma) \leq -c/8(t - \sigma).$$

Using all these in (4.3.4), one obtains

$$\dot{V}(t, \sigma) \leq -\frac{c}{4}V(t, \sigma) + Qv(t)V(t, \phi);$$

and with $r(t) = V(t, x_t(\sigma, \phi))$, we have

$$\dot{r}(t) \leq \left(-\frac{c}{4} + Qv(t)\right)r(t).$$

The solution of this inequality satisfies

$$r(t, \sigma) \leq e^{-c/8(t-\sigma)} r_0.$$

Since (4.2.4b) is valid, this last inequality leads to

$$|D(t)x_t| \leq e^{-c/8(t-\sigma)} M\|\phi\| \equiv M(t).$$

Because $D$ is assumed uniformly stable, Lemma 4.2.2 can be invoked to ensure that $x(\sigma, \phi) \to 0$ as $t \to \infty$, uniformly with respect to $\sigma \in [\tau, \infty)$, and $\phi$ in closed bounded sets. To conclude the proof we verify uniform stability as follows:

Because $D$ is uniformly stable, there are positive constants $a_1$, $a_2$, $a_3$, $a_4$ such that

$$\|x_t(\sigma, \phi)\| \leq e^{-a_1(t-\sigma)} \left( a_2\|\phi\| + a_3 \sup_{\sigma \leq s \leq t} |M(s)| \right) + a_4 \sup_{\sigma \leq s \leq t} |M(s)|.$$

For any $\epsilon > 0$ choose $\delta = (a_2 + a_3 M + a_4 M)^{-1}\epsilon$, and observe that if $\|\phi\| < \delta$, then

$$\|x_t(\sigma, \phi)\| < a_2\delta + a_3 M\delta + a_4 M\delta < \epsilon,$$

for $t \geq \sigma$. This proves that $x = 0$ is uniformly stable, so that the zero solution of (4.1.3) is WUASL. $\qquad\square$

**Remark 4.3.1**    Condition (4.3.1) can be replaced by the assumption

$$V \equiv \int_\sigma^\infty v(s)ds < \infty \qquad\qquad (4.3.5)$$

to deduce the same results.

When $f(t, \phi) \equiv A(t, \phi)$ is linear in $\phi$, one can obtain a generalization of the famous theorem of Lyapunov on the stability with respect to the first approximation. It is the global stability result that is most useful when dealing with controllability questions of neutral systems with limited power.

Consider the linear system

$$\frac{d}{dt}D(t)x_t = A(t, x_t), \qquad\qquad (4.3.6)$$

and its perturbation

$$\frac{d}{dt}[D(t)x_t - G(t, x_t)] = A(t, x_t) + f(t, x_t), \qquad\qquad (4.3.7)$$

where $A(t, \phi)$, $G(t, \phi)$ are linear in $\phi$. The following result is valid:

**Theorem 4.3.2**    *In (4.3.6) and (4.3.7) assume that*

(i) *the linear system (4.3.6) is uniformly asymptotically stable, so that for some $k \geq 1$, $\alpha > 0$ the solution $x(\sigma, \phi)$ of (4.3.6) satisfies*

$$\|x_t(\sigma, \phi)\| \leq ke^{-\alpha(t-\sigma)}\|\phi\|, \qquad t \geq \sigma \; \phi \in C.$$

(ii) *The function $F = F_1 + F_2$ satisfies*

$$|F_1(t, \phi)| \leq v(t)\|\phi\|, \quad \|F_2(t, \phi)\| \leq \epsilon\|\phi\|, \quad \forall \, t \geq \sigma, \quad \phi \in C,$$

*where*

$$\epsilon = \frac{\alpha}{2k}\frac{1 - k(\lambda + M_0)}{1 + k(\lambda + M_0)}$$

*and where if $\xi = \alpha/4$, then for each $p > 0$*

$$\frac{1}{p} \int_t^{t+p} v(s)ds < \xi \,, \quad \forall\, t \geq \sigma \,. \tag{4.3.8}$$

(iii) *The function $G = G_1 + G_2$ satisfies*

$$|G_1(t, \phi)| \leq \pi(t)\|\phi\| \,, \qquad |G_2(t, \phi)| \leq \lambda\|\phi\| \,,$$

*for $t \geq \sigma$, $\phi \in C$ where $\pi(t)$ is continuous and bounded with a bounded $M_0$ such that*

$$\pi(t) \leq M_0 \,, \quad t \geq \sigma \,, \quad k(\lambda + M_0) < 1 \,.$$

*Then every solution $x(\sigma, \phi)$ of (4.3.7) satisfies*

$$\|x_t(\sigma, \phi)\| \leq N\|\phi\|e^{-\alpha/4(t-\sigma)} \,, \quad t \geq \sigma \,,$$

*for some $N$.*

The proof is contained in [1, p. 866].

**Remark 4.3.2** The requirement of (4.3.8) can be replaced by the hypothesis

$$\int_\sigma^\infty v(s)ds < \infty \,.$$

It is possible to weaken the conditions on $F$, to deduce global eventual weak stability. Because of its importance it is now stated.

**Theorem 4.3.3** *For the system (4.3.6) and (4.3.7) assume that*

(i) *Equation (4.3.6) is uniformly asymptotically stable.*
(ii) *The function $F$ in (4.3.7) can be written as*

$$F = F_1 + F_2 + F_3 \,,$$

*where*

$$|F_1(t, \phi)| \leq v(t)\|\phi\| \,, \quad |F_2(t, \phi)| \leq \epsilon\|\phi\| \,, \quad |F_3(t, \phi)| \leq g(t) \,,$$

*for all $t \geq \sigma \geq T_0$ for some sufficiently large $T_0$ and all $\phi \in C$, where*

$$\lim_{\tau \to \infty} \frac{1}{\tau} \int_\sigma^{\sigma+\tau} v(s)ds < \xi \,,$$

*$\xi$ is a sufficiently small number, and where*

$$\epsilon = \frac{\alpha}{4k} \frac{1 - k(\lambda + M_0)}{1 + k(\lambda + M_0)} \,.$$

*Also*

$$H(t) = \int_t^{t+1} g(s)ds \to 0 \quad as \quad t \to \infty \,.$$

(iii) *The function $G = G_1 + G_2$ satisfies*

$$|G_1(t,\phi)| \leq \pi(t)\|\phi\|, \qquad |G_2(t,\phi)| \leq \lambda\|\phi\|,$$

*for $t \geq \sigma$, $\phi \in C$, where $\pi(t)$ is continuous and bounded with a bound $M_0$ such that*

$$\pi(t) \leq M_0, \quad t \geq \sigma, \quad k(\lambda + M_0) < 1.$$

*Then every solution of* (4.3.7) *with initial data* $(\sigma, \phi)$ *satisfies*

$$\|x_t(\sigma, \phi)\| \to 0 \quad as \quad t \to \infty$$

*for some $T_0$, $\sigma \geq T_0$. As a consequence the trivial solution of* (4.3.7) *is* (EVUASL).

For the proof consult [1, p. 869].

The above result deals with the linear case $f(t,\phi) = A(t,\phi)$. A similar result holds for general nonlinear $f$. Indeed, we have the following:

**Theorem 4.3.4**    *Assume that*

(i) $|g(t,\phi)| \leq N\|\phi\|$, $|f(t,\phi) - f(t,\psi)| \leq N\|\phi - \psi\|$, $\forall\, \phi,\, \psi \in C$, $t \in [\tau,\infty)$ *and $D$ is uniformly stable.*

(ii) *The trivial solution of* (4.3.6) *is uniformly asymptotically stable.*

(iii) *The function $F$ in* (4.3.7) *is such that $F = F_1 + F_2 + F_3$ where*

$$|F_1(t,\phi)| \leq v(t)|D(t)\phi|, \qquad |F_2(t,\phi)| \leq \epsilon|D(t)\phi|,$$

$$F_3(t,\phi)| \leq |\,M(t), \qquad \forall\, \phi \in C, \quad t \geq \sigma \geq T_0$$

*for some $T_0$, where*

$$\lim_{\tau \to \infty} \frac{1}{\tau} \int_\sigma^{\sigma+\tau} v(s)ds < \xi,$$

*with $\epsilon, \xi$ sufficiently small and for $\sigma$ large. Also*

$$M(t) = \int_t^{t+1} m(s)ds \to 0 \quad as \quad t \to \infty.$$

*Then there exists some $T$ such that if $\sigma > T$,*

$$\|x_t(\sigma, \phi)\| \to 0 \quad as \quad t \to \infty.$$

*Thus the solution $x = 0$ of* (4.3.7) *is* (EVWUASL).

A proof is indicated in [1, p. 872].

**Example 4.3.1**    Delay Logistic Equation of Neutral Type [5].

Consider the system

$$\dot{N}(t) = rN(t)\left[1 - \frac{N(t-h_1)}{K} - 2c\frac{\dot{N}(t-h_2)}{1+\dot{N}^2(t-h_2)}\right], \qquad (4.3.9)$$

where $h_1$, $h_2$, $K$, $c$, $r$ are constants $\geq 0$, with $r$, $h_2$, $K > 0$. We assume initial conditions of the type

$$x(s) = \phi(s) > 0, \qquad s \in [-h,0], \qquad x(0) > 0,$$

$$\phi \in C([-h_1,0]) \cap C^1((-h_2,0)).$$

If we let

$$N(t) \equiv K[1 + x(t)],$$

$$A(t) = r[1 + x(t)],$$

$$B(t) = 2ckA(t)|[1 + k^2\dot{x}^2(t-h_2)],$$

then (4.3.9) is equivalent to

$$\frac{d}{dt}\left[x(t) - \int_{t-h}^{t} A(s+h_1)x(s)ds\right] = -A(t+h_1) + B(t)\dot{x}(t-h_2). \qquad (4.3.10)$$

Set

$$a^* = (1+c)\exp[r(1+c)h_1],$$

$$b^* = (1-c)\exp[h_1r(1-c)],$$

$$\mu_1 = 2 - \left[2a^*rh_1 + \frac{2cka^*}{b^*\exp(-a^*)} + 2ckra^*h_1 + \frac{4rck(a^*)^2}{b^*\exp(-a^*)}\right],$$

$$\mu_2 = \frac{1}{ra^*} - \left[1 + ra^*h_2 + \frac{2(2cka^*r)^2}{rb^*\exp(-a^*)}\right].$$

For this we use the Lyapunov functional

$$V(t) = V(t,x(t))$$

$$= \left[x(t) - \int_{t-h}^{t} A(s+h_1)x(s)ds\right]^2 + \int_{t-h}^{t} \frac{B(s+h_2)}{A(s+h_2)}x^2(s)ds$$

$$+ 2\int_{t-h}^{t} \frac{B(s+h_1+h_2)}{A(s+h_1+h_2)}A^2(s+h_1)^2x^2(s)ds$$

$$+ \int_{t-h}^{t} A(s+2h_1)\left(\int_{s}^{t} A(u+h_1)x^2(u)du\right)ds$$

$$+ \int_{t-h}^{t} B(s + h_1) \left( \int_{s}^{t} A(u + h_1) x^2(u) du \right) ds \,,$$

and calculate the upper right derivative $\frac{D^+ V}{dt}$ of $V$ along solutions of (4.3.10).

**Proposition 4.3.1**   *Assume that*

$$r, h_1, k \in (0, \infty); \qquad h_2 \in [0, \infty)\,, \tag{4.3.11}$$

$$\mu_1 > 0\,, \qquad \mu_2 > 0\,, \tag{4.3.12}$$

$$2ckra^* < 1; \qquad 0 < c < 1\,. \tag{4.3.13}$$

*Then every positive solution of* (4.3.9) *satisfies*

$$N(t) \to 0 \quad as \quad t \to \infty\,.$$

**Proof**   We need only show that if (4.3.11) and (4.3.12) hold, an arbitrary solution of (4.3.10) satisfies

$$x(t) \to 0 \quad as \quad t \to \infty\,. \tag{4.3.14}$$

The needed details are contained in [5].   $\square$


## 4.4   Perturbations of Nonlinear Neutral Systems

Consider a nonlinear system that is more general than (4.1.3), namely

$$\frac{d}{dt} D(t, x_t) = f(t, x_t) + g(t, x_t)\,. \tag{4.4.1}$$

We assume the following as basic: Let $\Lambda \subset C$ be an open set, $I_1 \subset E$ an open interval, $P = I_1 \times \Lambda$. Denote by $E^{n^2}$ the set of $n \times n$ real matrices. Suppose $D$, $f : \Gamma \to E^n$ are continuous functions. We assume also that $0 \in \Lambda$. The main result in this section is stated as follows:

**Theorem 4.4.1**   *Assume*

(i)  *$D$, $f$ in* (4.4.1) *satisfy the following hypothesis:*

  (a)  *The Fréchet derivatives of $D, f$ with respect to $\phi$, denoted by $D_\phi$, $f_\phi$ respectively exist and are continuous on $\Gamma$ as well as $D_{\phi\phi}$, and*

  (b)  *For each $(t, \phi) \in \Gamma$, $\psi \in C$ write*

$$D_\phi(t, \phi)[\psi] = A(t, \phi)\psi(0) - \int_{-h}^{0} d_\theta \mu(t, \phi, \theta)[\psi(\theta)]$$

  *for some functions*

$$A : \Gamma \to E^{n^2}, \ \mu : \Gamma \times [-h, 0] \to E^{n^2}$$

with $A$ continuous, $\mu(t, \phi, \cdot)$ of bounded variation on $[-h, 0]$ and such that the map $(t, \phi) \to \int_{-h}^{0} d_\theta \mu(t, \phi, \theta)[\psi(\theta)]$ is continued for each $\psi \in C$.

(c) Assume for each $(t, \phi) \in \Gamma$ $D$ is uniformly atomic at zero on compact sets $K \subset \Gamma$, i.e.,

$$\det A(t, \phi) \neq 0,$$

$$\left| \int_{-s}^{0} d_\theta \mu(t, \phi, \theta)[\psi(\theta)] \right| \leq \ell(s)\|\psi\|$$

$s \in [0, h]$ for all $(t, \phi) \in K$ for some function $\ell : [0, h] \to [0, \infty)$ that is continuous and nonincreasing, $\ell(0) = 0$.

(ii) $g$ is Lipschitzian in $\phi$ on compact subsets of $\Gamma$.
(iii) $f(t, 0, 0) = 0$ for all $t \in I$.
(iv) For each $(\sigma, \phi) \in \Gamma$, we have

$$\|T(t, \sigma, \phi)\| \leq \exp(a(t) + b(\sigma)), \qquad t \geq \sigma,$$

where $a(t)$, $b(t)$ are continuous functions from $E^+ = [0, \infty)$ into $E^+$, i.e., are elements of $C(E^+, E^+)$. Here $T(t, \sigma; \phi)$ is the solution operator associated with the variational equation

$$\frac{d}{dt} D_\phi(t, x_t(\sigma, \phi))[Z_t] = f_\phi(t, x_t(\sigma, \phi))[z_t], \qquad t \in [\sigma, \sigma + a). \qquad (4.4.2)$$

(v) For each $(\sigma, \phi) \in \Gamma$,

$$|g(t, \psi)| \leq \sum_{i=1}^{N} c_i(t)\|\psi\|^{r_i} + C_{N+1}(t), \qquad t \geq \sigma,$$

where $r_i \in (0, 1]$, $c_j(t) \in C(E^+, E^+)$, $j = 1, 2, \ldots, N + 1$.
(vi) $c_j(s)e^{r_j a(s) + b(s)}$, $c_{N+1}^{(s)} e^{b(s)} \in L^1(\tau, \sigma)$, $j = 1, \ldots, N$.
(vii) $a(t) \to -\infty$ as $t \to \infty$.

*Then every solution of* (4.1.1) *satisfies* $\|x_t(\sigma, \phi)\| \to$ *as* $t \to \infty$.

The proof is contained in Chukwu and Simpson in [6, p. 57]. It uses the nonlinear variation of parameter equation

$$x_t(\sigma, \phi) = y_t(\sigma, \phi) + \int_{\sigma}^{t} T(t, s; x_s(\sigma, \phi)X_0 \cdot A^{-1}(s, x_s(\sigma, \phi))X_0 g(s, x_s(\sigma, \phi))ds$$

$$(4.4.3)$$

where $y_t(\sigma, \phi)$ is the solution of

$$\frac{d}{dt} D(t, y_t) = f(t, y_t). \qquad (4.4.4)$$

This formula (4.4.3) holds when the solution of (4.4.1) is unique.

## References


1. E. N. Chukwu, "Global Asymptotic Behavior of Functional Differential Equations of the Neutral Type," *Nonlinear Analysis Theory Methods and Applications* **5** (1981) 853–872.
2. M. A. Cruz and J. K. Hale, "Stability of Functional Differential Equations of Neutral Type," *J. Differential Equations* **7** (1970) 334–355.
3. J. K. Hale, *Ordinary Differential Equations*, Interscience, New York, 1969.
4. J. K. Hale, "Theory of Functional Differential Equations," *Applied Mathematical Sciences* **3**, Springer-Verlag, New York, 1977.
5. K. Gopalsamy, "Global Attractivity of Neutral Logistic Equations," in *Differential Equations and Applications*, edited by A. R. Aftabizadeh, Ohio University Press, Athens, 1989.
6. E. N. Chukwu and H. C. Simpson, "Perturbations of Nonlinear Systems of Neutral Type," *J. Differential Equations* **82** (1989) 28–59.

# Chapter 5

# Synthesis of Time-Optimal and Minimum-Effort Control of Linear Ordinary Systems

## 5.0   Control of Ordinary Linear Systems

The material in this section is introductory and deals with ordinary differential equations. Our aim is to solve the optimal problem for linear ordinary systems that will then become the basis for generalization to hereditary systems. We shall first introduce in this section the theory of controllability of linear ordinary differential equations

$$\dot{x}(t) = A_0(t)x(t) + B(t)u(t)\,, \tag{5.0.1}$$

where $A_0(t)$, $B(t)$ are analytic $n \times n$, $n \times m$ matrices defined on $[0, \infty)$. The admissible controls $u \in \Omega^*$ are at first required to be square integrable on finite intervals. We need the following controllability concepts:

**Definition 5.0.1**   System (5.0.1) is controllable on an interval $[t_0, t_1]$ if, given $x_0, x_1 \in E^n$, there is a control $u \in \Omega^*$ such that the solution $x(t, t_0, x_0, u)$ of (5.0.1) with $x(t_0, t_0, x_0, u) = x_0$ satisfies $x(t_1, t_0, x_0, u) = x_1$. It is called controllable at time $t_0$ if it is controllable on $[t_0, t_1]$ for some $t_1 > t_0$. If it is controllable at each $t_0 \geq 0$, we say it is controllable. System (5.0.1) is said to be fully controllable at $t_0$ if it is controllable on $[t_0, t_1]$ for each $t_1 > t_0$. It is said to be fully controllable if it is fully controllable at each $t_0 \geq 0$.

If $u$ is a control, the solution of (5.0.1) corresponding to this $u$ is given by the variation of constant formula

$$x(t, t_0, x_0, u) = X(t)X^{-1}(t_0)x_0 + X(t) \int_{t_0}^{t} X^{-1}(s)B(s)u(s)ds\,,$$

where $X(t)$ is a fundamental matrix solution of

$$\dot{x}(t) = A_0(t)x(t)\,. \tag{5.0.2}$$

Define

$$Y(s) = X^{-1}(s)B(s)\,, \tag{5.0.3}$$

113

and define the operator $\Gamma$ by

$$\Gamma = -A_0 + D, \tag{5.0.4}$$

where $\frac{d}{dt} = D$. Also define the expression $M$ as follows:

$$M(t_0, t) = \int_{t_0}^{t} Y(s)Y^T(s)ds = \int_{t_0}^{t} X^{-1}(s)B(s)B^T(s)X^{-1T}(s)ds, \tag{5.0.5}$$

where $X^T$ is the transpose of $X$. Then the following theorem is well known:

**Theorem 5.0.1**    *The following are equivalent:*

  (i)  $M(t_0, t)$ *is nonsingular for all $t_1 > t_0$ and all $t_0 \geq 0$.*
  (ii) *System (5.0.1) is fully controllable.*

*Also if $A_0(t)$ and $B(t)$ are analytic on $(t_0, t_1)$, $t_1 > t_0 \geq 0$, then (5.0.1) is fully controllable at $t_0$ if and only if for each $t_1 > t_0$ there exists $t \in (t_0, t_1)$ such that*

$$\mathrm{rank}[B(t), \Gamma B(t), \ldots, \Gamma^{n-1}B(t)] = n. \tag{5.0.6}$$

*Furthermore, if $A_0$ and $B$ are constant, then (5.0.1) is fully controllable if and only if*

$$\mathrm{rank}[B, A_0 B, \ldots, A_0^{n-1}B] = n. \tag{5.0.7}$$

From the definition we note full controllability of $t_0$ implies that one can start at $t_0$, and using an admissible control, reach any point $x_1$ in an arbitrarily short time (any $t_1 > t_0$). If the matrices are constant and the system is full controllable, then (5.0.7) holds no matter the time, and this contrasts with the situation in the delay case. (Sec. 6.1.)


## 5.1   Synthesis of Time-Optimal Control of Linear Ordinary Systems

Minimum-Time Problem
    Consider the following problem:
    Find a control $u$ that minimizes $t_1$ subject to

$$\dot{x}(t) = Ax(t) + Bu(t),$$
$$x(0) = x_0, \qquad x(t_1) = 0, \tag{5.1.1}$$

where $A$ is an $n \times n$ and $B$ an $n \times m$ matrix, and where

$$u \in \mathcal{U} = \{u \text{ measurable } u(t) \in E^m, |u_j(t)| \leq 1, \ j = 1, \ldots, m\}.$$

Thus we want explicitly to find a function $f(\cdot, x_0) : [0, t_1] \to \mathcal{U}$, $f(t, x_0) = u(t)$ that drives System (5.1.1) from $x_0$ to 0 in minimum $t_1$. The following is well known:

**Theorem 5.1.1**    [1] *In (5.1.1) assume that*

(i) $\text{rank}[B, AB, \ldots, A^{n-1}B] = n$.

(ii) *The eigenvalues of $A$ have no positive real part.*

(iii) *For each $j = 1, \ldots, m$, the vectors $[b_j, Ab_j, \ldots, A^{n-1}b_j]$ are linearly independent. Then there exists a unique optimal control $u^*$ of the form*

$$u^*(t) = u(t, x_0) = \text{sgn}[c^T e^{-At} B] \tag{5.1.2}$$

*almost everywhere, which drives $x_0$ to zero in minimum time. Thus if $y(t)$ is the $E^m$-vector, $c^T \cdot e^{-At} B$, the so-called index of the control system (5.1.1), then $u_j(t) = \text{sgn } y_j \ (j = 1, \ldots, m)$, where*

$$\text{sgn } y_j = \begin{cases} 1 & \text{if } y_j > 0 \\ -1 & \text{if } y_j < 0. \end{cases}$$

*It is undefined when $y_j(t) = 0$. Because (5.1.1) is normal in the sense that (iii) holds, $y_j(t)$ has only isolated zeros, where the control switches in the following sense:*

**Definition 5.1.1**  If $t > 0$ and $u_j^*(t - 0; x_0)u_j^*(t, \cdot, x_0) < 0$, then $t$ is a switch time of $u^*$ (or $u$ switches at $t$). Thus each point of discontinuity of $u^*$ is called a switch time, and $u^*(s) = u_1$ on some interval $[t_1, t)$ and $u^*(s) = u_2$ on some $(t, t_2)$, $t_1 < t < t_2$, where $u_1^* \neq u_2^*$ in $\mathcal{U}$, then we say that $u^*$ switches from $u_1^*$ to $u_2^*$ at time $t$. It is clear that $u^*$ has only a finite number of switch points on $[0, t_1]$. If (5.1.2) is valid, then $c$ is said to generate the control $u^*$ and $-u^*$ (which is the optimal control function for $-x_0$). If system (5.1.1) satisfies (iii) of Theorem 5.1.1, then it is said to be normal.

Let

$$\mathbb{R}(t) = \left\{ \int_0^t e^{-As} Bu(s)ds : u(s) \in \mathcal{U}, \ u \text{ measurable} \right\}$$

be the reachable set at $t \geq 0$, and $\mathbb{R} = \bigcup_{t \geq 0} \mathbb{R}(t)$ the reachable set. Note that $\mathbb{R}(t)$ is related to $\mathcal{C}(t)$, the set of points that can be steered to zero by controls as follows:

$$\mathcal{C}(t) = e^{At}\mathbb{R}(t).$$

If $x$ is in $\mathcal{C} = \bigcup_{t \geq 0} \mathcal{C}(t)$ and $u^*$ is its unique time-optimal control function, then the function

$$f(x) = u^*(0; x), \qquad x \neq 0, \qquad f(0) = 0$$

defines a map

$$f : \mathcal{C} \to \mathcal{U}$$

that is called the time-optimal feedback control or the synthesis function. Once $f$ is found, the time-optimal problem is completely solved. For example, consider the system

$$\dot{x}(t) = Ax + Bf(x),$$
$$x(0) = x_0. \tag{5.1.3}$$

If $z(x_0, x_0)$ is a solution of (5.1.3), then $z$ describes the optimal trajectory from $x_0$ to the origin, and $u^*(t, x^0) = f(z(t, x_0))$ is the time-optimal control function for $x_0$.

We shall now describe how Theorem 5.1.1 helps us to construct optimal feedback control for the simple harmonic oscillator.

**Example 5.1.1**    The system

$$\ddot{x} + x = u, \qquad |u| \le 1, \tag{5.1.4a}$$

is equivalent to

$$\dot{\underline{x}} = A\underline{x} + Bu, \tag{5.1.14b}$$

where

$$A = \begin{bmatrix} 0 & 1 \\ -1 & 0 \end{bmatrix}, \qquad B = \begin{bmatrix} 0 \\ 1 \end{bmatrix}, \qquad \underline{x} = \begin{bmatrix} x_1 \\ x_2 \end{bmatrix}.$$

It is controllable, normal, and satisfies condition (ii) of Theorem 5.1.1. Because

$$e^{At} = \begin{bmatrix} \cos t & \sin t \\ -\sin t & \cos t \end{bmatrix}, \qquad e^{-At} B = \begin{bmatrix} -\sin t \\ \cos t \end{bmatrix},$$

$$c^T e^{-At} B = -c_1 \sin t + c_2 \cos t, \qquad c = (c_1, c_2),$$

where $c_1^2 + c_2^2 \ne 0$. Therefore the (unique) optimal control is given by

$$u^*(t) = \operatorname{sgn}(\sin(t + \delta)), \qquad \tan \delta = \frac{-c_2}{c_1}$$

for some $-\pi \le \delta \le \pi$, since $-c_1 \sin t + c_2 \cos t = a \sin(t + \delta)$, $a > 0$.

We now derive optimal feedback control that drives $(x_0, y_0) \in E^2$ to $(0,0)$ in minimum $t_1$. For each initial $(x_0, y_0) \in E^2$ there exists an optimal control $u^*(t)$ uniquely determined by $u^*(t) = \operatorname{sgn}[\sin(t + \delta)]$. When $u = \pm 1$, the trajectory in $E^2$ is a circle of center $(\pm 1, 0)$ with a clockwise direction of motion of increasing $t$. (In time $t$ the trajectory moves through an arc of the circle of angle $t$). Indeed, when $+1$,

$$\dot{x}_1 = x_2, \qquad \dot{x}_2 = -x_1 + 1,$$
$$\frac{dx_1}{dx_2} = \frac{x_2}{-x_1 + 1}.$$

Hence $(1 - x_1)dx_1 = x_2 dx_2$, and $x_2^2 + (1 - x_1)^2 = a^2$. This is a circle centered at $(1, 0)$. Note that $1 - x_1 = a \cos t$, $x_2 = a \sin t$. When $u = -1$, $x_2^2 + (-1 - x_1)^2 = a^2$, which is a circle centered at $(-1, 0)$, where $x_1 = a \cos t - 1$, $x_2 = -a \sin t$.

One way of discovering an optimal control law is to begin at zero and move backward in time until we hit $(x_0, y_0)$ at time $-t_1$. Since we are moving in circles, the motion is counterclockwise. Begin with $u = 1$, when $0 < \delta \leq \pi$, and move with decreasing $t$ away from the origin in a counterclockwise direction along a semicircle center $(1, 0)$. At $P_1$ on this semicircle, $t = -\delta$, and $\sin(t + \delta)$ changes sign so that $u$ switches to $-1$. With this value the trajectory is a circle with center $(-1, 0)$ that passes through $P_1$. For exactly $\pi$ seconds we move counterclockwise along this circle, and describe a semicircle. (Recall that a full circle takes $2\pi$ seconds.) After $\pi$ seconds, $P_2$ is reached. This point is a reflection of $P_1$ onto the circle with radius 1 and center $(-1, 0)$. At $P_2$ the control becomes $u = +1$, and optimal trajectory becomes a circle with center $(1, 0)$ that passes $P_2$. After $\pi$ seconds on this circle we reach $P_3$ and the control switches to $u = -1$. This is the way that an optimal trajectory that passes through $(x_0, y_0)$ is generated. Because the system is normal, this trajectory is unique.

There is another route. Begin at zero with $u = -1$, when $-\pi \leq \delta < 0$ until at $Qt = -\delta - \pi$, and $\sin(t + \delta) = 0$. This is a circle with center $(-1, 0)$, and motion is counterclockwise until we switch at $Q$ to $u = 1$. For $\pi$ seconds the motion is a semicircle with center $(1, 0)$ and at $Q_2$ we switch to $u = -1$, etc. Clearly, the switching occurs along two arcs $\Gamma_+$ and $\Gamma_-$. The plane is divided into two sets $M_1$ and $M_2$, and then $M_0 = \{0\}$. Optimal feedback control is described as follows:

$$f(x, y) = \begin{cases} -1 & \text{if } (x, y) \text{ is above the semicircles or the arc } \Gamma_- : M_2\,, \\ +1 & \text{if } (x, y) \text{ is below the semicircles or on } \Gamma_+ : M_1\,. \end{cases}$$

Thus for

$$\ddot{x} + x = u\,, \qquad |u| \leq 1\,, \qquad u^*(t) = \text{sgn}[ce^{-At}B]\,.$$

The optimal trajectories are just solutions of $\ddot{x} + x = f(x, \dot{x})$ with initial point $(x_0, y_0)$ and final point $(0, 0)$ at time $t_1$. See Fig. 5.2.1.

From this analysis we deduce that there exists a unique function $f : E^2 \to E$ such that

$$\underline{\dot{x}}(t) = A\underline{x}(t) + Bf(\underline{x}(t)) \tag{5.1.5}$$

describes an optimal trajectory of reaching 0 in minimum time. In (5.1.5), $A$ and $B$ are as described following (5.1.4).

The semicircles constitute the switching locus: $\Gamma_+$ are arcs of circles of radius 1, centers $(1, 0), (3, 0) \ldots$; $\Gamma_-$ are arcs of circles of radius 1, centers $(-1, 0), (-3, 0) \ldots$. They move counterclockwise. These are described as "two one-manifolds". Above the semicircle is the subset $M_2$ of $E^2$, and below is the subset $M_1$ of $E^2$. These portions of the plane $E^2$ in some $\epsilon$ neighborhood of zero are called "terminal two-manifolds". The synthesis of optimal feedback control is realized as follows: Given any initial state $(x_0, \dot{x}_0)$ in some terminal manifold $M_k$, the optimal feedback function remains constant and optimal solution remains in $M_k$ until the instant the optimal solution enters $M_{k-1}$, when a switching occurs. After this it continues to move within $M_{k-1}$ until it reaches the origin, $M_0$, and the motion terminates.

For multidimensional controls ($m > 1$) the feedback function can be quite complicated. But Yeung [2] has proved that there are precisely $2m^{n-1}$ "terminal $n$-manifolds", and the "switching locus consists of a finite number of (analytic) $k$-manifolds" $1 \leq k \leq n-1$); one for $k = 0$ and at least two for, $k \geq 1$. Just as before, motion continues within $M_{k-1}$ until it reaches $M_{k-2}$ and the optimal feedback function switches again. The process terminates when the optimal solution reaches the origin, $M_0$. This description of the general situation is valid for controllable, strictly normal systems in an $\epsilon$-neighborhood of origin, i.e., in Int $\mathbb{R}(\epsilon)$, $\epsilon > 0$. In the case of the simple harmonic oscillator, $\epsilon$ can be taken to be $\pi$, and $\mathbb{R}(\pi)$ contains a circle of radius 2 about the origin. If $(x_0, \dot{x}_0) \in M_2$, one uses the control $u = -1$ to describe a circle with center $(-1, 0)$. This motion hits $\Gamma_+$ at some point. The control switches to $u = 1$ and motion continues on $\Gamma_+$, which is part of $M_1$, until it terminates at the origin, which is $M_0$. An analogous situation holds for $(x_0, \dot{x}_0) \in M_1$. Note carefully that in our circle of radius $\epsilon = 2$ there is at most one switch. In the next section we shall describe Yeung's description of the general situation.


## 5.2     Construction of Optimal Feedback Controls

In this section, in an $\epsilon$-neighborhood of the origin, we construct an optimal feedback control for controllable autonomous strictly normal systems. Using the switching times of the controls, which are among the roots of the index of the control system $y(t) = c^T e^{At} B$ in the interval $[0, \epsilon]$, we define terminal manifolds. Optimal feedback is constructed by identifying its value as a constant unique vertex of the unit cube in each manifold.

**Definition 5.2.1**     The system

$$\dot{x}(t) = Ax(t) + Bu(t) \tag{5.2.1}$$

is strictly normal if for any integers $r_j \geq 0$ satisfying $\sum_{j=1}^{m} r_j = n$, the vectors $A^i b_j$ with $j = 1, \ldots, m$, and $0 \leq i \leq r_j - 1$ are linearly independent (whenever $r_j = 0$, there are no terms $A^i b_j$).

**Remark 5.2.1**     If the system (5.2.1) is strictly normal, then rank $B = \min[m, n]$. Also if $m = 1$ and $B$ is an $n \times 1$ matrix, strict normality is equivalent to normality.

The number $\epsilon > 0$ that determines a neighborhood where the optimal feedback is constructed is obtained from the following fundamental lemma of Hájek:

**Lemma 5.2.1**     *System* (5.2.1) *is strictly normal if and only if there exists an $\epsilon > 0$ with the following property: For every complex nonzero $n$-vector $c \neq 0$ and in any interval of length less than or equal to $\epsilon$, the number of roots, counting multiplicities of the coordinates of the index $y(t) = c^T e^{-At} B$ of* (5.2.1)*, is less than $n$.*

With $\epsilon$ determined by the above lemma, define the reachable set

$$\mathbb{R}(\epsilon) = \left\{ \int_0^\epsilon e^{-As} B u(s) ds : u \in \mathcal{U} \right\} .$$

It is in Int $\mathbb{R}(\epsilon)$, the interior of $\mathbb{R}(\epsilon)$ relative to $E^n$ that the construction of a feedback will be made. In Int $\mathbb{R}(\epsilon)$ we shall identify the terminal manifolds. But first recall that optimal controls are bang-bang and are the vertices of the unit cube $\mathcal{U}$. We need some definitions.

**Definition 5.2.2** Let $0 < \theta < \epsilon$, and suppose $k$ is an integer $1 \leq k \leq n$. For any optimal control $u : [0, \theta] \to \mathcal{U}$ defined by $u(s) = u_j$ for $t_{j-1} \leq s < t_j$ with $0 = t_0 < t_1 < \cdots < t_k < \epsilon$ and $u_{j-1} < u_j$ in $U$, where $u_j$ is a vertex, the finite sequence $(u_1 \to \cdots \to u_k)$ is called an optimal switching sequence. If $u$ is not optimal and $u_j$ is not necessarily a vertex, then the finite sequence is called a switching sequence. The numbers $t_j$ $1 \leq j \leq k$ are points of discontinuities of $u$, and they are called switch times of $u$. The number of discontinuities of optimal controls is important. The following statement is valid:

**Corollary 5.2.1** *Let $\epsilon$ be given by the Fundamental Lemma, and consider the interval $[0, \epsilon]$. Every optimal control $u : [0, \epsilon] \to \mathcal{U}$ has at most $n - 1$ discontinuities.*

**Proof** Let $u : [\theta_1, \theta_2] \to \mathcal{U}$ be optimal control, where $0 \leq \theta_1 < \theta_2$ and $\theta_2 - \theta_1 < \epsilon$. Then by Theorem 5.1.1, $u = \operatorname{sgn} c^T e^{-As} B$ a.e. on $[\theta_1, \theta_2]$, for some $c \neq 0$ in $E^n$. Since the discontinuities of $u$ are among the roots of the index $y(s) = c^T e^{-As} B$, and this, by Hájek's Lemma, is less than $n$, the assertion follows. $\qquad\square$

**Definition 5.2.3** For each $k = 1, \ldots, n$, let $M_k$ denote the points in Int $\mathbb{R}(\epsilon)$ whose corresponding optimal controls have exactly $k - 1$ discontinuities. Assume $M_0 = \{0\}$. We are now ready to identify and define "terminal manifolds."

**Definition 5.2.4** For each $k = 1, \ldots, n$, and each $k$-tuple $i = \{u_1 \to \cdots \to u_k\}$ of vertices of $\mathcal{U}$ with $u_{j-1} \neq u_j$, define as terminal manifold the set $M_{ki}$ of points in Int $\mathbb{R}(\epsilon)$, whose optimal controls have $\{u_1 \to \cdots \to u_k\}$ as the optimal sequence. (Note that $i$ ranges over a finite set $I_k$.) We will say that the optimal switching sequence $\{u_1 \cdots u_k\}$ corresponds to $M_{ki}$. The terminal manifolds $M_{ki}$ are disjoint since optimal controls are unique.

Next we shall relate the terminal manifolds to the reachable sets. It is contained in the following statement:

**Proposition 5.2.1** *For terminal manifolds, the following disjoint unions are valid:*

$$M_k = \bigcup_{j \in I_k} M_{kj}, \qquad \text{Int } \mathbb{R}(\epsilon) = \bigcup_{k=0} M_k .$$

*For $k = n$, there are precisely $2m^{n-1}$ nonvoid terminal manifolds $M_{ki}$. The set $M_{ni}$ is open and connected (in $E^n$). Furthermore, each $M_{ki}$ is the image of the set*

$Q_k$ of all $t = (t_1 \cdots t_k) \in E^k$ with $0 = t_1 < t_1 < \cdots < t_k < \epsilon$ of a diffeomorphism $F : M_{ki} = F(Q_k)$, where $F$ is defined as the map $F : Q_k \to \text{Int } \mathbb{R}(\epsilon)$, by

$$F(t) = \sum_{j=1}^{k} \int_{t_{j-1}}^{t_j} e^{-As} B u_j ds \,,$$

a function that is analytic in its variables, and whose Jacobian has rank $k$ at each point of $Q_k$. $F^{-1}$ is continuous.

The description of $M_{ki}$ contained in Proposition 5.2.1 asserts that each $M_{ki}$ is an analytic $k$-submanifold of $E^n$. The following is a consequence of these results:

**Corollary 5.2.2**   *If $\{u_1 \to \cdots \to u_k\}$ is an optimal switching sequence corresponding to $M_{ki}$, then $\{u_{j+1}, \ldots, u_k\}$ is an optimal switching sequence corresponding to $M_{k-j}, i$ for $j = 0, \ldots, k-1$ $M_{k-j}, i$ is nonvoid whenever $M_{ki}$ is nonvoid. Moreover, if we let $M_{0,i} = M_0$, and if an optimal solution in $\text{Int } \mathbb{R}(\epsilon)$ meets $M_{k,i}$, then it meets only $M_{k-1,i}, \ldots, M_0$ thereafter in this order.*

These assertions are used to prove the next theorem.

**Proposition 5.2.2**   *The set $\text{Int } \mathbb{R}(\epsilon)$ is made up of a union of $2m^{n-1}$ disjoint connected nonvoid open sets and a finite number of analytic $k$-manifolds, $0 \le k \le n - 1$; one for $k = 0$ and at least two for $k \ge 1$.*

**Proof**   From Proposition 5.2.1, $\text{Int } \mathbb{R}(\epsilon) = \bigcup_{k=0}^{n} M_k = M_n \bigcup_{k \le n-1} M_k$, and $M_n$ is the union of $2m^{n-1}$ disjoint connected nonvoid open sets. We easily see that the union $\bigcup_{k \le n-1} M_k$ validates the last assertion. But Proposition 5.2.1 states that $M_{ni}$ is nonvoid, so that $M_{n-1}, \ldots, M_{1i}$ are also nonvoid, and by symmetry $-M_{n-1}, \ldots, -M_{11}$ are nonvoid as well. Because we have observed before that $M_{ki}$ are distinct analytic manifolds in $E^n$ by Proposition 5.2.1, the proof is complete.                                                   $\square$

The promised synthesis of optimal feedback control is contained in the proof of the following theorem of Yeung [8].

**Theorem 5.2.1**   *For the system (5.2.1) where $A$ is an $n \times n$ matrix and $B$ is an $n \times m$ matrix, assume that*

    (i)  *the system is strictly normal.*
    (ii)  $\text{rank}[B, \ldots, A^{n-1}B] = n$.
    (iii)  *No eigenvalues of $A$ have a positive real part.*

*Then there exists $\epsilon > 0$, and a function*

$$f : \text{Int } \mathbb{R}(\epsilon) \to E^m \,,$$

*where*

$$\mathbb{R}(\epsilon) = \left\{ \int_0^\epsilon e^{-As} B u(s) ds : u \in \mathcal{U} \right\} ,$$

*such that in* Int $\mathbb{R}(\epsilon)$*, the set of solutions of*

$$\dot{y}(t) = Ay(t) + Bf(y(t)) \tag{5.2.2}$$

*coincides with the set of optimal solutions of* (5.2.1)*. Furthermore,* $f(0) = 0$*; for* $x \neq 0$*,* $f(x)$ *is among the vertices of the unit cube* $U$ *and* $f(x) = -f(-x)$*. If* $m \leq n$*, then* $f$ *is uniquely determined by the condition that optimal solutions solve* (5.2.2)*. Also the inverse image under* $f$ *of an open set is an* $F_\sigma$*-set, that is: the union of at most countably many closed sets.*

**Proof of Theorem 5.2.1** An $\epsilon$ is obtained from Hájek's Lemma. Let $f(0) = 0$. If $x_0 \neq 0$, and $x_0 \in$ Int $\mathbb{R}(\epsilon)$, find a terminal manifold $M_{ki}$ that contains $x_0$. Suppose $\{u_1 \to \cdots \to u_k\}$ is its corresponding optimal control, set $f(x) = u_1$, and note that $f(x_0) = -f(-x_0)$. Because of the disjointedness of the union in Proposition 5.2.1, $f$ is well defined. Let $x_0 \neq 0$, and $x_0 \in$ Int $\mathbb{R}(\epsilon)$, we can let $x(\cdot, x_0)$ be the optimal solution of (5.2.1) in Int $\mathbb{R}(\epsilon)$ through $x_0$. Suppose now the optimal control $u(s)$ of $x_0$ is defined by $u(s) = u_j$ for $t_{j-1} \leq s < t_j$, with $0 = t_0 < t_1 < \cdots < t_k = T(x_0) < \epsilon$, where $T(x_0)$ is the minimal time function defined by $T(x_0) = \text{Inf}\{t \geq 0 : x_0 \in \mathbb{R}(t)\}$. It follows that $x_0$ belongs to some $M_{ki}$, so that by Corollary 5.2.2, $f(x(x, x_0)) = u(s)$. As a consequence, every optimal solution of (5.2.1) in Int $\mathbb{R}(\epsilon)$ is a solution of (5.2.2).

The converse follows from the following assertion: If $y : [0, \theta] \to E^n$ is a solution of (5.2.2), and $0 < \theta < \epsilon$, and if $y(t) \in M_{ki}$ with $0 \leq t \leq \theta$, then also $y(s) \in M_{ki}$ for sufficiently small $s - t \geq 0$. Indeed, the assertion implies that $f(y(\cdot))$ is constant on any interval on which $y(\cdot)$ is differentiable, so that Corollary 5.2.2, $y(\cdot)$ coincides with the optimal solution through $y(t)$. The above assertion is proved by induction on $n - k$. Continuing the proof, we recall that

$$f : \text{Int } \mathbb{R}(\epsilon) \to E^m \,,$$

and consider any open set $O$ in $E^m$. Then $f^{-1}(O)$ is the union of some $\{0\}$, $M_{ki}$ which, because they are manifolds in $E^n$, are $\sigma$-compact. Thus the inverse image of an open set is a $\sigma$-compact set, the union of at most countably many compact spaces, an $F_\sigma$-set. Dugunji [4, pp. 74, 240]. To prove that $f$ is unique we let $g$ be another optimal feedback control. Because after reaching zero optimal solutions remain constant, $g(0) = 0 = f(0)$. If $x_0 \neq 0$, and $x_0 \in$ Int $\mathbb{R}(\epsilon)$, find $M_{ki}$, which contains $x_0$. Suppose $x(t) = x(\cdot, x_0)$ is an optimal solution through $x_0$, and $\{u_1 \cdots u_k\}$ is an optimal switching sequence corresponding to $M_{ki}$. Then $\dot{x}(t) = Ax(t) - Bu_1$ for sufficiently small $t \geq 0$, because of Proposition 5.2.1. Since $x(\cdot, x_0)$ also solves

$$\dot{y} = Ay - Bg(y) \,,$$

we have

$$\dot{x}(t) = Ax(t) - Bu_1 = Ax(t) - Bg(x(t))$$

on some $(0, \alpha]$, $\alpha > 0$. Because $\dot{x}(\cdot)$ is continuous on $(0, \alpha]$, it follows that $Bu_1 = Bg(t)$. Since the system is strictly normal and $m \leq n$, the columns of $B$ are linearly independent. As a result, $g(x) = u_1 = f(x)$. The theorem is proved. $\qquad\square$

**Example 5.2.1**   We apply Theorem 5.2.1 to our system (5.1.4). From Hájek's Lemma [3], $c^T e^{-At} b$ has $n - 1$ discontinuities on $[0, \epsilon]$ where $\epsilon < \frac{\pi}{\omega}$ and $\alpha_i + i\omega_i$ are the roots of the characteristic equation $|A - \lambda I| = 0$, $\omega = \max_i(\omega_i)$. In our case, $\omega_1 = 1$, $\omega_i = -1$, so that $\epsilon < \pi$. Since in (5.1.4) $n = 2$, Int $\mathbb{R}(\epsilon) = M_1 \cup M_2 \cup \Gamma_+ \cup \Gamma_-$. The switching curves are shown below in Fig. 5.2.1.

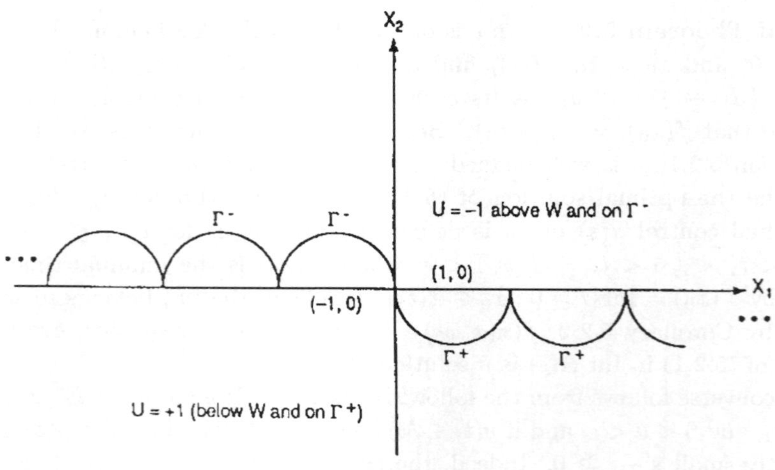


Fig. 5.2.1

We may define our terminal manifolds by moving backward in time from the origin (considering all possible control sequences as being moved backwards in time). The terminal manifolds, $M_k$'s, is the set of points in Int $\mathbb{R}(\epsilon)$ whose optimal controls have exactly $(k - 1)$ discontinuities. If our final control is $u_2$, the only possible sequence as we approach the origin is $\{u_1 \to u_2\}$. If we move backward in time (ccw) in our phase plane, we can describe a terminal manifold $M_1$. Control $u_1 = +1$ corresponds to a circle centered at $(1, 0)$, so if we vary the radius of the circle from 1 to 3 (and move backward in time until we reach the switching curve again) we may describe that particular terminal manifold. See Fig. 5.2.2.

Using Definition 5.2.3 we see that $M_1$ is defined above. Once the system is within $M_1$, there is exactly one discontinuity before reaching the origin. Terminal manifold $M_2$ is constructed similarly, but with optimal control sequences $\{u_1 \to u_2\}$. See Fig. 5.2.3.

We will now look at the possible control before $M_1$ (the manifolds). Our final control is $\{u_1 \to u_2\}$. Moving backward in time, the only possible control is $u_2$. So our optimal control sequence will be $\{u_2 \to u_1 \to u_2\}$. Now we will construct the corresponding manifold. Control of $u_2 = -1$ corresponds to a circle centered

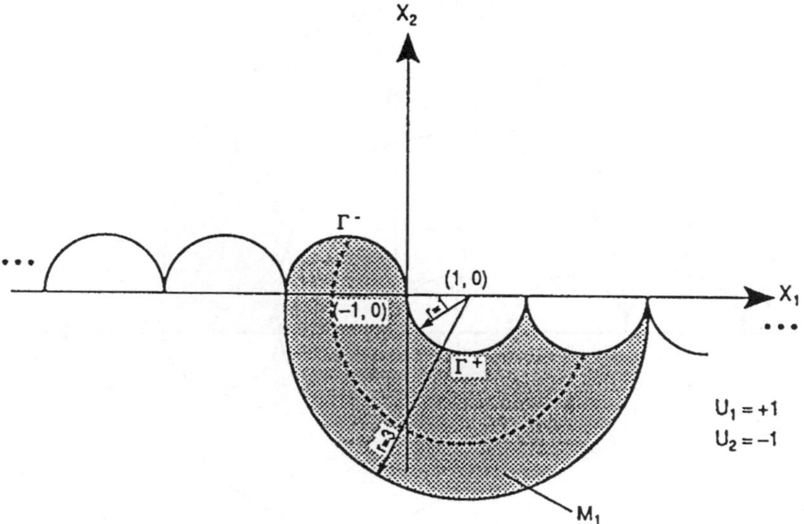


Fig. 5.2.2

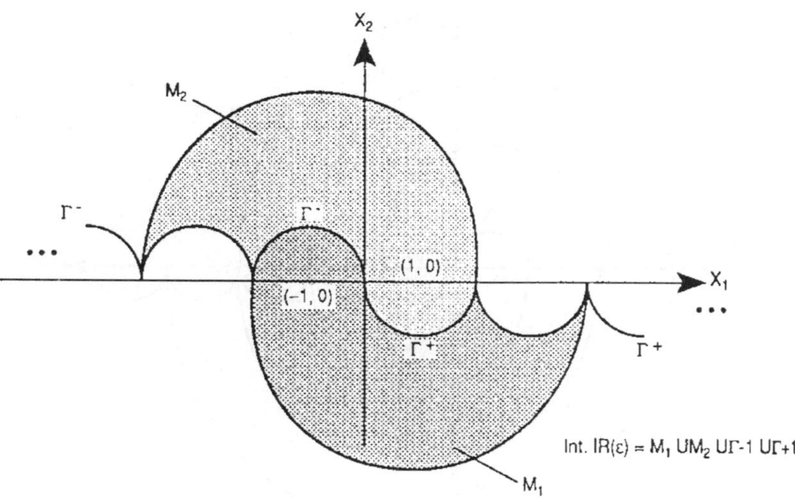


Fig. 5.2.3

at $(-1, 0)$. So if we vary the radius of this circle (moving backward in time, counterclockwise) from $r = 3$ to $r = 5$ until we reach our switching curve again, the manifold will be defined. See Fig. 5.2.4. So the optimal control sequence of the system, once it reaches the manifold above, is $\{u_2 \to u_1 \to u_2\}$. The rest of the manifolds may be found in a similar manner. The result is shown in Fig. 5.2.5.

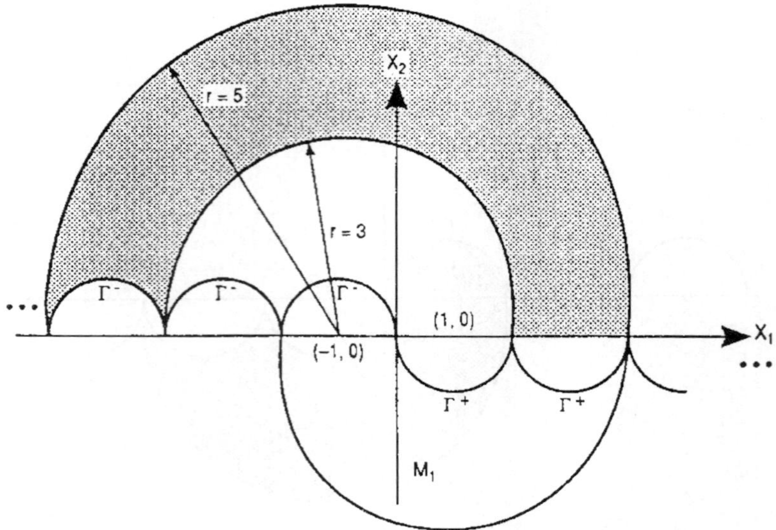



Fig. 5.2.4

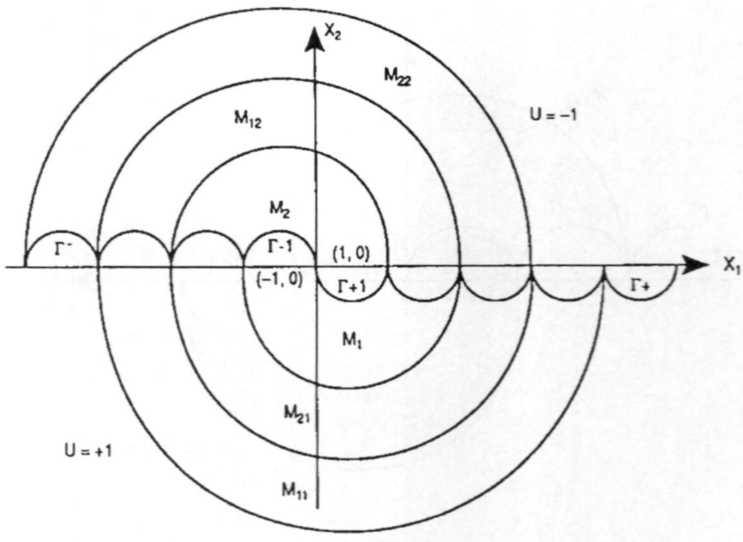



Fig. 5.2.5

Now that our terminal manifolds have been described and drawn and our $\epsilon$ computed, our time-optimal feedback control is known.

**Example 5.2.2**    Our initial conditions will be represented by the "$x$" in our phase plane in Fig. 5.2.6. The initial conditions place it in $M_{11}$, so our initial control will

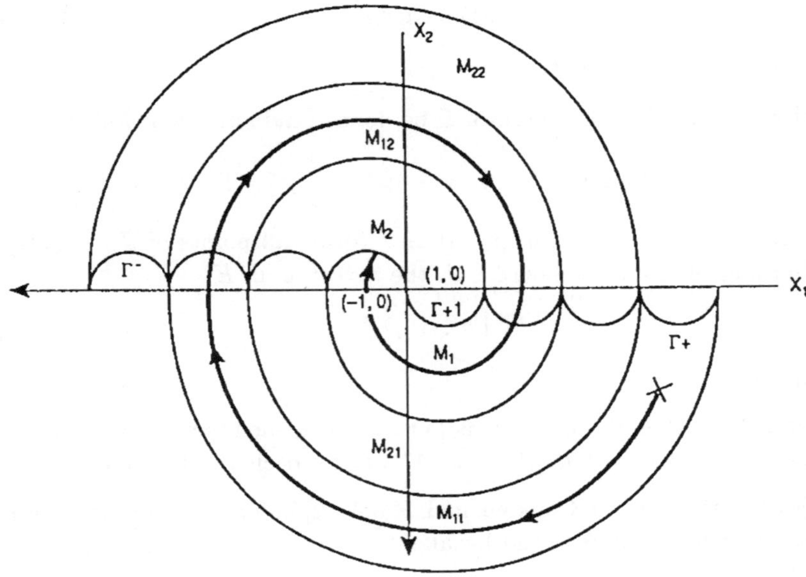


Fig. 5.2.6

be $u_1 = +1$ until our trajectory takes the states to $\Gamma_-$. Here we enter $M_{12}$ so the control will be $u_2 = -1$. Our trajectory then reaches $\Gamma^+$, at which time the states will be in $M_{11}$. The control here will be $u_1 = +1$. The states now enter $\epsilon$ so there will be at most $n - 1 = 1$ more switches. We reach $\Gamma^-$ where the control switches to $u = -1$. This control takes the states to the origin.

## 5.3    Time-Optimal Feedback Control of Nonautonomous Systems

In this section we extend Yeung's construction [8] of time-optimal control feedback to cover the time varying $n$-dimensional differential equation

$$\dot{x}(t) = A(t)x(t) + B(t)u(t), \qquad (5.3.1)$$

where $A$ is $n \times n$, $B$ is $n \times m$ matrices, and where $t \to A(t)$ has $n - 1$ continuous derivatives and $t \to B(t)$ has $n$ continuous derivatives on the interval $[t_0, t_1]$. If $X(t, t_0)$ is the fundamental matrix solution of

$$\dot{x}(t) = A(t)x(t), \qquad (5.3.2)$$

then the function $g : [t_0, t_1] \to E^m$ defined by $g(t) = c^T X^{-1}(t, t_0)B(t)$, with $c \in E^n$ a constant, is the index of the control system. If we assume normality (which is defined for (5.3.1) below), then $g$ is the switching function of the optimal controls in the sense that

$$u(t) = \operatorname{sign} g(t)\,. \tag{5.3.3}$$

By convention, $X(t,0) = X(t)$.

**Definition 5.3.1**    Let the operator $\Gamma$ be defined inductively as follows:

$$\Gamma^0 = \text{the identity}\,, \qquad \Gamma = -A(t) + \frac{d}{dt}\,, \qquad \Gamma^{k+1} = \Gamma\Gamma^k\,.$$

The system (5.3.1) is normal in $[t_0, t_1] \equiv J$ (or at all points of $J \equiv [t_0, t_1]$) if for each column $b_j$ of $B$ and at each $t \in J$, the $n$ vectors in $E^n$,

$$b_j, \Gamma b_j, \ldots, \Gamma^{n-1} b_j\,, \tag{5.3.4}$$

are independent.

**Remark 5.3.1**    If $A$ and $B$ are constant matrices, then normality is equivalent to the linear independence of $b_j,\ Ab_j, \ldots, A^{n-1} b_j$ for each $j = 1, \ldots, m$.

The following theorem of Chukwu and Hájek [2] relates the above definition of normality to those of Hermes and Lasalle [6].

**Proposition 5.3.1**    *System (5.3.1) is normal in $[t_0, t_1]$ if and only if the following conditions are satisfied:*

(i) *Each index $g = c^T X^{-1} B$ of (5.3.1) satisfies the so-called indicial equation,*

$$y^n(t) = \sum_{k=0}^{n-1} y^k(t) D_k(t)\,, \tag{5.3.5}$$

   *with continuous diagonal matrices $D_k(t)$; and*

(ii) *for each $c \neq 0$ and each column $b_j$ of $B$, the set*

$$G_j(c) = \{t \in [t_0, t_1] : c^T X^{-1}(t, t_0) b_j(t) = 0\}$$

   *is finite.*

**Proof**    Assume system (5.3.1) is normal, and consider any column $b_j$ of $B$, and $t \in [t_0, t_1]$. Then at $t \in [t_0, t_1]$, $\Gamma^n b_j$ must be a linear combination of vectors from the basis in (5.3.4). Thus

$$\Gamma^n b_j = \sum_{k=0}^{n-1} d_k \Gamma^k b_j\,,$$

with real $d_k$ depending continuously on $t$. If we collect the $d_k$ for $i = 1, \ldots, m$ into a diagonal matrix, we find

$$\Gamma^n B = \sum \Gamma^k B D_k\,. \tag{5.3.6}$$

Because the $k$th derivative of each index $g = c^T X^{-1} B$ is $g^k = c^T X^{-1} \Gamma^k B$ (5.3.6) leads to (5.3.4), for any $c \in E^n$, so that (i) holds.

Assume that (ii) fails and that $s_0 \in [t_0, t_1]$ is an accumulation point of zeros of the component $g_j = c^T X^{-1} b_j$. Then at $s$,

$$0 = (d^k/dt^k) c^T X^{-1} b_j = c^T X^{-1} \Gamma^k b_j \quad \text{for} \quad k = 0, \ldots, n-1. \tag{5.3.7}$$

Because $X^{-1}$ is nonsingular and $c \neq 0$, the vector $(c^T X^{-1})^T \neq 0$ and is perpendicular to $b_j, \ldots, \Gamma^{n-1} b_j$, which contradicts normality.

To prove the converse, we assume that normality fails, so that $\text{rank}[b_j, \ldots, \Gamma^{n-1} b_j] < n$, for some $b_j$ at some $s \in [t_0, t_1]$. If we choose a nonzero vector of the form $(c^T X^{-1}(s))^T$ that is perpendicular to all the columns above, then (5.3.7) holds at $s$. With this $c$ form the index

$$g(t) = c^T X^{-1}(t) B(t)$$

that satisfies (5.3.5). Because $D_k$ is diagonal, post multiplication by $e_j$ yields that $\eta = c^T X^{-1} b_j$ satisfies the equation $\eta^{(n)} = \sum_{k=0}^{n-1} \eta^k d_{kj}(t)$. Clearly this equation has (5.3.7) as initial condition. Hence $\eta \equiv 0$, so that $G_j(c) = [t_0, t_1]$. Gathering results, if normality fails, then (i) implies that (ii) is not valid. This concludes the verification of Proposition 5.3.1. $\qquad\square$

We need the notion of strict normality for nonautonomous systems.

**Definition 5.3.2**   The system (5.3.1) is strictly normal on $J$ (i.e., on each $t \in J$) if for any integers $r_j \geq 0$ with $\sum_{j=1}^{m} r = n$, the vectors

$$\Gamma^k b_j(t), \qquad 1 \leq j \leq m, \qquad 0 \leq k \leq r_{j-1},$$

are independent for each $t \in J$. Strict normality is equivalent to the "disconjugacy" of the indices of the system at each point. For an elaboration of this concept, see Chukwu and Hájek [1]. With these concepts, one follows the treatment in Sec. 2 and obtains the following extension of Theorem 5.2.1:

**Theorem 5.3.1**   *Suppose that*

(i)   *$A$ has $k-2$ continuous derivatives and $B$ has $k-1$ continuous derivatives on the interval $[0, t_1]$. Assume that*

(ii)   $\text{rank}[B(t), \Gamma B(t), \ldots, \Gamma^{k-1} B(t)] = n$ *on* $[0, t_1]$.

(iii)   *The system*

$$\dot{x}(t) = A(t)x \tag{5.3.8}$$

*is uniformly asymptotically stable.*

(iv)   *The system (5.3.1) is strictly normal at $t = 0$.*

*Then there exists $\epsilon > 0$ and a function*

$$f : \text{Int } \mathbb{R}(\epsilon) \to E^m,$$

*such that in* $\text{Int } \mathbb{R}(\epsilon)$ *the set of solutions of*

$$\dot{y}(t) = A(t)y(t) + B(t)f(y(t)) \tag{5.3.9}$$

*coincides with the set of optimal solutions of* (5.3.1). *Also* $f(0) = 0$; *and for* $x \neq 0$, $f(x)$ *is among the vertices of the unit cube* $U$. *Also,* $f(x) = -f(-x)$. *If* $m \leq n$, $f$ *is uniquely determined by the condition that optimal solution of* (5.3.1) *solve* (5.3.9).

Consider the autonomous system

$$\dot{x} = Ax + Bu\,. \tag{S}$$

The construction of time-optimal controls that is implied by Theorem 5.2.1 and Theorem 5.3.1 is dependent on the value of $\epsilon$. On that interval $[0, \epsilon)$ every function

$$g(t) = c^T e^{At} b_j\,, \qquad j = 1, \ldots, m\,, \tag{5.3.10}$$

has at most $n - 1$ roots. Here $A$ is an $n \times n$ real constant matrix and $b_j \equiv b$ is an $n \times 1$ matrix. It is important to know what is the largest value of $\epsilon$ such that on the interval of length less than $\epsilon$, $g(t)$ has at most $n - 1$ roots. The solution given by Hájek [3] will be stated after introducing the following notation. Let $r$ be the number of real eigenvalues $\alpha_i$ of $A$, and $2k$ that of the complex ones $\beta_j \pm iw_j$ (counting multiplicities, $w_j > 0$), so that $n = r + 2k$. It is possible to write

$$g(t) = \sum_{i=1}^{\ell} \alpha_{it} p_i(t) + \sum_j^m e^{\beta_j t}[q_j(t) \cos w_j(t) + r_j(t) \sin w_j t]\,, \tag{5.3.11}$$

where $p_i$, $q_j$, $r_j$ are real polynomials of appropriate degrees.

Set $w = \max w_j$, with $w > 0$ chosen arbitrarily if $k = 0$. The smallest integer $M$ greater than $x$ is denoted by $[x]^*$. Observe that $[x]^* \geq 1$ whenever $x > 0$.

**Theorem 5.3.2**    *Consider* $g(t)$ *in* (5.3.11). *Then either* $g(t) \equiv 0$ *or in any interval* $[0, T]$, $T > 0$, *the function* $g(t)$ *has at most* $N$ *distinct zeros, where*

$$N \leq (n - 1)[Tw/\pi]^*\,.$$

*Indeed,*

$$N \leq n - 1$$

*if all the eigenvalues of* $A$ *are real, and*

$$N \leq r + (2k - 1)[Tw/\pi]^* \tag{5.3.12}$$

*if all the eigenvalues are not real, in which case we have the following special case: If* $g(t) = a \cos wt + b \sin wt$ *(i.e.,* $r = 0$, $k = 1$, $n = 2$), *a closed interval of length* $T = \ell\pi/w$, *in which case the estimate in* (5.3.12) *(i.e.,* $N \leq \ell$) *should be replaced by* $N \leq \ell + 1$.

We have the following:

**Corollary 5.3.1**    *Suppose in* (5.3.10) $g(t) \not\equiv 0$. *Then* $g(t)$ *has at most* $n - 1$ *zeros, counting multiplicities in every nonclosed interval of length* $\pi/w$.

**Proof**   If $[0, T]$, $T > 0$, and $T < \pi/w$, then $\left[\frac{Tw}{\pi}\right]^* = 1$.
    Note that $w = \max Imag$ (eigenvalues of $A$).   $\square$

A somewhat better estimate on the number $N$ of the zeros of (5.3.11) is given in the next result that is due to Voorhoeve.

**Theorem 5.3.3**   *Let $g(t)$ be as given in (5.3.11). For $T > 0$, let $N$ be the number of zeros on $[0\,T]$. Put $J = \max_{1 \leq j \leq m}(Imag\, w_j)$. Then*

$$N \leq (n-1) + TJ/2\pi\,.$$

*In general, consider*

$$f(z) = \sum_{j=1}^{m} p_j(z)\exp(w_j z)\,, \tag{5.3.13}$$

*where $p_j$ are complex polynomials, $p_j \neq 0$, $w_j$ are complex numbers such that $w_i \neq w_j$ if $i \neq j$. If $I = \min_{1 \leq k \leq m}(Imag\, w_j)$; $J = \max_{1 \leq j \leq m}(Imag\, w_j)$, then*

$$N \leq n - 1 + T(J - I)/2\pi\,.$$

**Remark**   Note that $f(z)$ in (5.3.13) satisfies linear ordinary differential equations.

**Example 5.3.1**
1. The Simple Harmonic Oscillator
    Equation of motion:

$$m\ddot{x} + kx = u\,. \tag{5.3.14}$$

For $\frac{k}{m=1}$, $w = \sqrt{\frac{k}{m}} = 1$. This is the equation treated in (5.1.4a) and (5.1.4b). See Figs. 5.3.1 and 5.3.2.

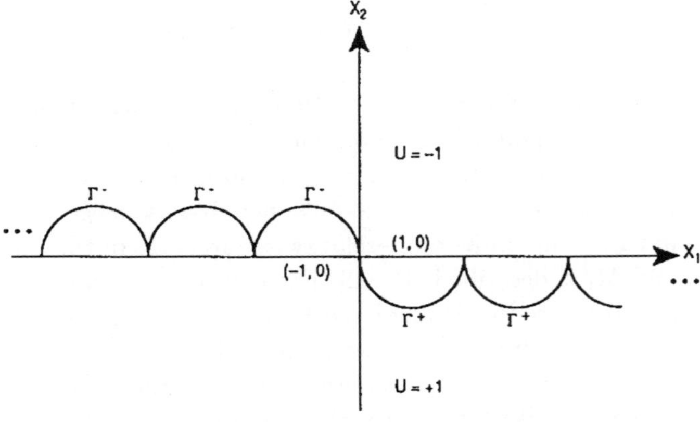


Fig. 5.3.1

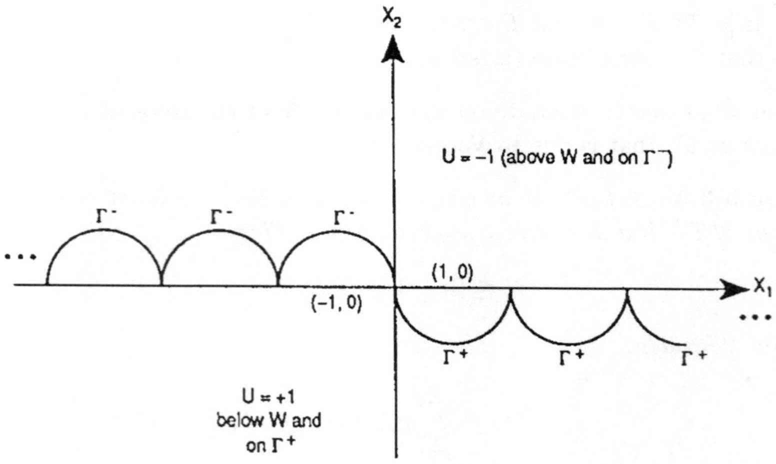


Fig. 5.3.2

It is stable, controllable, and normal, since it is the system

$$\dot{x} = Ax + Bu, \qquad (5.3.15)$$

where $A = \begin{bmatrix} 0 & 1 \\ -1 & 0 \end{bmatrix}$, $B = \begin{bmatrix} 0 \\ 1 \end{bmatrix}$. Optimal controls are given by $\text{sgn}[g(t,c)] = \text{sgn}(c^T e^{-At} B) : u^*(t) = \text{sgn}[c^T e^{-At} B] = \text{sgn}[\sin(t + \delta)]$, where $\tan \delta = -\frac{c_2}{c_1}$, $c = (c_1, c_2)$. The simple harmonic oscillator is strictly normal, so that there exists $\epsilon > 0$ such that the number of roots (counting multiplicities) of the coordinates of the index of the control system $g(t,c) = c^T e^{-At} B$ in $[0, \epsilon)$ is less than 2 and is at most 1.

By Corollary 5.3.1,

$$\epsilon < \frac{\pi}{\omega}, \qquad \text{where } \omega = \max_i \omega_i .$$

Since the eigenvalues of $A$ are $\pm i$, $\omega = 1$. Hence $\epsilon < \frac{\pi}{1}$. Recall that the *terminal manifold* $M_k$ is the set of points in Int $\mathbb{R}(\epsilon)$ whose optimal controls have exactly one discontinuity. If we set $u_1 = 1$, and $u_2 = -1$, and if our final control that steers any initial point to zero is $-1$, then the only possible sequence is $\{u_1 \to u_2\}$. If we move backward in time in a counterclockwise direction in the phase plane, the terminal manifold $M_1$ is described. Recall that with $u_1 = +1$ we described a circle centered at $(1, 0)$. If the radius is increased from 1 to 3 (and we steer backward in time until we hit the switching curve), we obtain all of $M_1$. See Fig. 5.3.3.

In $M_1$, there is exactly one discontinuity for the optimal control that drives any point in $M_1$ to the origin. With the optimal sequence $\{u_2 \to u_2\}$ we can analogously describe the terminal manifold in Fig. 5.3.4.

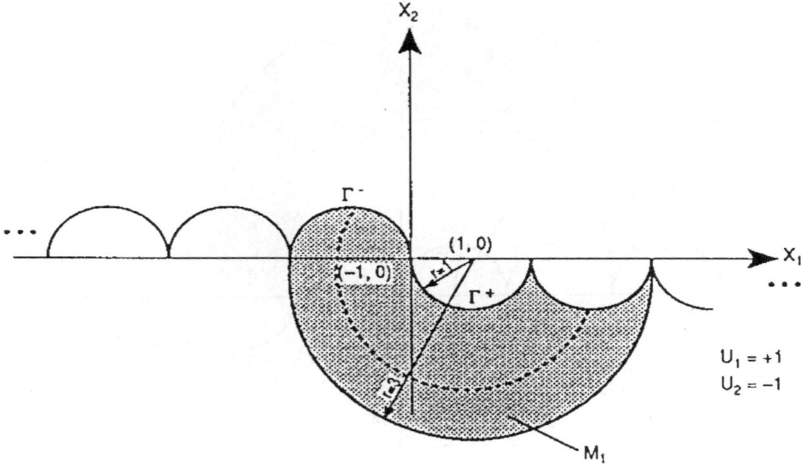



Fig. 5.3.3

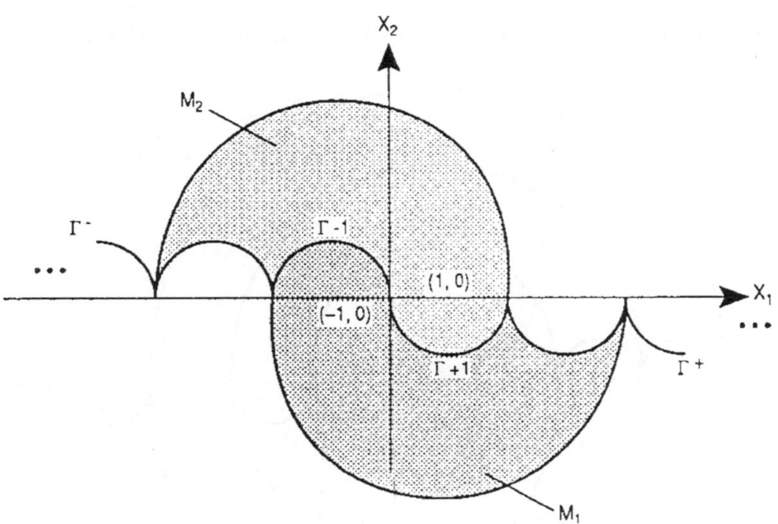



Fig. 5.3.4   Terminal manifolds $M_1$, $M_2$.

Going beyond terminal manifolds to other manifolds before $M_1$ or $M_2$ we can obtain a strategy as follows: Assume our final optimal sequence is $\{u_1 \to u_2\}$. Then if we move backward in time, the only possible sequence is $\{u_2 \to u_1 \to u_2\}$. To construct the corresponding manifold we note that $u_2 = -1$ corresponds to a circle centered at $(-1, 0)$. If we vary the radius of this circle (moving counterclockwise and backward in time) from $r = 3$ to $r = 5$ until we hit the switching curve again, the manifold $M_{12}$ in Fig. 5.3.5 is determined.

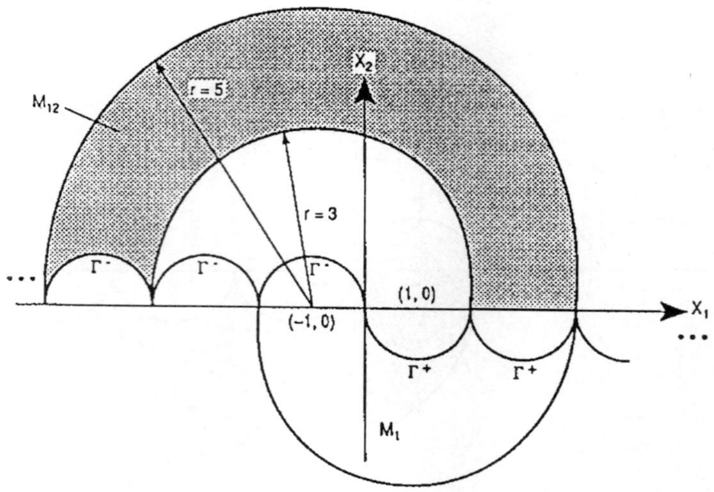


Fig. 5.3.5

Other manifolds may be determined in a similar manner. See Fig. 5.3.6.

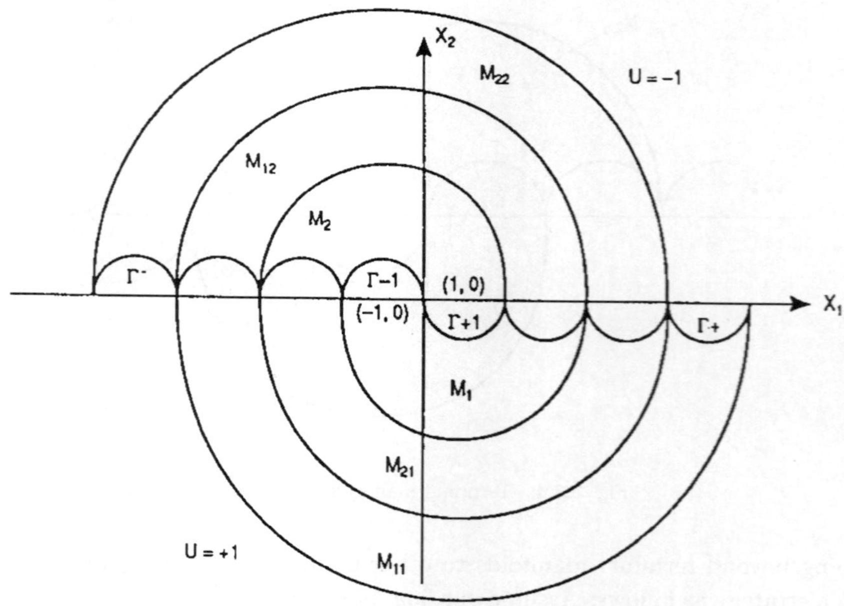


Fig. 5.3.6

With terminal manifolds described and $\epsilon$ computed, our time-optimal control law is known. For example, if our initial condition is point $x_0$ in Fig. 5.3.7.

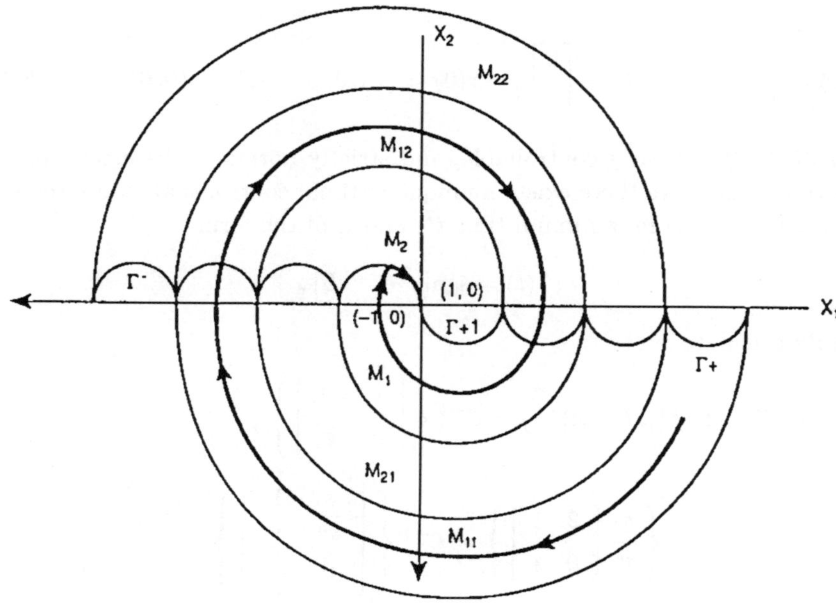


Fig. 5.3.7

$x_0 \in M_{11}$ so that the initial control is $u_1 = +1$, until we enter $M_{12}$ and switch to $u_2 = -1$ at the switching locus $\Gamma_-$. With this control we travel until $\Gamma_+$ is hit and enter $M_1$, where the control is $u_1 = +1$. Since we are within $\mathbb{R}(\epsilon)$, we will have at most one more switch at $\Gamma_-$, where the control is $u = -1$. With this control we enter zero optimally.

## Example 5.3.2    Rest-to-Rest Rigid Body Maneuver

Consider a rigid body such as a crane or a trolley whose mass is $m$, which moves without friction along a track. If $u(t)$ is an external controlling force that is applied to the rigid body, and if $x(t)$ represents the position at time $t$, we have the following equation of motion:

$$m\ddot{x}(t) = u(t), \qquad t > 0.$$

(5.3.16)

Assuming initial displacement $x(0) = x_0$ and initial velocity $\dot{x}(0) = 0$, we want to construct a control $u$ that brings the trolley to rest at the origin in minimum time: $x(t^*) = 0$, $\dot{x}(t^*) = 0$. This is the problem of the rest-to-rest maneuver. We assume $|u(t)| \leq 1$, $m = 1$, and the dynamics become

$$\dot{x} = Ax + Bu,$$

(5.3.16a)

where

$$A = \begin{bmatrix} 0 & 1 \\ 0 & 0 \end{bmatrix}, \quad B = \begin{bmatrix} 0 \\ 1 \end{bmatrix}, \quad x(0) = (x_0, 0), \quad \underline{x}(t^*) = (0, 0). \qquad (5.3.16b)$$

System (5.3.16) is clearly controllable, and strictly normal. The eigenvalues of $A$ are all zero. Therefore there exists a unique optimal control that steers the system from $(x_0, 0)$ to $(0, 0)$ in minimum time $t^*$, and is of the form

$$u^*(t) = \text{sgn}[c^T e^{-At} B].$$

We easily calculate

$$e^{-At} = \mathcal{L}^{-1}\{[sI - A]^{-1}\} = \mathcal{L}^{-1}\left\{\begin{bmatrix} s & -1 \\ 0 & s \end{bmatrix}\right\},$$

$$= \mathcal{L}^{-1}\left\{\frac{1}{s^2}\begin{bmatrix} 2 & 1 \\ 0 & s \end{bmatrix}\right\} = \mathcal{L}^{-1}\left\{\begin{bmatrix} \dfrac{s}{s^2} & \dfrac{1}{s} \\ 0 & \dfrac{s}{s^2} \end{bmatrix}\right\} = \begin{bmatrix} 1 & -t \\ 0 & 1 \end{bmatrix}.$$

Therefore $e^{-At} = \begin{bmatrix} 1 & -t \\ 0 & 1 \end{bmatrix}$,

$$c^T e^{-At} B = \frac{1}{m}[-c_1 t + c_2] = g(t, c),$$

where $g(t, c)$ is a straight line in $t$ and has only one switch at $t = \frac{c_2}{c_1}$. We examine two possible controls $u = +1$, $u = -1$. If $u = +1$,

$$\dot{x}_1 = x_2,$$

$$\dot{x}_2 = u(t) = 1, \qquad \frac{dx_1}{dx_2} = \frac{x_2}{1},$$

$$x_2 dx_2 = dx_1,$$

$$\frac{x_2^2}{2} - x_1 = c_1, \qquad \text{or}$$

$$x_2^2 = 2(x_1 - c_1), \qquad \text{say}.$$

In the phase plane this is a parabola on its side that opens to the right and is shifted to the right by $c_1$. See Fig. 5.3.8. This parabola passes through the origin if $c_1 = 0$. For $u = -1$, $x_2^2 = -2(x_1 - c_2)$. This is a parabola that opens to the left and is shifted to the right by $c_2$. See Fig. 5.3.9. To obtain the switching curves, we move backward in time counterclockwise from our final target $(0, 0)$. With $u = +1$, $x_2 = \pm\sqrt{2x_1}$. With $u = -1$, $x_2 = \pm\sqrt{-2x_1}$. Inspection of Fig. 5.3.8 with $u_1 = +1$ shows that $x_2$ is always increasing and we must use the lower half of the parabola in Fig. 5.3.10.

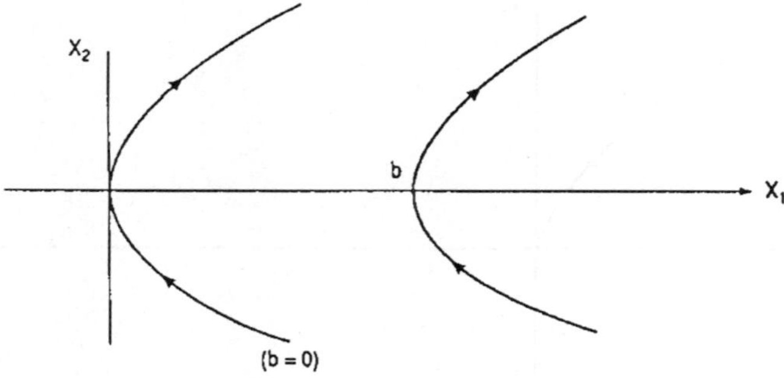



Fig. 5.3.8

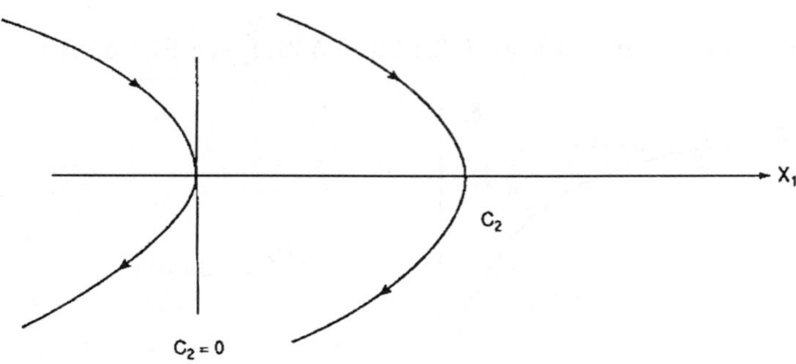



Fig. 5.3.9

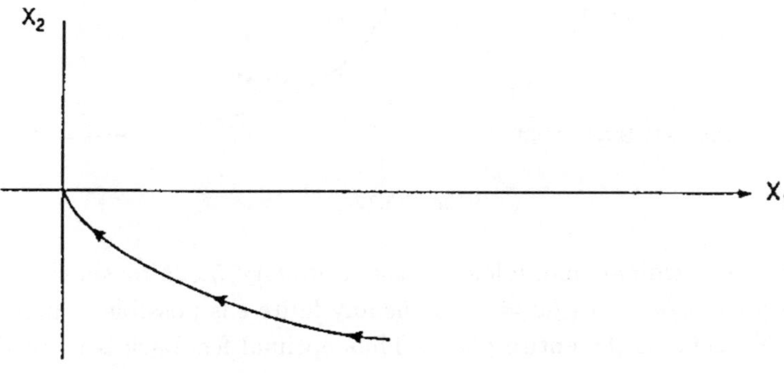



Fig. 5.3.10

In an analogous situation for $u = -1$, $x_2$ is always decreasing so that we must use the upper half of the parabola in Fig. 5.3.11.

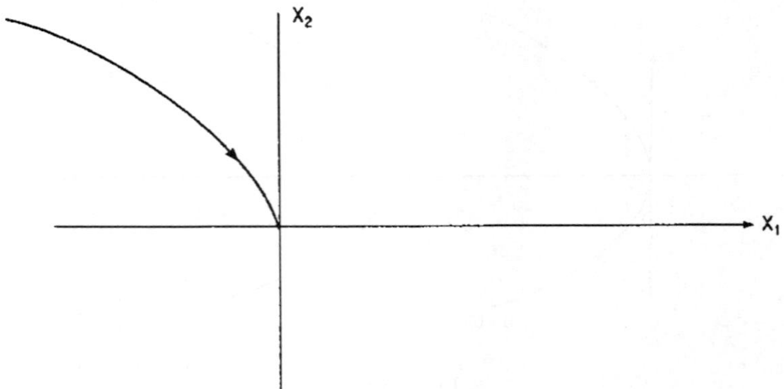


Fig. 5.3.11

The switching locus combines Figs. 5.3.10 and 5.3.11, see Fig. 5.3.12.

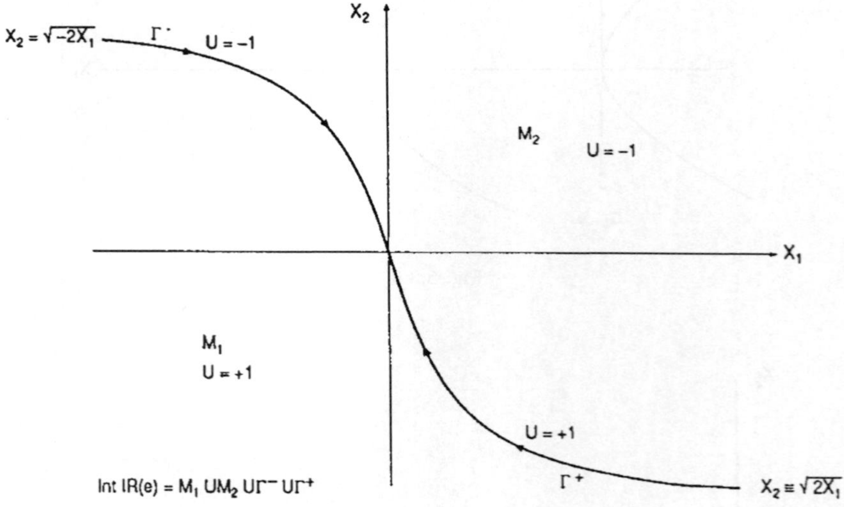


Fig. 5.3.12

To identify the terminal manifolds we use Corollary 5.3.2 to show that for one discontinuity $\epsilon < \pi/\omega = \infty$, $(\omega = 0)$. Thus any finite $\epsilon$ is possible. Thus Int $\mathbb{R}(\epsilon) = M_1 \cup M_2 \cup \Gamma_+ \cup \Gamma_+$ is the entire plane. Thus optimal feedback is defined by

$$f(x_1, x_2) = \begin{cases} -1 & \text{if } (x_1, x_2) \text{ lies on } M_2 \text{ or on } \Gamma_-, \\ +1 & \text{if } (x_1, x_2) \text{ lies on } M_1 \text{ or on } \Gamma_+. \end{cases}$$

Optimal solutions are trajectories of the system $M\ddot{x}(t) = f(x_1, \dot{x}_1)$.

**Example 5.3.3**　Damped Harmonic Oscillator

Many problems in engineering give rise to a damped harmonic oscillator equation. For example, the control surface of a flexible structure that is designed for space exploration is required to rest in a fixed position. If disturbed by an external force, and if nothing is done to restore it, the surface may behave as a damped harmonic oscillator. Let $x(t)$ be a displacement from the desired position. Then the surface satisfies the differential equation

$$m\ddot{x}(t) + 2b\dot{x}(t) + k^2 x = 0 \,,$$

with initial conditions $x(0) = x_0$ and $\dot{x}(0) = \dot{x}_0$. The value $x_0$ is an initial displacement, and $\dot{x}_0$ is the initial velocity imparted by the external force. The aerospace engineers aim to damp out the oscillatory behavior of $x(t)$ by applying a restoring torque $u$ that should bring the surface to rest $(0,0) = (x(t^*), \dot{x}(t^*))$ in minimum time. Thus the dynamics is given by

$$m\ddot{x}(t) + 2b\dot{x}(t) + k^2 x(t) = u(t) \,, \qquad x(0) = x_0 \,, \qquad \dot{x}(0) = \dot{x}_0 \,,$$
$$x(t^*) = 0 \,, \qquad \dot{x}(t^*) = 0 \,, \tag{5.3.17a}$$

where $t^*$ is minimum. We assume that

$$|u(t)| \leq 1 \,. \tag{5.3.17b}$$

The state space representation of (5.3.17a) is

$$\underline{\dot{x}}(t) = A\underline{x}(t) + Bu(t) \,, \qquad \underline{x}(0) = \underline{x}_0 \,, \qquad \dot{x}(0) = \dot{x}_0 \,, \tag{5.3.17b}$$

where

$$A = \begin{bmatrix} 0 & 1 \\ -\omega^2 & -2\alpha \end{bmatrix} \,, \qquad B = \begin{bmatrix} 0 \\ 1/m \end{bmatrix} \,,$$

where $\frac{2b}{m} = 2\alpha$, $\frac{k^2}{m} = \omega^2$. We note that $m$ is the mass, $2b$ is the damping, and $k^2$ is the stiffness. All these constants are positive, so that (5.3.17) is stable. We solve the time-optimal problem using Theorem 5.1.1. We simplify our calculation by setting $k = 1$, $m = 1$, so that

$$A = \begin{bmatrix} 0 & 1 \\ -\omega^2 & -2\alpha \end{bmatrix} \,, \qquad B = \begin{bmatrix} 0 \\ 1 \end{bmatrix} \,.$$

Then (5.3.17) is controllable:

$$\text{rank}[B, AB] = \text{rank} \begin{bmatrix} 0 & 1 \\ 1 & -2\alpha \end{bmatrix} = 2 \,.$$

The system is also strictly normal since $m = 1$. The eigenvalues, $\lambda$, of $A$, are solutions of

$$\det(A - \lambda I) = 0 = \det \begin{bmatrix} -\lambda, & 1 \\ -\omega^2, & -2\alpha - \lambda \end{bmatrix} = \lambda^2 + 2\alpha\lambda + \omega^2 = 0\,,$$

$$\lambda = \frac{-2\alpha \pm \sqrt{4\alpha^2 - 4\omega^2}}{2} = -\alpha \pm \sqrt{\alpha^2 - \omega^2}\,.$$

Hence no eigenvalues of $A$ have a positive real part. The conditions of Theorem 5.1.1 are satisfied, and therefore there exists a unique time-optimal control that drives our system from $x_0$ to $0$ in minimum time. This control is given almost everywhere by

$$u^*(t) = \text{sgn}[c^T e^{-At} B]\,.$$

There are three cases:

### 5.3.1   *The underdamped case $(\alpha^2 - \omega^2 < 0)$*

Let $\beta^2 = \alpha^2 - \beta^2 < 0$. If we let $y(t) = c^T e^{-At}$, then $\dot{y}(t) = -y(t)A$, where

$$A = \begin{bmatrix} 0 & 1 \\ -\omega^2 & -2\alpha \end{bmatrix},$$

$$\begin{bmatrix} \dot{y}_1(t) \\ \dot{y}_2(t) \end{bmatrix}^T = -[y_1(t), -y_2(t)] \begin{bmatrix} 0 & 1 \\ -\omega^2 & -2\alpha \end{bmatrix},$$

$$\dot{y}_1 = \omega^2 y_2\,,$$

$$\dot{y}_2 = -y_1 + 2\alpha y_2\,,$$

which is equivalent to $\ddot{y}_2 - 2\alpha\dot{y}_2 - \omega^2 y_2 = 0$. The characteristic equation (set $y_2 = e^{st}$) is $s^2 - 2\alpha s + \omega^2 = 0$, where

$$s = \alpha \pm \sqrt{\alpha^2 - \omega^2} = \alpha \pm i\beta\,,$$

$$g(t, c) = y_2(t)$$

$$= e^{\alpha t}[c_1 \cos \beta t + c_2 \sin \beta t]$$

$$= D e^{\alpha t} \sin(\beta t + \delta)\,,$$

where $D = [c_1^2 + c_2^2]^{\frac{1}{2}}$ and $\delta = \tan^{-1}(-c_2/c_1)$. Thus the time-optimal control is

$$u^*(t) = \text{sgn}[y_2(t)] = \text{sgn}[D e^{\alpha t} \sin(\beta t + \delta)]\,, \quad D e^{\alpha t} \text{ is positive}\,.$$

Hence

$$u^*(t) = \text{sgn}[\sin(\beta t + \delta)] \,.$$

The zeros are the $t$, such that

$$e^{\alpha t}[c_1 \cos \beta t + c_2 \sin \beta t] = e^{\alpha t}[D \sin(\beta t + \delta)] = 0 \,,$$

i.e., $t$ such that

$$\beta t + \delta = k\pi \,, \qquad (k \text{ integers}) \,, \qquad t = \frac{k\pi - \delta}{\beta} \,,$$

so that controls switch at intervals $\pi/\beta$ after the first switch. We now determine the switching curves for the underdamped case:

$$\dot{x}_1 = x_2 \,,$$

$$\dot{x}_2 = -\omega^2 x_1 - 2\alpha x_2 \pm 1 \,,$$

$$\frac{dx_1}{dx_2} = \frac{x_2}{-\omega^2 x_1 - 2\alpha x_2 \pm 1} \,.$$

If $u = +1$, we obtain

$$(-\omega^2 x_1 - 2\alpha x_2 + 1)dx_1 = x_2 dx_2 \,,$$

$$-\frac{1}{2}\omega^2 x_1^2 - 2\alpha x_1 x_2 + x_1 = \frac{1}{2}x_2^2 + c_1 \,,$$

$$\omega^2 x_1^2 + 4a x_1 x_2 - 2x_1 + x_2^2 = a_1^2 \,,$$

$$\left(\omega x_1 - \frac{1}{\omega}\right)^2 + x_2^2 + 4\alpha x_1 x_2 = a^2 \,.$$

The curve is described in Fig. 5.3.13, and it is a spiral approaching $\left(\frac{1}{\omega^2}, 0\right)$ as $t \to \infty$. For $u = -1$, we deduce the curve

$$\left(\omega x_1 + \frac{1}{\omega}\right)^2 + x_2^2 + 4\alpha x_1 x_2 = a^2 \,.$$

See Fig. 5.3.13. If we let $x_2 = a \sin t$ in the case $u = +1$, then

$$x_1 = \frac{-a \cos t - 2\alpha a \sin t + 1}{\omega^2} \,.$$

In the case $u = -1$,

$$x_1 = \frac{-a \cos t - 2\alpha a \sin t - 1}{\omega^2} \,.$$

In this underdamped case $\beta^2 = \alpha^2 - \omega^2 < 0$,

$$\dot{x}_1 = x_2 \,,$$

$$\dot{x}_2 = -\omega^2 x_1 - 2\alpha x_2 \pm 1 \,,$$

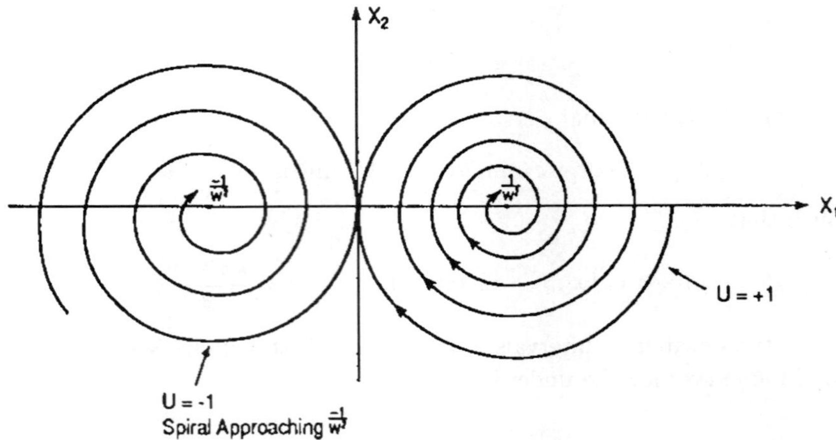


Fig. 5.3.13

$$x_1(t) = e^{-\alpha t}[c_1 \cos \beta t + c_2 \sin \beta t] + 1/k^2 \,,$$

$$x_2(t) = e^{-\alpha t}[\beta c_2 \cos \beta t - \beta c_1 \sin \beta t + 1/k^2 \,.$$

All solutions tend to $(1/k^2, 0)$ as $t \to \infty$. If $t$ is increasing, one of these spirals will be traversed clockwise (cw). If $t$ is decreasing, one of these spirals will be traversed counterclockwise. Let us move backward in time in order to find the switching curves. We know that our optimal control is of the form $u^*(t) = \text{sgn}[\sin(\beta t - \delta)]$. Control switches at intervals of $\pi/\beta$ after the first switch. Let $\beta = 1$. Following the same process as for the undamped oscillator, we see that the switching curves correspond with "unfolding" our spirals corresponding to $u = -1$ and $u = +1$ from the origin along the $x_2$ axis. The resulting switching curves are seen in Fig. 5.3.15.

Let $\omega = 1$. Let $0 < \delta \leq \pi$. Move $-\delta$ seconds counterclockwise around the arc (on our spiral) corresponding to $u = +1$. At $t = -\delta$, $\sin(t + \delta)$ changes sign so our control switches to $u = -1$. With $u = -1$, the optimum trajectory is centered at $(-1, 0)$ and passes through $P_1$. This arc is traversed for $\pi$ seconds ($\pi$ seconds between switches) until on a spiral $P_2$ is reached. At $P_2$, $t = -\delta - \pi$, so $\sin(t + \delta)$ changes sign and the control is once again $u = +1$. With $u = +1$, the optimum trajectory is a spiral centered at $(1, 0)$. If $\delta$ is varied from 0 to $\pi$ seconds along the arc shown in Fig. 5.3.14, the switching curves begin to take shape. The entire switching curve for our problem can be formed by moving backward in time, beginning with $u = +1$ as done above, then beginning with $u = -1$. The switch curves will follow. See Fig. 5.3.15.

We may now define our terminal manifolds by moving backward in time from the origin (considering all possible control sequences as we move back in time). The terminal manifolds, $M_k$'s, are the set of points in Int $\mathbb{R}(\epsilon)$ whose optimal controls have exactly $(n - 1)$ discontinuities.

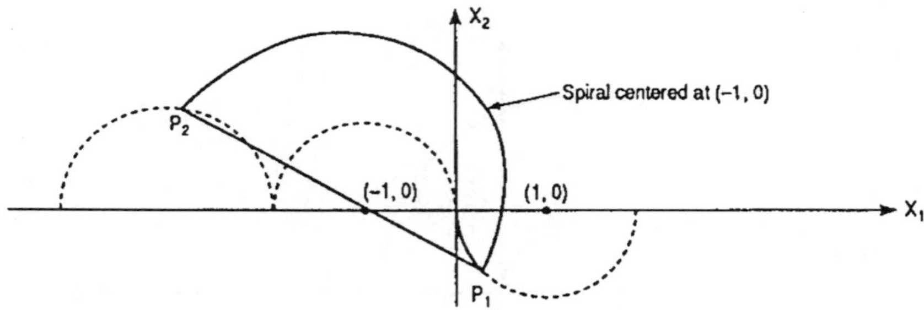


Fig. 5.3.14

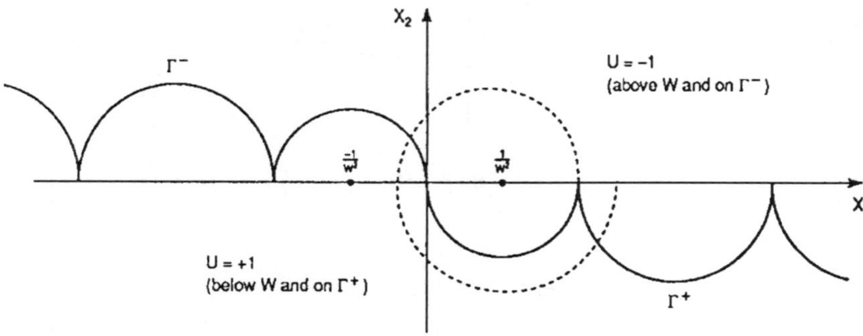


Fig. 5.3.15

The steps for determining terminal manifolds $M_1, M_2$ mirror those for the undamped harmonic oscillator.

See Fig. 5.3.16 for the Terminal Manifolds. The manifolds are shown in Fig. 5.3.17.

From Hájek's Lemma [3], there is for $n-1 = 1$, $(n = 2)$ discontinuity of $\eta^T e^{-At} B$ on $[0, \epsilon)$, $\epsilon < \frac{\pi}{\bar{\omega}}$. Indeed, $\bar{\alpha} + i\bar{\omega}_i$ roots of $|A - \lambda I|$, $\bar{\omega} = \max_i(\bar{\omega}_i)$

$$A = \begin{bmatrix} 0 & 1 \\ -\omega^2 & -2\alpha \end{bmatrix}, \qquad |A - \lambda I| = \begin{vmatrix} -\lambda & 1 \\ -\omega^2 & -2\alpha - \lambda \end{vmatrix} = 0,$$

$$\lambda(2\alpha + \lambda) + \omega^2 = 0, \qquad \lambda^2 + 2\alpha\lambda + \omega^2 = 0, \qquad \lambda = \frac{-2\alpha \pm \sqrt{4\alpha^2 - 4\omega^2}}{2}.$$

So $\bar{\omega} = \sqrt{\omega^2 - \alpha^2}$, $\omega > \alpha$, and $\epsilon < \frac{\pi}{\sqrt{\omega^2 - \alpha^2}}$.

We have

$$\text{Int } \mathbb{R}(\epsilon) = M_1 \cup M_2 \cup \Gamma_{+1} \cup \Gamma_{-1}.$$

See Fig. 5.3.17.

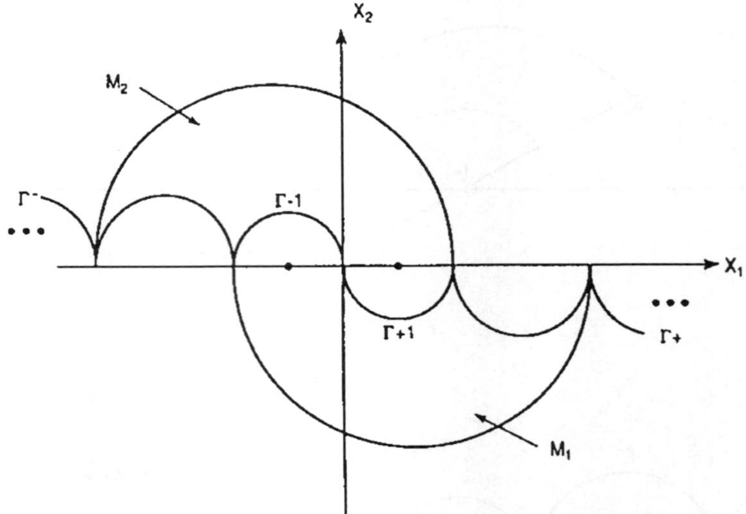

Fig. 5.3.16

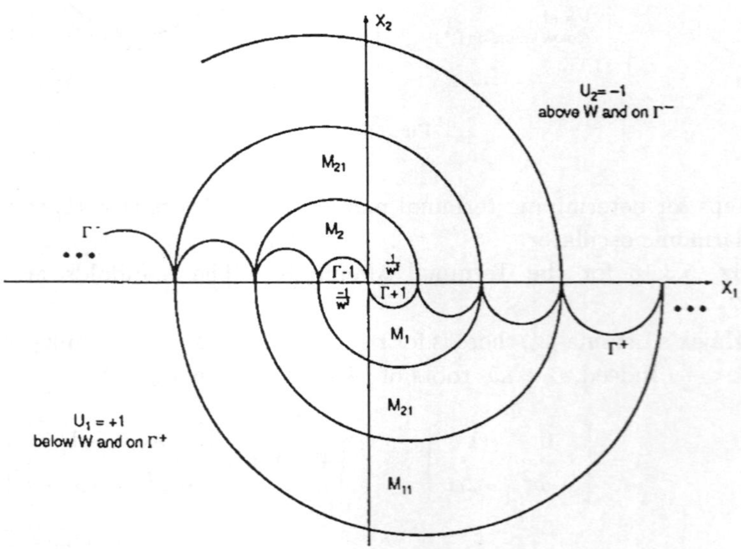

Fig. 5.3.17

So our optimal feedback control will be defined as

$$f(x_1, x_2) = \begin{cases} -1 & \text{if } (x_1, x_2) \text{ lies on } M_2 \text{ or on } \Gamma^-, \\ +1 & \text{if } (x_1, y_2) \text{ lies on } M_1^2 \text{ or on } \Gamma^+. \end{cases}$$

Our optimal trajectory is simply solutions of the equation

$$\ddot{x}_1 + \frac{c}{m}\dot{x}_1 + \frac{k}{m}x_1 = f(x_1, \dot{x}_1),$$

or

$$\ddot{x} + \frac{c}{m}\dot{x} + \frac{k}{m}x = f(x, \dot{x})$$

for underdamped system $\beta^2 = \alpha^2 - \omega^2 < 0$, $2\pi = \frac{c}{m}\omega^2 = \frac{k}{m}$.

### 5.3.2 *Critically damped system $(\alpha^2 - \omega^2 = 0)$*

In this case, $s = 0$

$$\ddot{y}_2 - 2\alpha\dot{y}_2 - \omega^2 y_2 = 0,$$

$$y_2(t) = e^{\alpha t}[c_1 + c_2 t],$$

$$\dot{y}_1(t) = \omega^2 y_2(t) = \omega^2 e^{\alpha t}[c_1 + c_2 t],$$

$$\dot{y}_2(t) = \alpha e^{\alpha t}[c_1 + c_2 t] + c_2 e^{\alpha t},$$

$$y_1(t) = -\dot{y}_2(t) + 2\alpha y_2(t) = -\dot{y}_2(t) + 2\alpha e^{\alpha t}[c_1 + c_2 t],$$

$$y_1(t) = \alpha e^{\alpha t}[c_1 + c_2 t] - c_2 e^{\alpha t}.$$

Because $B = \begin{bmatrix} 0 \\ 1 \end{bmatrix}$, we have

$$g(t, c) = c^T e^{-At} B = y(t)B = y_2(t) = e^{\alpha t}[c_1 + c_2 t].$$

Optimal control is given by $u^*(t) = \text{sgn}[y_2(t)]$. There is at most one switch when $c_1 + c_2 t = 0$, i.e., $t = -\frac{c_1}{c_2}$. To obtain the trajectory,

$$\dot{x}_1 = x_2, \qquad \dot{x}_2 = -\omega^2 x_1 - 2\alpha x_2 \pm 1.$$

This passes through the origin if $(u = -1)$,

$$\frac{dx_1}{dx_2} = \frac{x_2}{-\omega^2 x_1 - 2\alpha x_2 - 1} = \frac{x_2}{-\alpha^2 x_1 - 2\alpha x_2 - 1}, \qquad (\alpha^2 = \omega^2).$$

The trajectory describes the curve

$$\left(\alpha x_1 + \frac{1}{\alpha}\right)^2 + x_2^2 + 4\alpha x_1 x_2 = \frac{1}{\alpha^2}.$$

This corresponds to $\Gamma_-$ in Fig. 5.3.18.

For $u = 1$,

$$\left(\alpha x_1 - \frac{1}{\alpha}\right)^2 + x_2^2 + 4\alpha x_1 x_2 = 1/\alpha^2.$$

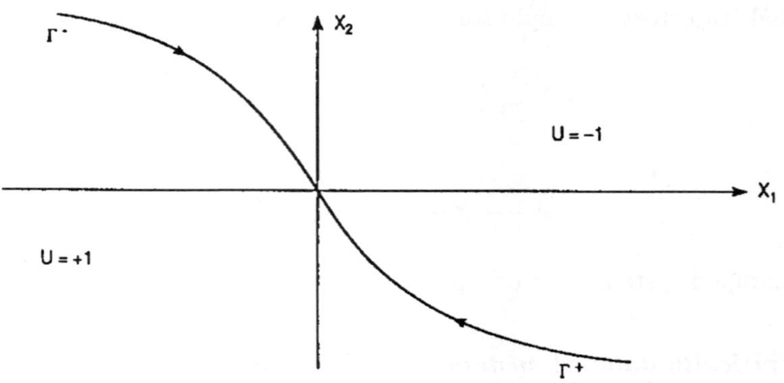


Fig. 5.3.18

This corresponds to $\Gamma_+$ in Fig. 5.3.18. For the overdamped case $\alpha^2 - \omega^2 > 0$,

$$s = -\alpha \pm \beta < 0\,,$$

$$y_2(t) = e^{\alpha t}[c_1 e^{\beta t} + c_2 e^{-\beta t}]\,.$$

This is at most one switch when $t = -\frac{1}{2\beta}\ln(-c_1/c_2)$. (Note that Hájek's Lemma yields $\epsilon = \pi/0 = \infty$).

For $u = +1$,

$$\left(\omega x_1 - \frac{1}{\omega}\right)^2 + x_2^2 + 4\alpha x_1 x_2 = \frac{1}{\omega^2}\,.$$

This corresponds to $\Gamma_+$ in Fig. 5.3.18.

For $u = -1$,

$$\left(\omega x_1 + \frac{1}{\omega}\right)^2 + x_2^2 + 4\alpha x_1 x_2 = \frac{1}{\omega^2}\,.$$

This corresponds to $\Gamma_-$ in Fig. 5.3.18.

The terminal manifolds are described in Fig. 5.3.19. The curves in both cases are hyperbolas.

**Example 5.3.4**　Consider the system

$$\dot{x}_1 = x_2 + u_1\,, \qquad \dot{x}_2 = -x_1 + u_2\,, \qquad |u_i| \le 1\,, \qquad i = 1, 2\,,$$

which is equivalent to $\dot{x} = Ax + Bu$, where

$$A = \begin{bmatrix} 0 & 1 \\ -1 & 0 \end{bmatrix}, \qquad B = \begin{bmatrix} 1 & 0 \\ 0 & 1 \end{bmatrix}\,.$$

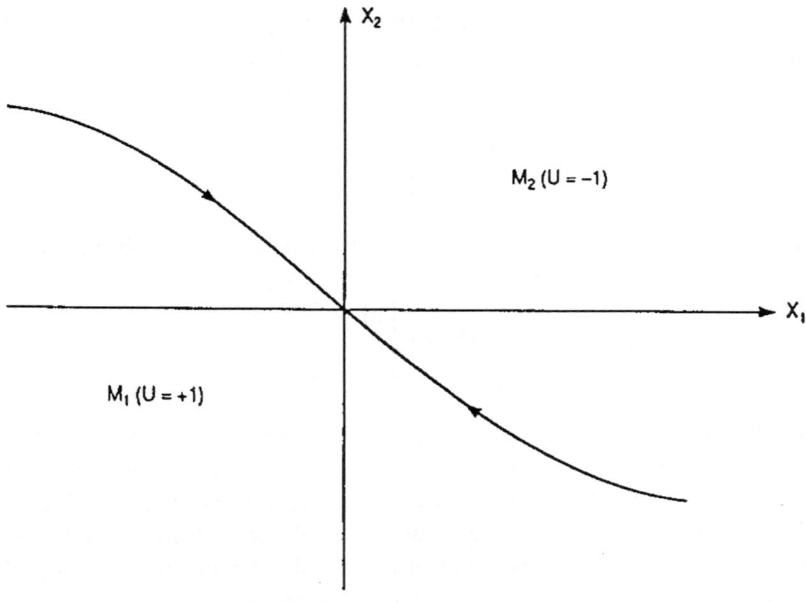


Fig. 5.3.19

The system is normal and controllable since

$$\operatorname{rank}[B, AB] = \operatorname{rank} \begin{bmatrix} 1 & 0 & 0 & 1 \\ 0 & 1 & -1 & 0 \end{bmatrix} = 2,$$

and

$$\operatorname{rank}[b_j, Ab_j] = 2, \qquad j = 1, 2.$$

It is also strictly normal. For any integers $r_j \geq 0$ with $2 = \sum_{j=1}^{2} r_j$, $(n = 2,\ m = 2)$, the vectors

$$A^i b_j, \qquad j = 1, 2, \qquad 0 \leq i \leq r_{j-1},$$

are linearly independent. (If $r_j = 0$, then there are no terms $A^i b_j$.) We have the following choices:

| $r_1$ | $r_2$ |
|---|---|
| 0 | 2 |
| 1 | 1 |
| 2 | 0 |

$r_1 = 0$ implies there are no terms in $A^i b_1$, since $r_1 - 1 = -1$ and $0 \leq i \leq r_{j-1}$. Also $r_1 = 1$ implies $i = r_1 - 1 = 0$. $r_1 = 1$ implies $r_2 = 1$, so $[A^i b_1, A^i b_2] =$

$[A^0 b_1, A^0 b_2] = [b_1, b_2]$, ($A^0$ is the identity matrix), $[b_1, b_2] = \begin{bmatrix} 1 & 0 \\ 0 & 1 \end{bmatrix}$, and this has rank 2. If $r_1 = 2$, $i = 0, 1$ and $r_2 = 0$; and we have that

$$[b_1, Ab_1] = \begin{bmatrix} 1 & 0 \\ 0 & -1 \end{bmatrix}.$$

This matrix has rank 2. If $r_2 = 2$, then $r_1 = 0$ and the corresponding vectors are

$$[b_2, Ab_2] = \begin{bmatrix} 0 & 1 \\ 1 & 0 \end{bmatrix}$$

with rank 2. We note that $r_2 = 0$ implies no terms in $A^i b_2$, since $r_2 - 1 = -1$ and $0 \le i \le r_j - 1$. If $r_2 = 1$, $i = r_2 - 1 = 0$, and $r_1 = 1$. An earlier argument yields linear independence. We have verified the strict normality condition. The equation $\dot{x} = Ax$ has eigenvalues $\lambda = \pm i$, which has zero real parts. All the conditions of the theorem are satisfied: There exists an optimal feedback control $f(x_1, x_2)$, which we shall now determine. The open loop control is given as

$$u^*(t) = \mathrm{sgn}[c^T e^{-At} B], \qquad 0 \le t \le t^*,$$

where $t^*$ is the minimum time. For our system

$$e^{At} = \begin{bmatrix} \cos t & \sin t \\ -\sin t & \cos t \end{bmatrix}, \qquad e^{-At} = \begin{bmatrix} \cos t & -\sin t \\ \sin t & \cos t \end{bmatrix},$$

$$c^T e^{-At} B = c^T e^{-At} = \begin{bmatrix} c_1 \cos t + c_2 \sin t \\ -c_1 \sin t + c_2 \cos t \end{bmatrix} = A \begin{bmatrix} \sin(t + \delta) \\ \cos(t + \delta) \end{bmatrix},$$

where $A = \sqrt{c_1^2 + c_2^2}$, $\delta = \tan^{-1}(-c_2/c_1)$. It follows that

$$u^*(t) = \mathrm{sgn} \begin{bmatrix} \sin(t + \delta) \\ \cos(t + \delta) \end{bmatrix}.$$

Because $\cos a$ and $\sin a$ differ by $\pi/2$, the sign changes of the optimal control components are $\pi/2$ apart. Each component switches every $\pi$ seconds. The control $u^*$ is bang-bang and are the extreme values

$$\left\{ \begin{bmatrix} 1 \\ 1 \end{bmatrix} \begin{bmatrix} 1 \\ -1 \end{bmatrix} \begin{bmatrix} -1 \\ 1 \end{bmatrix} \begin{bmatrix} -1 \\ -1 \end{bmatrix} \right\}$$

of the 2-dimensional unit cube. We now construct $f(x_1, x_2)$.

**Case 1:** $u = \begin{bmatrix} 1 \\ 1 \end{bmatrix}$.

$$\left.\begin{aligned} \dot{x}_1 &= x_2 + 1 \\ \dot{x}_2 &= -x_1 + 1 \end{aligned}\right\} \Rightarrow \frac{dx_1}{dx_2} = \frac{x_2 + 1}{-x_1 + 1},$$

$$(-x_1 + 1)dx_1 = (x_2 + 1)dx_2,$$

integrating both sides, we get

$$-\frac{1}{2}x_1^2 + x_1 = \frac{1}{2}x_2^2 + x_2 + c_1,$$

$$x_1^2 - 2x_1 + 1 + 1 + 2x_2 + x_2^2 = a^2,$$

$$(x_1 - 1)^2 + (x_2 + 1)^2 = a^2.$$

The equation describes a circle of radius $a$ centered at $(1, -1)$ in the phase plane.

**Case 2:** $u = \begin{bmatrix} 1 \\ -1 \end{bmatrix}$.

$$\left.\begin{aligned} \dot{x}_1 &= x_2 + 1 \\ \dot{x}_2 &= -x_1 - 1 \end{aligned}\right\} \Rightarrow \frac{dx_1}{dx_2} = \frac{x_2 + 1}{-x_1 - 1},$$

$$(-x_1 - 1)dx_1 = (x_2 + 1)dx_2,$$

so that

$$-\frac{1}{2}x_1^2 - x_1 = \frac{1}{2}x_2^2 + x_2 + c_1, \qquad (x_1 + 1)^2 + (x_2 + 1)^2 = a^2,$$

which is a circle of radius $a$ centered at $(-1, -1)$.

**Case 3:** $u = \begin{bmatrix} -1 \\ 1 \end{bmatrix}$.

$$\left.\begin{aligned} \dot{x}_1 &= x_2 - 1 \\ \dot{x}_2 &= -x_1 + 1 \end{aligned}\right\} \Rightarrow \begin{aligned} \frac{dx_1}{dx_2} &= \frac{x_2 - 1}{-x_1 + 1}, \quad (-x_1 + 1)dx_1 = (x_2 - 1)dx_2, \\ -\frac{1}{2}x_1^2 + x_1 &= \frac{1}{2}x_2^2 - x_2 + c_1(x_1 - 1)^2 + (x_2 - 1)^2 = a^2, \end{aligned}$$

which is a circle of radius $a$ centered at $(1, 1)$.

**Case 4:** $u = \begin{bmatrix} -1 \\ -1 \end{bmatrix}$.

$$\left.\begin{aligned} \dot{x}_1 &= x_2 - 1 \\ \dot{x}_2 &= -x_1 - 1 \end{aligned}\right\} \Rightarrow (x_1 - 1)^2 + (x_2 + 1)^2 = a^2.$$

This is a circle of radius $a$ centered at $(-1, 1)$.

### 5.3.3   *Forming the switching curves*

With increasing time $t$, the circles described on the previous page are traversed clockwise; with $t$ decreasing, the circles are traversed counterclockwise. Let us move backward in time to find the switching curve.

Let $\delta_1 > \frac{\pi}{2} > \delta_2 > 0$. Suppose our final control is $u_a = \begin{bmatrix} -1 \\ 1 \end{bmatrix}$. Move $-\delta_2$ seconds counterclockwise to $P_1$ around the arc corresponding to $u_1 = \begin{bmatrix} -1 \\ 1 \end{bmatrix}$ (circle centered at $(1, 1)$ that passes through the origin). Since we are moving backward in time, our only possible control is $u_b = \begin{bmatrix} -1 \\ -1 \end{bmatrix}$, because $\delta_1 > \pi/2 > \delta_2 > 0$ and $u^*(t) = \text{sgn}\begin{bmatrix} \sin(t - \delta_1) \\ \sin(t - \delta_2) \end{bmatrix}$. The control $u_b = \begin{bmatrix} -1 \\ -1 \end{bmatrix}$ corresponds to a circle centered at $(-1, 1)$ and passes through $P_1$ in the phase plane. We now traverse this circle counterclockwise for $\pi/2$ seconds ($\delta_1 - \delta_2 = \pi/2$). Traversing the circle $\pi/2$ seconds corresponds to moving a quarter circle counterclockwise from $P_1$ to $P_2$. At $P_2$, $t = -\delta_2 - \pi/2 (= -\delta_1)$, so $\sin(t - \delta_1)$ changes sign and the control will be $u_c = \begin{bmatrix} 1 \\ -1 \end{bmatrix}$, circle centered at $(-1, -1)$. See Fig. 5.3.20.

Looking at Fig. 5.3.20, if we vary $\delta_2$ from $t = 0$ to $t = \pi/2$ seconds, we see our switching curves for our problem take shape. The remaining switching curves for our problem can be formed as before by moving backward in time, starting with all possible controls. The switching curves are shown below in Fig. 5.3.21.

### 5.3.4   *Terminal manifolds and switching locus*

Recall the definition. For each $k = 1, \dots, n$, let $M_k$ be the set of points in Int $\mathbb{R}(\epsilon)$ whose corresponding optimal controls have exactly $k-1$ discontinuities. Set $M_0 = [0]$. For each $k = 1, \dots, n$ and each $k$-tuple $\underline{i} = \{u_1 \to \cdots \to u_k\}$ of vertices of $U$ with $u_{j-1} \neq u_j$, let $M_{k\underline{i}}$ be the set of points in $\mathbb{R}(\epsilon)$ whose optimal controls have $\{u_1 \to \cdots \to u_k\}$ as the optimal switching sequence ($M_{k\underline{i}}$ may be void; the indices $\underline{i}$ range over a finite set $I_k$). We may also say that the optimal switching sequence $\{u_1 \to \cdots \to u_k\}$ corresponds to $M_{k\underline{i}}$. The $M_{k\underline{i}}$ are called *terminal manifolds.*

It follows from the uniqueness of optimal controls that the terminal manifolds are disjoint.

For $k = n$ there are precisely $2m^{n-1}$ nonvoid sets $M_{k\underline{i}}$. The switching curves for this problem is below.

We may define our terminal manifolds by moving backward in time from the origin (considering all possible control sequences as we move backward in time). The terminal manifolds, $M_k$'s, are the set of points in Int $\mathbb{R}(\epsilon)$ whose optimal controls have exactly $(n-1)$ discontinuities ($n-1 = 2 - 1 = 1$ for this case). Suppose our

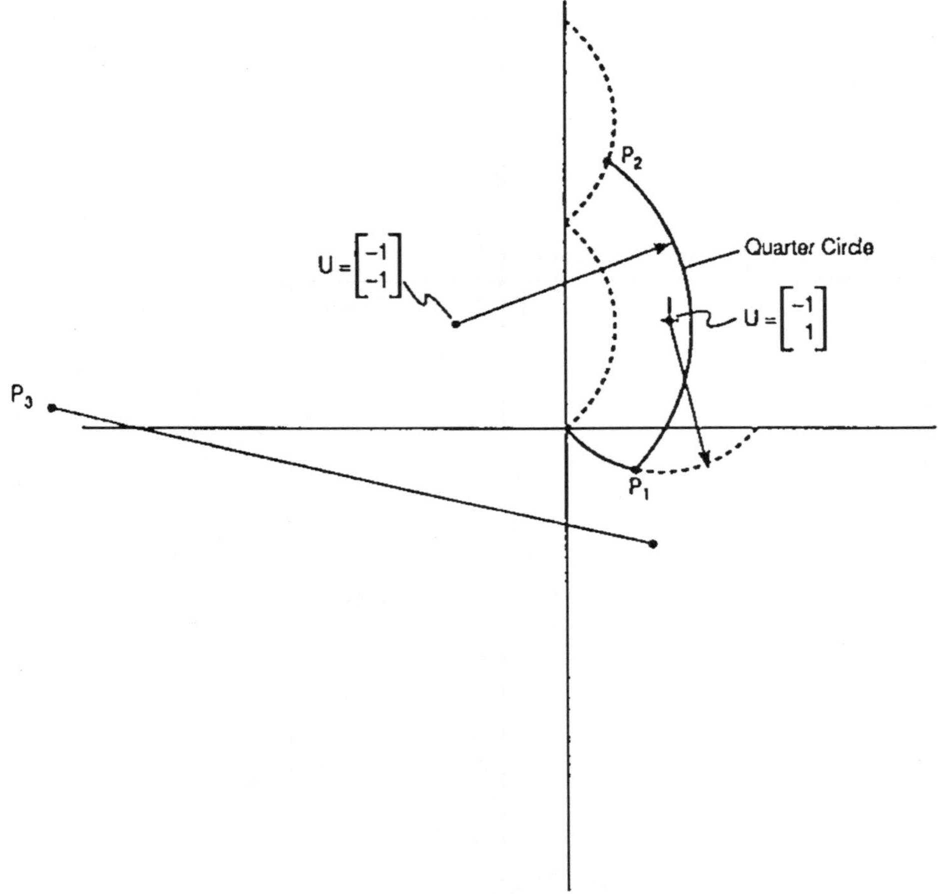


Fig. 5.3.20

final control is $u_c$ ($\Gamma^{+2}$ in Fig. 5.3.22). If we move backward in time (ccw) in our phase plane, we can describe the terminal manifold $M_2$. If our final control is $u_c$, the only possible control sequence as we approach the origin is $\{u_d \to u_c\}$. Control $u_d = \begin{bmatrix} 1 \\ -1 \end{bmatrix}$ corresponds to a circle centered at $(-1, -1)$; so we vary the radius of this curve from $\sqrt{2}$ to $\sqrt{10}$ (and move backward in time (ccw) until we reach the next switching curve) to describe that particular manifold. See Fig. 5.3.23.

Using the definition, we see that $M_2$ is defined above. Once the system is within $M_2$, there is exactly $n - 1 = 1$ discontinuity before reaching the origin. Terminal manifolds $M_3$, $M_4$, and $M_1$ are constructed in a similar manner but with control sequences $\{u_c \to u_b\}$, $\{u_b \to u_a\}$, and $\{u_a \to u_d\}$ respectively. See Fig. 5.3.24 for terminal manifolds.

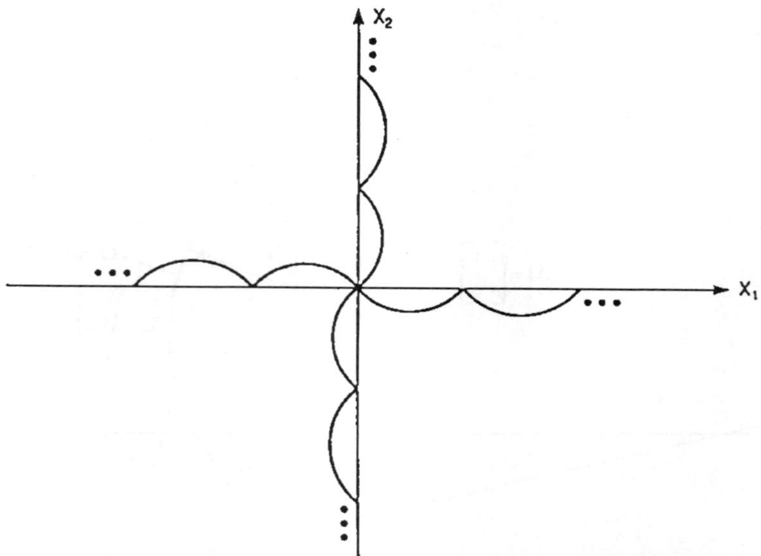

Fig. 5.3.21    Switching curves.

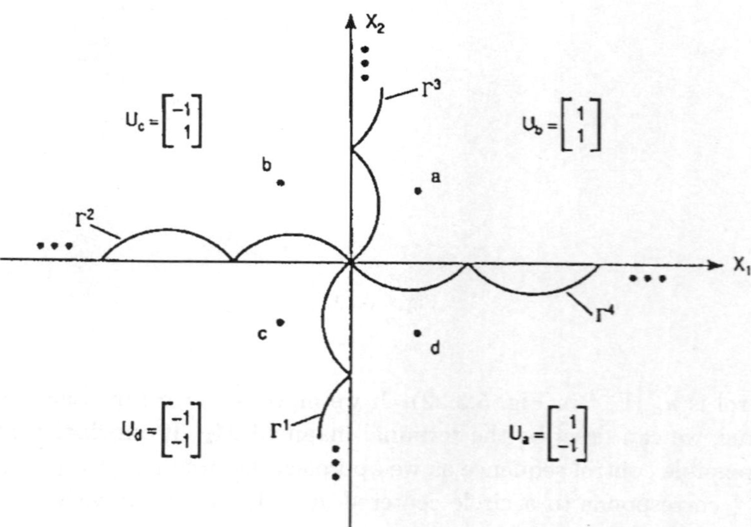


Fig. 5.3.22

Now we may define the other manifolds in our phase plane, by moving backward in time (ccw) from our terminal manifolds $M_k$'s. Suppose our final control sequence is $\{u_d \to u_c\}$ (terminal manifold $M_2$). If we move backward in time from $M_2$, the only possible control is $u_a$. So our optimal control sequence will be $\{u_a \to u_d$

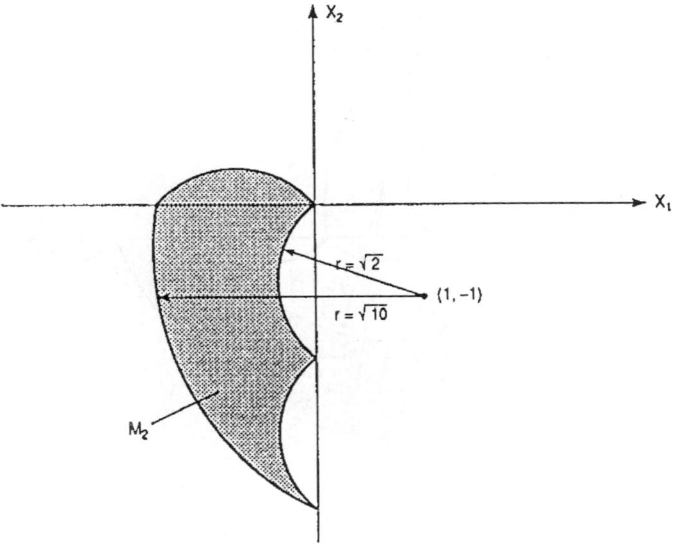


Fig. 5.3.23   Terminal manifold $M_2$.

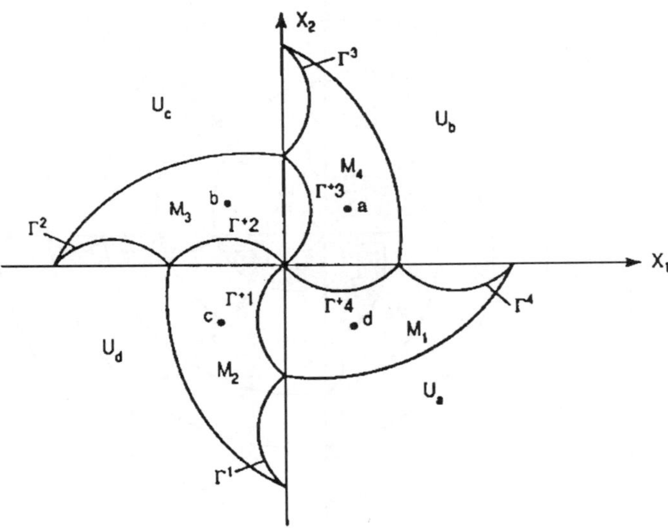


Fig. 5.3.24

$\rightarrow u_c\}$. Now we will construct the corresponding manifold. The control $u_a = \begin{bmatrix} 1 \\ -1 \end{bmatrix}$ corresponds to a circle centered at $(1, 1)$ in the phase plane. So if we vary the radius of this circle (moving backward in time, ccw) from $r = \sqrt{10}$ to $r = \sqrt{26}$, until we reach our switching curve again, the manifold will be defined. See Fig. 5.3.25.

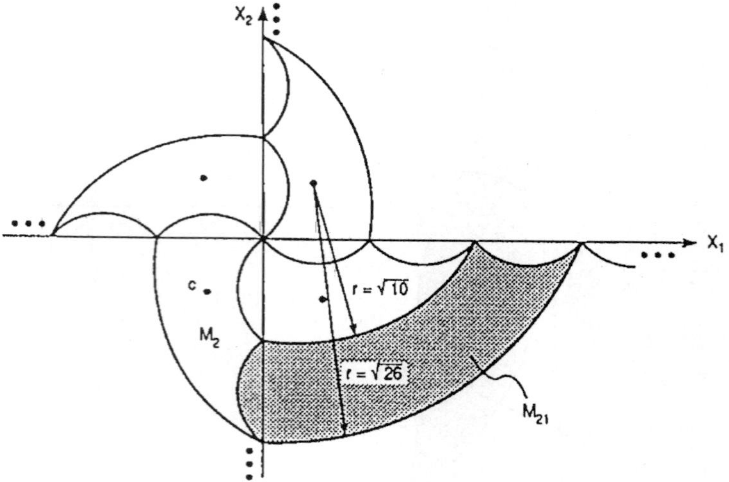

Fig. 5.3.25    Manifold $M_{21}$.

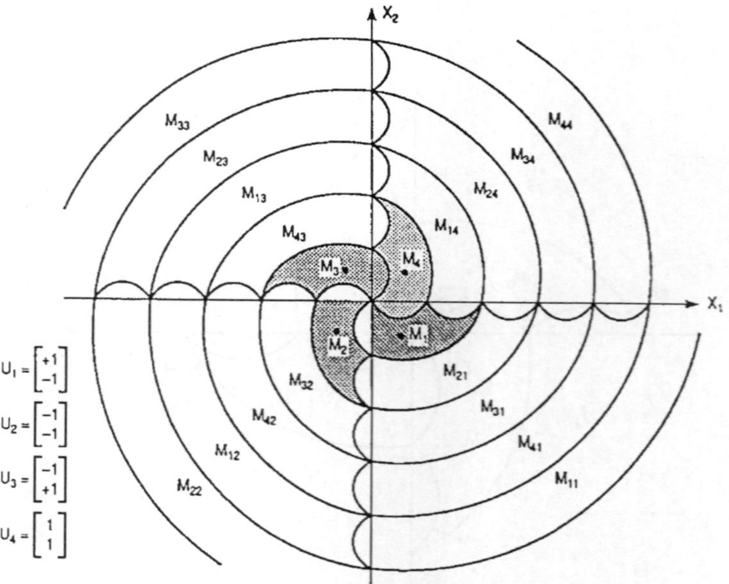

Fig. 5.3.26    Manifolds.

    The optimal control sequence of the system once it reaches $M_{21}$ in Fig. 5.3.25 is $\{u_1 \to u_d \to u_c\}$. The rest of the manifolds may be found in a similar manner. The result is shown in Fig. 5.3.26.

Since the system is strictly normal, there is $\epsilon > 0$ such that in Int $\mathbb{R}(\epsilon)$, $g(t) = c^T e^{-At} B$ has at most one discontinuity. Clearly $\epsilon < \pi$. Also Int $\mathbb{R}(\epsilon) = \bigcup_{j=1}^{4} M_j \cup \Gamma_{+j}$.

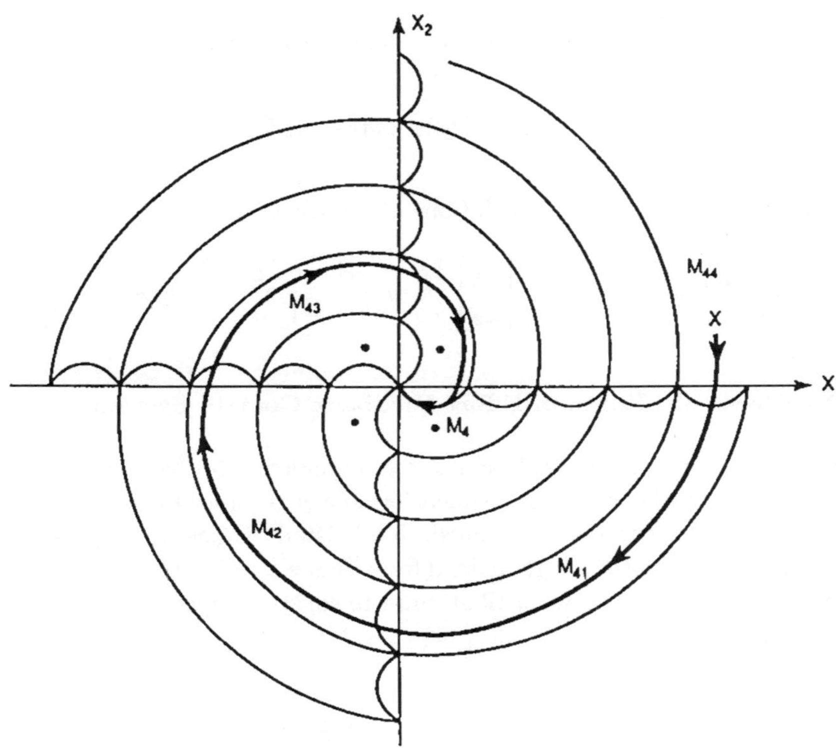


Fig. 5.3.27

Now, since we have our terminal manifolds described, our time-optimal feedback control is also described. As an example, suppose our initial conditions place the states in $M_{44}$ as shown by the "x" in Fig. 5.3.27. The trajectory of our states and the time-optimal feedback control necessary to reach the origin is known.

As seen from Fig. 5.3.27, our control will be $u_4 \to u_1 \to u_2 \to u_3 \to u_4 \to u_1$. The manifolds traversed will be $M_{44}$, $M_{41}$, $M_{42}$, $M_{43}$, $M_4$. Our control switches as we switch manifolds. Our final switch in control, once we are within the $\epsilon$-neighborhood of the origin, will be from $u_4$ to $u_1$.

Our time-optimal feedback control for this problem will be defined as:

$$
f(x_1, x_2) = \begin{cases}
\begin{bmatrix} 1 \\ -1 \end{bmatrix} & \text{if } x_1,\, x_2 \text{ lie between } \Gamma^4 \text{ and } \Gamma^1 \text{ or on } \Gamma^1\,, \\[2ex]
\begin{bmatrix} -1 \\ -1 \end{bmatrix} & \text{if } x_1,\, x_2 \text{ lie between } \Gamma^1 \text{ and } \Gamma^2 \text{ or on } \Gamma^2\,, \\[2ex]
\begin{bmatrix} -1 \\ 1 \end{bmatrix} & \text{if } x_1,\, x_2 \text{ lie between } \Gamma^2 \text{ and } \Gamma^3 \text{ or on } \Gamma^3\,, \\[2ex]
\begin{bmatrix} 1 \\ 1 \end{bmatrix} & \text{if } x_1,\, x_2 \text{ lie between } \Gamma^3 \text{ and } \Gamma^4 \text{ or on } \Gamma^4\,.
\end{cases}
$$

Our optimal trajectory is simply solutions of the equation

$$
\underline{\dot{x}} = A\underline{x} + Bf(x_1, x_2),\quad \underline{x} = \begin{bmatrix} x_1 \\ x_2 \end{bmatrix},\quad A = \begin{bmatrix} 0 & 1 \\ -1 & 0 \end{bmatrix},\quad B = \begin{bmatrix} 1 & 0 \\ 0 & 1 \end{bmatrix}.
$$

## 5.4   Synthesis of Minimum-Effort Feedback Control Systems

In this section we present the solution of the minimum-effort/energy control problem. For unrestrained controls, an explicit formula is given for the optimal feedback control. When the controls are restrained, the Neustadt/Hájek [4, 7] open-loop solution is presented. With the insight gained from Yeung's optimal feedback solution of the minimum-time problem, we shall attempt to construct optimal feedback control of the system.

Define the set $\mathcal{U}$ as follows:

$$
\mathcal{U} = \{u : [0, T] \to E^m,\ u \text{ measurable and integrable}\}\,.
$$

Suppose $J : \mathcal{U} \to E$, then $J(u(T))$ is the effort associated with $u(t)$. The minimum-effort problem is stated as follows:

$$
\text{Minimize } J(u(T))_{u \in \mathcal{U}}\,, \tag{5.4.0}
$$

$$
\text{subject to } \dot{x}(t) = Ax(t) + Bu(t)\,, \tag{5.4.1}
$$

$$
\text{where } x(0) = x_0\,,\quad x(T, x_0, u) = x_1\,. \tag{5.4.2}
$$

Here $x(\cdot, x_0, u)$ is the solution of (5.4.1) with $x(0, x_0, u) = x_0$. Implicit in this problem is the assumption that with $x_0, x_1 \in E^n$ arbitrarily given, there is an appropriate $T$ such that some control $u$ transfers $x$ from $x_0$ to $x_1$ in time $T$; that is, the solution satisfies $x(0, x_0, u) = x_0$ and $x(T, x_0, u) = x_1$. Thus we assume that system (5.4.1) is controllable on the interval $[0, T]$ with controls as specified.

We now isolate several cost functions $J(u(T))$ that describe efforts to be minimized, and then state the solutions. A general proof of these theories will be discussed in Sec. 5.5.

**Theorem 5.4.1** *Consider Problem (5.4.0) subject to (5.4.1) and (5.4.2), where*

$$J_0(u(T)) = \int_0^T u(t)^+ R(t) u(t) dt \,, \tag{5.4.3}$$

*and where $R(t)$ is a positive definite $m \times n$ matrix. (Here $+$ denotes transpose.) Suppose*

$$\mathrm{rank}[B, AB, \ldots, A^{n-1}B] = n \,. \tag{5.4.4}$$

*Let*

$$W(T) = \int_0^T (e^{-At}B)R^{-1}(t)B^+[e^{-At}]^+ dt \,. \tag{5.4.5}$$

*Then the control*

$$u^*(t) = R^{-1}(t)B^+(e^{-At})^+ W^{-1}(T)q, \ t \in [0, T] \,, \tag{5.4.6}$$

*where*

$$q = (e^{-AT}x_1 - x_0) \tag{5.4.7}$$

*is the optimal one.*

**Remark 5.4.1** Note carefully that there is no constraint on the controls. When $R(t) \equiv I$, the identity matrix, then

$$J_0(u(T)) = \int_0^T |u(t)|^2 dt \,, \tag{5.4.8}$$

and the following corollary is valid:

**Corollary 5.4.1** *Consider problem (5.4.0) with $J$ in (5.4.8) subject to (5.4.1) and (5.4.2). The optimal control is*

$$u^*(t) = B^+(e^{-At})^+ M^{-1}q \,, \tag{5.4.9}$$

*where*

$$M = \int_0^T e^{-At}BB^+(e^{-At})^+ dt \,, \tag{5.4.10}$$

$$q = (e^{-AT}x_1 - x_0) \,.$$

**Proof**   It is quite standard that condition (5.4.4) is equivalent to the nonsingularity of $W(T)$, so that (5.4.5) is well defined. Insert (5.4.5) into the solution of (5.4.1) that is given by the variation of parameter

$$x(t, x_0, u) = e^{At}\left[x_0 + \int_0^t e^{-As}Bu(s)ds\right]. \tag{5.4.11}$$

Then

$$x(T, x_0, u) = e^{At}\left[x_0 + \int_0^T e^{-As}B \cdot R^{-1}(s) \cdot B^+(e^{-As})^+ W^{-1}[e^{-AT}x_1 - x_0]\right]ds$$

$$= e^{AT}[x_0 + e^{-AT}x_1 - x_0] = x_1.$$

Thus $u^*$ transfers $x_0$ to $x_1$ in time $T$. Suppose $u(t)$ is any other control that transfers $x_0$ to $x_1$ in time $T$. We shall prove that

$$J_0(u^*(T)) \leq J_0(u(T)), \tag{5.4.12}$$

and also that

$$J(u^*(T)) = (q, W^{-1}q), \tag{5.4.13}$$

where $(\cdot, \cdot)$ denotes inner product. Because

$$x(T, x_0, u) = x(T, x_0, u^*),$$

we have

$$\int_0^T e^{-As}Bu(s)ds = \int_0^T e^{-As}Bu^*(s)ds. \tag{5.4.14}$$

If we subtract both sides and use inner product, we obtain

$$\left(\int_0^T e^{-As}B[u(s) - u^*(s)]ds, \ W^{-1}(T)[e^{-AT}x_1 - x_0]\right) = 0. \tag{5.4.15}$$

We use (5.4.6) and the properties of inner product to obtain

$$\int_0^T (u(s) - u^*(s), \quad u^*(s))ds = 0. \tag{5.4.16}$$

Using (5.4.16) and some easy manipulation, one deduces that indeed

$$\int_0^T u(t)^+ R(t)u(t)dt \geq \int_0^T u^{*+}R(t)u^*(t)dt.$$

But $J(u^*(T)) = \int_0^T u^{*+}(t)R(t)u^*(t)dt$, so that

$$J_0(u^*(T)) = \int_0^T \|R^{-1}(t)B^+(e^{-At})^+ W^{-1}(T)q\|^2 dt, \tag{5.4.17}$$

where $q = [e^{-AT}x_1 - x_0]$. Since $W(T)$ is symmetric, (5.4.17) yields

$$J_0(u^*(T)) = \int_0^T (q, W^{-1}(T)) \cdot [e^{-As}B]R^{-1}(s)[e^{-As}B]^+ W^{-1}(T)q)ds$$

$$= (q, W^{-1}q).$$

The optimal solution given in Theorem 5.4.1 is a feedback one. It is global. But the controls are "big" and unrestrained. When the effort function is defined to correspond to the "maximum thrust" available, there are hard limits that bound the controls. This situation is treated in the next theorem. $\qquad\square$

**Theorem 5.4.2**  *Consider the problem (5.4.0) with*

$$J_1(u(T)) = \max_{1 \le j \le m} \sup_{0 \le t \le T} |u_j(t)|, \tag{5.4.18}$$

*where*

$$u \in U_1 = \{u \in E^m,\ u\ measurable,\ |u_j(t)| \le 1,\ j = 1,\ldots,m, 0 \le t \le T\}, \tag{5.4.19}$$

*subject to (5.4.1) and (5.4.2), with $x_1 \equiv 0$. Define the function*

$$y(t) = e^{-At}x_1(t) - x_0, \tag{5.4.20}$$

$$g(t,c) = c \cdot e^{-At}B, \tag{5.4.21}$$

*where $c$ is an $n$-dimensional row vector, and $g$ the index of the system (5.4.1). Let*

$$F(c,T) = \sum_{j=1}^m \int_0^T |g_j(t,c)|dt. \tag{5.4.22}$$

*Assume that:*

  (i) $\mathrm{rank}[B, AB, \ldots, A^{n-1}B] = n$. $\hfill (5.4.23)$
  (ii) *For each $j = 1, \ldots, m$, the vectors*

$$\{b_j, Ab_j, \ldots, A^{n-1}b_j\} \tag{5.4.24}$$

   *are linearly independent.*
  (iii) *No eigenvalue of $A$ has a positive real part.*

*Then for each $x_0$, and some response time $T$, there exists a minimum-effort control that transfers $x_0$ to 0 in time $T$. If $y(T) \ne 0$, i.e., $x_0 \ne 0$, the minimum-effort $J_{1\,min} = J_1(u^*(T))$ is given by*

$$\frac{1}{J_{1\,min}} = \min_{c \in P} F(c,T), \tag{5.4.25}$$

*where $P$ is the plane $c \cdot y(T) = 1$. Furthermore, the optimal control $u^*(t)$ is unique almost everywhere and is given by*

$$u^*(t) = J_{1\,min}\ \mathrm{sgn}\ g(t,c^*), \tag{5.4.26}$$

*where $c^*$ is any vector in $P$ for which the minimum in (5.4.25) is attained. If $y(T) = 0$, i.e., $x_0 = 0$, then the control $u^*(t) = 0$ is the desired minimum-effort control.*

**Remark 5.4.2**   The statement and proof of Theorem 5.4.2 is given by Neustadt [7]. He suggests that the method of steepest descent can be used to calculate $\min F(c, T)$ $c \in P = \{c : c \cdot y(T) = 1\}$.

It is important to observe that for some fixed time $T$, minimum-effort optimal control $u^*$ is given by

$$u^*(t) = J_{1\min} \operatorname{sgn} g(t, c), \qquad t \in [0, T],$$

where $J_1 = J_{1\min}$ is a constant and $g$ is the index of the control system. For the time-optimal problem, the optimal control $v^*(t)$ is given by $v^*(t) = \operatorname{sgn} g(t, c)$; $t \in [0, T]$ with $T$ the minimum time. That is, the indices are switching functions of optimal controls. Thus, optimal solution of the minimum-effort control of (5.4.1) is a constant multiple of the minimum-time one. Therefore, the treatment of Secs. 5.4.2 and 5.4.3 can be appropriated for the construction in a neighborhood of zero, of optimal feedback control of minimum-effort problem. It is stated in the next result.

**Theorem 5.4.3**   *Consider the problem (5.4.0) with $J_1(u(T))$ defined in (5.4.19) subject to (5.4.1) and (5.4.2), where $x_1 = 0$. Assume that:*

  (i)   $\operatorname{rank}[B, AB, \ldots, A^{n-1}B] = n$.
  (ii)  *No eigenvalue of $A$ has a positive real part.*
  (iii) *The system (5.4.1) is strictly normal. Then there exists an $\lambda > 0$ and a function $f : \operatorname{Int} \mathbb{R}(\epsilon) \to E^n$, where $\epsilon = \min[\lambda, T]$, and*

$$\mathbb{R}(\epsilon) = \left\{ \int_0^\epsilon e^{-As} Bu(s)ds : u \in \mathcal{U} \right\},$$

*such that in $\operatorname{Int} \mathbb{R}(\epsilon)$, the set of solutions of*

$$\dot{z}(t) = Az(t) + Bf(z(t)) \tag{5.4.27}$$

*coincides with the set of optimal solutions of (5.4.1). Furthermore, $f(0) = 0$, and for $x \neq 0$, $f(x)$ is among the vertices of the cube $V$*

$$V = \{v \in E^m, \ |v_j| \leq J_{\min}, \ j = 1, \ldots, m\},$$

*and $f(x) = -f(-x)$. If $m \leq n$, then $f$ is uniquely determined by the condition that optimal solutions solve (5.4.27). Also, the inverse image of an open set in $E^m$ is an $F_\sigma$-set.*

**Remark 5.4.3**   If $\lambda \geq T$, the optimal feedback control constructed from Theorem 5.4.3 is a global one for the given $T$ so that the minimal control strategy $f(x)$ drives any $x_0$ to 0 in time $T$. If $\lambda < T$, only a local fuel-optimal feedback control is constructed for strictly normal systems.

**Proof**   The only modification needed in the proof of Theorem 5.4.3 is to replace the unit cube $U$ by the cube $V$ whose components have dimensions $\pm J_{\min}$. Thus, the optimal switching sequence is the finite sequence $\{v_1 \cdots v_k\}$, where $v_j$ are among the vertices of $V$, and optimal control $u$ on $[0, \theta]$ for $\theta < \epsilon$ is defined by $u(s) = v_j$ for $t_{j-1} \le s < t_j$, and $0 = t_1 < t_1 < \cdots t_k$ with $v_{j-1} \ne v_j$, where $k$ is an integer $1 \le k \le n$. As before, we set $M_0 = \{0\}$, and for $k = 1, \ldots, n$, we let $M_k$ be the set of points in $\operatorname{Int} \mathbb{R}(\epsilon)$ whose components have exactly $k - 1$ discontinuities. For each $k = 1, \ldots, n$, and each $k$-tuple $\underline{i} = \{v_1, \ldots, v_k\}$ of vertices of $V$ with $v_{j-1} \ne v_j$, let $M_{ki}$ be the set of points in $\operatorname{Int} \mathbb{R}(\epsilon)$ whose optimal controls have $\{v_1 \to \cdots \to v_k\}$ as the optimal switching sequence.

Because (5.4.1) is strictly normal, there exists $\epsilon > 0$ such that the indices $g(t, c)$ of (5.4.1) has at most $n - 1$ discontinuities on $[0, \epsilon]$. With this $\epsilon$, define $\mathbb{R}(\epsilon)$ and note that $\operatorname{Int} \mathbb{R}(\epsilon) = \bigcup_{k=0}^{n} M_k$, $M_k = \bigcup_{j \in I_k} M_{kj}$ as before. Set $f(0) = 0$, and for $0 \ne x \in \operatorname{Int} \mathbb{R}(\epsilon)$, find $M_{ki}$ containing $x$ and let $\{v_1 \to \cdots \to v_k\}$ be the optimal switching sequence corresponding to $M_{ki}$. Finally, set $f(x) = v_1$. The rest of the argument follows as in Theorem 2.1. $\qquad\square$

**Theorem 5.4.4**   *Consider the problem (5.4.0) with the cost function representing a pseudo fuel (energy if $p = 2$) :*

$$J_2(u(T)) = \left( \int_0^T \sum_{j=1}^{m} |u_j(t)|^p dt \right)^{\frac{1}{p}}, \qquad p > 1, \qquad (5.4.28)$$

*where admissible controls are*

$$u \in \mathcal{U}_2 = \left\{ u(t) \in E^m,\ u \text{ measurable} : \left( \int_0^T \sum_{j=1}^{m} |u_j(t)|^p dt \right)^{\frac{1}{p}} \le 1 \right\}. \qquad (5.4.29)$$

*As usual, the minimization is subject to (5.4.1) and (5.4.2).*
   *Assume that*

   (i) $\operatorname{rank}[B, AB, \ldots, A^{n-1}B] = n.$
   (ii) *No eigenvalue of $A$ has a positive real part.*
   (iii) *For each $j = 1, \ldots, m$, the vectors $b_j, Ab_j, \ldots, A^{n-1}b_j$ are linearly independent. Then for each $x_0$ and some $T$, there exists an optimal control $u^*(t)$ that transfers $x$ from $x_0$ to $0$ in time $T$.*

   *Furthermore, we let*

$$g(t, c) = ce^{-At}B, \qquad c \in E^n, \qquad a \ row \ vector,$$

$$F(c, T) = \left( \sum_{j=1}^{n} \int_0^T |g_j(t, c)|^q dt \right)^{\frac{1}{q}}, \qquad (5.4.30)$$

*where $\frac{1}{p} + \frac{1}{q} = 1$. If $x_0 \neq 0$, then the minimum-effort $J_{\min} = J(u^*(T))$ is given by*

$$\frac{1}{M_2(T)} = \frac{1}{J_{2\min}} = \min_{c \in P} F(c, T) , \tag{5.4.31}$$

*where*

$$P = \{c \in E^n : -cx_0 = 1\} . \tag{5.4.32}$$

*The optimal control is unique almost everywhere and is given by*

$$u_j^*(t) = \mu[g_j(t, c^*)]^{q/p} \, \mathrm{sgn}(g_j(t, c^*)) , \tag{5.4.33}$$

*where $\mu = J_{2\min}[F(c^*, T)]^{-\frac{q}{p}}$ and $c^* \in E^n$ is any vector in $P$ where the minimum in (5.4.31) is attained. If $x_0 = 0$, the optimal control is $u^*(t) = 0$.*

The last cost function considered is the measure of absolute fuel defined by

$$J_3(u(T)) = \int_0^T \sum_{j=1}^m |u_j(t)| dt , \tag{5.4.34a}$$

where the set of admissible controls is given by

$$U_3 = \left\{ u : u(t) \in E^m, \int_0^T \sum_{j=1}^m |u_j(t)| dt \leq 1 \right\} . \tag{5.4.34b}$$

This is the *so-called absolute fuel minimum problem*. There is no optimal solution for this cost function if we assume that $u \in U_3$ is integrable. Hájek [4] has proved the following result:

**Theorem 5.4.5**   *In problem (5.4.0), with cost (5.4.34) and constraints (5.4.1) and (5.4.2) with $x_1 = 0$, assume:*

   (i) $\mathrm{rank}[B, AB, \ldots, A^{n-1}B] = n$.
   (ii) $\mathrm{rank}[Ab_j, \ldots, A^n b_j] = n$                (5.4.35)
      *for each column $b_j$ of $B$ (this means that (5.4.1) is metanormal).*

*There is no optimal solution with $u$ integrable that steers $x_0$ to 0 in some $T$, while minimizing $J_3(u(T))$.*

We now admit, in absolute fuel minimization problems, as admissible controls those that are unbounded and *"impulsive"* in nature, the so-called Dirac delta functions. If such functions are considered optimal minimum, $-\|u\|_1$ controls exist.

**Theorem 5.4.6**   *In problem (5.4.0), with the cost function (5.4.34) and constraints (5.4.1) and (5.4.2) with $x_1 = 0$, suppose*

   (i) $\mathrm{rank}[B, AB, \ldots, A^{n-1}B] = n$.

(ii) $\mathrm{rank}[Ab_j, \ldots, A^n b_j] = n$ *for each* $j = 1, \ldots, m$. *Suppose*

$$g(t, c) = c \cdot e^{-At} B,$$

$$F(T, c) = \max_{1 \le j \le m} \max_{0 \le t \le T} |g_j(c, t)|. \tag{5.4.36}$$

*If* $x_0 \ne 0$, *and if we consider generalized (delta) functions to be admissible, then there exists a minimum-effort control* $u^*(t)$ *that transfers* $x_0$ *to* $0$ *in time* $T$. *The minimum fuel* $J_3(u^*(T)) = J_{3\,\min}$ *is given by*

$$\frac{1}{J_{\min}} = \min_{c \in P} F(T, c) = F(T, c^*),$$

*where* $P = \{-cx_0 = 1\}$. *The optimal control is given by* $\overline{u}^*(t) = J_{3\,\min}\overline{u}(t, c^*)$, *where the boundary control is given by*

$$\overline{u}_j(t) = \sum_{i=1}^{Nj} \mathrm{sgn}(g_j(c^*, \tau_{ji}))\delta(t - \tau_{ji}) \Big/ \sum_{j=1}^{m} N_j, \qquad 1 \le j \le m.$$

*Here the maximum in (5.4.36) can occur at multiple* $j$ *and at multiple instances* $\tau_{ji}$, $i = 1, 2, \ldots, N_j$, *where* $N_j$ *equals zero if* $g_j$ *does not contain the maxima.*

*If we replace the control set by*

$$U_4 = \left\{ u(t) \in E^m : \int_0^T |u_j(t)|dt \le 1, \; 1 \le j \le m \right\},$$

*then optimal control is given by*

$$u^*(t) = J_{\min}\overline{u}(t, c^*)$$

*where the boundary control is given by*

$$u_j(t) = \sum_{i=1}^{Mj} \frac{1}{M_j} \, \mathrm{sgn}(g_j(\tau_{ji}, c^*))\delta(t - \tau_{ji}), \qquad 1 \le j \le m,$$

*where*

$$\frac{1}{J_{\min}} = \min_{c \in P} F(T, c) = F(c^*)$$

*and*

$$F(T, c) = \sum_{j=1}^{m} \max_{0 \le t \le T} |g_j(c, t)|, \tag{5.4.37}$$

*with* $\tau_{ji}$, $i = 1, 2, \ldots, m_j$, *being where maximal in (5.4.37) may occur at multiple instances* $\tau_{ji}$ $i = 1, 2, \ldots, M_j$, $M_j \ge 1$.

## 5.5 General Method for the Proof of Existence and Form of Time-Optimal, Minimum-Effort Control of Ordinary Linear Systems

In the strings of results, Theorems 5.4.2–5.4.5, on the system

$$\dot{x}(t) = A(t)x(t) + B(t)u(t), \qquad x(0) = x_0, \tag{5.5.1}$$

one begins with the variation of parameters,

$$x(t, x_0, u) = X(t)\left[x_0 + \int_0^t X^{-1}(s)Bu(s)ds\right], \tag{5.5.2}$$

where $X(t)$ is the fundamental matrix solution of

$$\dot{x}(t) = A(t)x(t), \tag{5.5.3}$$

and defines the functions

$$y(t) = X^{-1}(t)x_1(t) - x_0, \tag{5.5.4}$$

$$g(t, c) = cX^{-1}(t)B(t), \qquad c \text{ a row vector}. \tag{5.5.5}$$

Also defined is the map

$$S_t(u) = \int_0^t X^{-1}(s)B(s)u(s)ds, \tag{5.5.6}$$

where $S_t : Y \to E^n$ is a continuous linear map with $S_0(Y) = 0$. Here $Y$ is the dual space $Z^*$ of a separable Banach space $Z$, or is a reflexive Banach space. It is the space of controls whose elements $u$ influence the state space of (5.5.1) that are elements of $E^n$. We assume a uniform bound $M$ in norm of admissible controls:

$$U = \{u \in X : \|u\| \leq M\}. \tag{5.5.7}$$

If $x_1(t)$ is a time-varying target that the system will hit, then $y(t)$ describes a reachable set — a desirable state of the system. Thus, at $t = 0$, $y(0) = x_1(0) - x_0 \neq 0$, and (5.5.1) is not at a desirable state. But if it hits the target at some $t$, then we have the coincidence

$$y(t) = X^{-1}(t)x_1(t) - x_0 = S_t(u) \tag{5.5.8}$$

for some $u \in U$. As a consequence, $y(t) \in \mathbb{R}(t)$, the reachable set, defined by

$$\mathbb{R}(t) = \left\{\int_0^t X^{-1}(s)B(s)u(s)ds : u \in U\right\}. \tag{5.5.9}$$

For the time-optimal problem, the minimal time for hitting the target is given by

$$t^* = \text{Inf}\{t \in [0, T] : S_t(u) = y(t) \text{ for some } u \in U\}. \tag{5.5.10}$$

The admissible control $u^* \in U$ that has $s_{t^*}(u^*) = y(t^*)$ is the time-optimal control. For the minimum fuel problem, the admissible control $u^*$ that has $S_t(u^*) = y(t)$ while minimizing $J(u(T))$ is the minimum fuel optimal control. As suggested by Hájek and Krabs [5], it is useful to study in abstract setting the mapping

$$S_t : Y \to E^n$$

and its adjoint

$$S_t^* : E^n \to Y$$

with the following insight

$$(S_t(u), c) = (c, S_t(u) = (S_t^*(c), u).$$

If $Y = L_\infty([0,T], E^m)$, then $Y = Z^*$ with $Z = L_1([0,T], E^m)$. $S_t$ is given in (5.5.6), and its adjoint $S_\tau^*$ is given by

$$S_\tau^*(c^T) = \begin{cases} 0, & \text{a.e. } \tau \in (t, T], \\ (X^{T-1}(\tau)B(\tau))^T c, & \text{a.e. } t \in [0, t], \end{cases} \tag{5.5.11}$$

which obviously maps $E^n$ into $L_1([0,T], E^m) \subseteq Y^*$ so that

$$\|S_t^*(c)\| = \int_0^t \|(X^{-1}(s)B(s))^T c\|_1 ds, \qquad c \in E^n, \tag{5.5.12}$$

where $\| \cdot \|_1$ denotes the $L_1$ norm in $E^m$. If $Y = L_2([0,T], E^m)$, then $Y = Y^*$ is a Hilbert space. For each $t \in [0,T]$, the adjoint operators $S_t^* : E^n \to Y$ of $S_t$ are given by (5.5.6) so that

$$\|S_t^*(c)\| = \left( \int_0^t \|(X^{-1}(s)B(s))^T c^T\|_2^2 ds \right)^{\frac{1}{2}}, \qquad \text{for } c \in E^n. \tag{5.5.13}$$

We now consider the continuous linear mapping $S_t : Y \to E^n$ under the following assumptions:

  (i)  $\mathbb{R}(t) = \{S_t(u) : u \in U\}$ is closed.
  (ii)  The mapping $t \to S_t$, $t \in [0,T]$ is continuous with respect to operator norm topology of $L(Y, E^n)$, the space of linear maps from $Y$ into $E^n$.
  (iii)  The function

$$y : [0, T] \to E^n$$

  is continuous.
  (iv)  $y : [0,T] \to E^n$ is constant and nonzero, i.e., $y(t) = y_1$, $\forall t \in [0,T]$, $y \neq 0$.
  (v)  For each $c \in E^n$, $c \neq 0$, the function

$$t \to \|S_t^*(c)\|$$

  is strictly increasing in $[0,T]$.

**Theorem 5.5.1** *Assume condition* (i), *and let* $t \in [0, T]$. *Then there is an admissible control* $u \in U$, *such that* $S_t(u) = y(t)$ *if and only if*

$$cy(t) \leq M\|S_t^*(c)\|, \qquad \forall\, c \in E^{n^*}, \qquad a\ row\ vector \qquad (5.5.14)$$

*where* $E^{n^*}$ *is the Euclidean space,* $c^T$ *is a row vector, and* $S_t^*$ *is the adjoint of* $S_t$ *mapping* $E^{n^*}$ *to* $Y^*$.

**Proof** From the assumptions there is a $u \in U$ with $S_t(u) = y(t)$, so that

$$cy(t) = c(S_t(u)) = S_t^*(c)(u) \leq M\|S_t^*(c)\|.$$

This proves that (5.5.14) is true. Conversely, recall the definition of the reachable set $\mathbb{R}(t)$, and note that it is convex since $S_t$ is linear and $U$ convex. Because it is a closed convex subset of $E^n$, if there is no $u \in U$ with $S_t(u) = y(t)$, then $y(t) \neq \mathbb{R}(t)$, and by the Separation Theorem of closed convex sets in $E^n$, [6, p. 33], there is a hyperplane which separates $\mathbb{R}(t)$ and $y(t)$. This means there is a $c \in E^{n^*}$ such that

$$cy(t) > \sup\{c^T(S_t(u)) : u \in U\} = \sup\{S^*(c)(u) : u \in U\}.$$

This contradicts (5.5.14). $\qquad\qquad\qquad\qquad\qquad\qquad\qquad\qquad\qquad\qquad\square$

**Theorem 5.5.2** *Suppose conditions* (i)–(iii) *are satisfied. Then there exists a* $c \in E^{n^*}$, *a row vector with* $\|c\| = 1$ *such that*

$$cy(t^*) = M\|S_{t^*}^*(c)\|. \qquad\qquad (5.5.15)$$

*The proof is contained in Antosiewicz* [9].

**Theorem 5.5.3**    (**Hájek–Krabs Duality Theorem**) [5] *Suppose* (i), (ii), (iv), *and* (v) *are valid. Let* $t^*$ *be the optimal time defined in* (5.5.10). *Then*

$$t^* = \max\{t \in (0, T] \quad such\ that\ cy(t) = M\|S_t^*(c)\| \quad for\ some\ c \in E^{n^*}$$

$$with\ \|c\| = 1\}.$$

**Proof** From Theorem 5.5.2 and its Corollary, the minimum time is a point in the set from which the maximum is taken. Suppose $t > t^*$ where $t \in (0, T]$, and where $cy_1 = M\|S_t^*(c)\|$ for some $c \in E^{n^*}$ with $\|c\| = 1$. Then by Theorem 5.5.1 and (v),

$$cy_1 \leq M\|S_{t^*}^*(c)\| < M\|S_t^*(c)\|.$$

This is a contradiction. $\qquad\qquad\qquad\qquad\qquad\qquad\qquad\qquad\qquad\qquad\square$

**Corollary 5.5.1** *Let* $u^*$ *be a time-optimal control and* $t^*$ *the minimum time. Then*

$$S_{t^*}^*(c)(u^*) = c(S_{t^*}(u^*)) = cy(t^*) = M\|S_{t^*}^*(c)\|.$$

Now, consider $U \subset L_\infty([0, t^*], E^m)$ as the control space. From the above there exists $u^*$ such that

$$cy(t^*) = \int_0^{t^*} (cX^{-1}(\tau)B(\tau))^T u^*(\tau)d\tau$$

$$= c \int_0^{t^*} X^{-1}(\tau)B(\tau)u^*(\tau)d\tau \qquad (5.5.16)$$

$$= M \int_0^t \|(X^{-1}(s)B(s))^T c\|_1 ds.$$

This implies that $u^*$ is of the form

$$u_j^*(t, c) = M \, \mathrm{sgn}(cX^{-1}(t)B(t))_j \qquad (5.5.17a)$$

when

$$(cX^{-1}(t)B(t))_j \neq 0$$

for each $c \neq 0$ and each $j = 1, \ldots, m$. Thus if the system is normal, optimal controls are given by

$$u^*(t) = M \, \mathrm{sgn}(cX^{-1}(t)B(t)), \qquad (5.5.17b)$$

when $U \subseteq L_\infty([0\,T], E^m)$.

Suppose

$$U = \{u \in L_p : \|u\|_p \leq M\}.$$

If $U \subset L_p([0, t^*], E^m)$, where $p > 1$, we obtain from (5.5.15) that

$$c^T y(t^*) = M\|S_{t^*}^*(c)\|_p,$$

$$= \int_0^{t^*} (cX^{-1}(\tau)B(\tau))^T u^*(\tau)d\tau, \qquad (5.5.18)$$

$$= M \left( \int_0^{t^*} \|cX^{-1}(\tau)B(\tau)\|^q d\tau \right)^{\frac{1}{q}}$$

where $\frac{1}{p} + \frac{1}{q} = 1$. Recall that

$$U = \left\{ u \in L_p : \left( \int_0^T |u(s)|^p ds \right)^{\frac{1}{p}} \leq M \right\}.$$

Let $g(s) = cX^{-1}(s)B(s)$. Then

$$cy(t^*) = \int_0^{t^*} \sum_{j=1}^m g_j(s)u_j(s)ds \leq \int_0^{t^*} \sum_{j=1}^m |g_j(s)u_j(s)|ds,$$

$$\leq \int_0^{t^*} \left( \sum_{j=1}^m |g_j(s)|^q \right)^{\frac{1}{q}} \left( \sum_{j=1}^m |u_j(s)|^p \right)^{\frac{1}{p}} ds,$$

$$\leq \left( \int_0^{t^*} \sum_{j=1}^m |g_j(s)|^q ds \right)^{\frac{1}{q}} \left( \int_0^{t^*} \sum_{j=1}^m |u_j(s)|^p ds \right)^{\frac{1}{p}},$$

$$\leq M \left( \int_0^{t^*} \sum_{j=1}^m |g_j(s)|^q ds \right)^{\frac{1}{q}},$$

since by (5.5.18)

$$cy(t^*) = M \left( \int_0^{t^*} \sum_{j=1}^m |g_j(s)|^q \right)^{\frac{1}{q}}$$

we have equality everywhere and the control $u^*$ that gives the equality is

$$u_j^*(s,c) = MK|g_j(s)|^{\frac{q}{p}} \, \text{sgn}(g_j(s)), \qquad j = 1,\ldots,m, \tag{5.5.19}$$

where

$$K = \left( \int_0^{t^*} \sum_{j=1}^m |g_j(s)|^q ds \right)^{-\frac{1}{p}}.$$

This is the time-optimal control. It holds when the system is normal.

To obtain expressions for the minimum fuel optimal controls, we note carefully that the controls $u^*(t,c)$ in (5.5.17) and (5.5.19) are boundary controls in the sense that if

$$z(T,c) = \int_0^T X^{-1}(s)B(s)u^*(s,c)ds, \tag{5.5.20}$$

then $z(T,c)$ is on the boundary of the reachable set $\mathbb{R}(t)$, so that

$$cz(T,c) = F(c,T) = M\|S_T^*(c)\|,$$

$$cz(T,c) > cy, \quad \forall\, y \in \mathbb{R}(T), \quad y \neq z(T,c). \tag{5.5.21}$$

If $y(T)$ is as defined in (5.5.4) and $y(T) \in \mathbb{R}(T)$, we can extend $y(T)$ to reach the boundary as follows: Let

$$\alpha = \max\{\beta : \beta y(T) \in \mathbb{R}(T)\}. \tag{5.5.22}$$

If $y(T) \neq 0$, $\alpha$ can be taken positive, and clearly $\alpha y(T)$ is a boundary point of $\mathbb{R}(T)$, so that

$$\alpha y(T) = z(T,c)$$

for some $c$ a row vector, where matters can be so arranged that

$$c \cdot y(T) = 1 \,. \tag{5.5.23}$$

We observe that

$$\bar{u}^*(t) = \frac{u^*(c,t)}{\alpha} \tag{5.5.24}$$

transfers $x_0$ to $x_1(T)$ in time $T$ while minimizing $J(u(T))$. Also

$$\frac{1}{\alpha} = \frac{1}{M(T)} = \min J(u(T)) \,, \tag{5.5.25}$$

where

$$\alpha = \min_{c \in P = \{c \in E^n : cy(T) = 1\}} F(c,T) \,. \tag{5.5.26}$$

As a consequence of these observations, minimum energy control is given by

$$\bar{u}^*(t) = \frac{u^*(c,t)}{M(T)} \,, \qquad 0 \le t \le T \,, \tag{5.5.27}$$

where $c^*$ is the minimizing vector in (5.5.26). These are the verifications of Theorem 5.4.2 and Theorem 5.4.4. In details for $L_\infty$ control, minimum fuel controls are

$$\bar{u}^*(t) = \frac{M \operatorname{sgn}(g(t,c^*))}{M(T)} \,, \tag{5.5.28}$$

where $c^*$ is the minimizing vector. For $L_p$ control, minimum pseudo-fuel controls are

$$\bar{u}_j^*(t) = \frac{K |g_j(s,c)|^{\frac{q}{p}} \operatorname{sgn}(g_j(s,c)) M}{M(T)} \,, \tag{5.5.29}$$

where

$$K = \left( \int_0^T \sum_{j=1}^M |g_j(s)|^q dt \right)^{-\frac{1}{p}} \,.$$

Thus for minimizing $L_p$ controls in $U$,

$$F(c,T) = \left( \int_0^T \sum_{j=1}^M |g_j(s)|^q ds \right)^{\frac{1}{q}} \,, \tag{5.5.30}$$

$$\alpha = \min_{c \in P} F(c,T) = F(c^*,T) = \frac{1}{M(T)} \,, \tag{5.5.31}$$

and optimal control (5.5.31) is given by (5.5.29) where $c^*$ is the minimizing vector in (5.5.30) and (5.5.31). It is illuminating to link up minimum time control with minimum energy problems. In some situations, we show that the optimal controls are the same.

**Definition 5.5.1**    For each $t \in (0, T]$, let

$$\frac{1}{M(t)} = \text{Inf}\{\|S_t^*(c)\| : c \in P\}, \tag{5.5.32}$$

where

$$P = \{c \in E^{n^*} : cy_1 = 1\}.$$

**Theorem 5.5.4**    *Suppose the basic assumptions* (i), (ii), (iv), *and* (v) *are valid. Then for each* $t \in (0, T]$,

$$M(t) \left\{\begin{matrix} \geq \\ = \\ < \end{matrix}\right\} M \Leftrightarrow t \left\{\begin{matrix} \leq \\ = \\ > \end{matrix}\right\} t^*, \tag{5.5.33}$$

*where $M$ is the bound on the corresponding control set.*

**Proof**    From the definition we quickly deduce that

$$cy_1 \leq M(t)\|S_t^*(c)\|, \qquad \forall\, c \in E^n.$$

Because $E^n$ is finite dimensional, it is routine to verify that for each $t \in (0, T]$ there exists some $c(t) \in P$ such that

$$\|S_t^*(c(t))\| = \frac{1}{M(t)} > 0, \tag{5.5.34}$$

and there is a $u_t \in Y$ such that

$$S_t(u_t) = y_1 \quad \text{with} \quad \|u_t\| = M(t).$$

Using this in (5.5.34), there exists $c(t) \in E^{n^*}$ such that

$$c(t)y_1 = M(t)\|S_t^*(c_t^T(t)\| = 1.$$

If we invoke Theorems 5.5.1 and 5.5.2, the result follows at once.    $\square$

**Theorem 5.5.5**    *Assume* (i), (ii), (iv), *and* (v). *The optimal control $u^*$ that transfers $x_0$ to 0 in minimum time $t^*$ while minimizing the effort $J(u(t^*))$ is given by*

$$u^*(t) = \text{sgn}[g(t, c^*)], \qquad 0 \leq t \leq t^*, \tag{5.5.35}$$

*when $U_1 \subseteq L_\infty$ with $J = J_1$ and $c$ minimizes* (5.5.26) *and* (5.4.22). *Also, $u^*$ is given by*

$$u_j^*(t) = K|g_j(t, c)|^{\frac{q}{p}}\, \text{sgn}\, g_j(t, c), \tag{5.5.36}$$

*where*

$$K = \left(\int_0^{t^*} \sum_{j=1}^{m} |g_j(s, c)|^q dt\right)^{\frac{1}{p}}$$

*if $U_2 \subset L_p$ with $J = J_2$, where $c$ in* (5.5.36) *minimizes* (5.5.31) *with $F$ in* (5.4.30).

**Proof** If $T = t^*$, the minimum time required to transfer $x_0$ to 0, then from Theorem 5.5.4,

$$\frac{1}{M} = \frac{1}{M(t^*)} = \inf_c\{\|S_{t^*}^*(c)\| : c \in P\},$$

where $P = \{c \in E^n : c^T y_1 = 0\}$, and where $M$ is the bound on $U$, i.e.,

$$U = \{u \in E^m : \|u\| \le M\}.$$

Since minimum fuel controls are given by (5.5.27), optimal controls are given by

$$\bar{u}^*(t) = \frac{u^*(t,c)}{M} \qquad 0 \le t \le t^*,$$

where $c$ is the minimizing vector of (5.5.26). But if we use $\bar{u}^*(t)$ in (5.5.28) or (5.5.29), the results (5.5.35) and (5.5.36) are deduced. Thus, when $M = 1$, time-optimal control for the system (5.5.1) is also a minimum fuel control.

### 5.5.1 *Remarks on the assumptions*

(i) The coincidence $y(t) = S_t(u)$ for some $u \in U$ is equivalent to controllability with constraints. If $y(t) = -x_0$, this is null controllability with constraints, which is ensured by the following assumptions:

    (a) $\operatorname{rank}[B, AB \cdots A^{n-1}B] = n$.

    (b) No eigenvalue of $A$ has positive real part.

    (c) The condition that for each $c \ne 0$, $t \to \|S_t^*(c)\|$ is strictly increasing is ensured by the system being proper, i.e., for each $t_1, t_2 \in [0, T]$ with $0 \le t_1 \le t_2 \le T$, $[ce^{-At}(t)B(t)]^T = 0$, $\forall\, t \in [t_1, t_2]$ if and only if $c = 0$. This is equivalent to the controllability condition (a).

(ii) In (5.5.17) the optimal controls are uniquely determined if for each $j = 1, \ldots, m$, and each $c^{\ne 0} \in E^n$, $\quad [ce^{-At}(t)B(t)]_j \ne 0$.

We note that theorems on minimum fuel problems, Theorems 5.4.2–5.4.4, depend on the assumption that ensures that the reachable set is closed. Because the reachable sets are also convex, optimal controls are controls that generate points on the boundary of $\mathbb{R}(t)$. If, however, the problem is to minimize the $L_1$-cost function

$$J_3(u(T)) = \|u\|_1 = \int_0^T \sum_{j=1}^M |u_j(t)|dt \qquad (5.5.37a)$$

subject to $x(0) = x_0$ and $x(T) = 0$ where $U = U_3$ is given in (5.5.34), the reachable set

$$\mathbb{R} = R(t) = \left\{ \int_0^t e^{-As}Bu(s)ds : \|u\|_1 \le 1 \right\} \qquad (5.5.37b)$$

is open, if we assume that

$$\operatorname{rank}[Ab_j, \ldots, A^n b_j] = n,$$

for each column $b_j$ of $B$. This is the content of the following Lemma by Hájek, [4].                                                                    □

**Lemma 5.5.1**  *Suppose $\dot{x}(t) = Ax + Bu$ (5.5.1) is metanormal, i.e., $\operatorname{rank}[AB_j \ldots A^n b_j] = n$ for each $j = 1, \ldots, m$. Then the reachable set $\mathbb{R}(t)$ of (1) is open, bounded, convex, and symmetric.*

**Proof**  Since boundedness, convexity, and symmetry are obvious, we give only the proof that $\mathbb{R}(t)$ is open. Suppose it is not open, so that

$$x_0 = \int_0^T e^{-As} B u_0(s) ds$$

is on a boundary point. Let $c \neq 0$ be an outernormal to (the closure of) $\mathbb{R}(t)$ at $x_0$, and use $c$ to define the index

$$y(t) = c \cdot e^{-At} B.$$

It is clear that control $u_0$ maximizes

$$\int_0^T y(s)u(s) \leq \int_0^T y(s)u_0(s), \quad \text{whenever } \|u\|_1 \leq 1. \tag{5.5.38}$$

But the mapping

$$u(\cdot) \to \int_0^T y(s)u(s)ds$$

is a linear functional on $L_1[0, T]$ with norm

$$\|y\|_\infty = \max_{1 \leq j \leq m} \max_{s \in [0, T]} |y_j(s)|.$$

Since (5.5.38) implies that $\|y\|_\infty$ is attained at the element $u_0(\cdot)$ of the unit ball, we have

$$\|y\|_\infty = \int_0^T y u_0 \leq \int_0^T |y_j| \, |u_{0j}| \leq \int_0^T \sum_{j=1}^M \|y\|_\infty |u_j| = \|y\|_\infty \|u_0\|_1 \leq \|y\|_\infty.$$

$$\tag{5.5.39}$$

This shows that we have equality throughout, so that

$$\int_0^T \sum_{j=1}^M (\|y\|_\infty - |y_j|) |u_{0j}| = 0.$$

But Hájek has observed [4, p. 417] that metanormality implies $|y_j|$ is nonconstant a.e. This implies $u_0 = 0$ a.e. With this $u_0 = 0$, our boundary point is $x_0 = 0$. This contradicts the following containment:

$$R_\alpha(T) \subset \text{Int } \mathbb{R}_\beta(T), \quad \text{Int}(1/\alpha)\mathbb{R}_\alpha(T) \supset \left(\frac{1}{\beta}\right) R_\beta,$$

whenever $0 < \alpha < \beta \leq mT$, which is a consequence of metanormality. Thus the set of points $-x_0$ which can be steered to 0 at $T$ by using controls with $L_1$ bound $\|u\|_1 \leq 1$ is the open set $\mathbb{R}$.

If the controls have $\|u\|_1 \leq M$, the set is $M\mathbb{R}$ for $M > 0$. From this it follows that minimal $M$ can never be attained unless $x_0 = 0$. This proves Theorem 5.5.5. $\qquad\square$

Because the reachable set is open, its boundary points can only be "reached" by convex combination of delta functions. We can therefore admit delta functions as controls when we consider absolute fuel minimization. Optimal controls are impulsive in nature. If such functions are considered, optimal minimum $-\|u\|_1$ controls exist.

Because $\mathbb{R}$ is open we can now prove Theorem 5.4.6 as follows:

**Proof of Theorem 5.4.6**  Let

$$\mathbb{R}_\alpha = R_\alpha(T) = \left\{ \int_0^T e^{-As} Bu(s)ds : \|u\|_\infty \leq 1 \text{ and } \|u\|_1 \leq \alpha \right\}.$$

Then by Hájek [4, p. 435],

$$\mathbb{R} = \bigcup_{\alpha>0} 1/\alpha\mathbb{R}_\alpha = \left\{ \int_0^T e^{-As} Bu(s)ds : \|v\| \leq 1, \ \|v\|_\infty < \infty \right\},$$

where $1/\alpha u = v$, and $\mathbb{R}$ is open.

Recall that the set of admissible controls is

$$U = \left\{ u(t) \in E^m : \int_0^T \sum_{j=1}^1 |u_j(t)|dt \leq M \right\}, \tag{5.5.40}$$

and the reachable set

$$\mathbb{R} = \mathbb{R}(T) = \left\{ S_T(u) = \int_0^T e^{-As} Bu(s) : u \in U \right\}$$

is open. If $g(s) = ce^{-As}B(s)$, then $cy(T) = \int_0^T g(c,t)u(t)dt$. The control that realizes a reachable set on the boundary of $\mathbb{R}$ maximizes (5.5.38) over $U$.

We note that

$$cy(t) = \int_0^T \sum_{j=1}^m g_j u_j dt \leq \int_0^T \sum_{j=1}^m |g_j| |u_j| dt \leq M \max_{1 \leq j \leq m} \max_{0 \leq t \leq T} |g_j(c,t)|. \tag{5.5.41}$$

By inspection, we observe for time-optimal controls

$$cy(t^*) = M\|S_{t^*}^*(c)\|\,,\tag{5.5.42}$$

where

$$\|S_{t^*}^*(c)\|_\infty = \max_{1\le j\le m}\ \sup_{0\le t\le T}\ |g_j(c,t)|\,.\tag{5.5.43}$$

If equality holds everywhere in (5.5.41), then impulsive functions, applied at the points where $g_i$ is largest, maximizes (5.5.38). Note that the maxima in (5.5.43) can occur at multiple $j$ and at multiple instances of $\tau_{ji}$, $i = 1, 2, \ldots, N_j$, where $N_j$ is taken to be zero if $g_j$ does not contain the maximum. As a consequence, impulsive optimal controls are

$$u_j(t) = M\sum_{j=1}^{N_j}\mathrm{sgn}(g_j(c,\tau_{ji})\delta(t-\tau_{ji})\sum_{j=1}^{m}\frac{1}{N_j}\,,\qquad 1\le j\le m\,.\tag{5.5.44}$$

For the minimum absolute fuel problem, we have

$$F(c) = \max_{1\le j\le m}\ \sup_{0\le t\le T}\ |g_j(c,t)|\,,\qquad \alpha = \min_{c\in P}F(c) = F(c^*)\,,\tag{5.5.45}$$

and optimal minimum fuel control is $\overline{u}^* = \frac{u(t,c^*)}{\alpha}$, where $c^*$ is the minimizing vector in (5.5.45), and where the boundary controls $u(t,c^*)$ are given in (5.5.44).

If we replace the constraint set $U$ in (5.5.40) by

$$U = \left\{u(t)\in E^n : \int_0^T |u_j(t)|dt \le 1,\ j = 1,\ldots,m\right\}\,,\tag{5.5.46}$$

then from (5.5.46) and (5.5.38) we have

$$cy(T) = \int_0^T\sum_{j=1}^{m}g_ju_j\,dt \le \sum_{j=1}^{m}\sup_{0\le t\le T}|g_i| = \sum_{j=1}^{m}\max_{0\le t\le T}|g_j(t,c)|\,.\tag{5.5.47}$$

The maximum in (5.5.47) occurs as before at multiple instances of time $\tau_{ji}$, $i = 1, 2, \ldots, M_j$ where $M_j \ge 1$. The maximizing impulsive controls that approximate reachable points on the "boundary" of $\mathbb{R}(t)$ are

$$u_j(t,c) = \sum_{i=1}^{M_j}\frac{1}{M_j}\mathrm{sgn}(g_j(\tau_{ji},c)\delta(t-\tau_{ji})\,,\qquad 1\le j\le m\,.\tag{5.5.48}$$

Therefore, if

$$F(c^*) = \sum_{j=1}^{m}\max_{1\le t\le T}|g_j(c,t)|\,,\qquad \overline{\alpha} = \min_{c\in P}F(c) = F(c^*)\,,\tag{5.5.49}$$

then optimal control that minimizes the absolute fuel is given by $u^* = \frac{u(t,c^*)}{\alpha}$, where $c^*$ and $\alpha$ are determined as in (5.5.49).

**Example 5.5.1**   Simple Harmonic Oscillator

$$x(0) = x_0, \qquad \dot{x}(0) = 0$$

$$\ddot{x} + x = u$$

$$x(T) = 0, \qquad \dot{x}(T) = 0$$

$$\dot{\underline{x}} = A\underline{x} + Bu$$

$$A = \begin{bmatrix} 0 & 1 \\ -1 & 0 \end{bmatrix}, \qquad B = \begin{bmatrix} 0 \\ 1 \end{bmatrix}.$$

Appropriate the earlier treatment in Example 5.1.1.

$$y(T) = e^{-AT}\underline{x}(T) - \underline{x}(0) = -\underline{x}(0) = \begin{bmatrix} -x_0 \\ 0 \end{bmatrix}.$$

The hyperplane $P$ is defined by $P = \{c : cy(T) = 1\}$. This implies that $c_1 = -1/x_0$ where $c_2$ is free. The index of the control system is

$$g(t, c) = -c_1 \sin t + c_2 \cos t = \sqrt{c_1^2 + c_2^2}\, \sin(t + \delta),$$

where $\delta = \tan^{-1}(-\frac{c_2}{c_1})$. Therefore

$$g(t, c) = \left( \frac{1}{x_0^2} + c_2^2 \right)^{\frac{1}{2}} \sin(t + \delta).$$

With cost function $J_1$,

$$J_1(u(T)) = \max_{1 \leq j \leq m} \sup_{0 \leq t \leq T} |u_j(t_1)| = \sup_{0 \leq t \leq T} |u(t)|,$$

and $U = \{u : |u(t)| \leq 1\}$,

$$F(c, T) = \int_0^T \sum_{j=1}^m |g_j(t, c)|dt = \left( \frac{1}{x_0^2} + c_2^2 \right)^{\frac{1}{2}} \int_0^T |\sin(t + \delta)|dt,$$

$$\frac{1}{J_{1\min}} = \overline{\alpha} = \min_{c \in P} F(c, T) = \min_{c \in P} \left( \frac{1}{x_0} + c_2^2 \right)^{\frac{1}{2}} \int_0^T |\sin(t + \delta)|dt.$$

Clearly $c_2 = 0$ is the minimizer, and this forces $\delta = 0$. Thus the minimizing vector $c^* = (-\frac{1}{x_0}, 0)$. It follows from Theorem 5.4.2 that the unique optimal open-loop strategy is

$$u^*(t) = J_{1\min} \operatorname{sgn}[g(t, c^*)] = \frac{|x_0|}{\int_0^T |\sin t|dt} \operatorname{sgn}[\sin t].$$

Thus $u^* = J_{1\min}$ or $u^* = -J_{1\min}$. For $u = +J_{1\min}$, $x_2^2 + (J_{1\min} - x_1)^2 = a^2$. This is a circle of radius $a$ centered at $(J_{1\min}, 0)$. For $u = -J_{1\min}$, $(x_1 + J_{\min})^2 + x_2^2 = a^2$. This is a circle of radius $a$ centered at $(-J_{\min}, 0)$. Terminal manifolds are constructed as before. Note that $\epsilon < \pi$. Int $\mathbb{R}(\epsilon) = M_1 \cup M_2 \cup \Gamma_+ \cup \Gamma_-$. But since $u$ is not $\pm 1$, but $\pm J_{1\,\min}$, the circles on the switching loci have centers $(J_{1\,\min}, 0)$ and $(-J_{\min}, 0)$. See Figs. 5.5.1 and 5.5.2. Optimal feedback control is

$$f(x_1, x_2) = \begin{cases} +J_{1\,\min} & \text{if } (x_1, x_2) \in M_1, \Gamma_+ , \\ -J_{1\,\min} & \text{if } (x_1, x_2) \in M_2, \Gamma_- . \end{cases}$$

The switching curves may be found as in Fig. 5.2.1. See Fig. 5.5.1 below.

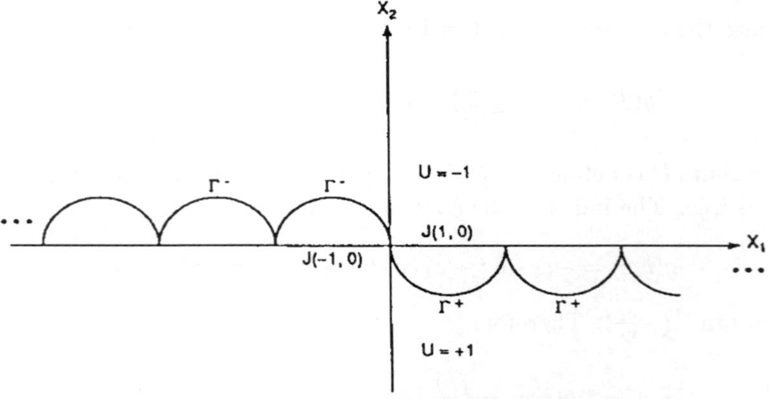


Fig. 5.5.1   Switching Curves $W$.

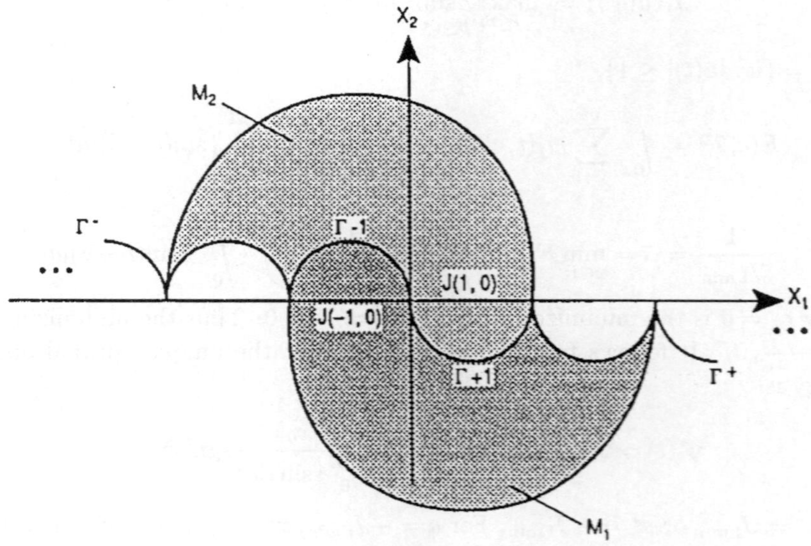


Fig. 5.5.2

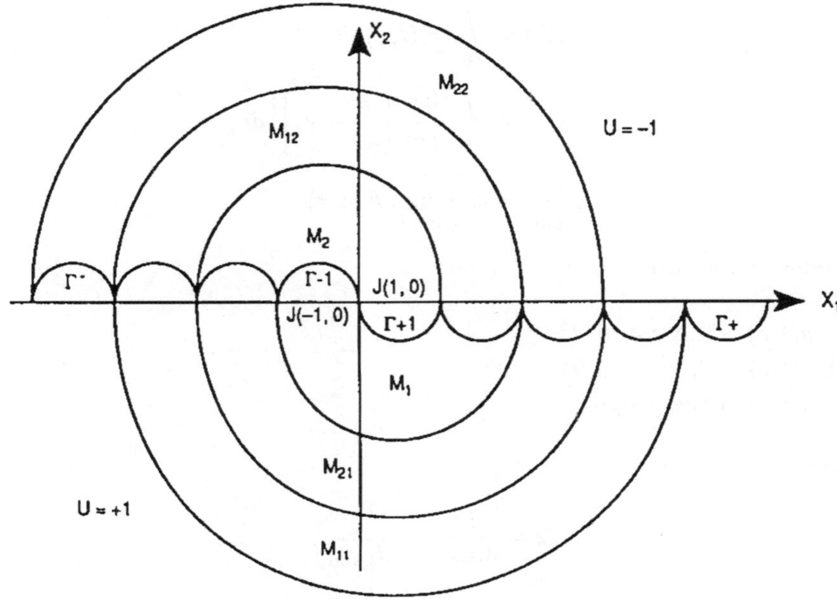


Fig. 5.5.3

The associate manifolds are constructed as before in Example 5.3.1.   See Fig. 5.3.5.

**Example 5.5.2**   Rigid Body Maneuver:
$m\ddot{x} = u.$

$$\underline{x}(0) = \begin{bmatrix} x_0 \\ 0 \end{bmatrix}, \qquad \underline{x}(T) = \begin{bmatrix} 0 \\ 0 \end{bmatrix}.$$

We use earlier calculations.

$$y(T) = e^{-At}\underline{x}(T) - \underline{x}(0) = -\underline{x}_0.$$

The hyperplane $P$ is defined to be $P_1 = cy(T) = 1$, $[c_1, c_2]\begin{bmatrix} -x_0 \\ 0 \end{bmatrix} = 1$, i.e., $c_1 = -1/x_0$, where $c_2$ is free.

The index of the control system is defined by

$$g(t, c) = c^T e^{-At} B = [c_1, c_2] \begin{bmatrix} 1 & -t \\ 0 & 1 \end{bmatrix} \begin{bmatrix} 0 \\ 1/m \end{bmatrix} = \frac{1}{m}\left[\frac{t}{x_0} + c_2\right],$$

on $P$. Using the notations of Theorem 5.4.2,

$$J_1(u(T)) = \sup_{0 \le t \le T} |u(t)|,$$

$$F(c, T) = \int_0^T |g(t, c)| \, dt \, ,$$

$$= \int_0^T \left| \frac{1}{m} \left[ \frac{t}{x_0} + c_2 \right] \right| dt \, ,$$

$$\frac{1}{J_{1\,\min}} = \overline{\alpha} = \min_{c \in P} F(c, \tau) \, .$$

If we consider the three cases

  (i)  $g(t, c) > 0, \; \forall \, t \in [0, T]$.
  (ii)  $g(t, c) < 0, \; \forall \, t \in [0, T]$.
  (iii)  $g(t, c)$ changes sign,

we find that

$$\overline{\alpha} = \frac{T^2}{4mx_0} = \frac{1}{J_{1\,\min}} \, .$$

By Theorem 5.5.2, the unique (a.e.) optimal control is

$$u^*(t) = \frac{4mx_0}{T^2} \operatorname{sgn} \left[ \frac{t}{x_0} - \frac{T}{2x_0} \right] \quad \text{for all } x_0 \neq 0, \quad 0 \leq t \leq T \, .$$

Thus

$$u^* = \frac{+4mx_0}{T^2} \, , \qquad u^* = \frac{-4mx_0}{T^2} \, .$$

Just as in Example 5.3.2 of (5.3.16), we consider the dynamics with these two values of $u^*$. For $u = +J_{\min}$,

$$x_2^2 = 2J_{\min}(x_1 - b_1) \, ;$$

$u = -J_{\min}$,

$$x_2^2 = -2J_{\min}(x_1 - b_2) \, .$$

Note that if $J_{\min} > 1$, our parabola will be spread out more; but if $J_{\min} = 1$ ($m = 1$, $x_0 = 1$, $T = 2$), the strategy is the same as the time-optimal control one. But if $J_{\min} < 1$, the parabola is more condensed. The switching curve is as described in Fig. 5.5.4. Note that $\epsilon = \infty$, as we saw before. The terminal manifolds $M_1$ and $M_2$ are as shown in Fig. 5.5.4, $M_2$ is the space above the switching curve, and $M_1$ below it. Optimal controls law is $f(x_1, x_2)$:

$$f(x_1, x_2) = \begin{cases} -J_{\min} & \text{on } M_2 \cup \Gamma_+ \, , \\ +J_{\min} & \text{on } M_1 \cup \Gamma_+ \, . \end{cases}$$

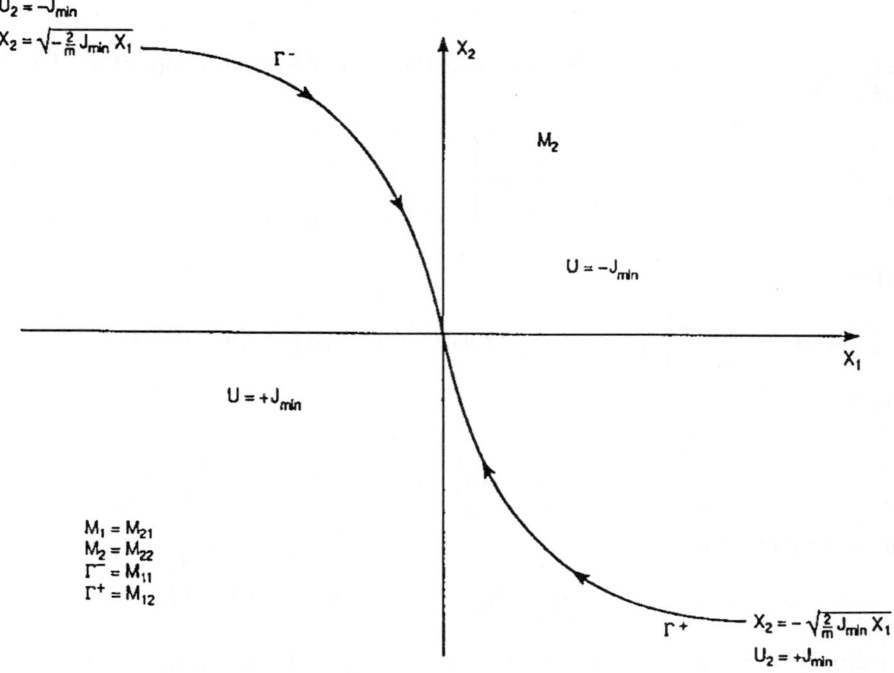


Fig. 5.5.4

Switching locus $W$:

$$x_2(t) = \begin{cases} -\sqrt{2J_{\min} x_1} & \text{for } x_1 \geq 0 \quad \Gamma_+ \,, \\ \sqrt{-2J_{\min} x} & \text{for } x_1 < 0 \quad \Gamma_- \,. \end{cases}$$

**Example 5.5.3**    Damped Harmonic Oscillator

$$\ddot{x}(t) + 2\alpha\dot{x}(t) + \omega^2 x(t) = u(t) \,.$$

$$|u| \leq 1 \qquad \left( \frac{b}{m} = 2\alpha, \ \frac{k}{m} = \omega^2 \right) \,.$$

$$\underline{x}(t_1) = \begin{bmatrix} 0 \\ 0 \end{bmatrix} \,, \qquad \underline{x}(0) = \begin{bmatrix} x_0 \\ 0 \end{bmatrix}$$

$$\underline{x} = \begin{bmatrix} x \\ \dot{x} \end{bmatrix} \,, \qquad A = \begin{bmatrix} 0 & 1 \\ -\omega^2 & -2\alpha \end{bmatrix} \,, \qquad B = \begin{bmatrix} 0 \\ 1/m \end{bmatrix} \,.$$

Recall the following:

$$y(t_1) = e^{-At_1}\underline{x}(t_1) - \underline{x}(0) = -\underline{x}(0) = \begin{bmatrix} -x_0 \\ 0 \end{bmatrix} \,,$$

$$P = \{c : c^T y(t_1) = 1\}, \qquad c = (c_1, c_2).$$

Hence $c_1 = -\frac{1}{x_0}$, $c_2$ is free. The index of the control system is $g(t, c) = c^T e^{-At} B$. If $\alpha^2 - \omega^2 < 0$, $\beta^2 = \omega^2 - \alpha^2$,

$$e^{-At} B = \frac{1}{m} e^{\alpha t} \begin{bmatrix} -\dfrac{1}{\beta} \sin \beta t \\[2mm] \cos \beta t + \dfrac{\alpha}{\beta} \sin \beta t \end{bmatrix},$$

so that

$$g(t, c) = \frac{1}{m} e^{\alpha t} \left[ \frac{1}{\beta} (-c_1 + \alpha c_2) \sin \beta t + c_2 \cos \beta t \right] = D e^{\alpha t} \sin(\beta t + \delta),$$

where

$$D = \frac{1}{m} \left[ \left( -\frac{c_1 + c_2 \alpha}{\beta} \right)^2 + c_2 \right]^{\frac{1}{2}}, \qquad \delta = \tan^{-1} \left[ \frac{c_2 \beta}{-c_1 + c_2 \alpha} \right].$$

The cost function is

$$J_1(u(t_1)) = \sup_{0 \le t \le t_1} |u(t)|, \qquad \mathcal{U} = \{u : |u(t)| \le 1\}.$$

Our aim is to find optimal feedback control $u^*(t) \in \mathcal{U}$ that transfers $x(0)$ to $x(t_1)$ in time $t_1$ while minimizing $J_1(u(t_1))$.

$$F(c_1, t_1) = D \int_0^{t_1} |e^{\alpha t} \sin(\beta t + \delta)| dt, \qquad \frac{1}{J_{1\,\min}} = \min_{c \in P} F(c, t_1) = F(c^*, t_1).$$

This minimum occurs when $c_2 = 0$, $c_1 = \begin{bmatrix} -1/x_0 \\ 0 \end{bmatrix}$,

$$D = \frac{1}{m} \left[ -\frac{\frac{-1}{x_0}}{\beta} + 0 \right]^{\frac{1}{2}} = \frac{1}{m|\beta x_0|}, \qquad \delta = 0.$$

$$\frac{1}{J_{1\,\min}} = \frac{1}{m|\beta x_0|} \int_0^{t_1} |\sin \beta t| dt.$$

By Theorem 5.4.2, the optimal control is $u^*(t) = J_{1\,\min} \operatorname{sgn}[\sin \beta t]$. Thus $u^* = J_{1\,\min}$ or $-J_{1\,\min}$. With these controls we have that if $u = J_{1\,\min}$,

$$\left( \omega x_1 - \frac{J_{1\,\min}}{\omega} \right)^2 + x_2^2 = a^2,$$

which is a spiral approaching $(\frac{J_{1\,\min}}{\omega^2}, 0)$ as $t \to \infty$. The drawing similar to Fig. 5.3.13. If $u = -J_{1\,\min}$,

$$\left( \omega x_1 + \frac{J_{1\,\min}}{\omega} \right)^2 + x_2^2 = a^2,$$

which is a spiral approaching $\left(\frac{-J_{1\,\min}}{\omega^2}, 0\right)$ as $t \to \infty$. The switching curves are similar to those given in Figs. 5.3.15 and 5.3.17. We now use Hájek's Corollary to deduce that

$$\epsilon < \frac{\pi}{\sqrt{\omega^2 - \alpha^2}} \,.$$

In Int $\mathbb{R}(\epsilon)$ optimal controls have at most one discontinuity, and

$$\text{Int } \mathbb{R}(\epsilon) = M_1 \cup M_2 \cup \Gamma_+ \cup \Gamma_- \,.$$

Our optimal strategy is

$$f(x_1, x_2) = \begin{cases} -J_{1\,\min} & \text{in } M_2 \cup \Gamma_- \,, \\ J_{1\,\min} & \text{in } M_1 \cup \Gamma_+ \,. \end{cases}$$

**Example 5.5.4**

$$\dot{x}_1 = x_2 + u_1 \,, \qquad \dot{x} = Ax + Bu \,,$$

$$\dot{x}_2 = -x_1 + u_2 \,, \qquad A = \begin{bmatrix} 0 & 1 \\ -1 & 0 \end{bmatrix} \,, \qquad B = \begin{bmatrix} 1 & 0 \\ 0 & 1 \end{bmatrix} \,,$$

$$x(t_1) = \begin{bmatrix} 0 \\ 0 \end{bmatrix} \,, \qquad x(0) = \begin{bmatrix} x_0 \\ 0 \end{bmatrix} \,.$$

Just as before, $c_1 = -\frac{1}{x_0}$, $c_2$ is free.

$$e^{-At} = \begin{bmatrix} \cos t, & -\sin t \\ \sin t, & \cos t \end{bmatrix} \,, \qquad g(t, c) = D \begin{bmatrix} \sin(t + \delta) \\ \cos(t + \delta) \end{bmatrix} \,,$$

where

$$D = \sqrt{c_1^2 + c_2^2} \,, \qquad \delta = \tan^{-1} -\frac{c_2}{c_1} \,,$$

$$F(t_1, c) = \int_0^{t_1} \sum_{j=1}^{2} |g_j(t, c)| dt \,,$$

$$= D \int_0^{t_1} [|\sin(t + \delta)| + |\cos(t + \delta)|] dt \,,$$

$$\frac{1}{J_{1\,\min}} = \min_{c \in P} D \int_0^{t_1} [|\sin(t + \delta)| + |\cos(t + \delta)|] \,.$$

The minimum is attained when $c_2 = 0$, i.e., $\delta = 0$. Thus

$$\frac{1}{J_{1\,\min}} = \frac{1}{|x_0|} \int_0^{t_1} [|\sin(t)| + |\cos(t)|] dt \,.$$

Just as before,

$$u^*(t) = J_{1\min} \operatorname{sgn} \begin{bmatrix} \sin t \\ \cos t \end{bmatrix}.$$

It follows that optimal controls will be the cube of length $J_{1\min}$

$$u_b = \begin{bmatrix} J_{1\min} \\ -J_{1\min} \\ \text{Case 2} \end{bmatrix}, \quad u_a = \begin{bmatrix} J_{1\min} \\ J_{1\min} \\ \text{Case 1} \end{bmatrix}, \quad u_c = \begin{bmatrix} -J_c \\ J_m \\ \text{Case 3} \end{bmatrix}, \quad u_d = \begin{bmatrix} -J_{1\min} \\ -J_{\min} \\ \text{Case 4} \end{bmatrix}.$$

Just as before we have the four cases:

**Case 1:**

$$\dot{x}_1 = x_2 + J, \qquad u_a = \begin{bmatrix} J \\ J \end{bmatrix}, \qquad \dot{x}_2 = -x_1 + J, \qquad (x_1 - J)^2 + (x_2 + J)^2 = d^2,$$

which is a circle of radius $d$ centered at $(J, -J) \equiv d$, in the phase plane.

**Case 2:**

$$u = \begin{pmatrix} J \\ -J \end{pmatrix}, \quad \dot{x}_1 = x_2 + J, \quad \dot{x}_2 = -x_1 - J, \quad (x_1 + J)^2 + (x_2 + J)^2 = d^2.$$

This is a circle of radius $d$ center at $(-J, -J)$.

**Case 3:**

$$u = \begin{bmatrix} -J \\ J \end{bmatrix},$$

$$(x_1 - J)^2 + (x_2 - J)^2 = d^2.$$

This is a circle of radius $d$ center at $(J, J)$.

**Case 4:**

$$u = \begin{bmatrix} -J \\ -J \end{bmatrix},$$

$$(x_1 + J)^2 + (x_2 - J)^2 = d^2.$$

This is a circle of radius $d$ centered at $(-J, J)$. We observe that $\epsilon < \pi$, and with this we define the terminal manifold $\mathbb{R}(\epsilon) = \mathbb{R}(\pi)$ as follows:

$$\operatorname{Int} \mathbb{R}(\pi) = \bigcup_{k=0}^{2} M_k = M_1 \cup M_2,$$

where

$$M_k = \bigcup_{\underline{i} \in I_k} M_{k\underline{i}} \, ,$$

with $I_k$ as the permutation of the controls;

$$M_0 = \{0\} \, , \quad M_1 = M_{11} \cup M_{12} \cup M_{13} \cup M_{14} \, , \quad M_2 = \bigcup_{j=1}^{4} M_{2j} \, .$$

Recall that there are $2.2^{2-1}$ nonvoid sets $M_{2i} = 4$ terminal manifolds. Suppose $\left\{ \begin{bmatrix} J \\ J \end{bmatrix} \begin{bmatrix} J \\ -J \end{bmatrix} \begin{bmatrix} -J \\ J \end{bmatrix} \begin{bmatrix} -J \\ -J \end{bmatrix} \right\}$ is the optimal sequence. The corresponding centers of the circular arcs are $\begin{bmatrix} J \\ -J \end{bmatrix} \begin{bmatrix} -J \\ -J \end{bmatrix} \begin{bmatrix} J \\ J \end{bmatrix} \begin{bmatrix} -J \\ J \end{bmatrix}$. By definition, $M_1$ is the set of points that can be driven to the origin with no switch in control.

$$M_1 = \Gamma_{++} \cup \Gamma_{+-} \cup \Gamma_{--} \cup \Gamma_{-+} \equiv \sum_{j=1}^{4} M_{1j} \, ,$$

$M_2$ is the set of points that can be steered to zero with at most one switch. $M_{21}$ corresponds to the control $\begin{bmatrix} J \\ J \end{bmatrix}$. $M_{22}$ corresponds to $\begin{bmatrix} J \\ -J \end{bmatrix}$, $M_{23}$ to $\begin{bmatrix} -J \\ -J \end{bmatrix}$, and $M_{24}$ to $\begin{bmatrix} -J \\ J \end{bmatrix}$. Switching occurs at intervals of length $\pi/2$, after the first switch, not necessarily including the last switch. See Fig. 5.5.5. Optimal feedback control is

$$f(x_1, x_2) = \begin{cases} J \begin{bmatrix} 1 \\ 1 \end{bmatrix} & \text{in } M_{11} \cup M_{21} \, , \\[2ex] J \begin{bmatrix} 1 \\ -1 \end{bmatrix} & \text{in } M_{12} \cup M_{22} \, , \\[2ex] J \begin{bmatrix} 1 \\ -1 \end{bmatrix} & \text{in } M_{13} \cup M_{23} \, , \\[2ex] J \begin{bmatrix} -1 \\ 1 \end{bmatrix} & \text{in } M_{14} \cup M_{24} \, . \end{cases}$$

Note that $J$ depends on $x_0$ and on the final time. It has been observed by Ukwu (personal communication) that

$$J = \begin{cases} \dfrac{-x_0}{4k + 1 - \cos(T - k\pi) + \sin(T - k\pi)} & \text{if } k\pi \le T \le \dfrac{2k+1}{2}\pi \, , \\[3ex] \dfrac{-x_0}{4k + 3 - \cos(T - k\pi) - \sin(T - k\pi)} & \text{if } \dfrac{2k+1}{2}\pi < T \le (k+1)\pi \, . \end{cases}$$

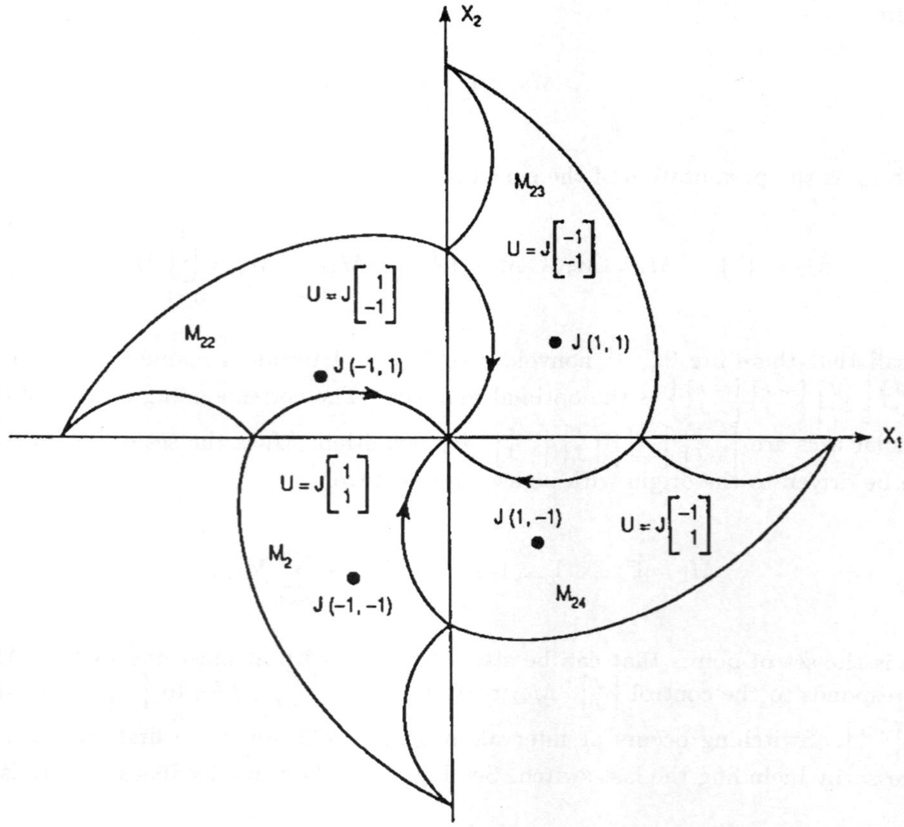


Fig. 5.5.5

## Example 5.5.5    Impulsive Controls

We consider the cost function $J_3(u(T))$ for the damped oscillator with $m = k = 1$, $x_0 = 1$, and $\alpha = 0.03$. Recall that $g(t, c) = De^{\alpha t} \sin(\beta t + \delta)$, where

$$D = \left( \left( -\frac{c_1}{\beta} + c_2 \alpha/\beta \right)^2 + c_2^2 \right)^{\frac{1}{2}} \bigg/ m \,,$$

$$\delta = \tan^{-1} \left[ \frac{c_2 \beta}{-c_1 + c_2 \alpha} \right] ,$$

$$F(c^*, T) = \max_{0 \le t \le T} g(c, T) \,,$$

$$\frac{1}{J_{\min}} = \min_{c \in P} F(c, T) \,.$$

The maximum occurs at $\tau = (\frac{11\pi}{2} - \delta)/\beta$ (approximately) (see Redmond and Silverberg [10]).

$$F(c, T) = De^{\alpha\tau}|\sin(\beta\tau + \delta)|.$$

$F(c, T)$ is minimized over $H$ by letting $c_2 = 0$, so that

$$\frac{1}{J_{3\,min}} = \frac{1}{\beta m x_0}e^{\alpha\tau},$$

in which $\tau = \frac{11\pi}{2\beta}$. Optimal strategy is $u^*(t) = \beta m x_0 e^{-\alpha\tau}\mathrm{sgn}(\sin\beta\tau)\delta(t - \tau)$. As observed by Redmond and Silverberg [10], Figs. 5.5.6A and 5.5.6B show that impulse control allows for a period of free oscillation during which the natural damping present in the system removes energy from the system. Then at the last instant, when the systems displacement is identically zero, an impulse of magnitude equal to the system's momentum is applied. This instantaneous change in velocity abruptly transfers the system to the origin at time $\tau$. The timing of the impulse is extremely important. In Fig. 5.5.6B, the impulse is applied when the potential energy is a minimum and the system's kinetic energy is a local maximum. This observation is of great value in the development of near-optimal control laws of large scale systems.

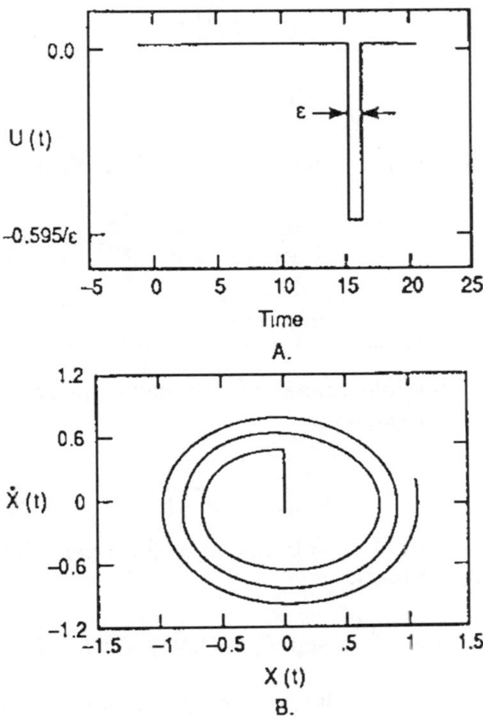


Fig. 5.5.6   Impulse control of a damped harmonic oscillator.

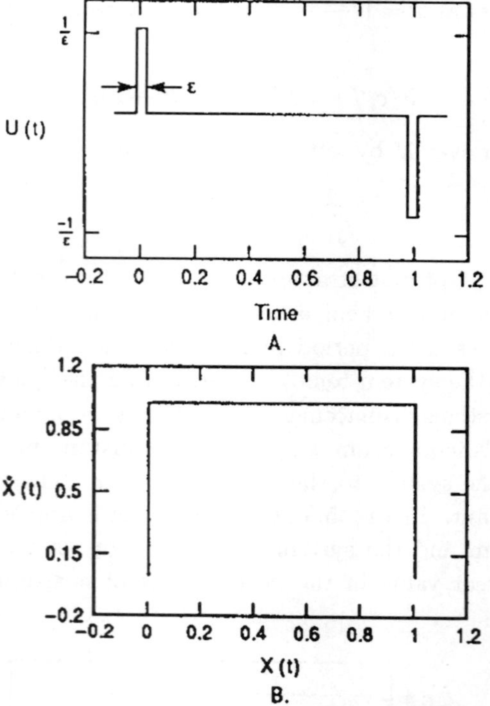


Fig. 5.5.7

We consider another example — the Rigid Body Maneuver. $m\ddot{x}(t) = u(t)$.

$$F(c, T) = \sup_{0 \leq t \leq T} \left| -\frac{t}{x_0 m} + \frac{c_2}{m} \right|,$$

$$\frac{1}{J_{3\,\min}} = \min_{c_2} \sup_{0 \leq t \leq T} \left| -\frac{t}{x_0 m} + \frac{c_2}{m} \right|.$$

Considering all three possible ranges of argument (positive, negative, and sign change) the minimum is attained at

$$\frac{1}{J_{3\,\min}} = \frac{T}{2x_0 M}.$$

This minimum occurs both at $\tau_1 = 0$, and $\tau_2 = T$. From (5.5.37) with $M = 2$, the open-loop optimal control law is

$$u^*(t) = \frac{x_0 M}{T} \operatorname{sgn}(T_2 - t)[\delta(t) + \delta(t - T)].$$

This control consists of one initial impulse to impart a velocity to the system. Then the system drifts until the desired time $T$, when a final impulse is applied to terminate the motion at rest. See Figs. 5.5.7A and 5.5.7B.

Redmond and Silverberg have determined that if fuel consumption in propulsive systems is defined by

$$Fu = \frac{1}{Isp} \int_0^T |u(t)|dt\,,$$

where $u(t)$ represents the thrust and $Isp$ the specific impulsive, which is set to be 1, there are big savings in fuel if impulse controls are used. The result is illustrated in the table below. Continuous controls are those of Theorem 5.4.4; bang-bang controls represent Theorem 5.4.2.

Table 5.5.1

| | Fuel Consumption | | |
| --- | --- | --- | --- |
| | Bang-Bang | Continuous | Impulse |
| Oscillator | 1.168 | 0.921 | 0.595 |
| Maneuver | 4.0 | 3.0 | 2.0 |

## References


1. E. N. Chukwu and O. Hájek, *A Connection Between Optimal Control and Disconjugacy in Dynamical Systems, Vol. II*, An International Symposium, Academic Press, 1976.
2. E. N. Chukwu and O. Hájek, "Disconjugacy and Optimal Control," *J. Optimization Theory and Applications* **27**(3) (1979) 333–356.
3. O. Hájek, "On the Number of Roots of Exp-Trig Polynomials," *Computing* **18** (1977) 177–183.
4. O. Hájek, "$L_1$-Optimization in Linear Systems with Bounded Controls," *J. Optimization Theory and Applications* **29** (1979) 409–436.
5. O. Hájek and W. Krabs, *On A General Method For Solving Time-Optimal Linear Control Problems*, Preprint No. 579, 1981, Technische Hochschule Darmstadt.
6. H. Hermes and J. P. LaSalle, *Functional Analysis and Time Optimal Control*, Academic Press, New York, 1969.
7. L. W. Neustadt, "Minimum Effort Control Systems," *SIAM Journal on Control* **1** (1962) 16–31.
8. D. Yeung, "Synthesis of Time-Optimal Control," Ph.D Thesis, Case Western Reserve University, Cleveland, Ohio, 1974.
9. H. A. Antosiewicz, "Linear Control Systems," *Arch. Rat. Mech. Anal.* **12** (1963) 313–324.
10. J. Redmond and L. Silverberg, "Fuel Consumption in Optimal Control," *AIAA Journal of Guidance Control and Dynamics* **15**(2) (1992) 424–430.
11. J. Dugunji, *Topology*, Allyn and Bacon, Boston, 1966.

## Chapter 6

# Control of Linear Delay Systems

In this section we study some controllability questions of the system

$$\dot{x}(t) = L(t, x_t) + B(t)u(t), \qquad (6.1.1)$$

where

$$L(t, x_t) = \int_{-h}^{0} d_\theta \eta(t, \theta) x(t + \theta), \qquad (6.1.1b)$$

and $\eta(t, \theta)$ is a matrix-valued function. We assume that $\eta$ is measurable in $t$ and of bounded variation in $\theta$ on $[-h, \theta]$, and that there is an integrable function $m :$ $(-\infty, \infty)$ such that

$$|L(t, \phi)| \le m(t)\|\phi\|, \quad t \in (-\infty, \infty), \quad \phi \in C.$$

Throughout this section we assume that $t \to B(t)$ is continuous.

## 6.1 Euclidean Controllability of Linear Delay Systems

In time-optimal control theory, one assumes that beginning from some initial point the target point can be reached in finite time by using some admissible control. The usual initial point is the point $\phi$ in the function space $C = C([-h, 0], E^n)$ of continuous functions from $[-h, 0]$ into $E^n$, or the Sobolev space $W_p^{(1)}([-h, 0], E^n) = W_{p,\,2\le p\le\infty}^{(1)}$, which is the space of absolutely continuous functions from $[-h, 0]$ to $E^n$ with the first derivative square integrable. We first consider the target to be a point in Euclidean space $E^n$. It is desirable to reach this point (from any initial function) in finite time using some admissible controls $u$ that are unlimited in magnitude. We now investigate this Euclidean controllability problem for the system

$$\dot{x}(t) = L(t, x_t) + B(t)u(t), \qquad (6.1.1)$$

where $L$ and $B$ satisfy all the conditions of Sec. 2.2.

187

**Definition 6.1.1**  The linear control process (6.1.1) is Euclidean controllable on the interval $[\sigma_0, t_1]$ if for each $\phi \in C$ and $x_1 \in E^n$, there is a square integrable controller $u$ such that $x_\sigma(\sigma\phi, u) = \phi$ and $x(t_1, \sigma, \phi, u) = x_1$. It is Euclidean null-controllable if in the definition $x_1 = 0$. It is Euclidean controllable if it is Euclidean controllable on every interval $[\sigma, t_1]$, $t_1 > \sigma$.

The conditions for Euclidean controllability have been well studied: See Kirillova and Churakova [11], Gabasov and Kirillova [8], Weiss [15], and Manitius and Olbrot [13, 14]. They are all conditions on matrices representing the system, and are a consequence of the following lemma:

**Lemma 6.1.1**  *In the system* (6.1.1), *the following are equivalent:*

(i)  *The matrix*

$$G(\sigma, t_1) = \int_\sigma^{t_1} U(t_1, s)B(s)B^*(s)U(t, s)ds \qquad (6.1.2)$$

*has rank $n$, where $U(t, s)$ is the $n \times n$ matrix solution of*

$$\dot{x}(t) = L(t, x_t) \, . \qquad (6.1.3)$$

(ii)  *The relation*

$$c^T Y(s, t_1)B(s) \equiv 0 \, , \qquad c \in E^n \, , \qquad s \in [\sigma, t_1]$$

*implies $c = 0$, where $Y(s, t_1) = U(t_1, s)$ a.e. in $s$ is an $n \times n$ matrix defined by*

$$Y(\sigma, t) = \begin{cases} 0, & \sigma > t \, , \\ I - \displaystyle\int_\sigma^t Y(\alpha, t)\eta(\alpha, \sigma - \alpha)d\alpha, & \sigma \leq t \, . \end{cases} \qquad (6.1.4)$$

(iii)  *The system* (5.1.1) *is Euclidean controllable on* $[\sigma_1, t_1], t_1 > \sigma$.

As a consequence of this lemma, the next characterization of Euclidean controllability in terms of the system's coefficients is available for the autonomous system

$$\dot{x}(t) = A_0 x(t) + A_1 x(t - h) + Bu(t) \, , \qquad (6.1.5)$$

where $A_0, A_1, B$ are constant matrices. First introduce the determining equations,

$$Q_k(s) = A_0 Q_{k-1}(s) + A_1 Q_{k-1}(s - h), \quad k = 1, 2, 3, \ldots, \quad s \in (-\infty, \infty) \, .$$

$$Q_0(s) = \begin{cases} B, & s = 0 \, , \\ 0, & s \neq 0 \, , \end{cases} \qquad (6.1.6)$$

and define

$$\overline{Q}_n(t_1) = [Q_0(s), Q_1(s), \ldots, Q_{n-1}(s), \quad s \in [0, t_1]] \, .$$

We have:

**Theorem 6.1.1**  *System* (6.1.5) *is Euclidean controllable on* $[0, t_1]$ *if and only if*

$$\operatorname{rank} \overline{Q}_n(t_1) = n.$$

**Remark 6.1.1**  Note that the nonzero elements of $Q_k(s)$ form the sequence:

$$
\begin{array}{lll}
s = 0 & h & 2h \\
Q_0(s) = B_0 & & \\
Q_1(s) = A_0 B & A_1 B_0 & \\
Q_2(s) = A_0^2 B & (A_0 A_1 + A_1 A_0) B_0 & A_1^2 B_0 \,.
\end{array}
$$

Note that if $t_1 \leq h$, the only elements in the sequence are the terms $[B, A_0 B, \ldots, A_0^{n-1} B]$, so that $E^n$-controllability on an interval less than $h$ implies the full rank of $[B, \ldots, A_0^{n-1} B]$. If this has less than full rank and $t_1 > h$, other terms can be added to $\overline{Q}_n(t_1)$ and the system may still be controllable on $[0, t_1]$. This contrasts with the situation in Sec. 5.1 for systems without delay, where if the system can be steered to some point of $E^n$, it can be steered to that point in an arbitrarily short time.

**Example 6.1.1**  Consider

$$\dot{x}(t) = A_0 x(t) + A_1 x(t-1) + B u(t),$$

where

$$
A_0 = \begin{bmatrix} -1 & 0 \\ 0 & -1 \end{bmatrix}, \quad
A_1 = \begin{bmatrix} 0 & 0 \\ 1 & 0 \end{bmatrix}, \quad
B = \begin{bmatrix} 1 \\ 0 \end{bmatrix},
$$

$$
Q_0 = \begin{bmatrix} 1 \\ 0 \end{bmatrix}, \quad Q_0(s) \equiv 0, \quad s \neq 0.
$$

$$
Q_1(0) = A_0 B = \begin{bmatrix} -1 \\ 0 \end{bmatrix}, \quad
Q_1(1) = A_1 B = \begin{bmatrix} 0 \\ 1 \end{bmatrix},
$$

$$
Q_1(s) \equiv 0, \quad s \neq 1 \quad \text{and} \quad s \neq 0.
$$

$$
\overline{Q}_2(2) = \begin{bmatrix} 1 & -1 & 0 \\ 0 & 0 & 1 \end{bmatrix}.
$$

Since $\operatorname{rank} \overline{Q}_2(2) = 2$, the system is Euclidean controllable on $[0, 2]$.

**Example 6.1.2**  Consider

$$\dot{x}(t) = A_0 x(t) + A_1 x(t-h) + B u(t),$$

$$
A_0 = \begin{bmatrix} 1 & 0 & 0 \\ 0 & 1 & 1 \\ 0 & 0 & 0 \end{bmatrix}, \quad A_1 = \begin{bmatrix} 0 & 0 & 1 \\ 0 & 0 & 0 \\ 0 & 0 & 0 \end{bmatrix}, \quad B = \begin{bmatrix} 0 \\ 0 \\ 1 \end{bmatrix},
$$

$$
A_0 B = \begin{bmatrix} 0 \\ 1 \\ 0 \end{bmatrix}, \quad A_1 B = \begin{bmatrix} 1 \\ 0 \\ 0 \end{bmatrix}, \quad B = \begin{bmatrix} 0 \\ 0 \\ 1 \end{bmatrix},
$$

$$
\overline{Q}(t_1) = [A_0 B, A_1 B, B], \quad t_1 > h,
$$

$$
\operatorname{rank} \overline{Q}(t_1), \quad t_1 > h \text{ is } 3.
$$

Hence this system is $E^3$-controllable.

**Example 6.1.3**    Consider the scalar differential-difference equation

$$
x^n(t) + \sum_{i=1}^{n-1} a_i x^i(t) + \sum_{i=0}^{n-1} b_i x^i(t-1) = cu(t),
$$

where $a_i$, $b_i$ and $c$ are constants and $x$, $u$ are scalar functions. Define $x_1 = x^{(0)}$, $x_2 = x^{(1)} \cdots x_n = x^{(n-1)}$

$$
\dot{x}_1(t) = x_2(t)
$$

$$
\dot{x}_2(t) = x_3(t)
$$

$$
\vdots
$$

$$
\dot{x}_n(t) = -\sum_{i=0}^{n-1} a_i x_{i+1}(t) - \sum_{i=0}^{n-1} b_i x_{i+1}(t-1) + cu(t).
$$

Written in matrix form

$$
\dot{x}(t) = A_0 x(t) + A_1 x(t-1) + Bu(t),
$$

we have

$$
A_0 = \begin{bmatrix} 0 & 1 & 0 & \cdots \\ 0 & 0 & 1 & \cdots \\ \vdots & \vdots & & \cdots \\ -a_0 & -a_1 & -a_2 & \cdots & -a_{n-1} \end{bmatrix},
$$

$$
A_1 = \begin{bmatrix} 0 & 0 & \cdots & 0 \\ -b_0 & -b_1 & \cdots & -b_n \end{bmatrix},
$$

$$
B = \begin{bmatrix} 0 \\ 0 \\ \vdots \\ 0 \\ 1 \end{bmatrix},
$$

$$
\overline{Q}_n(0) = [Q_0(s) \quad Q_1(s), \ldots, Q_{n-1}(s)]
$$
$$
= [B, A_0 B, \ldots, A_0^{n-1} B] .
$$

This has full rank. Hence the system is Euclidean space controllable on any $[0, t_1]$, $t_1 > 0$.


## 6.2 Linear Function Space Controllability

The true state of solutions of (6.1.1) is an element of some function space. Thus the state at time $t$ is denoted by $x_t$, and this is a segment of the trajectory $s \to x(s)$ $t - h \le s \le t$. Very often $C = C([-h, 0], E^n)$ is used. But $W_2^{(1)}([-h, 0], E^n) = W_2^{(1)}$, the state space of absolutely continuous functions from $[-h, 0]$ to $E^n$ with first derivative square integrable and in $L_2([-h, 0], E^n)$ is also natural. Indeed, if $\phi \in W_2^{(1)}([-h, 0], E^n)$ and $u \in L_2([0, t_1], E^n)$, then $x(\phi) = x : [0, t_1] \to E^n$ is absolutely continuous, and if $\dot{x}(t) = L(x_t) + u(t)$, then $\dot{x} \in L_2([0, t], E^n)$. Therefore $x \in W_2^{(1)}([-h, t_1], E^n)$, so that $x_t \in W_2^{(1)}([-h, 0], E^n)$ for all $t \in [0, t_1]$. In this section we explore controllability questions in $W_2^{(1)}$.

**Definition 6.2.1** The system (6.1.1) is controllable on the interval $[\sigma, t_1]$ if for each $\phi, \psi \in W_2^{(1)}$ there is a controller $u \in L_2([\sigma, t_1], E^m)$ such that $x_{t_1}(\sigma, \phi, u) = \psi$, $x_\sigma(\sigma, \phi, u) = \phi$. It is null controllable if $\psi \equiv 0$ in the above definition. It is said to be controllable (null controllable) if it is controllable (null controllable) on every interval $[\sigma, t_1]$ with $t_1 > \sigma + h$. The following fundamental result is given in Banks, Jacobs, and Langenhop [2, p. 616].

**Theorem 6.2.1** *In (6.1.1), in addition to the prevailing conditions of this section (for $L$ and $B$), assume that*

(i) *$t \to B^+(t)$, $t \in E$ is essentially bounded on $[t_1 - h, t_1]$, where $B^+(t)$ denotes the Moore–Penrose generalized inverse of $B(t)$ [11]. Then (6.1.1) is controllable on the interval $[\sigma, t_1]$ with $t_1 > \sigma + h$ if and only if*

(ii) *$\operatorname{rank} B(t) = n$ on $[t_1 - h, t_1]$.*

*When $t \to B(t)$ is constant, we have the following sharp result:*

**Corollary 6.2.1** *Suppose the $n \times m$ matrix $B$ in (6.1.1) is constant. A necessary and sufficient condition that (6.1.1) is controllable on $[\sigma, t_1]$ is that $\operatorname{rank} B = n$.*

The proof of Theorem 6.2.1 is available in [2] where a thorough discussion of the result is made. Manitius also discusses the result [14].

In the investigation of controllability of (6.1.1) in the state space $W_2^{(1)}$, with $L_2$ controls, it is impossible to handle the situation of controls with pointwise control constraints. In this case controls are taken in the space $L_\infty$, and the state space is either $W_\infty^{(1)}$ or $C$.

Though several researches are available on the optimal control of variants of (6.1.1) when the state space is $C$ and the controls are $L_\infty$, (see Banks and Kent [3] and Angell [1]) there seems to be no general results on the controllability of such systems, a necessary requirement for the existence of an optimal control. There are other problems connected with such treatment; for example, the attainable sets cannot be closed and the multipliers cannot in general be nontrivial. See Manitius [14, pp. 120, 128]. We begin to address some of these problems by first formulating function space controllability state conditions of (6.1.1) when the space is $C$ and the controls are $L_\infty$. The result is also valid for $W_\infty^{(1)}$ state space of initial functions and $C^1([-h,0], E^n)$ space of terminal conditions, with $L_\infty$ controls. We assume that the control matrix $B(t)$ is continuous.

**Theorem 6.2.2**  *In (6.1.1), namely,*

$$\dot{x}(t) = L(t, x_t) + B(t)u(t), \tag{6.1.1}$$

*suppose the state space is $C = C([-h,0], E^n)$, and $B$ is continuous. Assume:*

> (i) $\operatorname{rank} B(t) = n$ *on* $[t_1 - h, t_1]$.
> *Then (6.1.1) is controllable on* $[\sigma, t_1]$, $t_1 > \sigma + h$.

**Proof**  Assume (i). Then rank $B(t_1 - h) = n$, and this implies that

$$H(\sigma, t_1 - h) = \int_\sigma^{t_1 - h} X(t_1 - h, s)B(s)B^*(s)X^*(t_1 - h, s)ds \tag{6.2.1}$$

has rank $n$. Let $t_1 > \sigma + h$, $\phi, \psi \in C([-h,0], E^n)$. It is a consequence of Lemma 2.3 of Manitius [14], (or Lemma 6.1.1) that there is a $u \in L_\infty([\sigma, t_1 - h], E^m)$ such that $x(t_1 - h, \sigma, \phi, u) = \psi(-h)$. We extend $u$ and $x = x(\cdot, \sigma, \phi, u)$ to the interval $[\sigma, t_1]$ so that $\psi(t - t_1) = x(t)$, $t_1 - h \le t \le t_1$, and

$$\psi(t - t_1) = \psi(-h) + \int_{t_1-h}^{t} L(s, x_s)ds + \int_{t_1-h}^{t} B(s)u(s)ds, \tag{6.2.2}$$

on $[t_1 - h, t_1]$. Note that the integral form of (6.1.1) is

$$x(t) = x(0) + \int_0^t L(s, x_s)ds + \int_0^t B(s)u(s)ds,$$
$$x(t) = \phi(t), \quad t \in [-h, 0], \quad t \ge 0, \quad \phi \in C. \tag{6.2.3}$$

Because of (i),

$$\text{rank}[B(t)B^*(t)] = n, \quad \forall t \in [t_1 - h, t_1],$$

and

$$H(t) = \int_{t_1-h}^{t} B(s)B^*(s)ds, \quad t > t_1 - h$$

has rank $n$ for each $t \in (t_1 - h, t_1)$. As a consequence of this rank condition, define for each $t \in (t_1 - h, t_1]$ a control $v(s)$, $s \le t$,

$$v(s) = B^*(s)H^{-1}(t)\left[\psi(t - t_1) - \psi(-h) - \int_{t_1-h}^{t} L(s, x_s)ds\right],$$

where $x(\cdot)$ is that solution of (6.1.1) or (6.2.3) with

$$x_\sigma(\phi) = \phi, \quad x(t_1 - h, \phi, u) = \psi(-h).$$

That $v$ exists follows from the usual argument [9, p. 141] for the existence of a solution of a linear system (6.1.1). With this $v$ in the right-hand side of (6.2.2), we have

$$\psi(-h) + \int_{t_1-h}^{t} L(s, x_s)ds + \int_{t_1-h}^{t} B(s)B^*(s)H^{-1}(t)$$

$$\times \left[\psi(t - t_1) - \psi(-h) - \int_{t_1-h}^{t} L(s, x_s)ds\right]$$

$$= \psi(t - t_1), \quad t \in [t_1 - h, t_1]. \qquad \square$$

It is interesting to note the absence of condition $t \to B^+(t)$ being essentially bounded on $[t_1 - h, t_1]$, which is required in Banks, Jacobs, and Langenhop [2]. Indeed, if $B(t)$ has rank $n$ on $[t_1 - h, t_1]$ and is continuous, then $B^+(t)$ is continuous on $[t_1 - h, t_1]$. See Campbell [4, p. 225]. In this case $B^+(t)$ is uniformly bounded on $[t_1 - h, t_1]$.

**Remark 6.2.1**  Note that the control that does the transfer from $\phi$ to $\psi$ is

$$\bar{u} = \begin{cases} u \in L_\infty([\sigma, t_1 - h], E^m), \\ v : [t_1 - h, t_1], \quad \text{continuous}. \end{cases}$$

Thus we can assume that in our situation the subspace of admissible controls is

$$\mathcal{U} = \{u \in L_\infty([0, t_1])u_{t_1} \in C([-h, 0], E^m)\}.$$

**Remark 6.2.2**  Because the proof is carried out in an integrated form, it is easy to consider $\psi \in C$. If, as was done in [2], a "differentiated" version of (6.1.1) is used in the definition of control, $\psi$ will be naturally in $C^1([-h, 0], E^n)$. In both cases the full rank of $B(t)$ is needed as was formulated by Bank, Jacobs, and Langenhop [2].

The full rank of $B$ required by Condition (ii) of Theorem 6.2.1 is very strong: for $W_2^{(1)}$-controllability we must have as many controls as there are state variables. There are very many practical situations when this will fail. For example, the scalar $n$th order retarded equation with a single control of Example 6.1.3 is ruled out. We now study other types of controllability, which we must need to make progress in our study of time-optimal control.

Consider the system

$$\dot{x}(t) = L(t, x_t) + B(t)u(t), \qquad (6.1.1)$$

where $L$ is given in (2.2.4) with assumptions on $L(t, \phi)$ stated there. With these assumptions we may decompose $L(t, \phi)$ as follows:

$$L(t, \phi) = A_0(t)\phi(0) + H(t, \phi),$$

where

$$A_0(t) = \eta(t, 0^-),$$

$$H(t, \phi) = \int_{-h}^{0} d_\theta \overline{\eta}(t, \theta)\phi(\theta),$$

$$\overline{\eta}(t, \theta) = \begin{cases} 0, & \theta \geq 0, \quad \forall t, \\ \eta(t + \theta) + A_0(t), & \theta < 0, \quad \forall t. \end{cases} \qquad (6.2.4)$$

We call the operator $t \to H(t, \phi)$ the "strictly retarded" part of $L(t, \phi)$. In the system

$$\dot{x}(t) = A_0(t)x(t) + \sum_{i=1}^{N} A_i(t)x(t - w_k), \quad 0 < w_1 < w_2 < \cdots < W_N = h,$$

$$H(t, \phi) = \sum_{i=1}^{N} A_i(t)\phi(-w_k). \qquad (6.2.5)$$

We call system (6.1.1) a system with strict retardations if the mapping $\overline{\eta}$ in (6.2.4) satisfies the condition, there exists a $\delta$, $0 < \delta < h$ such that

$$\overline{\eta}(t, \theta) = 0 \quad \text{for} \quad -\delta \leq \theta \leq 0, \quad \forall t \in \theta.$$

Note that all systems of the form (6.2.5) are strictly retarded. We have:

**Proposition 6.2.1**   *Suppose system (6.1.1) has strict retardation, then (6.1.1) is Euclidean null controllable if and only if*
$G(\sigma, t_1 - h) = \int_{\sigma}^{t_1 - h} U(t_1 - h, s)B^*(s)B^*(s)U^*(t_1 - h, s) \cdot ds$ *has* rank $n$ *for every choice of* $\sigma, t_1$, *with* $t_1 > \sigma + h$.

With this one proves the following [2].

**Theorem 6.2.3**   *Suppose:*

(i) (6.1.1) *has strict retardation, and*

(ii) $t \to B^+(t)$, $t \in E$ *is essentially bounded on* $[t_1 - h, t_1]$ *for each* $t_1$ *with* $t_1 > \sigma + h$. *Then (6.1.1) is null controllable if and only if*

(iii) *Condition* (i) *of Proposition 6.2.1 is satisfied.*

(iv) $B(t)B^+(t)\overline{\eta}(t, \theta) = \overline{\eta}(t, \theta)$ *if* $-h \leq \theta \leq 0$ *for almost every* $t \in [t_1 - h, t_1]$ *for every* $t_1 > \sigma + h$.

*When* $L, B$ *are independent of time, the following sharp condition is deduced:*

**Corollary 6.2.2** *Suppose (6.1.1) is strictly retarded. Then (6.1.1) is null controllable if and only if*

(i) *Condition* (iv) *of Theorem 6.2.3 is satisfied, and*

(ii) $\operatorname{rank}[B, A_0 B, \ldots, A_0^{n-1} B] = n$.

*In particular, if we consider*

$$\dot{x}(t) = A_0 x(t) + \sum_{i=1}^{N} A_i x(t - h_i) + Bu(t), \tag{6.2.6}$$

*where* $0 < h_1 < \cdots < h_N$ *with* $A_i$, $i = 0, \ldots, n$ $n \times n$ *matrices, then one obtains:*

**Corollary 6.2.3** *The system (6.2.6) is null controllable if and only if*

$$BB^+ \sum_{i=1}^{N} A_i = \sum_{i=1}^{N} A_i,$$

*and*

$$\operatorname{rank}[B, A_0 B, \ldots, A_0^{n-1} B] = n.$$

*As a consequence, the nth-order scalar differential difference equation in Example 6.1.3 is null controllable.*

## 6.3 Constrained Controllability of Linear Delay Systems

In the last sections the controls are big. In this section we consider controllability of

$$\dot{x}(t) = L(t, x_t) + B(t)u(t), \tag{6.3.1}$$

when the controls are required to lie on a bounded convex set $\mathcal{U}$ with a nonempty interior. For ease of treatment, $\mathcal{U}$ will be assumed to be the unit cube

$$C^m = \{u \in E^m : |u_j| \leq 1, j = 1, \ldots, m\}. \tag{6.3.2}$$

Here $u_j$ denotes the $j$th component of $u \in E^m$. Consistent with our earlier treatment, the class of admissible controls is defined by

$$\mathcal{U}_{ad} = \{u \in L_\infty([0, t_1], E^m) : u(t) \in C^m \quad \text{a.e. on} \quad [0, t_1]\}.$$

It is easy to see that $\mathcal{U}_{ad}$ has a nonempty interior relative to $L_\infty$, and that $0 \in \mathcal{U}_{ad}$. The state space is either $W_\infty^{(1)}$ or $C$. Conditions on $L$ and $B$ of Secs. 6.1–6.3 are assumed to prevail. We need some definitions.

**Definition 6.3.1**   The system (6.3.1) is null controllable with constraints if for each $\phi \in C$, there is a $t_1 < \infty$ and a control $u \in \mathcal{U}_{ad}$ such that the solution $x(\cdot)$ of (6.3.1) satisfies

$$x_\sigma(\sigma, \phi, u) = \phi, \quad x_{t_1}(\sigma, \phi, u) = 0.$$

It is locally null controllable with constraints if there exists an open ball $\mathcal{O}$ of the origin in $C$ with the following property: For each $\phi \in \mathcal{O}$, there exists a $t_1 < \infty$ and a $u \in \mathcal{U}_{ad}$ such that the solution $x(\ )$ of (6.3.1) satisfies

$$x_\sigma(\sigma, \phi, u) = \phi, \quad x_{t_1}(\sigma, \phi, u) = 0.$$

We now study the null controllability with constraints of (6.3.1). Two preliminary propositions are needed.

**Proposition 6.3.1**   [5] *Suppose (6.3.1) is null controllable on the interval $[\sigma, t_1]$. Then for each $\phi \in C$, there exists a bounded linear operator $H : C \to L_\infty([\sigma, t_1], E^m)$ such that the control $u = H\phi$ has the property that the solution $x(\sigma, \phi, H\phi)$ of (6.3.1) satisfies*

$$x_\sigma(\sigma, \phi, H\phi) = \phi, \quad x_{t_1}(\sigma, \phi, H\phi) = 0.$$

*With this $H$, one proves the following:*

**Proposition 6.3.2**   [5] *Assume that (6.3.1) is null controllable. Then it is locally null controllable with constraints.*

**Proof**   Since (6.3.1) is null controllable, we have by Proposition (6.3.1) that there exists a bounded linear operator $H : C \to L_\infty([\sigma, t_1], E^m)$ such that for each $\phi \in C$, $u = H\phi$ and the solution $x(\sigma, \phi, H\phi)$ of (6.3.1) satisfies

$$x_\sigma(\sigma, \phi, H\phi) = \phi, \quad x_{t_1}(\sigma, \phi, H\phi) = 0.$$

Because $H$ is continuous, it is continuous at zero in $C$. Here for each neighborhood $V$ of zero in $L_\infty([\sigma, t_1], E^m)$ there is a neighborhood $M$ of 0 in $C$ such that $H(M) \subset V$. In particular, choose $V$ to be any open set in $L_\infty([\sigma, t_1], E^m)$ containing zero that is contained in $\mathcal{U}_{ad}$. This choice is possible since $\mathcal{U}_{ad}$ has zero in its interior. For this particular choice, we see that there exists an open set $\mathcal{O}$ around the origin in $C$ such that $H(\mathcal{O}) \subset V \subset L_\infty([\sigma, t_1], E^m)$. Every $\phi \in \mathcal{O}$ can be steered to zero by the control $u = H\phi$. Hence (6.3.1) is locally null controllable with constraints.   $\square$

**Theorem 6.3.1**   *Assume that:*

(i) *The system (6.3.1) is null controllable.*

(ii) *The system*

$$\dot{x}(t) = L(t, x_t) \qquad (6.3.3)$$

*is uniformly asymptotically stable, so that there are constants $k > 0$, $\alpha > 0$ such that for each $\sigma \in E$ the solution $x$ of (6.3.3) satisfies*

$$\|x_t(\sigma, \phi)\| \le k\|\phi\|e^{-\alpha(t-\sigma)}.$$

*Then (6.3.1) is null controllable with constraints.*

**Proof**   Condition (i) and Proposition 6.2.2 guarantee the existence of an open ball $\mathcal{O} \subset C$ such that every $\phi \in \mathcal{O}$ can be steered to the zero function with controls in $\mathcal{U}_{ad}$ in time $t_1 < \infty$. Condition (ii) assures us that every solution of (6.3.3), i.e., every solution of (6.3.1) with $u = 0$, satisfies

$$x_t(\sigma, \phi, 0) \to 0 \quad \text{as} \quad t \to \infty.$$

Thus, using $u = 0 \in U_{ad}$, the system rolls on as Eq. (6.3.3), and there is a finite $t_0$, such that $\psi = x_{t_0}(\sigma, \phi, 0) \in \mathcal{O}$. With this initial data $(t_0, \psi)$ there exists a $t_1 > t_0$ such that for some $u \in \mathcal{U}_{ad}$, $x_{t_0}(\sigma, \psi, u) = \psi$, $x_{t_1}(\sigma, \psi, 0) = 0$. Thus with the control, $v$,

$$v = \begin{cases} 0 & \text{in } [\sigma, t_0], \\ u & \text{in } [t_0, t_1], \end{cases}$$

which is contained in $\mathcal{U}_{ad}$, $\phi$ is indeed transferred to 0 in a time $t_1 < \infty$. This concludes the proof. $\qquad\qquad\square$

As a consequence we have the following sharp result:

**Corollary 6.3.1**   *Consider*

$$\dot{x}(t) = L(x_t), \qquad (6.3.4)$$

*where for an $n \times n$ matrix $\xi(\theta)$ $-h \le \theta \le 0$, $L$ is given by*

$$L(\phi) = \int_{-h}^{0} [d\xi(\theta)]\phi(\theta), \quad \phi \in C.$$

*Let $\Delta(\lambda) = \lambda I - \int_{-h}^{0} e^{\lambda\theta} d\xi(\theta)$.*

(i) *Let the roots of the characteristic equation*

$$\det \Delta(\lambda) = 0$$

*have negative real parts.*

(ii) *Suppose*

$$\dot{x}(t) = L(x_t) + Bu(t) \tag{6.3.5}$$

*is null controllable.*

*Then (6.3.5) is null controllable with constraints.*

**Corollary 6.3.2**    *Consider*

$$\dot{x}(t) = A_0 x(t) + \sum_{i=1}^{N} A_i x(t - h_i) + Bu(t), \quad 0 < h_1 < \cdots < h_N = h. \tag{6.3.6}$$

*Assume that:*

(i) *There exists a positive definite symmetric matrix $H$ such that the matrix $G$ is negative semidefinite, where*

$$G \equiv HA_0 + A_0^T H + \sum_{k=1}^{N} 2qnM_k H,$$

*and where $q \geq 1$ is a constant, $M_k = \max|A_k ij|$.*

(ii) $\mathrm{rank}[\Delta(\lambda), G] = n$ *for each complex $\lambda$ where $\Delta(\lambda) = \lambda I - A_0 - \sum_{k=1}^{N} A_k e^{-\lambda h_i}$.*

(iii)

$$\mathrm{rank}
\begin{bmatrix}
A_0 - \lambda I & \Delta_1 \cdots A_N & G \\
A_1 & AN & 0 \\
A_N & 0 & 0
\end{bmatrix}
= n + \mathrm{rank}
\begin{bmatrix}
A_1 & \cdots & A_N \\
A_N & 0 & 0
\end{bmatrix},$$

*or in place of (ii), (iii) we have that $G$ is negative definite.*

(iv) $BB^+ \sum_{i=1}^{N} A_i = \sum_{i=1}^{N} A_i.$

(v) $\mathrm{rank}[B, A_0 B, \ldots, A_0^{n-1} B] = n.$

*Then (6.3.6) is null controllable with constraints.*

**Proof**    For the system (6.3.6) Conditions (i), (ii), and (iii) are the requirements of uniform asymptotic stability of Corollary 3.2.3 or of Chukwu [6]. Hypotheses (iv) and (v) are needed for null controllability.      □

**Example 6.3.1**    Consider

$$y^n(t) = \sum_{j=0}^{n-1} b_i y^i(t - h) + \sum_{i=0}^{n-1} a_i y^i(t) + u(t),$$

where $a_i, b_i$ are constants. If the homogeneous system is uniformly asymptotically stable, then Example 6.3.1 is null controllable with constraints.

The following result is implied by the main contribution in Chukwu [5, Corollary 4.1].

**Theorem 6.3.2**   *Consider the system*

$$\dot{x}(t) = L(t, x_t) + B(t)u(t) \,, \tag{6.1.1}$$

*where $L$ is given in (2.2.4) with assumptions on $L(t, \phi)$ stated there and with $B$ continuous. Assume that the controls $u$ are $L_2(L_\infty)$ functions with values in a compact convex subset $P$ of $E^m$ and $0 \in P$. Assume that the trivial solution of*

$$\dot{x}(t) = L(t, x_t) \tag{6.3.3}$$

*is uniformly stable, and (6.1.1) function-space $W_2^{(1)}$ controllable (Euclidean $E^n$ controllable) on some finite interval $[\sigma, t_1]$ with $L_2(L_\infty)$ controls.*

*Then (6.1.1) is function space (Euclidean) controllable with constraints. As a consequence, uniform stability of (6.3.3) and controllability of (6.1.1) on some interval $[\sigma, t_1]$ suffice for global null controllability of (6.1.1), where the controls are measurable with values in $P$.*

# References


1. T. S. Angell, "Existence Theorems for Optimal Control Problems Involving Functional Differential Equations," *J. Optimization Theory Appl.* **7** (1971) 149–169.
2. H. T. Banks, M. Q. Jacobs, and C. E. Langenhop, "Characterization of the Control States in $W_2^{(1)}$ of Linear Hereditary Systems," *SIAM J. Control* **13** (1975) 611–649.
3. H. T. Banks and G. A. Kent, "Control of Functional Differential Equations of Retarded and Neutral Type to Target Sets in Function Space," *SIAM J. Control* **10** (1972) 567–594.
4. S. L. Campbell and C. D. Meyer, *Generalized Inverses of Linear Transformations*, Pitman, London, 1979.
5. E. N. Chukwu, "Controllability of Delay Systems with Restrained Controls," *J. Optimization Theory Appl.* **29** (1979) 301–320.
6. E. N. Chukwu, "Function Space Null Controllability of Linear Delay Systems with Limited Power," *J. Math. Anal. Appl.* **124** (1987) 193–304.
7. E. N. Chukwu, "Global Behavior of Linear Retarded Functional Differential Equations," *J. of Math. Anal. and Appl.* **162** (1991) 277–298.
8. R. Gabasov and F. Kirillova, *The Qualitative Theory of Optimal Processes*, Marcel Dekker, New York, 1976.
9. J. Hale, *Theory of Functional Differential Equations*, Springer-Verlag, New York, 1977.
10. H. Hermes and J. P. LaSalle, *Functional Analysis and Time Optimal Control*, Academic Press, New York, 1969.
11. F. M. Kirillova and S. V. Churakova, "On the Problem of Controllability of Linear Systems with After Effect," *Differential Nye Uravneniya* **3** (1967) 436–445.
12. D. G. Luenberger, *Optimization by Vector Space Methods*, John Wiley, New York, 1969.
13. A. Manitius and A. W. Olbrot, "Controllability Conditions for Linear Systems of Linear Hereditary Systems," *SIAM J. Control* **13** (1975) 611–649.
14. A. Manitius, "Optimal Control of Hereditary Systems," in *Control Theory and Topics in Functional Analysis* **III**, International Atomic Energy Agency, Vienna, 1976.
15. L. Weiss, "An Algebraic Criterion for Controllability of Linear Systems with Time-Delay," *IEEE Trans. on Autom. Control* **AC-15** (1970) 443.

## Chapter 7

# Synthesis of Time-Optimal and Minimum-Effort Control of Linear Delay Systems

## 7.1 Linear Systems in Euclidean Space

Our aim here is to consider the time-optimal control of the linear system

$$\dot{x}(t) = A_0(t)x(t) + \sum_{i=1}^{N} A_i(t)x(t - h_i) + B(t)u(t)\,, \tag{7.1.1}$$

where $0 < h_1 < h_2 < \cdots < h_N = h\ A_i$, $i = 0, \ldots, N$ are $n \times n$ analytic matrix functions, and $B$ is an $n \times m$ real analytic matrix function. The controls are locally measurable $L_\infty$ functions that are constrained to lie in the unit cube

$$C^m = \{u \in E^m : |u_j| \le 1,\ j = 1, \ldots, m\}\,. \tag{7.1.2}$$

Thus the class of admissible controls is defined by

$$U_{ad} = \{u \in L_\infty C[\sigma, t_1]\,,\ u(t) \in C^m \text{ a.e. on } [0, t_1]\}\,.$$

In $E^n$ we consider the following problem: Minimize

$$J(t, x(t)) = t\,, \tag{7.1.3}$$

where

$$x_\sigma = \phi\,, \qquad (x(t, \sigma, \phi, u), t) = \{0\} \times [0, \infty)\,. \tag{7.1.4}$$

Here $x(\cdot, \sigma, \phi, u)$ is a solution of (7.1.1) with $x_\sigma = \phi$. This is the time-optimal problem. It is old [17] and very important [13, 16, 18] and continues to be interesting even for linear ordinary differential systems [19] and [20]. In the economic applications, for example, it is desirable to force the value of capital to reach some target in minimum time while maximizing some welfare function. For example, we may want the Gross Domestic Product to grow as fast as possible. In Example 1.4, one would like to control the epidemic of AIDS. It is known [7, p. 183] that, associated with the delay equations

201

$$\frac{dS(t)}{dt} = \wedge - \lambda C(T)S(t)\frac{I(t)}{T(t)} - \mu S(t) \equiv L_1\,,$$

$$\frac{dI(t)}{dt} = \lambda\left[C(T)(t)S(t)\frac{I(t)}{T(t)} - C(T(t-h))S(t-h)\frac{I(t-h)}{T(t-h)}\right] - \mu I(t)$$

$$\equiv L_2\,,$$

that model AIDS as a progressive disease is the reproductive number $R$ given by

$$R = \lambda C\left(\frac{\wedge}{\mu}\right)\frac{(1-e^{-\mu h})}{\mu}\,.$$

This system has a unique positive endemic state if and only if $R > 1$. If $R < 1$, the infection-free state is globally asymptotically stable, while the endemic state is locally asymptotically stable whenever $R > 1$. Since $R$ is the number of secondary infections produced by an infectious individual in a purely susceptible population, to control the spread of AIDS as rapidly as possible, $R$ should be reduced to less than 1 in minimum time. This is a time-optimal problem. Theoretically this can be done by introducing control strategies that will decrease the number $\lambda C$, the transmission rate per unit time per infected partner, as fast as possible. Conceivably the control

$$u(t) = b(t)\lambda C(T)S(t)\frac{I(t)}{T(t)}$$

is introduced, so that the system becomes

$$\frac{dS(t)}{dt} = L_1 + u(t)\,,$$

$$\frac{dI(t)}{dt} = L_2 - u(t)\,.$$

The aim is to drive this control system to the infection-free state $(\frac{\wedge}{\mu}, 0)$ in minimum time. With this as a motivation, we now study (7.1.1), a linear approximation, by constructing an optimal feedback control in a way similar to the development in [10].

Recent advances in the theory of functional differential equations in Sec. 7.3, [10], [11] and [14], and in the constrained controllability questions [9], [15], have made it possible to attempt to solve the time-optimal control problem of linear systems (7.1.1) within the classical theory. The development of the theory was long abandoned because of lack of progress in the directions which have now taken place. We shall illustrate our theory with several examples of (7.1.1) from technology, and one from ecology. If $x(\sigma, \phi, 0)$ is a solution of

$$\dot{x}(t) = A_0 x(t) + \sum_{i=1}^{N} A_i(t)x(t-h_i)\,,$$

$$x_\sigma = \phi \in C\,,$$

$$\tag{7.1.5}$$

in $E^n$, the solution of (7.1.1) is given by

$$x(t, \sigma, \phi, u) = x(t, \sigma, \phi, 0) + \int_\sigma^t U(t, s)B(s)u(s)ds, \qquad (7.1.6)$$

where $U(t, s)$ is the fundamental solution of (7.1.5).

**Definition 7.1.1**  The Euclidean reachable set of (7.1.5) at time $t$ is defined by

$$\mathbb{R}(t, \sigma) = \left\{ \int_\sigma^t U(t, s)B(s)u(s)ds : u \in U_{ad} \right\}.$$

This is the set of all points in $E^n$ that can be attained from initial point $\phi = 0$ using all admissible controls. The following properties of $\mathbb{R}(t, \sigma)$ are easily proved:

**Proposition 7.1.1**  $\mathbb{R}(t, \sigma)$ *is convex and compact, and* $0 \in \mathbb{R}(t, \sigma)$, $\forall t \geq \sigma$. *Consider the subset of* $U_{ad}$ *defined by*

$$U_{ad}^0 = \{u \in E^m, \ u \text{ measurable } |u_j| = 1, \ j = 1, 2, \ldots, m\}.$$

*The following bang-bang principle is valid in* $E^n$ :

**Theorem 7.1.1**  *Let*

$$\mathbb{R}(t, \sigma) = \left\{ \int_\sigma^t U(t, s)B(s)u(s) : u \in U_{ad} \right\},$$

*and*

$$\mathbb{R}^0(t, \sigma) = \left\{ \int_\sigma^t U(t, s)B(s)u(s) : u \in U_{ad}^0 \right\}.$$

*Then* $\mathbb{R}^0(t, \sigma) = \mathbb{R}(t, \sigma)$, $\forall t \geq \sigma$.

**Proof**  Because $U(t, s) \in L_\infty([\sigma, t], E^{n \times n})$, and $B(t)$ analytic, we have

$$U(t, s)B(s) \in L_\infty([\sigma, t], E^{n \times m}).$$

It now follows from Corollary 8.2 of Hermes and LaSalle [4] that

$$\mathbb{R}(t, \sigma) = \mathbb{R}^0(t, \sigma), \qquad \forall t \geq \sigma.$$

$\square$

**Remark 7.1.1**  Theorem 7.1.1 states that the bang-bang principle of LaSalle is valid for (7.1.1); that if, of all bang-bang steering functions, there is an optimal one relative to $U_{ad}^0$, then it is optimal relative to $U_{ad}$. Also if there is an optimal steering function, there is always a bang-bang steering function that is optimal.

**Definition 7.1.2**    Let $r^n$ denote the metric space of all nonempty compact subsets of $E^n$ with the metric $\rho$ defined as follows: The distance of a point $x$ from $\mathcal{O}$ is

$$d(x, \mathcal{O}) = \inf\{|x - a| : a \in \mathcal{O}\},$$

$$N(\mathcal{O}, \epsilon) = \{x : E^n : d(x, \mathcal{O}) \le \epsilon\},$$

$$\rho(\mathcal{O}_1, \mathcal{O}_2) = \inf\{\epsilon : \mathcal{O}_1 \subset N(\mathcal{O}_2, \epsilon) \quad \text{and} \quad \mathcal{O}_2 \subset N(\mathcal{O}_1, \epsilon)\}.$$

**Proposition 7.1.2**    *The set $\mathbb{R}(\cdot, \sigma) : [\sigma, \infty) \to r^n$ is continuous.*

*Because $t \to U(t, s)$, $t \ge s$ is continuous the usual ideas such as are contained in Lee and Markus [5, p. 69] can be adapted to prove this using the variation of parameter. Associated with $\mathbb{R}(t, \sigma)$ is the Euclidean space attainable set*

$$\mathcal{A}(t, \sigma) = \{x(t, \sigma, \phi, u) : u \in U_{ad}, \ x \text{ is a solution of } (7.1.1)\}.$$

*The same type of proof of Proposition 7.1.2 yields that $t \to \mathcal{A}(t, \sigma) = x(t, \sigma, 0) + \mathbb{R}(t, \sigma)$ is continuous in the Hausdorff metric, and it is compact and convex.*

*The following properties of $\mathbb{R}(t, \sigma)$ as a subset of $E^n$ are available:*

**Lemma 7.1.1**    $\mathbb{R}(t, \sigma)$ *is compact and convex and satisfies the monotonicity relation:*

(i)  $0 \in \mathbb{R}(t, \sigma)$ *for each $t \ge \sigma$.*
(ii)  $W(t, s)\mathbb{R}(s, \sigma) \subset \mathbb{R}(t, \sigma)$, $\sigma \le s \le t$.
(iii)  *The mapping $t \to \mathbb{R}(t, \sigma)$ is continuous with respect to the Hausdorff metric.*

**Remark 7.1.2**    Relations (i) and (ii) are proved as in [21].

To motivate the discussion for the form of optimal controls, we note that if an admissible control $u$ drives $\phi$ to zero in time $t_1$, then

$$0 = x(t_1, \sigma, \phi, u) = T(t_1, \sigma)\phi(0) + \int_\sigma^{t_1} U(t_1, s)B(s)u(s)ds,$$

so that

$$-T(t_1, \sigma)\phi(0) \in \mathbb{R}(t_1, \sigma).$$

Note that $U(t_1, \sigma) = T(t_1, \sigma)I$ where $I$ is the identity (see [8, p. 146]). For example, for the system

$$\dot{x}(t) = A_0 x(t) + A_1 x(t - h), \qquad x_0 = \phi,$$

with solutions

$$x(\phi, 0)(t) = U(t)\phi(0) + A_1 \int_{-h}^0 U(t - s - h)\phi(s)ds,$$

we may designate

$$W(t) : C \to E^n$$

by

$$W(t)\phi \equiv x(\phi, 0)(t)\,.$$

If $x_\sigma = \phi$, then $W(t) \equiv W(t, \sigma)$. Also recall that $T(t, \sigma)(\phi) \equiv x_t(\sigma, \phi, 0)$. The coincidence

$$W(t_1)\phi \in \partial\mathbb{R}(t_1, \sigma)\,,$$

where $\partial\mathbb{R}$ is the boundary of $\mathbb{R}$, is very important in characterizing the form of optimal controls. We need the following definitions.

**Definition 7.1.3** The control $u \in U_{ad}$ is an extremal control if the associated solution $x(\sigma, \phi, u)$ corresponding to $(\sigma, \phi)$ lies on the boundary of $\mathcal{A}(t, \sigma)$, that is,

$$x(t, \sigma, \phi, u) = \partial\mathcal{A}(t, \sigma)\,, \qquad \sigma \leq t \leq t^*\,.$$

**Definition 7.1.4** System (7.1.5) is pointwise complete at time $t_1$, if and only if the solution operator defined by $W(t, \sigma)\phi = x(\sigma, \phi, 0)(t)$ is such that

$$W(t_1, \sigma) : C \to E^n$$

is a surjection: $W(t_1, \sigma)C = E^n$. Conditions for pointwise completeness are found in [6]. These conditions are stated in Lemma 7.1.2.

**Lemma 7.1.2** *For System* (2.1.12), *namely*

$$\dot{x}(t) = A_0 x(t) + A_1 x(t - 1)\,, \tag{S}$$

*a necessary and sufficient condition for the pointwise completeness for $t = t_1$ is that, for every nonzero $\eta \in E^n$ there exists a $t \geq t_1$ such that $\eta^T U(t) \neq 0$ where $U$ is the fundamental matrix solution of (2.1.12). If we define $F_k = Z_k(0)A_1$, where $Z_k$ is identified in* (2.1.16)–(2.1.21), *then* (2.1.12) *is pointwise complete for $t_1 \in [k, k+1)$, $k = 0, 1, 2, \ldots$ whenever the matrix*

$$M(t_1) = [E_{k-1}F_{k-1}, \ldots, E_{k-1}A_{k-1}^{n\,k-1}F_{k-1}, E_k Z_k(0)]$$

*has rank $n$.*

**Example 7.1.1** $A_0 = \begin{bmatrix} 0 & 2 & 0 \\ 0 & 0 & -1 \\ 0 & 0 & 0 \end{bmatrix}$, $A_1 = \begin{bmatrix} 0 & 0 & 0 \\ 1 & 0 & 0 \\ 0 & 2 & 0 \end{bmatrix}$, $U(2) = E_2 Z_2(0) =$

$\begin{bmatrix} 2 & 4 & -4 \\ 1 & 1 & -2 \\ 0 & 2 & 0 \end{bmatrix}$, and this is a singular matrix

$$M(2) = \begin{bmatrix} 2 & -2 & 0 & 2 & -4 & 0 & 0 & -4 & 0 \\ 1 & -2 & 0 & 0 & -2 & 0 & 0 & 0 & 0 \\ 0 & 2 & 0 & 2 & 0 & 0 & 0 & -4 & 0 \end{bmatrix}\,.$$

This has rank less than 3 because $\eta^T = [1, -2, -1]$ satisfies $\eta^T M(\eta) = 0$. The example is not pointwise complete for $t_1 \geq 2$.

The following is also true.

**Lemma 7.1.3**    *The system is pointwise complete for all $t_1 \in [0, \infty)$ whenever $A_0 A_1 = A_1 A_0$. The system is pointwise complete for all $t_1 \in [0, 2)$.*

**Theorem 7.1.2**    *The following are equivalent:*

   (i)  *The system (7.1.1) is Euclidean controllable on $[\sigma, t]$, $t > \sigma + h$.*
  (ii)  $0 \in \text{Int } \mathbb{R}(t, \sigma)$, $t > \sigma + h$.
 (iii)  $W(t, s)\mathbb{R}(s, \sigma) \subset \text{Int } \mathbb{R}(t, \sigma)$, $\sigma < s < t$.

**Proof**    Suppose (i) is valid. Then the mapping

$$H : L_\infty([\sigma, t], E^m) \to E^n$$

given by $Hu = x(t, \sigma, \phi, u)$ is surjective. Clearly $H(U_{ad}) = \mathbb{R}(t, \sigma)$. Also

$$H(\mathbb{B}) \subset H(U_{ad}) = \mathbb{R}(t, \sigma),$$

where $\mathbb{B}$ is an open ball containing $0$ such that $0 \in \mathbb{B} \subset U_{ad}$. Because $H$ is a continuous linear transformation of $L_\infty$ onto $E^n$, it is an open map. Hence $H(\mathbb{B})$ is open and contains $0$. Thus $0 \in \text{Int } \mathbb{R}(t, \sigma)$. Conversely, assume that $0 \in \text{Int } \mathbb{R}(t, \sigma)$. Then $0 \in \text{Int } \mathbb{R}(t, \sigma) \subset H(L_\infty([\sigma, t], E^n))$. Because the last set in the inclusion is a subspace, containing a full neighborhood of zero, we have $H(L_\infty([\sigma, t], E^m)) = E^n$. Thus (7.1.1) is controllable.

Assume that $0 \in \text{Int } \mathbb{R}(t, \sigma)$ and $W(t, s)q \in \partial \mathbb{R}(t, \sigma)$ where $q \in \mathbb{R}(s, \sigma)$, since by Lemma 7.1.1, $W(t, s)q \in \mathbb{R}(t, \sigma)$, $W(t, s) = T(t, s)I$, and aim at a contradiction. Because of the point in the boundary, we have

$$(c, \rho) \leq (c, W(t, s)q), \qquad \forall \rho \in \mathbb{R}(t, \sigma),$$

where $(\cdot, \cdot)$ is the inner product in $E^n$. We now observe that

$$\left( c, \int_\sigma^s W(t, s)U(s, \tau)B(\tau)u(\tau)d\tau \right) = \left( c, \int_\sigma^s U(t, \tau)B(\tau)u(\tau)d\tau \right)$$

because of the semigroup property of the solution operator.

Define

$$u^*(\tau) = \begin{cases} u(\tau), & 0 \leq \tau \leq s, \\ \text{sgn}[c^T U(b, \tau)B(\tau)], & s < \tau \leq t. \end{cases}$$

Then this admissible $u^*$ determines a $\rho \in \mathbb{R}(t, \sigma)$, and

$$(c, \rho - W(t, s)q) - \left( c, \int_s^t U(t, \tau)B(\tau)u^*(\tau)d\tau \right) = \int_s^t |c^T U(t, \tau)B(\tau)|d\tau > 0.$$

Thus $(c, \rho) > (c, W(t, s)q)$, contradicting the assumption that $W(t, s)q$ is the boundary of $\mathbb{R}(t, \sigma)$. Hence $W(t, s)\mathbb{R}(s, \sigma) \subset \text{Int } \mathbb{R}(t, \sigma)$, for $\sigma < s < t$. This completes the proof of Theorem 7.1.2. $\qquad\square$

**Proposition 7.1.3** *Suppose (7.1.5) is pointwise complete and $w$ is a time-optimal control for (7.1.1). Then $w$ is extremal.*

**Proof** We first show that if $w$ is a time-optimal control at the first instant $t_1$ of arrival at $0 \in E^n$, then the solution will lie on $\mathcal{A}(t_1, \sigma)$, the boundary of the attainable set. We shall then show that if $x(t, \sigma, \phi, \omega)$ lies on $\mathcal{A}(t, \sigma)$ at any fixed $t_*$, then

$$x(t, \sigma, \phi, \omega) \in \partial\mathcal{A}(t, \sigma), \qquad \sigma \leq t \leq t_*.$$

Suppose that $\omega$ is the optimal control that steers $\phi$ to $0 \in E^n$ in minimum time $t_1$, i.e., $x(t_1) = x(t_1, \sigma, \phi, \omega) = 0$, and assume that $x(t_1) = 0$ is not in the boundary, $\partial\mathcal{A}(t_1, \sigma)$. Then there is a ball $\mathcal{O}(0, \rho)$ of radius $\rho$ about $0$ such that $\mathcal{O}(0, \rho) \in \mathcal{A}(t_1, \sigma)$. Since $\mathcal{A}(t, \sigma)$ is continuous in $t$, we can preserve this inclusion at $t$ near $t_1$ by reducing the size of $\mathcal{O}(0, \rho)$, i.e., if there exists a $\delta > 0$ such that $\mathcal{O}(0, \rho/2) \subset \mathcal{A}(t, \sigma)$, $t_1 - \delta \leq t \leq t_1$. Thus we can reach $0$ at time $t_1 - \delta$. This contradicts the optimality of $t_1$. The conclusion is valid that $0 = x(t_1, \sigma, \phi, \omega) \in \partial\mathcal{A}(t, \sigma)$. We now prove the second part by assuming that $x^* = x(t_*, \sigma, \phi, \omega) \in \text{Int } \mathcal{A}(t_*, \sigma)$ for some $0 < t_* < t_1$. We claim that

$$x(t, \sigma, \phi, \omega) \in \text{Int } \mathcal{A}(t, \sigma) \quad \text{for} \quad t > t_*.$$

Since $x^* \in \text{Int } \mathcal{A}(t_*, \sigma)$, there is a ball $\mathcal{O}(x^*, \delta) \subset \mathcal{A}(t_*, \sigma)$ such that $c \in \mathcal{O}(x^*, \delta)$ can be reached from $\phi$ at time $t_*$ using control $u_0$. Now introduce the new system

$$z(t) = \sum_{i=0}^{N} A_i(t)z(t - h_i) + B(t)w(t), \qquad t \geq t^*, \qquad (7.1.7)$$

$z_{t^*} = \phi_0$ with $z(t^*) = c = \phi_0(0)$, $t^* < t$ with $\omega$ fixed. We observe that for the coincidence $\phi_0 = x^*$, we have from uniqueness of solutions that $z(t, \sigma, \phi_0, \omega) = x(t, \sigma, \phi, \omega)$, $t \geq t^*$. This solution is given in $E^n$ by

$$z(t, \sigma, \phi_0, \omega) = W(t, \sigma)\phi_0 + \int_{t^*}^{t} U(t)s)B(s)u(s)ds,$$

that is,

$$z(t, \sigma, \phi_0, \omega) = W(t, \sigma)\phi_0(0) + d(t),$$

where $d(t) = \int_{t_*}^{t} U(t, s)B(s)w(s)ds$. Since (7.1.5) is complete so that $W(t, \sigma)C = E^n$, $W(t, \sigma)$ is an open map. Hence $\phi_0 = c \to z(t, \sigma, \phi_0, \omega)$ is an open map and takes open sets into open sets. As a consequence the image of the open ball $\mathcal{O}$ lies in the interior of $\mathcal{A}(t, \sigma)$. Thus the image of $x^*$, which is just $x(t, \sigma, \phi, \omega)$, lies in Int $\mathcal{A}(t, \sigma)$. $\qquad\square$

**Theorem 7.1.3**   *Suppose* (7.1.5) *is pointwise complete and* (7.1.1) *is Euclidean controllable. Let $u^*$ be an optimal control and $t^*$ the minimum time. Then $\omega \equiv -T(t^*, \sigma)\phi(0) \in \partial \mathbb{R}(t^*, \sigma)$, the boundary of the reachable set $\mathbb{R}(t^*, \sigma)$ on $E^n$ if and only if*

$$u^*(s) = \text{sgn}[y(s)B(s)], \quad y(s) \not\equiv 0, \quad s \in [\sigma, t^*],$$

*where $y : [\sigma, t^*]$ is an $n$-row vector of bounded variation that satisfies the adjoint equation*

$$\frac{dy(s)}{ds} = -\sum_{i=0}^{N} y(s + h_i)A_i(s + h_i), \tag{7.1.8}$$
$$h_0 = 0 < h_1 < \cdots < h_N = h.$$

**Proof**   For the system

$$\dot{x}(t) = A_0(t)x(t) + \sum_{i=1}^{N} A_i(t)x(t - h_i) + B(t)u(t),$$

$$x(t) = \phi(t) \; t \in [-h, 0), \quad x(0) = \phi(0) = c, \quad t \geq \sigma,$$

by definition the $E^n$-reachable set is

$$\mathbb{R}(t^*, \sigma) = \left\{ \int_{\sigma}^{t^*} U(t^*, s)B(s)u(s)ds : u \in U_{ad} \right\}.$$

If (7.1.1) is null controllable with constraints so that optimal control $u^*$ exists, then $x(t^*, \sigma, \phi, u^*) = 0$, and

$$\omega \equiv -T(t^*, \sigma)\phi(0) = \int_{\sigma}^{t^*} U(t^*, s)B(s)u(s)ds.$$

Consequently $\omega \in \mathbb{R}(t^*, \sigma)$. Because $u^*$ is optimal and (7.1.5) is complete, the solution $x(\sigma, \phi, u^*)$ is extremal. Therefore $-T(t^*, \sigma)\phi(0) = \omega \in \partial \mathbb{R}(t^*, \sigma)$. But by Proposition 7.1.1, $\mathbb{R}(t^*, \sigma)$ is compact and convex, and because (7.1.1) is controllable, Theorem 7.1.2 yields that $0 \in \text{Int } \mathbb{R}(t^*, \sigma)$. It follows from the separation theorem of closed convex sets [1, p. 418] that there exists a nontrivial $d \in E^n$, a row vector, such that

$$(d, \rho) \leq (d, \omega), \qquad \forall \rho \in \mathbb{R}(t^*, \sigma),$$

where $(\cdot, \cdot)$ defines the scalar product in $E^n$. Thus

$$\left( d, \int_{\sigma}^{t^*} U(t^*, s)B(s)u(s)ds \right) \leq \left( d, \int_{\sigma}^{t^*} U(t^*, s)B(s)u^*(s)ds \right),$$

so that

$$\int_{\sigma}^{t^*} d^T U(t^*, s) B(s) u(s) ds \leq \int_{\sigma}^{t^*} d^T U(t^*, s) B(s) u^*(s) ds \,. \qquad (7.1.9)$$

Recall the definition $Y(s, t)$ as the matrix solution of the adjoint equation (7.1.8) as was treated in Hale [8, pp. 147–153] and the coincidence $Y(s, t) = U(t, s)$ a.e. Then clearly (7.1.9) yields the inequality

$$\int_{\sigma}^{t^*} d^T Y(s, t^*) B(s) u(s) ds \leq \int_{\sigma}^{t^*} d^T Y(s, t^*) B(s) u^* \,.$$

So that

$$\int_{\sigma}^{t^*} y(s, t^*) B(s) u(s) ds \leq \int_{\sigma}^{t^*} y(s, t^*) B(s) u^*(s) ds \,,$$

where $y(\cdot, t^*) : E \to E^{n^*}$ vanishes on $[t^*, \infty)$, satisfies (7.1.8) on $(-\infty, t^* - h]$ and is such that $y(t^*, t^*) = d \neq 0$. As a consequence,

$$u^*(s) = \text{sgn}[y(s, t^*) B(s)] \,, \quad y(s, t^*) \not\equiv 0 \,.$$

The argument can be reversed. $\qquad \square$

**Remark 7.1.3** Note that $y(t^*, t^*) = d \neq 0$, so that $y(s, t^*) \not\equiv 0$.

**Remark 7.1.4** The form of optimal control in Theorem 7.1.3 asserts that each component $u_j^*$ of $u^*$ is given by

$$u_j^*(s) = \text{sgn}[y(s, t^*, \psi) b_j(s)]$$

on $[\sigma, t^*]$, $j = 1, \ldots, n$, where $b_j(s)$ is the $j$th component of $B(s)$. Provided we assume completeness, it uniquely defines an optimal control if (7.1.1) is normal in the following sense:

**Definition 7.1.5** Define $g_j(d) = \{s : y(s, t^*, d) b_j(s) = 0, \ s \in [0, t^*]\}$. System (7.1.1) is normal on $[\sigma, t^*]$ if $g_j(d)$ has measure zero for each $j = 1, \ldots, m$, where $y(t^*, t^*, d) = d \neq 0$. If (7.1.1) is normal for each $[\sigma, t^*]$, we say (7.1.1) is normal.

It follows from the above definition that (7.1.1) is normal on $[\sigma, T]$, $T > \sigma$ if for each $j = 1, \ldots, m$, $d^T U(T, s) b_j = 0$ almost everywhere for each $j = 1, \ldots, m$, implies $d = 0$. Here $b_j$ is the $j$th component of $B$.

Recalling the definition of determining equations for constant coefficient delay systems,

$$\dot{x}(t) = A_0 x(t) + A_1 x(t - h) + B u(t) \,,$$

$$Q_k = A_0 Q_{k-1}(s) + A_1 Q_{k-1}(s - h) \ k = 1, 2, \ldots, \ s \in (-\infty, \infty), \qquad (7.1.10)$$

$$Q_0(s) = \begin{cases} I & s = 0, \ I \text{ (Identity)} \,, \\ 0 & s \neq 0 \,, \end{cases}$$

and defining

$$\overline{Q}_n(t_1) = \{Q_0(s), Q_1(s), \dots, Q_{n-1}(s), \ s \in (0, t_1]\},$$

one proves in a way similar to [3, p. 79] that the following proposition is valid.

**Theorem 7.1.4**   *A necessary and sufficient condition that a complete system* (7.1.10) *is normal on* $[0, T]$ *is that for each* $j = 1, \dots, m$, *the matrix*

$$\overline{Q}_{jn}(T) = \{Q_0(s)b_j, \dots, Q_{n-1}(s)b_j, \ s \in [0, T]\}$$

*has* rank $n$.

## 7.2   Geometric Theory and Continuity of Minimal Time Function

We note that if (7.1.1) is null controllable with constraints, then $x(\sigma, \phi, u)(t) = 0$ for some $t \geq \sigma$ and some $u \in U_{ad}$, that is,

$$w \triangleq -T(t, \sigma)\phi(0) = \int_\sigma^t U(t, s)B(s)u(s)ds.$$

Consequently

$$w = -T(t, \sigma)\phi(0) = x(t, \sigma, \phi, 0) \in \mathbb{R}(t, \sigma)$$

if and only if $\phi$ can be steered to zero at time $t$ by some admissible control. If the system is Euclidean null controllable with constraints, then

$$A(\sigma) = \{\phi : x(\sigma, \phi, u)(t) = 0 \text{ for some } u \in U_{ad} \text{ and some } t\},$$

$$= C = \{\phi : T(t, \sigma)\phi(0) = x(t, \sigma, \phi, 0) \in \mathbb{R}(t, \sigma)\},$$

where $x(t, \sigma, \phi, 0)$ solves (7.1.5).

**Definition 7.2.1**   The minimal time function is the function $M : A(\sigma) \to E$ defined by

$$M(\phi) = \text{Inf}\{t \geq \sigma : -T(t, \sigma)\phi(0) \in \mathbb{R}(t, \sigma)\}.$$

Thus $\sigma \leq M(\phi) \leq +\infty$ with $M(\phi) < \infty$ if and only if $-T(t, \sigma)\phi(0) \in \mathbb{R}(t, \sigma)$.

The following result is fundamental.

**Theorem 7.2.1**   *Let* (7.1.1) *be Euclidean controllable.   Then the minimal time function is continuous.*

**Proof**   First we show that $M$ is upper semicontinuous. Let $\phi_k \to \phi$ as $k \to \infty$ so that $-T(t, \sigma)\phi_k \to -T(t, \sigma)\phi$ as $k \to \infty$. Suppose $M(\phi) < \infty$, and for some fixed $t$, $t > M(\phi)$. We know that

$$-T(t, \sigma)\phi(0) \in \mathbb{R}(t, \sigma).$$

Also, since (7.1.1) is Euclidean controllable and Lemma 7.1.1(ii) is valid, there is some $s$, $t < s$ such that

$$T(s,t)[-T(t,\sigma)\phi(0)] \in \text{Int}(\mathbb{R}(s,\sigma)).$$

Therefore, for sufficiently large $k$,

$$T(s,t)[-T(t,\sigma)\phi_k(0)] \in \mathbb{R}(s,\sigma).$$

Because $T(s,t)[-T(t,\sigma)\phi_k(0)] = -T(s,\sigma)\phi_k(0)$, we have $M(\phi_k) \leq s$. Therefore for all $s > M(\phi)$, $\limsup M(\phi_k) \leq s$. If we let $s = M(\phi) + \frac{1}{j}$, then

$$\limsup M(\phi_k) \leq M(\phi).$$

This inequality is still valid if $M(\phi) = +\infty$, proving upper semicontinuity. $\square$

For the proof of lower semicontinuity, suppose

$$M(\phi) > \liminf M(\phi_k).$$

Since $M(\phi_k)$ cannot be $+\infty$, there is a subsequence such that $M(\phi_k)$ converges:

$$M(\phi_k) \to s < M(\phi).$$

Therefore, for $k$ large,

$$M(\phi_k) < s.$$

From the definition of $M(\phi_k)$, $M(\phi_k) \leq t$ whenever $-T(t,\sigma)\phi_k(0) \in \mathbb{R}(t,\sigma)$. There is some $\phi_{ki}$ such that

$$T(t_{ki(0)},\sigma)\phi_{ki}(0) \in \mathbb{R}(t_{ki},\sigma), \qquad t_{ki} < s,$$

$$-T(s,t_{ki})T(t_{ki},\sigma)\phi_{ki}(0) \in T(s,t_{ki})\mathbb{R}(t_{ki},\sigma) \in \mathbb{R}(s,\sigma).$$

Since $\mathbb{R}(s,\sigma)$ is closed,

$$-T(s,\sigma)\phi_k(0) \to T(s,\sigma)\phi(0) \in \mathbb{R}(s,\sigma),$$

so that

$$M(\phi) \leq s < M(\phi).$$

This contradiction proves that $M$ is lower semicontinuous. We conclude that $M$ is continuous. Note that $M$ is always lower semicontinuous.

**Corollary 7.2.1** *Let (7.1.1) be controllable and (7.1.5) be pointwise complete. For $t \geq \sigma$ we have the following relations:*

$$\{-T(t,\sigma)\phi(0) : M(\phi) \leq t\} = \mathbb{R}(t,\sigma).$$

$$\{-T(t,\sigma)\phi(0) : M(\phi) = t\} = \partial\mathbb{R}(t,\sigma), \text{ the boundary of } \mathbb{R}.$$

**Proof**   We now observe that

$$\mathbb{R}(t,\sigma) \subset \{-T(t,\sigma)\phi(0) : M(\phi) \le t\} \subset \bigcap_{p=1}^{\infty} \mathbb{R}\left(t+\frac{1}{p},\sigma\right).$$

We now show that

$$\mathbb{R}(t,\sigma) \supset \bigcap_{p=1}^{\infty} \mathbb{R}\left(t+\frac{1}{p},\sigma\right).$$

Let $\phi(0) \in \bigcap_{p=1}^{\infty} \mathbb{R}(t+\frac{1}{p},\sigma)$. Then there is an admissible control $u_p : [\sigma, t+\frac{1}{p}]$, such that

$$\phi(0) = \int_{\sigma}^{t+\frac{1}{p}} U\left(t+\frac{1}{p},s\right) B(s)u_p(s)ds$$

$$= \int_{\sigma}^{t} U\left(t+\frac{1}{p},s\right) B(s)u_p(s)ds + \int_{t}^{t+\frac{1}{p}} U\left(t+\frac{1}{p},s\right) B(s)u_p(s)ds$$

$$\equiv x_p + y_p\,.$$

We note that

$$x_p = \int_{\sigma}^{t} U\left(t+\frac{1}{p},t\right) U(t,s)B(s)u_p(s)ds$$

$$= U\left(t+\frac{1}{p},t\right) \int_{\sigma}^{t} U(t,s)B(s)u_p(s)ds\,.$$

Thus $x_p \in U(t+\frac{1}{p},t)\mathbb{R}(t,\sigma)$, or

$$x_p = T\left(t+\frac{1}{p},t\right) \phi_p(0)\,, \qquad \phi_p(0) \in \mathbb{R}(t,\sigma)\,.$$

Hence $x_p \to T(t,t)\phi(0) = \phi(0)$ as $p \to \infty$, since $y_p \to 0$ as $p \to \infty$, so that $\phi(0)$ is in the closed set $\mathbb{R}(t,\sigma)$.

For the second formula we note that since (7.1.1) is Euclidean controllable and (7.1.5) is pointwise complete, we have that

$$W(t,\sigma) : C \to E^n\,, \qquad (W(t,\sigma)\phi = T(t,\sigma)\phi(0))$$

is an open map, and $T$, $M$ are continuous. Hence

$$\{-T(t,\sigma)\phi(0) : M(\phi) < t\}$$

is open, so that

$$\{-T(t,\sigma)\phi(0) : M(\phi) < t\} \subset \text{Int } \mathbb{R}(t,\sigma)\,.$$

From the proof of Proposition 7.1.1, we know that if

$$-T(t^*, \sigma)\phi(0) \in \text{Int } \mathbb{R}(t^*, \sigma), \qquad \text{for } \sigma < t^* < \infty,$$

then

$$-T(t, \sigma)\phi(0) \in \text{Int } \mathbb{R}(t, \sigma), \qquad \text{for } t > t^*.$$

From the above we have

$$M(\phi) \leq t^* < t.$$

This completes the proof. $\qquad\qquad\square$

As an immediate by-product of the continuity of the minimal time function, we construct an optimal feedback control for the system (7.1.1). In the spirit of Hájek [10], we present some needed preliminaries.

We identify $B_0$ with the conjugate space of $C$ and define a subset of $B_0$, which we describe as the cone of unit outward normals to support hyperplanes to $\mathbb{R}(t, \sigma)$ at a point $T(t, \sigma)\phi(0)$ on the boundary of $\mathbb{R}(t, \sigma)$.

**Definition 7.2.2**  For the controllable system (7.1.1), let

$$\mathcal{A}(\sigma) = C = \{\phi : T(t, \sigma)\phi(0) \in \mathbb{R}(t, \sigma)\}.$$

For each $\phi \in \mathcal{A}(\sigma)$, let

$$K(\phi) = \{\psi \in B_0 : \|\psi\| = 1 \text{ and such that } (\psi(0), \rho) \leq \langle \psi(0), T(M(\phi), \sigma)\phi(0)\rangle,$$

$$\forall \, \rho \in \mathbb{R}(M(\phi), \sigma)\}.$$

Note that $T(M(\phi), \sigma)\phi(0) \in \partial\mathcal{R}(M(\phi), \sigma)$, the boundary of $\mathbb{R}(M(\phi), \sigma)$.

**Lemma 7.2.1**  *Suppose (7.1.1) is Euclidean controllable. Then $K(\phi)$ is nonvoid and $K(-\phi) = -K(\phi)$. Also*

$$\limsup K(\phi_n) \subset K(\phi) \quad as \quad \phi_n \to \phi \text{ in } \mathcal{A}(\sigma).$$

*That is, $K$ is upper semicontinuous at $\phi$.*

**Proof**  From the controllability assumption,

$$0 \in \text{Int } \mathbb{R}(t, \sigma), \qquad t > \sigma + h,$$

by Theorem 7.1.2. Consider $K(\phi)$. Since $T(M(\phi), \sigma)\phi(0) \in \partial\mathbb{R}(M(\phi), \sigma)$ by the usual separation theorem [1, p. 148], there exists a $\psi \in B_0$ the conjugate space of $C$, $\psi(0) \not\equiv 0$ such that

$$\langle \psi(0), \rho \rangle \leq \langle \psi(0), T(M(\phi), \sigma)\phi(0)\rangle \ \forall \rho \in \mathbb{R}(M(\phi), \sigma).$$

It is clear that $\psi$ can be chosen such that $\|\psi\| = 1$. Hence $K(\phi)$ is nonvoid.

For the second assertion, observe that $\mathbb{R}(M(\phi),\sigma)$ is symmetric about zero. Because $t \to \mathbb{R}(t,\sigma)$ is continuous [21] and closed, and because $\phi \to M(\phi)$ is continuous, a simple argument proves the last assertion.

Because $K(\phi)$ is a nonvoid subset of $B_0$, for each $\phi$ we can choose a $y(\phi) \in K(\phi)$ whenever (7.1.1) is Euclidean controllable. We now prove that this selection can be made in a measurable way. $\qquad\square$

**Lemma 7.2.2**   *Assume that (7.1.1) is Euclidean controllable. There exists a measurable function $y : \mathcal{A}(\sigma) \to B_0$ such that*

$$y(\phi) \in K(\phi) \ for \ all \ \phi \in \mathcal{A}(\sigma).$$

**Proof**   We observe that $C$ is a measurable space. Its conjugate $B_0$ is a Hausdorff space. Also $C$ is separable. Let $k : C \to B_0$ be the mapping given by the collection

$$K(\phi) = \{k(\phi) = \psi : \psi \in B_0 , \quad \|\psi\| = 1 \ (\psi(0),\rho) \le (\psi(0), T(M(\phi),\sigma)\phi(0))$$

$$\forall \ \rho \in \mathbb{R}(M(\phi),\sigma)\}.$$

The continuity of $k$ is a simple direct consequence of its definition as well as the continuity of $t \to \mathbb{R}(t,\sigma)$ and of $\phi \to M(\phi)$. In McShane and Warfield [23] we identify

$$M = C = \mathcal{A}(\sigma), \qquad A = B_0, \quad \psi \in B_0,$$

(and invoke the continuum hypothesis). Theorem 4 of [23] asserts that there exists a measurable function $y : \mathcal{A}(\sigma) \to B_0$ such that $y(\phi) \in K(\phi)$. This proves Lemma 7.2.2. $\qquad\square$

We are now ready for our main result.

**Theorem 7.2.2**   *In (7.1.1) assume that $A_i$, $B_i$ are real analytic. Assume that (7.1.1) is both Euclidean controllable and normal. Let (7.1.5) be pointwise complete. Then there exists a measurable function $f : \mathcal{A}(\sigma) \to E^m$ that is optimal feedback control for (7.1.1) in the following sense:*
*Consider the system*

$$\dot{z}(t) = \sum_{i=0}^{N} A_i(t)z(t - h_i) + B(t)f(z_t), \qquad z_\sigma = \phi. \tag{7.2.1}$$

*Each optimal solution of (7.1.1) is a solution of (7.2.1). As a partial converse, each solution of (7.2.1) is a solution (possibly not optimal) of (7.1.1). Also the function $f$ is given by*

$$f(\phi) = \lim_{t \to s^+} -\mathrm{sgn}[g(t,s,y(\phi))B(s)], \tag{7.2.2}$$

*where $g(t,s,\cdot) : B_0 \to E^n$ is defined by*

$$g(t,s,\psi) = y(t,s,\psi), \qquad s \in [\sigma,t],$$

*where $y : [\sigma, t]$ satisfies*

$$\frac{\partial y(t, s, \psi)}{ds} = -\sum_{i=1}^{N} y(t, s + h_i) A_i(s + h_i)\,,$$

$$y(t, t, \psi) = \psi(0)\,,$$

$$y_t = \psi\,.$$

**Proof**   Let $y : \mathcal{A}(\sigma) \to B_0$ be the measurable selection described in Lemma 7.2.2. For each $\phi \in \mathcal{A}(\sigma)$ set $y(\phi) = \psi \in B_0$, and define $f(\phi)$ as in (7.2.2). We recall that

$$g(t, s, \psi) = \int_{-h}^{o} [d\psi(\theta)][U(t + \theta, s)]\,,$$

where $U$ solves (7.1.5).

We recall from Tadmor [25, Theorem 4.1] (see also Corollary 10.2 of [18]) that $s \to U(t, s)$ is piecewise analytic for each $s \in [\sigma, M(\phi)]$, since $A_i(t)$ is analytic in $t$. It follows from the analyticity of $B(s)$ that each coordinate of $g(t, s, \psi)B(s)$ is analytic on $[s_{i-1}, s_i]$ $i = 1, 2, \ldots, v$ for each partition $\sigma \leq s_0 \leq s_1 \leq \cdots \leq s_v = M(\phi)$. Thus $\mathrm{sgn}[g(t, s, \psi)B(s)]$ is piecewise constant on $[s_{i-1}, s_i]$ and therefore on $[\sigma, M(\phi)]$. Hence the limit in (7.2.2) exists. Because $y(\phi)$ is a measurable selection of $\phi$, so is $f(\phi)$.

Let $x : [\sigma, M(\phi)] \to C$ be an optimal solution of (7.1.1) with $x_\sigma = \phi$. We now show that $x$ satisfies (7.2.1). Indeed, take an arbitrary $s$, $0 \leq s < M(\phi)$, then $x_s \in \mathcal{A}(\sigma)$. There is an optimal control that transfers $x_s$ to zero. We take it to be

$$u_s(t) = -\mathrm{sgn}[g(t, s, y)B(s)]$$

for any choice of $y \in K(x_s)$, i.e., $-y \in K(-x_s)$. This choice follows from Theorem 7.1.3. We choose $y = y(x_s)$. Because (7.1.1) is assumed normal, optimal control is unique almost everywhere and it uniquely determines a response on $[\sigma, M(\phi)]$; that is, the response $x$. Thus for almost all $t \geq s$ and each $s$ we have

$$u_\sigma(s) = u_s(t) = -\mathrm{sgn}[g(t, s, y(x_s))B(s)]\,.$$

Now take the limit as $t \to s^+$ ($\sigma \leq s \leq t$). We now have

$$u_\sigma(s^+) = f(x_s)\,, \qquad s \in [\sigma, M(\phi)]\,.$$

Since $u_\sigma$ is piecewise constant, $u_\sigma(s) = f(x_s)$ for almost all $s$. Therefore, the response $x$ to $u_\sigma$ satisfies

$$\dot{x}(s) = \sum_{i=0}^{N} A_i(s)x(s - h_i) + B(s)f(x_s)\,,$$

$$x_\sigma = \phi$$

almost everywhere. Because $x$ is absolutely continuous, it is a solution of (7.1.1).

Finally, let $x$ be a solution of (7.2.1) that is a response to $v(t) = f(x_t)$. Clearly $v \in L_\infty([\sigma, M(\phi)], E^n)$ and $v \in U_{ad}$. $\qquad\qquad\qquad\square$

## 7.3   The Index of a Control System

From our analysis it is clear that the properties of the function

$$k(s) = g(t, s, \psi)B(s) = y(t, s)B(s) \,,$$

with $y_s(t, \cdot) = \psi$, $\psi \in B_0$, the Banach space of functions $\psi : [-h, 0] \to E^{n*}$ of bounded variation on $[-h, 0]$, continuous on the left on $(-h, 0)$ and vanishing at zero with norm $\mathrm{Var}\ \psi_{[-h,0]}$, are important in determining optimal strategies of delay equations (7.1.1). We call $k$ the index of the control system. To determine it explicitly, one first finds the fundamental matrix $U(t, s)$ of (7.1.1). We consider its autonomous version, namely

$$\dot{x}(t) = A_0 x(t) + \sum_{i=1}^{N} A_i x(t - ih) \,,$$
$$x(0) = x_0 \in E^n \,; \qquad x(s) = \phi(s) \,, \qquad s \in [-hN, 0) \,, \qquad h > 0 \,. \tag{7.3.1}$$

The determination of the fundamental matrix $U$ of (7.3.1) was treated in Sec. 2.1. With the explicit formula for $U$, one determines the index of the control system using $g$, given as

$$
\begin{aligned}
g(t, s, \psi) &= \int_{-h}^{0} d\psi(\theta) U(t + \theta, s) \\[2mm]
&= \int_{-h}^{0} [d\psi(\theta)][U(t + \theta - s)] \\[2mm]
&= \int_{-h}^{0} [d\psi(\theta)] \left[ \sum_{i=1}^{k} V_i(t - s(i - 1)h + \theta) \right] \\[2mm]
&= \sum_{i=1}^{k} \int_{-h}^{0} [d\psi(\theta)][V_i(t - s + \theta - (i - 1)h)] \,.
\end{aligned}
\tag{7.3.2a}
$$

Thus,

$$k(s) = \sum_{i=1}^{k} \int_{-h}^{0} [d\psi(\theta)][V_i(t - s + \theta - (i - 1)h)]B(s) \,,$$

$$s \leq t, t - s \in [(k - 1)h, kh] \,. \tag{7.3.2b}$$

Consider $N = 1$, that is, the system

$$\dot{x}(t) = A_0 x(t) + A_1 x(t - h) + B(t)u(t) \,. \tag{7.3.3}$$

The index of this control system is given as follows: On $[0, 2h]$

$$k(s) = \int_{-h}^{0} [d\psi(\theta)] e^{A_0(t+\theta-s)} B(s), \quad t - s \in [0, h],$$

$$U(t) = e^{A_0 t} + \int_{h}^{t} e^{A_0(t-r)} A_1 e^{A_0(r-\tau)} d\tau, \quad t \in [h, 2h],$$

(7.3.4a)

$$g(t, s, \psi) = \int_{-h}^{0} [d\psi(\theta)] \left[ e^{A_0(t+\theta-s)} \right.$$
$$\left. + \int_{h}^{t+\theta-s} e^{A_0(t+\theta-s-\tau)} A_1 e^{A_0(\tau-h)} d\tau \right], \qquad t - s \in [h, 2h].$$

(7.3.4b)

We now observe that $k(s)$ is an $m$-vector function, the sign of whose components can be determined and thus an open loop control constructed. Thus, once the disconjugacy properties of $k$ are analyzed, an optimal feedback control of (7.3.1) can be constructed, as was done in [20].

### 7.3.1 *Applications*

For linear autonomous systems, we have developed some theory about optimal control that, when applied to simple examples, will give us insight to the construction of optimal control laws. Since optimal control exists when the complete system is Euclidean controllable with constraints, and is of the form $\text{sgn}(d^T U(t-s)) = u(s)$, the usual methods of ordinary differential equations can be applied. Using the values of $U$ on $[kh, (k+)h]$ beginning at the origin, integrate backward with controls $u(s)$, and find all optimal trajectories to the Euclidean origin. Since in the first interval $[0, h]$, $U(t) = e^{A_0 t}$, the analysis is relatively routine, though complicated. The application of this method, though feasible, is definitely limited, because of the complicated nature of $U$. It does, however, provide insight into possible general methods of construction of optimal controls of systems with limited controls. Very many numerical methods which have so far been developed [28, 29] are for $L_2$ controls, and can reasonably be modified to help solve the synthesis problem of time-optimal control systems with $L_\infty$ controls whose components are bounded, a problem long abandoned by researchers more brilliant than I. This book is an affirmation of hope that they will return!

**Example 7.3.1** We want to find a time-optimal control $u^*(t)$ such that

$$x(t) = \psi(t), \qquad t \in [-1, 0],$$

and $x(t^*) = 0$, with $t^*$ minimum when

$$\dot{x}(t) = -x(t) + x(t-1) + u(t), \qquad |u(t)| \le 1.$$

(7.3.5)

It is easily proved that optimal control exists. We use the fundamental matrix solution

$$\dot{x}(t) = -x(t) + x(t-1),$$

of Problem 2.1.1

$$U(t) = \begin{cases} e^{-t}, & t \in [0,1], \\ e^{-t}[1 + e(t-1)], & t \in [1,2], \\ e^{-t}\left[1 + e(t-1) + \dfrac{1}{2}e^2(t^2 - 4t + 4)\right], & t \in [2,3] \end{cases}$$

to deduce the index of the control system,

$$g(s) = g(t,s,\psi), \qquad (\psi(0) = d \neq 0)$$

$$= \begin{cases} dU(t-s) = de^{s-t^*}, & s \in [t^* - 1, t^*], \\ de^{s-t^*}[1 + e(t^* - s - 1)], & s \in [t^* - 2, t^* - 1], \\ de^{s-t^*}\left[1 + e(t^* - s - 1) + \dfrac{1}{2}e^2(t^* - s)^2 - 4(t^* - s) + 4\right], & s \in [t^* - 3, t^* - 2]. \end{cases}$$

Optimal control is given by

$$\operatorname{sgn} d,$$

since all factors of $d$ are positive. The time-optimal control will be either 1 or $-1$ on the entire interval $[0, t^*]$.

**Example 7.3.2**    Antirolling Stabilization of a Ship

A ship is rolling in the waves. The linear dynamics of the angle of tilt $x$ from the normal upright position is given by

$$\dot{x}(t) = y(t), \qquad \dot{y}(t) = -by(t) - qy(t-h) - kx(t) + u_2(t), \tag{7.3.6}$$

$$\dot{x}(t) = y(t), \qquad \dot{y}(t) = -by(t) - qy(t-h) - kx(t) + u_2(t), \tag{7.3.7}$$

System (7.3.7) is obtained when a servomechanism is introduced and designed to reduce $(x, y)$ to $(0, 0)$ as fast as possible. What the contrivance does is to introduce an input to the natural damping of the rolling ship, a term proportional to the velocity at an earlier instant $t - h : qy(t - h)$. Also introduced is a control with components $(0, u_2)$ yielding the equation above. Thus

$$\underline{\dot{x}}(t) = A_2\underline{x}(t) + A_1x(t-h) + Bu(t), \tag{7.3.8}$$

$$A_0 = \begin{bmatrix} 0 & 1 \\ -k & -b \end{bmatrix}, \quad A_1 = \begin{bmatrix} 0 & 0 \\ 0 & -q \end{bmatrix}, \quad B = \begin{bmatrix} 0 \\ 1 \end{bmatrix},$$

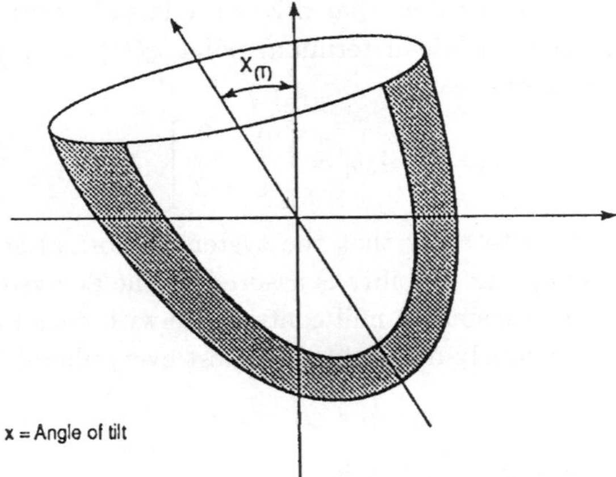


Fig. 7.3.1

with $k = 2$, $b = 3$, $q = 1$. The parameter values correspond to one particular operating point. In this example,

$$U(t) = U_1 = e^{A_o t} = \begin{bmatrix} 2e^{-t} - e^{-2t} & e^{-t} - e^{-2t} \\ -2e^{-t} + 2e^{-2t} & -e^{-t} + 2e^{-2t} \end{bmatrix}, \quad t \in [0, h]. \tag{7.3.9}$$

If $h = 1$ on $[1, 2]$, then

$$U_2 = U(t)$$

$$= \begin{bmatrix} e^{-t}(e(2t - 6)) + e^{-2t}(2 + e^2(2t)), & e^{-t}(e(t - 4)) + e^{-2t}(1 + e^2(2t)) \\ e^{-t}(2e(2 - t)) + e^{-2t}(-4 + e^2(-6 - 4t)), & e^{-t}(e(5 - t)) + e^{-2t}(-2 + e^2(2 - 4t)) \end{bmatrix}$$

$$\tag{7.3.10}$$

and if $h$ is arbitrary,

$$U_2 = U(t) = [e^{-t} + (t - h)e^{-(t-h)}] \begin{bmatrix} 2 & 1 \\ -2 & -1 \end{bmatrix}$$

$$+ [-e^{-(t-h)} + e^{-2(t-h)}] \begin{bmatrix} 0 & -1 \\ -2 & 0 \end{bmatrix} \tag{7.3.11}$$

$$+ [e^{-2t} - 2(t - h)e^{-2(t-h)}] \begin{bmatrix} -1 & -1 \\ 2 & 2 \end{bmatrix}$$

on $[h, 2h]$, and so on. The index of the control system is

$$g(s) = g(t^*, s, \psi) = d^T U(t^* - s),$$

$$\psi(0) = d = (d_1, d_2).$$

Our problem is to find a control $u^*$ that drives the initial position $\underline{x}_0 = (x_0, y_0) = \phi(s)$, $s \in [-h, 0]$ to the Euclidean terminal point $\underline{x}(t^*) = (x(t^*), y(t^*)) = 0$ in minimum time $t^*$. Because

$$Q = [b, A_0 b] = \begin{bmatrix} 0 & 1 \\ 1 & -2 \end{bmatrix},$$

we invoke Theorem 7.1.4 to show that the system is normal and Euclidean controllable. Uniform asymptotic stability is assured by the analysis of Example 2.4.1 since $3 = b > q = 1$. The system is null controllable with constraints, and optimal controls exist and are uniquely determined (almost everywhere) by

$$u^*(s) = \operatorname{sgn} g(s),$$

where the index of the control system is

$$g(s) = \begin{cases} d^T U_1(t^* - s), & s \in [0, h], & 0 \le s \le t^* \le h, \\ d^T U_2(t^* - s), & s \in [0, h], & s + h \le t^* \le 2h, \\ d^T U_3(t^* - s), & s \in [0, h], & s + 2h \le t^* \le 3h, \end{cases} \tag{7.3.12}$$

and so on.

In general,

$$g(s) = d^T U_k(t^* - s), \qquad s \in [0, h], \qquad s + (k-1)h \le t^* \le kh,$$

where $U_k = U$ is as defined in (2.1.8) or (2.1.10). We now use $U$, which has been determined. In $[0, h]$,

$$g(s) = g_1(s) = (2(d_1 - d_2)e^{-s} + (2d_2 - d_1)e^{-2s}, \ (d_1 - d_2)e^{-s} + (2d_2 - d_1)e^{-2s}).$$
$$\tag{7.3.13}$$

In $[h, 2h]$,

$$g(s) = g_2(s) = (2(d_1 - d_2), (d_1 - d_2))(e^{-(t^* - s)} + (t - s - h)e^{-(t-s-h)})$$
$$+ (2d_2, -d_1)(e^{-(t^* - h - s)} - e^{-2(t^* - s - h)})$$
$$+ (2d_2 - d_1)(1, 1)(e^{-2(t^* - s)} - 2(t^* - s - h)e^{-2(t-s-h)}),$$
$$s \in [0, h], \qquad s + h \le t^* \le 2h. \tag{7.3.14}$$

To obtain optimal trajectories in $E^2$, we start at the origin and integrate backwards in time. To do this we replace $t$ by $-\tau$ in our system (7.3.7) and obtain the dynamics

$$S(\tau): \quad \begin{aligned} \dot{x}(\tau) &= -y(\tau), \\ \dot{y}(\tau) &= 3y(\tau) + y(\tau + h) + kx(\tau) + u_2, \end{aligned} \tag{7.3.15}$$

where

$$u_2(\tau) = \operatorname{sgn}[(d_1 - d_2) + e^{\tau}(2d_2 - d_1)]. \tag{7.3.16}$$

Taking $d_1 < 0$, $d_2 < 0$, we begin at the origin ($x(0) = y(0) = 0$) along the trajectory $S(\tau)$ with $u_2 = -1$. This control may change sign at some $\tau_1 \in (0, h]$ where

$$\tau_1 = \text{In}\left(-\frac{(d_1 - d_2)}{2d_2 - d_1}\right).$$

Therefore, starting at $\tau = 0$ at the point $\underline{x}(\tau_1)$ of $S(\tau) = \alpha$ with $u_2 = 1$, we leave $\alpha$ along a new curve $\beta(\tau)$ that begins at $\underline{x}(\tau_1)$. There is no further change of sign if $\tau \in [0, h]$ until $\tau_2 = \tau \in [h, 2h]$, and then the control is $u(s) = \text{sgn } g_2(0)$. Also, if $\tau_1 \in [h, 2h]$, the control $u(s) = \text{sgn } g_2(s)$ is used. In both situations, and under the assumption $d_1 < 0$, $d_2 < 0$,

$$u_2(s) = \text{sgn}[(d_1 - d_2)(e^{-(t^*-s)} + (t - s - h)e^{(t-s-h)}$$

$$- d_1(e^{-(t^*-s-h)} - e^{-2(t^*-s-h)}) \tag{7.3.17}$$

$$+ (2d_2 - d_1)(e^{-(t^*-s)} - 2(t^* - s - h)e^{-2(t-s-h)})].$$

The new control then is

$$u_2(s) = \text{sgn}[(d_1 - d_2)(e^{-t^*} + (t^* - h)e^{-(t^*-h)}) - d_1(e^{-(t^*-h)} - e^{-2(t^*-h)})$$

$$+ (2d_2 - d_1)(e^{-t^*} - 2(t^* - h)e^{-2(t-h)})]$$

$$= \text{sgn}\{[d_2 e^{-t^*} + e^{-(t^*-h)}[(d_1 - d_2)(t^* - h)] - d_1$$

$$+ e^{-(t^*-h)}(4d_2 - 3d)]\}. \tag{7.3.18}$$

Thus we begin at the point $\beta(\tau_2)$ with $u_2(s) = $ the second component of $g_2(0)$, where $d_1 < 0$ $d_2 < 0$, and move along the trajectory $S(-\tau)$, which is $C(\tau)$, with this control until possible switches at the zero of the second component of $g_2(s)$ on $[h, 2h]$. The earlier process is repeated. Next we start at the origin. With $d_1 < 0$, $d_2 > 0$, we have $u_2 = 1$. The process is repeated, and the control law is deduced, by considering the situation $d_1 > 0$, $d_2 > 0$ and $d_2 > 0$, and $d_2 < 0$ as well.

**Example 7.3.3**  We consider a two-dimensional retarded system

$$\dot{x}_1(t) = -2x_1(t) - x_2(t - h) + u_1(t) + u_2(t),$$
$$\dot{x}_2(t) = -x_2(t) + x_1(t - h) + u_1(t) + 2u_2(t), \qquad |u_1| \le 1, \; |u_2| \le 1, \tag{7.3.19}$$

where $x_1$, $x_2$ stand for population densities of two species (and so they are nonnegative). This system may describe the linear dynamics of a predator–prey system around the equilibrium $(X, Y)$. If $\bar{x}$ and $\bar{y}$ are the population densities of the prey and predator respectively, then

$$x_1 = \bar{x} - X, \qquad x_2 = \bar{y} - Y.$$

The controls $\underline{u} = (u_1, u_2)$ are harvesting/seeding strategies. We want an optimal $u^*$ that will drive the populations to the equilibrium $(0, 0) \in E^2$ in minimum time starting from any initial population density function $(x_0, y_0) \equiv \phi \in C([-h, 0], E^2)$. The system can be written in matrix form as

$$\dot{x}(t) = A_0 x(t) + A_1 x(t - h) + Bu(t),$$

where

$$A_0 = \begin{bmatrix} -2 & 0 \\ 0 & -1 \end{bmatrix}, \qquad A_1 = \begin{bmatrix} 0 & -2 \\ 1 & 0 \end{bmatrix}, \qquad B = \begin{bmatrix} 1 & 1 \\ 1 & 2 \end{bmatrix}.$$

We observe that

$$\dot{x}(t) = A_0 x(t) + A_1 x(t - h)$$

is Exponentially Asymptotically Stable (EAS) for all $h \geq 0$. Indeed, we can use a result in a very recent book [27, pp. 98–99] that the system

$$\begin{aligned} \dot{x}_1(t) &= -a_{11} x_1(t) - b_{12} x_2(t - h), \\ \dot{x}_2(t) &= -a_{22} x_2(t) + b_{21} x(t - h) \end{aligned} \tag{7.3.20}$$

is EAS for all $h \geq 0$ if

$$a_{11} a_{22} > b_{12} b_{22} \quad \text{and} \quad a_{11} + a_{22} > \frac{b_{12} b_{21}}{\sqrt{a_{11} a_{22} - b_{12} b_{21}}}, \tag{7.3.21}$$

or if

$$a_{11} a_{22} > -b_{12} b_{21} \quad \text{and} \quad 2h < \frac{a_{11} + a_{22}}{b_{12} b_{21}}. \tag{7.3.22}$$

We can also use Proposition 3.3.3. The first condition is $3 > 1$, while the second is $3 > \frac{1}{\sqrt{2}}$ or $3 > -1$ and $2h < \frac{3}{2}$. The system is also Euclidean controllable, since by Theorem 6.1.1,

$$\overline{Q}_n(t) = [B, A_0 B] = \begin{bmatrix} 1 & 1 & -2 & -2 \\ 1 & 2 & -1 & -2 \end{bmatrix},$$

and this has rank 2 for any $t > 0$. Theorem 6.3.1 assures us that (7.3.19) is null controllable with constraints. Optimal controls exist, and since it is normal for each $t > 0$, the optimal control is uniquely determined by

$$\text{sgn}(d^T U_k(t - s)B), \qquad s \in [0, h], \qquad k \leq t \leq (k + 1)h,$$

where $U_k$ is the fundamental matrix which we now calculate the methods of Sec. 2.1:

$$U_0(t) = e^{A \cdot t} = \begin{bmatrix} e^{-2t} & 0 \\ 0 & e^{-t} \end{bmatrix} \quad \text{on} \quad [0, h]. \tag{7.3.23}$$

$$U_1(t) = e^{A_0 t} + \int_h^t e^{A_0(t-s_1)} A_1 e^{A_0(s-h)} ds_1$$

$$= \begin{bmatrix} e^{-2t} & 0 \\ 0 & e^{-t} \end{bmatrix} + \int_h^t \begin{bmatrix} 0 & -2e^{-2t+s+h} \\ e^{-(t+s-2h)} & 0 \end{bmatrix} ds$$

$$= \begin{bmatrix} e^{-2t} & 0 \\ 0 & e^{-t} \end{bmatrix} + \begin{bmatrix} 0, & -2(e^{-(t-h)} - e^{-2(t-h)}) \\ e^{-(t-h)} - e^{-2(t-h)} & 0 \end{bmatrix}$$

$$= \begin{bmatrix} e^{-2t} & -2(e^{-(t-h)} - e^{-2(t-h)}) \\ e^{-(t-h)} - e^{-2(t-h)} & e^{-t} \end{bmatrix} \quad \text{on} \quad [h, 2h].$$

$$U_2(t) = U_1 + \int_0^{t-2h} e^{A_0(t-h-s)} A_1 \int_0^{s_1} e^{A_0(s_1-s)} A_1 e^{A_0 s} ds)ds_1,$$

$$U_2 = U_1 + V_1, \tag{7.3.24}$$

where

$$V_1 = \begin{bmatrix} -2e^{-t} + 2e^{-2(t-h)}(t+1-2h), & 0 \\ 0, & -2e^{-(t-h)}(t-2h) \end{bmatrix}.$$

Hence

$$U_2(t) = \begin{bmatrix} e^{-2t} - 2e^{-t} + 2e^{-(t-h)}(t+1-2h), & -2(e^{-(t-h)} - e^{-2(t-h)} \\ e^{-(t-h)} - e^{-2(t-h)}, & e^{-t} - 2e^{-(t-h)}(t-2h) \end{bmatrix}, \tag{7.3.25}$$

$$t \in [2h, 3h].$$

For each point of the Euclidean state space (the plane), there is an optimal control. We first assume that the delay $h$ is sufficiently large. Thus

$$d^T U_k(t-s) B, \qquad s \in [0, h], \qquad k \le t \le (k+1)h$$

is given as follows:

$$d^T U_0(t-s) B = (d_1 e^{-2(t-s)} + d_2 e^{-(t-s)}, \; d_1 e^{-2(t-s)} + 2d_2 e^{-(t-s)})$$

$$\text{on } [0, h], \quad 0 \le s \le h,$$

$$d^T U_1(t-s) B = (d_1 \{ e^{-2(t-s)} - 2(e^{-(t-s-h)} - e^{-2(t-s-h)}) \}$$

$$+ d_2 \{ e^{-(t-s)} + e^{-(t-s-h)} - 2e^{-2(t-s-h)} \},$$

$$d_1 \{ e^{-2(t-s)} - 4(e^{-(t-s-h)} - e^{-2(t-s-h)}) \}$$

$$+ d_2\{e^{-(t-s-h)} - e^{-2(t-s-h)} + 2e^{-(t-s)}\}),$$

$$0 \le s \le h, \qquad t \in [h, 2h], \tag{7.3.26}$$

$$d^T U_2(t-s)B = (d_1\{e^{-2(t-s)} - 2e^{-(t-s)} + 2e^{-2(t-s-h)}(t - s + 1 - 2h)$$

$$+ d_2\{e^{-(t-s-h)} - e^{-2(t-s-h)}\},$$

$$d_1\{-2(e^{-(t-s-h)} - e^{-2(t-s-h)})$$

$$+ d_2\{e^{-(t-s)} - 2e^{-(t-s-h)}(t - s - 2h)\},$$

$$0 \le s \le h, \qquad t \in [2h, 3h]. \tag{7.3.27}$$

To obtain the optimal trajectories, we start at the origin and as usual integrate backwards. This is done by replacing $t$ with $-\tau$ and considering the dynamics

$$\dot{x}_1(\tau) = 2x_1(\tau) + x_2(\tau + h) - u_1(\tau) - u_2(\tau),$$
$$\dot{x}_2(\tau) = x_2(\tau) - x_1(\tau + h) - u_1(\tau) - 2u_2(\tau), \tag{7.3.28}$$

where

$$u_1(\tau) = \text{sgn}[d_1 e^{2(t-\tau)} + d_2 e^{(t-\tau)}]$$
$$= \text{sgn}[d_1 e^{t-2\tau} + d_2 e^{-\tau}],$$
$$u_2(\tau) = \text{sgn}[d_1 e^{2(t-\tau)} + 2d_2 e^{(t-\tau)}] \tag{7.3.29}$$
$$= \text{sgn}[d_1 e^{t-2\tau} + 2d_2 e^{-\tau}], \qquad \text{on } [0, h].$$

We now assume that the optimal time $t$ is known and that $d_1 e^t + d_2 < 0$, $d_1 e^t + 2d_2 < 0$, and $d_2 > 0$ and begin at $(0, 0)$ with controls $u_1 = -1$, $u_2 = -1$ and move along the parabola defined by (7.3.28), which is

$$\alpha: \ x_1(\tau) = e^{2\tau} - 1 \qquad \text{on } [0, h],$$
$$x_2(\tau) = 3(e^\tau - 1). \tag{7.3.30}$$

Clearly $u_2(\tau)$ will switch to a new value at

$$\tau_1 = t - \ell n\left(\frac{-2d_2}{d_1}\right) = t + \ell n\left(\frac{-d_1}{2d_2}\right),$$

and $u_1(\tau)$ will change sign at $\tau_2 = t + \ell n(\frac{-d_1}{d_2})$. Hence $\tau_2 - \tau_1 = \ell n\, 2$. It follows then that if we begin at $\tau = 0$ at the point $(x_1(\tau_1), x_2(\tau_1))$ of the curve $\alpha$ with $u_1 = -1$ and $u_2 = 1$, we depart from $\alpha$ along the parabola

$$x_1(\tau) = x_1(\tau_1)e^{2\tau}, \qquad x_2(\tau) = x_2(\tau_1)e^\tau = (e^\tau - 1).$$

Clearly $u_1$ changes signs at

$$x_2(\ln 2) = 4x_1(\tau_1), \qquad x_2(\ln 2) = 2x_2(\tau_1) - 1.$$

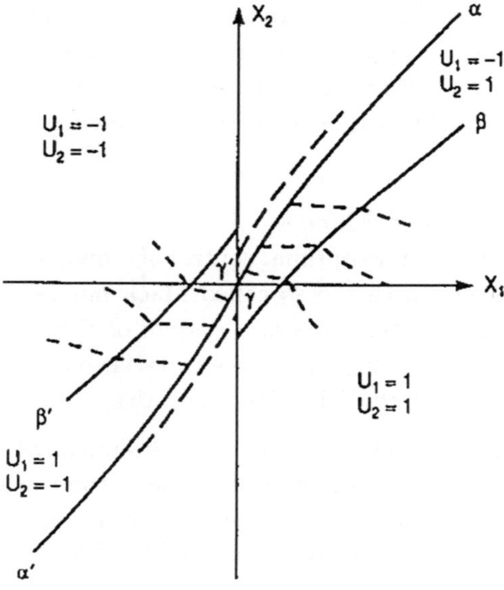


Fig. 7.3.2

These are the calculations of Hermes and LaSalle [4, p. 83] for ordinary differential systems. They are valid here because we have assumed that $h$ is so large that on $[0, h]$ all the indicated switches occur, and the dynamics is that of $e^{A_0\tau}$. Thus the last equations above tell us how to transform $\alpha$ to obtain the curve $\beta$ where $u_1$ will change sign. For each component there is only one sign change. We complete the analysis by assuming now that

$$d_1 e^t + d_2 < 0, \qquad d_1 e^t + 2d_2 \geq 0.$$

Now begin at $(0, 0)$, with control $u_1 = -1$, $u_2 = 1$, with a sign change of $u_1$ not later than $\tau = \ln 2$ apart. The trajectory is along the line $(0, 0)$ to $(-1, 0)$, which is the switching curve $\gamma$. The others are obtained by symmetry. Thus within $[0, h]$, $h$ large, the complete optimal control law is deduced. This is given in the Fig. 7.3.2.

If the switch time $\tau > h$, we have to go to the next interval $[h, 2h]$ before switching. Thus we consider $u(t)$, $t \in [h, 2h]$ in $[h, 2h]$; and then

$$u(\tau) = \text{sgn}[d^T U_1(s - t)B], \quad t \in [h, 2h] \quad \text{with } s = 0, \text{ i.e.,}$$

$$u_2(\tau) = \text{sgn}[d_1\{e^{+2t} - 4(e^{+(t-h)} - e^{+2(t-h)})\}$$

$$+ d_2\{e^{+(t-h)} - e^{+2(t-h)} + 2e^{+t}\}$$

$$u_2(\tau) = \text{sgn}[d_1\{e^{2t} - 2(e^{+t-h} - e^{+2(t-h)})\}]$$

$$+ d_2\{e^{+t} + e^{t-h} - 2e^{2(t-h)}\}],$$

where $d_1e^t + d_2 < 0$, $d_1e^t + 2d_2 < 0$, and $d_2 > 0$, $t \in [h, 2h]$. Thus we begin with this control at the point $\alpha(\tau_1)$ and move along this trajectory until switches occur at the zero of $g_1$ and the zero of $g_2$. The earlier process is repeated, after going to the next interval, $[2h, 3h]$; the analysis is completed. In this way the optimal control law is deduced.

## Example 7.3.4　Wind Tunnel Model

The problem concerns a time-optimal control of a high-speed closed-circuit wind tunnel. We are concerned with the so-called Mach number control. A linearized model of the Mach number dynamics is a system of three state equations with a delay. The state variables $x_1$, $x_2$, $x_3$ represent deviations from a chosen operating point (equilibrium point) for the following quantities:

$x_1 = $ Mach number, $x_2 = $ actuator position (guide vane angle in a driving fan), and $x_3 = $ actuator rate. The delay represents the time of the transport between the fan and the test section. We assume that the control variable has three components that constitute an input to the rate of the Mach number. The model has an equation of the form

$$\dot{x}_1(t) = -ax_1(t) + akx_2(t - h),$$

$$\dot{x}_2(t) = x_3(t), \qquad\qquad (7.3.31)$$

$$\dot{x}_3(t) = -w^2x_2(t) - 2\xi wx_3(t) + w^2u_3(t),$$

with $a = \frac{1}{1.964}$, $k = 0.117$, $w = 6$, $\xi = 1.6$, and $h = 0.33s$.

The parameter values correspond to one particular operating point. We write the equation as

$$\dot{x}(t) = A_ox(t) + A_1x(t - h) + Bu(t),$$

where

$$A_0 = \begin{bmatrix} -a & 0 & 0 \\ 0 & 0 & 1 \\ 0 & -w^2 & -2\xi w \end{bmatrix}, \quad A_1 = \begin{bmatrix} 0 & a_k & 0 \\ 0 & 0 & 0 \\ 0 & 0 & 0 \end{bmatrix} \quad B = \begin{bmatrix} 0 \\ 0 \\ k_3 \end{bmatrix}, \quad k_3 = 36.$$

The eigenvalues of $A_0$ are $\lambda_1 = -0.509165$, $\lambda_2 = -17.0939$, and $\lambda_3 = -2.106024$. The fundamental matrix is

$$e^{A_0 t} = \begin{bmatrix} e^{-\lambda_1 t} & 0 & 0 \\ 0 & 12e^{-\lambda_2 t} + 0.88e^{-\lambda_3 t} & 0.05(-e^{-\lambda_2 t} + e^{-\lambda_3 t}) \\ 0 & -2e^{-\lambda_2 t} + 2e^{-\lambda_3 t} & 0.89e^{-\lambda_2 t} + 0.11e^{-\lambda_3 t} \end{bmatrix}.$$

The system is Euclidean controllable on $[0, t]$, $\forall \ t > h$, since

$$\overline{Q}_3(h) = \begin{bmatrix} 0 & 0 & 0 & 0 & 2.144603 \\ 0 & 36 & -691.2 & 0 & 0 \\ 36 & -6912 & 11975.05 & 0 & 0 \end{bmatrix}$$

and rank $\overline{Q}_3(h) = 3$.

The system

$$\dot{x}(t) = A_0 x(t) + A_1 x(t - h)$$

is uniformly asymptotically stable since the eigenvalue $\lambda$ of the characteristic equation

$$\Delta(\lambda) = \begin{bmatrix} \lambda + \dfrac{1}{1.964} & -\dfrac{0.117}{1.964}e^{-\lambda h} & 0 \\ 0 & \lambda & -1 \\ 0 & 36 & \lambda + 19.2 \end{bmatrix}$$

$$= \left(\lambda + \frac{1}{1.964}\right)(\lambda^2 + 19.22\lambda + 36) = 0,$$

and

$$\lambda_1 = -0.50916, \qquad \lambda_2 = -17.094, \qquad \lambda_3 = -2.106.$$

It follows that the system is controllable with constraints. The fundamental matrix is

$$U(t)$$

$$= \begin{bmatrix} e^{-0.5092t} & 0 & 0 \\ 0 & 1.1405e^{-2.106t} - 0.1405e^{-17.094t} & 06672(e^{-2.106t} - e^{-17.044t}) \\ 0 & -2.40192(e^{-2.106t} - e^{-17.084t}) & 0.1405e^{-2.106t} + 1.1405e^{-17.044t} \end{bmatrix},$$

on $[0, 0.33]$.

On $[0.33, 0.66]$,

$$U(t) = e^{A_0 t} + \int_{0.33}^{t} e^{A_0(t-s)} A_1 e^{A_0(t-s)} ds.$$

Hence,

$U(t)$

$$
= \begin{bmatrix}
e^{-5092t} & 0.03153 - 0.0786e^{-2.6t} + 0.168e^{-17.6t} & 0.00129 - 0.003603e^{-2.6152t} \\
& & +0.07527e^{-17.6032t} \\
0 & 1.1405e^{-2.106t} - 0.1405e^{-17.94t} & 0.06672(e^{-2.106t} - e^{-17.094t}) \\
0 & -2.40192(e^{-2.106t} - e^{17.084t}) & -0.1405e^{-2.106t} + 1.11405e^{-17.084t}
\end{bmatrix}.
$$

The optimal controls are

$$
u^*(s) = \text{sgn}[0.06672 d_2 (e^{-2.106(t-s)} - e^{-17.094(t-s)})
$$
$$
+ d_3(-0.1405 e^{-2.106(t-s)} + 1.1405 e^{-17.044(t-s)})] \text{ on } [0, h],
$$
$$
u^*(s) = \text{sgn}[d_1(0.00129 - 0.003603 e^{-2.6152(t-s)} + 0.07527 e^{-17.6032(t-s)})
$$
$$
+ d_2(0.06672(e^{-2.106(t-s)} - e^{-17.044(t-s)}))
$$
$$
+ d_3(-0.1405 e^{-2.106t} + 1.1405 e^{-17.044t})].
$$

## 7.4　Time-Optimal Feedback Control of Autonomous Delay Systems

In this section we complete the investigation that is reported in Theorem 7.2.2 of the problem of the construction of an optimal feedback control needed to reach the Euclidean space origin in minimum time for the linear system

$$
\dot{x}(t) = A_0 x(t) + \sum_{j=1}^{N} A_j x(t - \tau j) + Bu(t). \tag{7.4.1}
$$

Here $0 < \tau < 2\tau < \cdots < \tau N = h$; $A_i$ are $n \times n$ constant matrices, and $B$ is an $n \times m$ constant matrix. The controls are $L_\infty$ functions whose values on any compact interval lie in the $m$-dimensional unit cube

$$
C^m = \{u \in E^m : |u_j| \le 1, \ j = 1, \ldots, m\}.
$$

We shall show that the time-optimal feedback system

$$
\dot{x}(t) = \sum_{j=0}^{N} A_j x(t - \tau j) + Bf(x(t)) \tag{7.4.2}
$$

executes the time-optimal regime for (7.4.1) in the spirit of Hájek [30] and Yeung [32, 33]. The construction of $f$ provides a basis for direct design, and it is done for strictly normal systems, which we now define.

**Definition 7.4.1**　Let

$$
J_0 = \{t = j\tau, \ j = 0, 1, 2, \ldots\},
$$

and assume $J_0$ is finite. Suppose $U(\epsilon, t)$ is the fundamental matrix solution of

$$\dot{x}(t) = A_0 x(t) + \sum_{j=1}^{N} A_j x(t - \tau j) \tag{7.4.3}$$

on some interval $[0, \epsilon]$, $\epsilon > 0$. Note that $U(\epsilon, t)$ is piecewise analytic, and its analyticity may break down only at points of $J_0$ (see Tadmor [25]). System (7.4.1) is strictly normal on some interval $[0, \epsilon)$ if for any integers

$$r_j \geq 0 \text{ satisfying } \sum_{j=1}^{M} r_j = n,$$

the vectors

$$Q_{ij}(s), \quad j = 1, \ldots, m, \qquad s \in [0, \epsilon] - J_0, \qquad 0 \leq i \leq r_{j-1} \tag{7.4.4}$$

are linearly independent. (If $r_j = 0$, there are no terms $Q_{ij}$ in (7.4.4).) Here, as before in (7.1.10),

$$Q_{kj}(s) = \sum_{j=0}^{N} A_i Q_{k-1j}(s - I_j), \qquad k = 1, 2, \qquad s \in (-\infty, \infty),$$

$$Q_{0j}(s) = \begin{cases} b_j & s = 0, \\ 0 & s \neq 0, \end{cases}$$

$$j = 1, \ldots, m,$$

and $B = (b_1 \ldots b_m)$.

It follows from Theorem 7.1.4 that a complete, strictly normal system is normal, and has rank $B = \min[m, n]$. Indeed, choose any column $b_j$ of $B$ and set $r_j = n$, $r_j = 0$ for $i \neq j$ in the definition. Clearly

$$Q_{0j}(s), \qquad Q_{n-1j}(s), \qquad s \in [0, \epsilon] - J_0$$

are linearly independent. Because of Theorem 7.1.4 we have normality since $b_j$ is arbitrary. The second assertion is obvious since $b_j$ is linearly independent and we can take $r_j = 1$ or $r_j = 0$. If the system is an ordinary one,

$$\dot{x}(t) = A_0 x(t) + B u(t), \tag{7.4.5}$$

our definition reduces to that of Hájek [31].

**Lemma 7.4.1 (Fundamental Lemma).** *System (7.4.1) is strictly normal if there exists $\epsilon > 0$ with the following property: for every $n$-vector $c \neq 0$ and in any interval of length $\leq \epsilon$, the sum of the number of roots counting multiplicities of the coordinates of the index $g$ of the control system (7.4.1)*

$$g(t,c) = c^T U(\epsilon, t) B$$

*is less than $n$.*

**Proof**   Suppose that such an $\epsilon > 0$ exists and (7.4.1) is not strictly normal on $[0, \epsilon) - J_0$. Then there exist integers $r_j \geq 0$ such that $\sum_{j=1}^{m} r_j = n$, and

$$Q_{kj}(i\tau), \quad (1 \leq j \leq m, \ 0 \leq k \leq r_{j-1}), \quad \epsilon - it > 0, \quad i = 1, 2, \cdots$$

are linearly dependent. Hence there exists $c \neq 0$, $c \in E^n$ such that

$$c^T Q_{kj}(i\tau) = 0, \qquad (1 \leq j \leq m, \ 0 \leq k \leq r_{d-1}). \tag{7.4.6}$$

Suppose $a$ is a zero of $g_j(t,c) = c^T U(\epsilon, t) b_j$, $(j = 1, \ldots, m)$ on $[0, \epsilon) - J_0$. Then analyticity yields

$$g_j(t,c) = \sum_{k=1}^{\infty} g_j^{(k)}(a^+, c) \frac{(t-a)^k}{k!},$$

$$g_j(t,c) = \sum_{k=0} g_j^{(k)}(a^-) \frac{(t-a)^k}{k!},$$

where $g_j$ is expressed as the Taylor series at a particular time ($t = a^+$, or $t = a^-$), and

$$g_j^{(k)}(a^+, c) \equiv \frac{d^k g}{dt^k}(a^+, c), \qquad k = 0, 1, 2, \ldots,$$

etc. It now follows that

$$0 = \sum_{k=0}^{\infty} \Delta g_j^{(k)}(a, c) \frac{(t-a)^k}{k!},$$

where $\Delta g_j^{(k)} \equiv g_j^{(k)}(a^-, ) - g_j^k(a^+, )$. If $a = \epsilon - ih > 0$, $i = 1, 2, \ldots$, then

$$0 = \sum_{k=0}^{\infty} \Delta g_j^{(k)}(\epsilon - i\tau) \frac{(t - (\epsilon - i - T))^k}{k!},$$

$$0 = \sum_{k=0}^{\infty} (-1)^k c^T Q_{kj}(i\tau) \frac{(t - (\epsilon - ih))}{k!}, \tag{7.4.7}$$

where we have used (7.4.7). It follows from (7.4.6) that (7.4.7) becomes

$$0 = [t - (\epsilon - i\tau)]^{r_j} \sum_{k=0}^{\infty} c^T Q_{k+r_j, j}(i\tau) \frac{(t - (\epsilon - \tau i))^k}{(k + r_j)!}.$$

Hence for each $j = 1, \ldots, m$, $t = \epsilon - iT$ is a zero of $g_j(t)$ of multiplicity $\geq r_j$, and consequently the sum of the number of roots in $[0, \epsilon] - J_0$ counting multiplicities of the coordinates of $g(t) = c^T U(\epsilon, t) B$ is at least $n$. This completes the proof.    $\square$

It is very desirable to have information on the largest $\epsilon$ that has the property of the Fundamental Lemma. Results on this do not seem to be currently available. It is conjectured that the converse of the Fundamental Lemma is valid.

In all that follows, we assume as basic that (7.4.1) is strictly normal and (7.4.3) complete in the sense that the operator $W(t, \sigma) : C \to E^n$, defined by $W(t, \sigma)\phi = x(\sigma, \phi, 0)(t)$, where $x(\sigma, \phi, 0)(t)$ is a solution of (7.4.3), is a surjection:

$$W(t, \sigma)C = E^n .$$

We shall also retain $\epsilon$ as that described by the Fundamental Lemma.

**Corollary 7.4.1**  *Let $k \le n$. Suppose there are distinct times $t_1 < \cdots < t_k$ with $t_k - t_1 \le \epsilon$ and integers $\ell_i$ among $1, \ldots, m$. If (7.4.1) is strictly normal, then the vectors*

$$U(\epsilon, t_1)b_{\ell_1}, \ldots, U(\epsilon, t_k)b_{\ell_k}$$

*are linearly independent.*

**Proof**  Since a subcollection of linearly independent vectors are also linearly independent, we shall prove the statement for $k = n$. Suppose the assertion is false, there exists $c \ne 0$, $c \in E^n$ such that

$$g_i(t_i, c) \equiv c^T U(\epsilon, t_i)b_{\ell_i} = 0, \qquad 1 \le i \le n .$$

Thus the sum of the number of roots, counting multiplicities of the coordinates of $g_i$, is at least $n$ in $[t_1, t_1 + \epsilon]$; and this is a contradiction.  $\square$

**Corollary 7.4.2**  *Suppose (7.4.1) is strictly normal, and $0 < t_1 < \cdots < t_n < \epsilon$. Then any $n$ terms among*

$$\int_0^{t_i} U(\epsilon, t)b_j dt, \qquad (1 \le i \le n,\ 1 \le j \le n)$$

*are linearly independent.*

**Proof**  Suppose that statement is invalid. Then there exists a nonzero vector $c \in E^n$ that is perpendicular to $n$ terms among

$$\int_0^{t_i} c^T U(\epsilon, t)b_j dt \equiv M(t), \qquad 1 \le i \le n, \qquad 1 \le j \le m .$$

Observe that $t = 0$ is a root of $M(t)$, so that the sum of the number of roots in

$$[0, \epsilon] \quad \text{of} \quad \int_0^t c^T U(\epsilon, t), B dt$$

is at least $n + m$. The first derivative of this integral is $c^T U(\epsilon, t)B$, so that its coordinates have at least $n$ zeros in $[0, \epsilon]$.  $\square$

**Corollary 7.4.3**   *Assume that (7.4.1) is pointwise complete and strictly normal. Then every optimal control on an interval of length less than $\epsilon$ has at most $n - 1$ discontinuities almost everywhere.*

**Proof**   Consider the interval $[\theta_1, \theta_2]$, with $0 \leq \theta_1 < \theta_2$ and $\theta_2 - \theta_1 < \epsilon$. If $u$ is an optimal control on $[\theta_1, \theta_2]$, then by Theorem 7.1.3, for some $c \neq 0$, $c \in E^n$,

$$U(t) = \text{sgn } c^T U(\epsilon, t) B$$

almost everywhere on $[\theta_1, \theta_2]$. It follows that the points of discontinuity of $u$ are among the roots of the coordinate of $c^T U(\epsilon, t) B$. This is at most $n - 1$ because of the Fundamental Lemma.                                                      $\square$

**Definition 7.4.2**   Let $k$ be an integer with $1 \leq k \leq n$, and let

$$Q_k = \{t = (t_1 \cdots t_k)^T \in E^k : 0 = t_0 < t_1 < \cdots < t_k < \epsilon\}, \qquad (7.4.8a)$$

and for the $k$-tuple $i = (u_1 \ldots u_k)$ of vertices of $C^m$ with $u_{j-1} \neq u_j$, define the mapping

$$F_{ki} : Q_k \to \text{Int } \mathbb{R}(\epsilon)$$

by

$$F_{ki}(t) = \sum_{j=1}^{k} \int_{t_{j-1}}^{t_j} U(\epsilon, t) B u_j \, dt. \qquad (7.4.8b)$$

**Definition 7.4.3**   The minimum time function $M : \mathbb{R} \to E^1$ is defined by

$$M(x) = \text{Inf}\{t \geq 0 : x = x(t, \sigma, \phi, 0) \in \mathbb{R}(t)\},$$

where $x(t, \sigma, \phi, 0)$ is the solution of

$$\dot{x}(t) = \sum_{j=0}^{N} A_j x(t - \tau_j), \qquad x_0 = \phi, \qquad (7.4.9)$$

and $\mathbb{R} = \bigcup_{t \geq 0} \mathbb{R}(t)$. Suppose

$$x = \int_{0}^{t_k} U(t_k, t) B u(t) \, dt = \sum_{i=1}^{k} \int_{t_{i-1}}^{t_i} U(t_k, t) B u_i(t) \, dt,$$

$0 = t_0 < t_1 < \cdots < t_k < \epsilon$. Then $M(x) = t_k$.

**Definition 7.4.4**   Let $k$ be any integer $1 \leq k \leq n$. For any (optimal) control $u$ defined by $u(s) = u_j$ on $[t_{j-1}, t_j)$ for $(t_1, \ldots, t_k) \in Q_k$ and $u_{j-1} \neq u_j$ in $C^m$, we define the sequence $\{u_1 \to \cdots \to u_k\}$ its (optimal) switching sequence.

**Corollary 7.4.4**   *Suppose $1 \leq k \leq n$. Suppose that $\phi \in C$, $\phi \neq 0$ is steered to the Euclidean origin in time $t_k$ by an optimal control $u$ whose switching sequence is given by $\{u_1 \to \cdots \to u_k\}$. Let*

$$v_i = B(u_i - u_{i+1}) \quad \text{for} \quad i = 1, \ldots, k-1, \qquad v_k = Bu_k \,.$$

*Then the vectors*

$$U(t_k, t_i)v_i \,, \qquad (1 \leq i \leq k)$$

*are linearly independent.*

**Proof**   The statement is trivially valid for $k = 1$. We assume $k \geq 2$ and use induction on $k$ and Corollaries 7.4.1 and 7.4.2 to prove the corollary using the arguments corresponding to [32, pp. 12–13].   $\square$

Now recall that in Sec. 7.2 we proved the following.

**Theorem 7.4.1a**   *The minimal time function*

$$M : E^n \to E$$

*is continuous if (7.4.1) is Euclidean controllable. Consequently,*

$$M : \mathbb{R} \to E$$

*is continuous.*

**Lemma 7.4.2**   *Let $Q_k$ be as defined in (7.4.8a), and $F_{ki}$ as defined in (7.4.8b). If $i = \{u_1 \to \cdots \to u_k\}$ is an optimal switching sequence, then the Jacobian matrix of $F_{ki}(t) = F_{ki}(t_0 \cdots t_k)$ has rank $k$ at each point of $Q_k$. $F_{ki}$ is an analytic function of its variables.*

**Proof**   We observe that

$$F_{ki} = \sum_{j=1}^{k} \int_{t_{j-1}}^{t_j} U(t_k, t)Bu_j ds$$

$$= \int_0^{t_1} U(t_k, t)Bu_1 ds + \int_0^{t_2} U(t_k, t)B(u_2 - u_3)ds$$

$$+ \cdots + \int_0^{t_k} U(t_k, t)Bu_k ds \,,$$

which is analytic. It is clear that the column vectors of its Jacobian matrix are constant multiples of $U(t_k, t)Bu_k$, or

$$U(t_k, t_j)B(u_j - u_{j+1}) \,, \qquad (1 \leq j \leq k-1) \,.$$

These vectors are linearly independent since Corollary 7.4.4 is true. Therefore the rank is $k$.

If $u$ is an optimal control on an interval $[0, t_1]$, $t_1 \in \epsilon$, then its values are extreme points of $C^m$, the $m$-dimensional unit cube. It is therefore bang-bang and piecewise constant. Since Corollary 7.4.3 asserts that $u$ has at most $n - 1$ discontinuities, if there are exactly $k$ discontinuities, $k \leq n - 1$ at $t_1 \cdots t_k$ with $t_0 = 0 < t_1 < \cdots t_k < 0$, $u(t) = u_k$ for $t \in [t_{k-1}, t_k)$, and if these switching times are changed, but the switching sequence $\{u_1 \to \cdots u_k\}$ is retained, then the resulting control is still optimal. This is the content of the next theorem.    $\square$

**Theorem 7.4.1b**    *Let $i = \{u_1 \to \cdots \to u_k\}$ be a switching sequence, and such that for some $t \in Q_n$, $x \in E^n$ is given by*

$$x = Fn_i(t) = \sum_{j=1}^{n} \int_{t_{j-1}}^{t_j} U(t_n, s) B u_j ds$$

*with $t_0 = 0 < t_1 < \cdots < t_n < \epsilon$; $M(x) = t_n$.*
    *Then for any*

$$s \in Q_n, \quad y = F_{n_i}(s) = \sum_{j=1}^{n} \int_{s_{j-1}}^{s_j} U(s_n, s) B u_j ds$$

*has $T(y) = s_n$.*

**Proof**    Let

$$F = \left\{ \tau \in Q_n : z = \sum_{j=1}^{n} \int_{\tau_{j-1}}^{\tau_j} U(\tau_n, s) B u(s) ds \text{ has } M(z) = \tau n \right\}.$$

By assumption, $t = (t_1, \ldots, t_n) \in F$, so that $F$ is nonempty. If we prove that $F$ is both open and closed in $Q_n$, then $F = Q_n$. Let a sequence $\tau_p$ in $E$ converge to $\tau \in Q_n$. Suppose

$$x_p = \sum_{j=1}^{n} \int_{\tau_{p,j-1}}^{\tau_{p,j}} U(t_{pn}, s) B u_j ds = F_{ni}(\tau_p),$$

$$z = \sum_{j=1}^{n} \int_{\tau_{j-1}}^{\tau_j} U(\tau_n, s) B u_j ds = F_{ni}(\tau).$$

Since $\tau_p \to \tau$, we have $x_p \to z$. Now $\tau_p \in F$, so that $M(x_p) = \tau_{p,n}$ and from continuity of $M$, $M(z) = \tau_n$. Hence $\tau \in F$. We have proved that $F$ is closed in $Q_n$. We now assume that $F$ is not open in $Q_n$ and deduce a contradiction. Because of our assumption, there is a sequence $\tau_p \in Q_n - F$ and a $\tau \in F$ such that $\tau_p \to \tau$. Let

$$x_p = \sum_{j=1}^{n} \int_{\tau_{p,j-1}}^{\tau_{p,j}} U(t_{p,n} s) B u_j ds,$$

$$z = \sum_{j=1}^{n} \int_{\tau_{j-1}}^{\tau_j} U(t_n, s) B u_j ds \,.$$

From $\tau_p \to \tau$ we have $x_p \to z$ in Int $\mathbb{R}(\epsilon)$. As a consequence of this, $x_p$ can be driven to the Euclidean origin by an appropriate optimal control

$$x_p = \sum_{j=1}^{kp} \int_{t_{p,j-1}}^{t_{p,j}^j} U(t_p, k_p, s) B u_{pj} ds \,, \tag{7.4.10}$$

$$k_p \leq n \,, \quad t_{p,0} = 0 < t_{p1}, \ldots, t_{p,k_p} < \epsilon \,, \quad M(x_p) = \tau_p, k_p \,.$$

Clearly $v_{pj}$ are among the vertices of $C^m$. Because $C^m$ has finitely many vertices and $1 \leq k_p \leq n$, we can take $k_p = k$ and $v_{pj} = v_j$, independent of $p$. Also since $0 < t_{p,j} < \epsilon$, by taking subsequences we can assume $t_{p,j} \to \sigma_j$. Because $x_p \to z$, we conclude that

$$z = \sum_{\alpha=1}^{k} \int_{\sigma_{j-1}}^{\sigma_j} U(\sigma_n, s) B v_j ds \,.$$

Because optimal controls are unique, we have $k = n$, $v_j = u_j$, and $\sigma_j = \tau_j$. Therefore

$$F_{ni}(\tau_p) = x_p = F_{ni}(t_p) \,, \qquad \tau_{p,j} \to \tau_j \leftarrow t_{p,j} \,.$$

Invoke (7.4.10) to obtain

$$M(x_p) = t_{p,n} < \tau_{p,n} \,. \tag{7.4.11}$$

But if $F_{ni}$ corresponds to the optimal switching sequence $\{u_1 \to \cdots \to u_n\}$, then $F_{ni}$ is locally one-to-one by Lemma 7.4.2. This is true in a neighborhood of $\tau$. But then $\tau_{p,j} \to \tau_j \in t_{p,j}$. Therefore for sufficiently large $p$, $\tau_p = t_p$, so that in particular $\tau_{p,n} = t_{p,n}$. This equality contradicts (7.4.11). The proof is complete. $\qquad \square$

We shall now follow the development of the report in Sec. 5.2 and generalize Yeung's thesis in [32] and [33] as follows:

**Theorem 7.4.2**  *Consider the system*

$$\dot{x}(t) = A_0 x(t) + \sum_{j=1}^{N} A_j x(t - \tau_j) + B u(t) \,, \tag{7.4.1}$$

*and assume:*

(i) *System* (7.4.1) *is Euclidean controllable.*
(ii) *System*

$$\dot{x}(t) = A_0 x(t) + \sum_{j=1}^{N} A_j x(t - \tau j) \tag{7.4.2}$$

*is uniformly asymptotically stable.*

(iii) *System* (7.4.1) *is strictly normal.*

*Then there exists an $\epsilon > 0$ and a function $f : \text{Int } \mathbb{R}(\epsilon) \to E^m$ that is an optimal feedback control of* (7.4.1) *in the following sense: If*

$$\dot{z}(t) = \sum_{j=0}^{N} A_j z(t - \tau j) + B f(z(t)), \qquad z \in \text{Int } \mathbb{R}(\epsilon), \qquad (7.4.12)$$

*then the set of solutions of* (7.4.12) *coincides with the set of optimal solutions of* (7.4.1) *in* $\text{Int } \mathbb{R}(\epsilon)$. *Also $f(0) = 0$ for $x \neq 0$, $f(x)$ is among the vertices of the unit cube $U$. Furthermore, $f(x) = -f(-x)$. If $m \leq n$, $f$ is uniquely determined by the condition that optimal solutions of* (7.4.1) *solve* (7.4.12).

**Proof**    We follow closely the treatment of Sec. 5.2, and define terminal manifolds $M_{ki}$ as follows. $\qquad\qquad\qquad\qquad\qquad\qquad\qquad\qquad\qquad\qquad\qquad\qquad\square$

**Definition 7.4.5**    For each $k = 1, \ldots, n$ and each switching sequence $i = \{u_1 \to \cdots \to u_k\}$ of vertices of $C^M$ with $u_{j-1} \neq u_j$, the terminal $M_{ki}$ is defined to be the set of points in $\text{Int } \mathbb{R}(\epsilon)$ whose optimal controls have $i$ as optimal switching sequence. We set $M_0 = \{0\}$. The set of points in $\text{Int } \mathbb{R}(\epsilon)$ whose optimal controls have exactly $k - 1$ discontinuities is designated as $M_k$, and the switching locus is $\bigcup_{k=0}^{n-1} M_k$.

The following proposition is easily proved as is done for ordinary systems in Proposition 5.2.1.

**Proposition 7.4.1**

(i) $M_k = \bigcup_{j \in I_k} M_{kj}$, $\text{Int } \mathbb{R}(\epsilon) = \bigcup_{k=0}^{n} M_k$, *and these are disjoint unions.*
(ii) *There are exactly $2M^{n-1}$ nonvoid sets $M_{ni}$.*
(iii) *Each nonvoid $M_{ki}$ is an analytic $k$-manifold of $E^n$, and $M_{ni}$ is open and connected.*
(iv) $\text{Int } \mathbb{R}(\epsilon)$ *is an open and dense set that is the union of $2m^{n-1}$ disjoint connected nonempty open sets, and a finite number of analytic $k$-manifolds, $0 \leq k \leq n - 1$, one for $k = 0$, and at least two for $k \geq 1$.*

With this proposition, one proves the following:

**Proposition 7.4.2**    *Suppose on an interval $[0, \theta] - J_0$ with $0 < \theta < \epsilon$, $i = \{u_1 \to \cdots \to u_k\}$ is an optimal switching sequence. If $v$ has the same switching sequence $i$, but not necessarily the same switching times, then $v$ is also optimal. Also if $i = \{u_1 \to \cdots \to u_k\}$ is an optimal switching sequence corresponding to $M_{k,i}$ and $\{U_{j+1} \to \cdots \to u_k\}$ corresponds to $M_{k-j,i}$ $j = 0, \ldots, k - 1$, with $M_{0j} = M_0$, then if an optimal solution in $\text{Int } \mathbb{R}(\epsilon)$ meets $M_{k,i}$ then thereafter it meets only $M_{k-1,i}, \ldots, M_0$ in this order.*

We now use these preliminary results to prove Theorem 7.4.2. Set $f(0) = 0$. Let $\phi \neq 0$ be an initial point that is driven to $0 \in E^n$ in time $t \leq \epsilon$, so that

$$x = x(t, \phi, 0) \in \text{Int } \mathbb{R}(\epsilon),$$

where $x(t, \phi, 0)$ is a solution of (7.4.3). Now find $M_{ki}$ containing $x$. Suppose the corresponding optimal switching sequence is $\{u_1 \to \cdots \to u_k\}$, set $f(x) = u$. Because $M_k = \bigcup_{j \in Ik} M_{kj}$, $\text{Int } \mathbb{R}(\epsilon) = \bigcup_{k=0}^{n} M_k$, are disjoint unions, $f$ is well defined. Let $x \neq 0$ be in $\text{Int } \mathbb{R}(\epsilon)$, and let $x(\psi)$ be the optimal solution of (7.4.1) in $\text{Int } \mathbb{R}(\epsilon)$ through $x$. Suppose the optimal control $u(s)$ of $x$ is given by

$$u(s) = u_j \qquad \text{on } [t_{j-1}, t_j),$$

with

$$0 = t_0 < t_1 < \cdots < t_k = M(x) < c.$$

Then $x$ belongs to some $M_{ki}$. It follows from Proposition 7.4.2 that $f(x(s)) = u(s)$, proving that any optimal solution of (7.4.1) in $\text{Int } \mathbb{R}(\epsilon)$ is also a solution of (7.4.2) or (7.4.12).

We indicate a proof of the converse. Just as in [32], if $y : [0, \theta] \to E^n$ is a solution of (7.4.12) with $0 < \theta < \epsilon$, and if $y(t) \in M_{ki}$ and $0 \leq t < \theta$, then for $s - t \geq 0$ sufficiently small $y(s) \in M_{ki}$. The proof of this assertion is given by induction on $n - k$. For $k = n$, the assertion is true because $y(\cdot)$ is continuous and $M_{n_i}$ is open in $E^n$. Details of the consequence of the inductive assumption is as outlined in [7, p. 79] for ordinary systems. With this proved, it will follows that $f(y(\cdot))$ is constant on any interval on which $y(\cdot)$ is differentiable. Because of Proposition 7.4.2, $y(\cdot)$ coincides with the optimal solution through $y(t)$. The rest of the proof is as in Theorem 5.2.1.

**Remark**  In [32, Theorem 4.1] Hájek's Proposition 12 [10] was used. An analogous result is valid in our situation when (7.4.1) is normal:

**Proposition 7.4.3**  *Every optimal solution of a normal system (7.4.1), on $[0, +\infty)$ can be extended to an optimal solution on $(-\infty, \infty)$. The proof is analogous to Hájek [10, p. 346]. See Hale [8, p. 68].*

## 7.5  Minimum-Effort Control of Delay Systems

In Secs. 7.2–7.4 we studied the time-optimal control of linear delay systems. In this section we study the minimum-effort control of the system

$$\dot{x}(t) = A_0(t)x(t) + \sum_{i=1}^{N} A_i(t)x(t - h_i) + B(t)u(t), \qquad (7.5.1)$$

where $0 < h_1 \leq h_2 \leq \cdots h_N = h$, $A_i\ i = 0, \ldots, N$ are $n \times n$ analytic functions, and $B$ is an $n \times m$ real analytic matrix function. The controls $u$ are bounded measurable functions. We consider the following problem: Find a $u$ (subject to some constraint, i.e., $u \in U$) that minimizes an "effort" function $E(u(t_1))$ subject to (7.5.1), where

$$x_\sigma = \phi, \qquad x(t_1, \sigma, \phi, u) = x_1 \in E^n. \tag{7.5.2}$$

Here $x = x(\cdot, \sigma, \phi, u)$ is the solution of (7.5.1) with $x_\sigma = \phi$.

We identify the following effort functions:

$$E_0(u(t_1)) = \int_\sigma^{t_1} u^T(t)R(t)u(t)dt, \tag{7.5.3a}$$

where $R(t)$ is a continuous positive definite $m \times m$ matrix. When $R(t) \equiv I$, the identity matrix, then

$$E_0(u(t_1)) = \int_\sigma^{t_1} |u(t)|^2 dt. \tag{7.5.3b}$$

$$E_1(u(t_1)) = \max_{1 \leq j \leq m} \sup_{0 \leq t \leq t_1} |u_j(t)|, \tag{7.5.4}$$

where

$$u \in U = \{u \in E^m, \ u \text{ measurable } |u_j(t)| \leq 1, \ j = 1, \ldots, m, \ \sigma \leq t \leq t_1\}. \tag{7.5.5}$$

In (7.5.4) $E_1(u(t_1))$ describes the maximum thrust available to the system

$$E_2(u(t_1)) = \left( \int_\sigma^{t_1} \sum_{j=1}^M |u_j(t)|^p dt \right)^{\frac{1}{p}} \equiv \|u\|_p, \tag{7.5.6}$$

where $p > 1$, and

$$u \in U = \{u \in E^m, \ u \text{ measurable } \|u\|_p = E_2(u(t_1)) \leq 1\}. \tag{7.5.7}$$

If $p = 2$ in (7.5.6), then $E_2(u(t_1)) = \|u\|_2$ represents the energy or power of the system, and this is to be minimized.

$$E_3(u(t_1)) = \int_\sigma^{t_1} \sum_{j=1}^m |u_j(t)| dt = \|u\|_1. \tag{7.5.8}$$

In (7.5.8) we may sometimes assume a constraint

$$u \in U = \{u \text{ measurable}, \ u \in E^m, \ |u_j(t)| \leq 1\}. \tag{7.5.9}$$

Without constraints (7.5.9) in $E_3$, the optimal controls are impulsive in nature. With constraints (7.5.9), the optimal controls are "bang-of-bang", as we shall see.

The solution of (7.5.1) in the state space $E^n$ is given by the variation of parameter

$$x(t, \sigma, \phi, u) = x(t, \sigma, \phi, 0) + \int_\sigma^t U(t, s)B(s)u(s)ds, \tag{7.5.10}$$

where $U(t,s)$ is the fundamental matrix solution of

$$\dot{x}(t) = A_0 x(t) + \sum_{j=1}^{N} A_i(t)x(t-h_i)\,,$$

$$x_\sigma = \phi \in C\,,$$

(7.5.11)

with

$$U(t,s) = I\,, \qquad t = s\,, \qquad U(t,s) = 0 \ s > t\,.$$

In (7.5.10), set

$$Y(t,s) = U(t,s)B(s)\,.$$

(7.5.12)

This is an $n \times m$ matrix function which, because $B$ is analytic, is piecewise analytic in $t, s$ [11] and at least measurable in $t, s$ [8, p. 145], and continuous in $t$ for $t \geq s$ for each fixed $s$.

We now state the solutions of the minimum-effort problems for the various efforts.

**Theorem 7.5.1** *Assume that*

(i) *(7.5.1) is Euclidean controllable on* $[\sigma, t_1]$, *and this holds if*

$$\text{rank } \overline{Q}_n(t_1) = n\,,$$

(7.5.13)

*where*

$$\overline{Q}_n(t_1) = \{Q_0(s,t_1), \ldots, Q_{n-1}(s,t_1)\,, \qquad s \in (\sigma, t_1]\}$$

(7.5.14)

*with $Q_i$ defined by the determining equations*

$$Q_k(s,t) = \sum_{j=0}^{N} A_j(t)Q_{k-1}(s-h_j, t-h_j) - \frac{d}{dt}Q_{k-1}(s,t)\,,$$

$$t \in [\sigma, t_1]\,, \qquad k = 1, 2, \ldots, n-1\,.$$

(7.5.15)

$$Q_0(s,t) = \begin{cases} B(t), & t \in [\sigma, t_1] \quad for \ s = h_i\,, \\ 0 & otherwise\,. \end{cases}$$

(ii) *Let*

$$W = \int_\sigma^{t_1} Y(t_1,s)R^{-1}(s)[Y(t_1,s)]^T ds\,.$$

(7.5.16)

*Then the control $u^*$ which is defined by*

$$u^*(t) = R^{-1}(t)(Y(t_1,t)^T)W^{-1}q\,, \qquad t \in [\sigma, t_1]\,,$$

(7.5.17)

*with*

$$q = [x_1 - x(t_1, \sigma, \phi, 0)],  \qquad (7.5.18)$$

*is the optimal control that minimizes $E_0(u(t_1))$, i.e.,*

$$E_0(u^*(t_1)) \le E_0(u(t_1)), \qquad \forall \, u.$$

**Proof** Because (7.5.13) holds, $W^{-1}$ exists and $u^*(t)$ is well defined. If we use $u^*(t)$ in (7.5.10), we obtain

$$x(t_1, \sigma, \phi, u^*) = x(t_1, \sigma, \phi, 0) + \int_\sigma^{t_1} Y(t_1, s) R^{-1}(s)(Y(t_z, s))^T W^{-1} q \, dt,$$

$$= x(t_1, \sigma, \phi, 0) + q = x_1.$$

Thus, indeed $u^*$ transfers to $x_1$ in time $t_1$. That $u^*$ minimizes $E_0$ follows standard arguments. Indeed, let $\bar{u}$ be any other control that transfers $\phi$ to $x_1$ at time $t_1$. Since $u^*$ in (7.5.17) also steers the system from $\phi$ to $x_1$, we have the equality

$$\int_\sigma^{t_1} U(t_1, s) B(s) u^*(s) ds = \int_\sigma^{t_1} U(t_1, s) B(s) \bar{u}(s) ds.$$

Using the inner product on both sides of this equality, we obtain

$$\left( \int_\sigma^{t_1} Y(t_1, s)(\bar{u}(s) - u^*(s)) ds, W^{-1} q \right) = 0.$$

Using (7.5.17) and the properties of the inner product, we obtain

$$\int_\sigma^{t_1} (\bar{u}(s) - u^*(s)), \qquad u^*(s)) ds = 0.$$

We now use this equality to derive

$$E_0(u^*(t_1)) = \int_\sigma^{t_1} u^*(s) R(s) u^*(s) ds \le \int_\sigma^{t_1} \bar{u}(s) R(s) \bar{u}(s) ds = E(\bar{u}(t_1)).$$

This completes the proof. $\qquad\qquad\qquad\qquad\qquad\qquad\qquad\qquad\qquad\qquad\qquad$ $\square$

**Remark 7.5.1** We can easily show that

$$E_0(u^*(t_1)) = (q, W^{-1} q).$$

Observe that there are no constraints on $u$ except that it is measurable and integrable. The controllability assumption enables one to infer that $\phi$ is steered to $x_1$ in time $t_1$. The rank condition is available in Manitius [6].

If the controls are constrained to lie on a bounded set $U$, then some stability condition on (7.5.11) is required.

For the solution of the minimum-effort problem with effort defined by $E_1(t)$ in (7.5.4), set

$$y(t) = x_1 - x(t, \sigma, \phi, 0) \,, \tag{7.5.19}$$

and call it the reachable state. Let

$$g(t_1, c) = c^T Y(t_1, t) = c^T U(t_1, t) B(t) \,, \tag{7.5.20}$$

where $c$ is an $n$-dimensional vector. Note that $g$ is an $m$-vector. Suppose

$$F_1(c, t_1) = \sum_{j=1}^{M} \int_{\sigma}^{t_1} |g_j(t, c)| dt \,. \tag{7.5.21}$$

**Theorem 7.5.2**  *In (7.5.1) assume:*

(i) *that (7.5.13) holds.*
(ii) *The system (7.5.11) is uniformly asymptotically stable.*
(iii) *Also (7.5.11) is complete, i.e., the map $T(t, \sigma) : C \to E^n$ defined by*

$$T(t, \sigma) = x(t, \sigma, \phi, 0) \,,$$

*where $x(t, \sigma, \phi, 0)$ is a solution of (7.5.11), satisfies*

$$T(t, \sigma)C = E^n \,. \tag{7.5.22}$$

(iv) *The system (7.5.1) is normal and this holds if for each $j = 1, \ldots, m$, the matrix*

$$\overline{Q}_{nj}(t_1) = \{Q_{0j}(s, t_1), \ldots, Q_{n-1j}(s, t_1), \ s \in (\sigma, t_1]\} \tag{7.5.23a}$$

*has rank $n$ where $Q_{ij}$ is defined by*

$$Q_{kj}(s, t) = \sum_{i=0}^{N} A_i(t) Q_{k-1j}(s - h_i, t - h_i) - \frac{d}{dt} Q_{k-1j}(s, t) \,,$$

$$k = 1, 2, \ldots, n-1 \,, \quad t \in [\sigma, t_1] \,,$$

$$Q_{0j}(s, t) = \begin{cases} b_j(t) \,, & t] \in [\sigma, t_1] \,, \quad for \ s = h_i \,, \\ 0 & otherwise \,, \end{cases}$$

*and $B = (b_1 \cdots b_j \cdots b_m) \,.$* $\tag{7.5.23b}$

*Then there exists a minimum-effort control $u^*(t)$ such that*

$$E_1(u^*(t_1)) \le E_1(u(t_1)) \,, \qquad \forall \, u \in U \,,$$

*with $U$ defined in (7.5.5) subject to (7.5.1) and (7.5.2), with $x_1 \equiv 0$. Furthermore, if $y(t_1) \neq 0$, then the minimum-effort $E_{1\,\min} = E_1(u^*(t_1))$ is given by*

$$\frac{1}{E_{1\,\min}} = \min_{c \in P} F_1(c, t_1), \tag{7.5.24}$$

*where $P$ is the plane $c^T y(t_1) = 1$. The optimal control $u^*(t)$ is unique almost everywhere and is given by*

$$u^*(t) = E_{1\,\min}\, \operatorname{sgn} g(t, c^*), \tag{7.5.25}$$

*where $c^*$ is any vector in $P$ for which the minimum in (7.5.24) is attained. If $y(t_1) = 0$, then the control $u^*(t) \equiv 0$ is the minimum-effort control.*

**Theorem 7.5.3**   *If the effort function $E_2(u(t_1))$ is as defined in (7.5.6), then we define*

$$F_2(t_1, c) = \left( \sum_{j=1}^{m} \int_{\sigma}^{t_1} |g_j(t, c)|^q dt \right)^{\frac{1}{q}}, \tag{7.5.26}$$

*where $\frac{1}{p} + \frac{1}{q} = 1$. For (7.5.1) assume conditions (i)–(iv) of Theorem 7.5.2. Then for each $\phi \in C$, $0 = x_1 \in E^n$, and some $t_1$, there exists an optimal control $u^*(t)$ that minimizes $E_2(u(t_1))$, i.e., $E_2(u^*(t_1)) \leq E_2(u(t_1))$ for all $u \in U$ defined in (7.5.7), subject to (7.5.1) and (7.5.2). Furthermore, if $y(t_1) \neq 0$, then the minimum-effort $E_{2\,\min} = E_2(u^*(t_1))$ is given by*

$$\frac{1}{E_{\min}} = \min_{c \in P} F_2(t_1, c), \tag{7.5.27}$$

*where $P = \{ c \in E^n : c^T y(t_1) = 1 \}$. The optimal control $u^*$ is unique almost everywhere and is given by*

$$u_j^*(t) = \mu |g_j(t, c^*)|^{1/p}\, \operatorname{sgn} g_j(t, c^*), \tag{7.5.28}$$

*where*

$$\mu = E_{2\,\min}[F_2(t_1, c^*)]^{-q/p},$$

*and $c^* \in E^n$ is any vector in $P$ where the minimum is attained. If $y(t_1) = 0$, then the minimum-effort control is $u^*(t) \equiv 0$.*

The solution of the minimum-effort problem when $E_3(u(t_1))$ is as defined in (7.5.8) (and it is not constrained) does not exist among integrable functions. If impulsive functions, the so-called Dirac functions, are admissible, then an optimal solution exists. These observations are contained in the next theorem.

**Theorem 7.5.4**   *Consider the minimum-effort problem with effort function $E_3(u(t_1))$ defined in (7.5.8), where $u \in U = \{ u : \|u\|_1 \leq 1 \}$ is defined in (7.5.9). Assume that for (7.5.1), the following conditions exist:*

(i) (i)–(iii) *of Theorem* 7.5.2.

(ii) *Assume metanormality for* (7.5.1), *i.e.,* rank $\overline{Q}_{n+1j} = n$, *for each* $j = 1, \ldots, m$, *where* $\overline{Q}_{n+1j} = \{Q_{0j}(s, t_1) \cdots Q_{nj}(s, t_1), \ s \in [\sigma, t_1]\}$ *with* $Q_{ij}$ *defined in* (7.5.23b). *Then there is no optimal solution with $u$ an integrable function that steers $\phi$ (unless $x(t_1, \sigma, \phi, 0) = 0$) to $0$ in some $t_1$, while minimizing $E_3(u(t_1))$. But if impulsive controls are admissible and*

$$F_3(t_1, c) = \max_{1 \leq j \leq m} \ \sup_{\sigma \leq t \leq t_1} |g_j(t, c)| \,, \tag{7.5.29}$$

*then there exists a minimum-effort control $u^*(t)$ if $y(t_1) \neq 0$.*
*The minimum-effort is given by $E_3(u^*(t_1)) = E_{3\,\min}$ and*

$$\frac{1}{E_{3\,\min}} = \min_{c \in P} F_3(t_1, c) \,, \tag{7.5.30}$$

*where $P = \{c \in E^n : c^T y(t_1) = 1\}$.*
*The optimal control is given by $u^*$, where*

$$u_j^*(t) = \frac{E_{\min}}{C_j} \left( \sum_{i=1}^{N_j} c_{ij} \, \text{sgn}(g_j(\tau_{ji}, c)) \delta(t, -\tau_{ji}) \right), \quad C_j = \sum_{i=1}^{N_j} c_{ij} \ 1 \leq j \leq m \,. \tag{7.5.31}$$

*Here $\delta(t, -\tau_{ji})$ is the so-called Dirac delta function. The suprema in* (7.5.29) *may occur at multiple $j$ and at multiple instances of times $\tau_{ji}$, $i = 1, 2, \ldots, N_j$, where $N_j$ equals zero if $g_j$ does not contain the suprema. Thus, $\tau_{ji} \in [\sigma, t_1]$ are the finite number of times at which $|g_j(t_1, c^*)| = F_3(t_1, c^*)$.*

In Theorem 7.5.4, the components of the controls are not bounded. If

$$U = \{u \text{ measurable } u(t) \in E^m \ |u_j(t)| \leq 1 \,, \ j = 1, \ldots, m \,, \ \|u\|_1 \leq \alpha\} \,,$$

then the optimal controls are "bang-of-bang."

**Theorem 7.5.5** *Assume in* (7.5.1) *the following:*

(i) *Conditions* (i) *and* (ii) *of Theorem* 7.5.4. *Suppose the problem is to minimize $\|u\|_1$ subject to $u \in U$ and* (7.5.1) *and* (7.5.2). *Then there exists a unique minimum fuel control $u^*(t)$. This control is "bang-of-bang" in the sense of having only the values $\pm 1$, and $0$, with no switches $+1$ to $-1$ or back (however, $1, 0, 1$ is possible) unless $\alpha = mt_1$ and $y(t_1) \leftarrow \partial \mathbb{R}(t_1)$ where*

$$\mathbb{R}(t_1) = \left\{ \int_\sigma^{t_1} U(t, s) B(s) u(s) ds : \|u\|_\infty \leq 1 \right\}. \tag{7.5.32}$$

*Optimal control is given by $u^*$ where*

$$u_j^*(t) = \begin{cases} \text{sgn } g_j(t, c) & \text{if } |g_j(t, c)|/\eta \geq 1 \,, \\ 0 & \text{otherwise} \,, \end{cases} \tag{7.5.33}$$

*where $\eta \geq 0$ is some constant.*

To prove the above theorems on minimization of effort, we recall the variation of parameter in (7.5.10) and $Y$ in (7.5.2), and define the function

$$S_t(u) = \int_\sigma^t Y(t,s)u(s)ds\,, \tag{7.5.34}$$

which maps the control space $L$ into the state space $E^n : S_t : L \to E^n$. This map is continuous and linear with $S_0(L) = 0$. If $U \subset L$ is the control constraint set, then the reachable set is the set

$$\mathbb{R}(t) = \{S_t(u) : u \in U\} = \left\{\int_\sigma^t Y(t,s)u(s)ds : u \in U\right\}\,. \tag{7.5.35}$$

Thus if $y(t)$ is a reachable state defined in (7.5.19), i.e., $y(t) = x_1 - x(t,\sigma,\phi,0)$, the coincidence

$$y(t) = x_1 - x(t,\sigma,\phi,u) = S_t(u)\,, \tag{7.5.36}$$

for some $u \in U$, implies that $y(t) \in \mathbb{R}(t)$, and therefore $\phi$ can be driven to $x_1$ by the control $u \in U$. We hit $x_1$ in minimum time $t^*$ if

$$t^* = \text{Inf}\{t \in [0,t_1] : S_t(u) = y(t) \text{ for some } u \in U\}\,. \tag{7.5.37}$$

The time-optimal control $u^* \in U$ is such that $y(t^*) = S_{t^*}(u^*)$. The minimum-effort control is the admissible control $u^* \in U$ such that $S_{t_1}(u^*) = y(t_1)$, with

$$E(u^*(t_1)) \leq E(u(t_1))\,, \qquad \forall\, u \in U\,.$$

In what follows we assume $L$ is either $L_\infty$ or $L_p$, $1 \leq p < \infty$. In this case, the map $S_t : L \to E^n$ has defined in (7.5.34) its adjoint as $S_t^* : E^n \to L$, represented by

$$S_\tau^*(c^T) = \begin{cases} 0, & \text{a.e. } \tau \in [t,t_1]\,, \\ Y(t,\tau)^T c^T, & \text{a.e. } \tau \in [0,t]\,. \end{cases} \tag{7.5.38}$$

If $L = L_\infty$, then $S_\tau^*$ maps $E^n$ into $L_1([0,t_1], E^m) \subset L_\infty$ so that

$$\|S_t^*(c^T)\| = \int_\sigma^t \|Y(t_1,s)^T c^T\|_1 ds\,, \tag{7.5.39}$$

$c \in E^n$, where $\|\cdot\|_1$ is the $L_1$ norm in $E^m$. If $L = L_p$, then $S_\tau^*$ is still given by (7.5.38) with

$$\|S_t^*(c^T)\| = \left(\int_\sigma^t \|Y(t,s)^T c^T\|_q^q ds\right)^{\frac{1}{q}} \tag{7.5.40}$$

for $c \in E^n$.

We now impose some conditions on $S_t : L \to E^n$ and then show what conditions on the system's coefficients ensure that the assumptions hold.

*Prevailing assumptions:*

I. The reachable set

$$\mathbb{R}(t) = \{S_t(u) : u \in U\}$$

is closed.

With $U$ in (7.5.5) or (7.5.7), the closure of $\mathbb{R}(t)$ is automatic and is proved using weak compactness argument. See also Proposition 7.1.1.

II. The map $t \to S_t$, $t \in [0, T]$ is continuous with respect to the operator norm topology of $\mathbb{B}(L, E^n)$, the space of bounded linear transformations from $L$ into $E^n$. See Proposition 7.5.2.

III. The function

$$y : [\sigma, t_1] \to E^n$$

defined by

$$y(t) = x_1(t) - x(t, \sigma, \phi, 0)$$

in (7.5.19) is continuous. Here $x_1(t)$ is a continuous point target. In the next assumption we maintain it is constant.

IV. $y : [\sigma, t_1] \to E^n$ is constant and not zero: $y(t) = y_1$ for all $t \in [\sigma, t_1]$.

V. For each $c \in E^n$, $c \neq 0$ the function $t \to \|S_t^*(c^T)\|$ is strictly increasing. This condition is guaranteed by the following condition:

$$\text{For each } \tau_1, \tau_2 \in [\sigma, t_1], \ c^T U(t_1, t) B(t) = 0, \ \forall \, t \in [\tau_1, \tau_2] \tag{7.5.41}$$
$$\text{if and only if } c = 0.$$

This condition (7.5.41) is equivalent to Euclidean controllability. See Manitius [37, pp. 77–86]. In terms of the system's coefficients, Euclidean controllability is assured by the rank condition (7.5.13).

VI. The system (7.5.1) is normal in the following sense: $S_t^*(c^T)$ is not identically zero for $t \in [\sigma, t_1]$, and $c \neq 0$.

In view of (7.5.38), we have that System (7.5.1) is normal on $[\sigma, t_1]$ if for each $c \in E^n$, $c \neq 0$, and each $j = 1, \ldots, m$, the set

$$\{t > \sigma : c^T U(t_1, t) b_j(t) \equiv 0, \ t \in [\sigma, t_1]\}$$

has measurable zero. In this case, $\|S_t^*(c^T)\| > 0$ in (7.5.39) or in (7.5.40), for each $c \neq 0$, $t > \sigma$. Conditions on the systems' coefficients for normality are given in (7.5.23). We state and prove this for the autonomous simple system

$$\dot{x}(t) = A_0 x(t) + A_1 x(t - h) + B u(t), \tag{7.5.42}$$

where the "determining equations" are given by

$$Q_{kj}(s) = A_0 Q_{k-1}(s) + A_1 Q_{k-1}(s-h), \quad k = 1,2,3,\ldots, \quad s \in [-\infty, \infty),$$

$$Q_{0j}(s) = \begin{cases} b_j, & s = 0, \\ 0, & s \neq 0. \end{cases} \tag{7.5.43}$$

Set

$$\overline{Q}_{nj}(t_1) = \{Q_{0j}(s), Q_{1j}(s), \ldots, Q_{n-1j}(s), \ s \in [0, t_1]\}. \tag{7.5.44}$$

**Theorem 7.5.6** *The system (7.5.42) is normal on $[0, t_1]$ if and only if for each $j = 1, \ldots, m$, rank $\overline{Q}_{nj}(t_1) = n$.*

**Proof** This is an easy adaptation of Manitius [37, p. 79]. Because of its utility in subsequent discussions we shall outline it in some detail. We note that $s \to U(t_1, s)$ is the fundamental matrix solution to the adjoint equation:

$$\frac{\partial}{\partial s} U(t_1, s) = -U(t_1, s)A_0 - U(t_1, s+h)A_1,$$

$$U(t_1, s) = \begin{cases} 0, & s > t_1, \\ I, & s = t. \end{cases} \tag{7.5.45}$$

The function

$$g_j(s, c) = c^T U(t_1, s) b_j,$$

the $j$th component of the index of the control system, has the following properties: $s \to g_j(s, c)$ is defined on $[0, \infty)$; it vanishes on $(t_1, \infty)$ and is piecewise analytic in $(0, \infty)$. The isolated exceptional points are at points $s = t_1, t_1 - h, t_1 - 2h, \ldots, t_1 - hi$, where $i$ is such that $t_1 - ih > 0$ (see Tadmor [11]). Define

$$\Delta g_j^k(t, c) = g_j^k(t - 0, c) - g_j^k(t + 0, c), \quad t \in (0, \infty), \tag{7.5.46}$$

and we have designated $g_j^k(t-0, c), g_j^k(t+0, c)$ as the left- (respectively right-) hand side limit of the $k$th derivative of $g_j$ at $s = t$, $k = 0, 1, 2, \ldots$. Clearly,

$$\Delta g_j^0(t, c) = \begin{cases} 0, & t = t_1, \\ c^T b_j, & t = t_1, \end{cases} \tag{7.5.47}$$

and the jump at $t_1$ occurs because of the jump discontinuity of the fundamental matrix $U(t_1, s)$ at $s = t_1$. Also

$$\Delta g_j^{(1)}(t, c) = \begin{cases} c^T A_0 b_j, & t = t_1, \\ c^T A_1 b_j, & t = t_1 - h, \\ 0, & \text{otherwise.} \end{cases} \tag{7.5.48}$$

We can prove by induction that, in general,

$$\Delta g_j^{(k)}(t_1 - ih) = (-1)^k c^T Q_{kj}(ih): \quad i = t_1 - ih > 0. \tag{7.5.49}$$

Assume that rank $\overline{Q}_{nj}(t_1) = n$, but (7.5.42) is not normal on $[0\ t_1]$. Then there is a $c \neq 0$ $c \in E^n$ such that for some $j = 1, \ldots, m$, $g_j(s, c) = 0$ on $[0, t_1]$. Because of this $g_j(s, c) \equiv 0$ on $[0, \infty]$, so that on differentiating,

$$0 = c^T \Delta g^k(t_1 - ih), \quad t_1 - ih > 0 = (-1)^k c^T Q_k(ih),$$

$$\text{for } k = 0, 1, 2, \ldots, \quad i = 0, 1, 2, \ldots, t_1 - ih > 0.$$

We deduce that $c \in E^n$ is orthogonal to all the vectors of $\overline{Q}_{nj}(t_1)$, which is a contradiction. We have proved normality. We now prove the necessity for this. The constant coefficient case is proved exactly the same way as was done in [37, p. 80]. For the general situation of (7.5.1), it is far from clear whether the rank condition is necessary.

The next prevailing assumption on (7.5.1) is that of metanormality. The notion was introduced by Hájek [35, p. 416].

VI. The system (7.5.1) is metanormal on $[\sigma, t_1]$ if and only if every index

$$g(t, c) = c^T U(t_1, t) B(t)$$

with $c \neq 0$ has each component $g_j(t, c)$ constant only on sets of measure zero: The set

$$\{t > \sigma : c^T U(t_1, t) b_j(t) = \alpha\}$$

has measure zero for each column $b_j$ of $B$, constant $\alpha \in E$, $c \neq 0$ in $E^n$.

Though metanormality is a condition on $S_t^*$, we characterize it as the full rank of some of the systems parameters. We restrict our discussion to (7.5.42).  $\square$

**Lemma 7.5.1**  *The system (7.5.42) is metanormal if and only if each $j = 1, \ldots, m$,* rank$\{Q_{1j}(s), \ldots, Q_{nj}(s), \ s \in [0, t_1)\} = n$ *where $Q_{kj}$ is as defined in (7.5.43).*

**Proof**  Suppose a $j$th coordinate

$$g_j(t, c) = c^T U(t_1, t) b_j \equiv \alpha,$$

$\alpha$ a constant on a set of positive measure. Then $t \to g_j(t, c) - \alpha \equiv M(t) = 0$ on a set of positive measure. Since it is piecewise analytic except at the points $s = t_1, t_1 - h, t_1 - 2h, \ldots, t_1 - ih$ where $i$ is such that $t_1 - ih > 0$, the function $m(t) \equiv 0$ on $[0, \infty)$. Hence it follows that

$$0 = \Delta M^k(t_1 - ih), \quad t_1 - ih > 0,$$
$$= (-1)^k c^T Q_{kj}(ih) \tag{7.5.50}$$

for $k = 1, 2, \ldots, n$, $i = 0, 1, \ldots, t_1 - ih > 0$. Hence $c$ is orthogonal to all vectors $Q_{1j}(s), \ldots, Q_{nj}(s)$, $s \in [0, t_1]$. Therefore (7.5.42) is not metanormal.    $\square$

We now prove necessity. We show that metanormality implies that for each $j = 1, \ldots, m$, rank $\hat{Q}_{\infty j} = n$ where $\hat{Q}_{\infty j}$ denotes a matrix with infinite number of columns given by $Q_{1j}(s), Q_{2j}(s), \ldots, s \in [0, t_1]$; and the assertion rank $\hat{Q}_{\infty j} = n$ means that there are $n$ linearly independent columns in the sequence

$$\{Q_{kj}(s), \; s \in [0, t_1), \; k = 1, 2, \cdots\}.$$

Suppose this is false. Then there is a $c \neq 0$, $c \in E^n$ such that

$$c^T Q_{kj}(s) = 0, \quad k = 1, 2, \ldots, \; s \in [0, t_1). \tag{7.5.51}$$

Since $A_0, A_1, B$ are constant, the function

$$t \to g_j(t, c) = c^T U(t_1, t) b_j$$

is piecewise analytic except at isolated points of $[0, t_1]$, which are seen to be $t_1, t_1 - h, t_1 - 2h$, etc. But $g_j(t, c)$ vanishes for $t > t_1$, hence $M(t) \equiv g_j(t, c) - \alpha \equiv -\alpha$ for $t > t_1$. It follows that $M^{(k)}(t_1 + 0) = 0$ for $k = 1, 2, \ldots$. Since (7.5.47)–(7.5.50) holds and $\Delta M^{(k)}(t) = M^{(k)}(t - 0) - M^{(k)}(t + 0)$, $t \in (0, \infty)$, we deduce that $M^{(k)}(t_1 - 0) = g_j^k(t_1 - 0, c) = 0$, $k = 1, 2$. Because $g_j(t, c)$ is piecewise analytic on $[0, t_1]$, $M(t) \equiv 0$ on $[t_1 - h, t_1]$. In this same way we obtain $M(t) \equiv 0$ on $[t_1 - 2h, t_1 - h]$, and by induction $M(t) \equiv 0$ on $[0, t_1]$, i.e., $g_j(t, c) = \alpha$ on $[0, t_1]$. This contradicts metanormality. To complete the proof it can be shown that rank $\hat{Q}_{\infty j} = n$ implies rank $\hat{Q}_n = n$, where

$$\hat{Q}_{nj} = [Q_{1j}(s), \ldots, Q_{nj}(s), \; s \in [0, t_1]].$$


## 7.6  Proof of Minimum-Effort Theorems

With the six general Assumptions I–VI stated, and conditions for their validity in terms of the system's coefficients deduced, we are now prepared to state three preliminary results on which the solution of the problem of the minimum-effort control strategies are based.

**Theorem 7.6.1**  *In (7.6.1), let $t \in [\sigma, t_1]$, and consider $U$ in (7.6.5) or (7.6.7). Then there exists an admissible control $u \in U$ such that*

$$S_t(u) = y(t) = x_1(t) - x(t, \sigma, \phi, 0)$$

*if and only if*

$$c^T y(t) \leq \|S_t^*(c^T)\|, \qquad \forall \, c \in E^n, \tag{7.6.1}$$

*where $S_t^*$ is the adjoint of $S_t$, a map of $E^{n^T}$ to $L$.*

**Proof**  It is assumed that there is a $u \in U$ such that $S_t(u) = y(t)$. It follows that

$$c^T y(t) = c^T(S_t(u)) = S_t^*(c^T)(u) \leq \|S_t^*(c^T)\| \, \|u\| \leq \|S_t^*(c^T)\| \, .$$

Therefore (7.6.1) is valid. To prove the converse, we recall that the Euclidean reachable set $\mathbb{R}(t)$ is closed. It is also convex, being a linear image of the convex set $U$. If there is no $u \in U$ such that $S_t(u) = y(t)$, then $y(t) \notin \mathbb{R}(t)$. Because $\mathbb{R}(t)$ is a closed and convex subset of $E^n$, the Separation Theorem [4, p. 33] asserts that there is a hyperplane that separates $\mathbb{R}(t)$ and $y(t)$: This means there exists a $c \in E^n$, such that

$$c^T y(t) \geq \sup\{c^T(S_t(u)) : u \in U\} = \sup\{S_t^*(c^T)(u) : u \in U\} \, .$$

This invalidates (7.6.1). $\qquad\qquad\qquad\qquad\qquad\qquad\qquad\qquad\qquad\qquad\qquad\square$

The assumption that there exists some $u \in U$ such that $S_t(u) = y(t)$ is a statement on the constrained controllability of (7.6.1) at some $t$. The optimal (minimum) time $t^*$ for hitting the target is defined as

$$t^* = \mathrm{Inf}\{t \in [\sigma, t_1] : S_t(u) = y(t) \text{ for some } u \in U\} \, . \qquad (7.6.2)$$

The admissible control $u^* \in U$ that ensures the coincidence $S_{t^*}(u^*) = y(t^*)$ is the time-optimal control. For the minimum-effort problem, the optimal strategy $u^*$ that ensures that $S_{t_1}(u^*) = y(t_1)$ while minimizing $E(u(t_1))$ is the (minimum) optimal control. Inspired by the ideas of Hájek and Krabs [36], we propose the following:

**Theorem 7.6.2**  *In (7.6.1) assume that the point target $x_1(t) \in E^n$ is continuous. Then there exists a $c \in E^n$ with $\|c\| = 1$ such that*

$$c^T y(t^*) = \|S_{t^*}^*(c^T)\| \, , \qquad (7.6.3)$$

*and*

$$S_{t^*}^*(c^T)(u^*) = c^T(S_{t^*}(u^*)) = c^T y(t^*) \, , \qquad (7.6.4)$$

*where $u^*$ is the time-optimal control, and*

$$\|u^*\| = 1 \, , \qquad (7.6.5)$$

*if the system is normal.*

**Remark 7.6.1**  Because $x_1(t)$ is continuous,

$$y(t) = x_1(t) - x(t, \sigma, \phi, 0)$$

is continuous. The proof as pointed out in [36, p. 5] is a little adaptation of the argument in the proof of Theorem 6 in [34].

Because the next result links the time-optimal control and the minimum fuel strategy, it is described by Hájek and Krabs in another setting [36] as the Duality Theorem. It is simple but fundamental.

**Theorem 7.6.3** *In (7.6.1) assume null controllability with constraints. This is satisfied if (7.6.11) is uniformly asymptotically stable and (7.6.1) Euclidean controllable. If $t^*$ is the minimum time, then*

$$t^* = \max\{t \in (0, t_1] \text{ such that } c^T y(t) = \|S_t^*(c^T)\| \text{ for some } c \in E^n : \|c\| = 1\}.$$

$$(7.6.6)$$

**Proof**    Because (7.6.1) is null controllable with constraints, we set $x_1(t) \equiv 0$; for each nontrivial $\phi \in C$, $y(t) = x_1(t) - x(t, \sigma, \phi, 0)$ becomes $y(t) = -x(t, \sigma, \phi, 0)$; and if $-x(t_1, \sigma, \phi, 0) \neq 0$ and if we set $y(t) \equiv -x(t_1, \sigma, \phi, 0) = y_1$, $\forall\, t \geq \sigma$, then (IV) is satisfied. Because of Euclidean controllability, assumption (V) is satisfied. Since Theorem 7.6.2 is valid, the minimum time $t^*$ is a point over which the maximum in (7.6.6) is assumed. Suppose there is a $t > t^*$ such that

$$c^T y_1 = \|S_t^*(c^T)\|$$

for some $c \in E^n$ with $\|c\| = 1$. Then

$$c^T y_1 \leq \|S_{t^*}^*(c^T)\| < \|S_t^*(c^T)\|\,.$$

The first inequality follows from Theorem 7.6.1, the second from assumption V. The obvious contradiction proves our assertion.      $\square$

We now designate $U$ to be as in (7.6.5) or (7.6.7), and derive from Theorems 7.6.1–7.6.3 optimal controls for minimizing effort and time.

**Theorem 7.6.4**    *In (7.5.1), assume:*

   (i) *(7.5.1) is Euclidean controllable.*
   (ii) *(7.5.11) is uniformly asymptotically stable.*
   (iii) *(7.5.1) is normal.*
   (iv) *System (7.5.11) is complete. If $U_\infty = L_\infty([0, t_1], C^m)$, $t^*$ is the minimum time, and $u^*$ is the time-optimal control, then $u^*$ is uniquely determined by*

$$u_j^*(t, c) = \text{sgn}[c^T U(t_1, t) b_j(t)], \qquad \sigma \leq t \leq t^*, \qquad (7.6.7)$$

*for some $c \neq 0$, and each $j = 1, \ldots, m$. If $U_p \subset L_p([\sigma, t_1], E^m)$, $p > 1$, $\frac{1}{p} + \frac{1}{q} = 1$ such that $u \in U$ implies $\|u\|_p \leq 1$, then the time-optimal controls are uniquely determined by*

$$u_j^*(t) = k|g_j(t, c)|^{q/p} \, \text{sgn}(g_j(t, c))\,, \qquad (7.6.8)$$

*where*

$$g(t, c) = c^T U(t^*, t) B(t)\,, \qquad (7.6.9)$$

$$k = \left( \int_\sigma^{t^*} \sum_{j=1}^M |g_j(t, c)|^q ds \right)^{-1/p}. \qquad (7.6.10)$$

**Proof**   From Theorem 7.6.3, if $u^*$ is a time-optimal control and $t^*$ the minimum time, then

$$S_{t^*}^*(c^T)(u^*) = c^T(S_{t^*}(u^*)) = c^T y(t^*) = \|S_{t^*}^*(c^T)\|\,.$$

But then the $u^* \in U_\infty$ is such that

$$c^T y(t^*) = \int_\sigma^{t^*} c^T U(t^*,t)B(t)u^*(t)dt$$

$$= c^T \int_\sigma^{t^*} U(t^*,t)B(t)u^*(t)dt$$

$$= \int_\sigma^{t^*} \|c^T U(t^*,t)B(t)\|_1 dt\,. \tag{7.6.11a}$$

This implies that $u^*$ is of the form a.e.

$$u_j^*(t) = \mathrm{sgn}[c^T U(t^*,t)b_j(t)]\,, \qquad \sigma \le t \le t^*\,,$$

when $c^T U(t^*,t)b_j(t) \neq 0$ for $c \neq 0$, and each $j = 1,\ldots,m$, which is true since the system is normal and (7.6.11) complete. For $U_p$, we obtain from (7.6.3) in Theorem 7.6.2 that

$$c^T y(t^*) = \|S_{t^*}^*(c^T)\|$$

$$= \int_\sigma^{t^*} c^T U(t^*,t)B(t)u^*(t)dt \tag{7.6.11}$$

$$= \int_\sigma^{t^*} \|c^T U(t^*,t)B(t)\|^q dt)^{\frac{1}{q}}\,,$$

where $\frac{1}{p} + \frac{1}{q} = 1$. Since $g(t,c) = c^T U(t^*,t)B(t)$,

$$c^T y(t^*) = \int_\sigma^{t^*} \sum_{j=1}^m g_j(t,c)u_j(t)dt$$

$$\le \int_\sigma^{t^*} \sum_{j=1}^m |g_j(t,c)u_j(t)|dt$$

$$\le \int_\sigma^{t^*} \left(\sum_{j=1}^m |g_j(t,c)|^q\right)^{\frac{1}{q}} \left(\sum_{j=1}^m |u_j(t)|^p\right)^{\frac{1}{p}} dt$$

$$\leq \left( \int_\sigma^{t^*} \sum_{j=1}^m |g_j(t,c)|^q dt \right)^{\frac{1}{q}} \left( \int_\sigma^{t^*} \sum_{j=1}^m |u_j(t)|^p dt \right)^{\frac{1}{p}}$$

$$\leq \left( \sum_\sigma^{t^*} \sum_{j=1}^m |g_j(t,c)|^q dt \right)^{\frac{1}{q}} .$$

But then (7.6.11) is valid, i.e.,

$$c^T y(t^*) = \left( \int_\sigma^{t^*} \sum_{j=1}^m |g_j(t,c)|^q \right)^{\frac{1}{q}} ,$$

and so we have equality everywhere in the above estimates. The control $u^*$ that gives the equality is (by inspection)

$$u_j^*(t) = k|g_j(t,c)|^{\frac{q}{p}} \operatorname{sgn}(g_j(t,c)), \qquad j = 1, \ldots, m, \qquad (7.6.12)$$

where $k = \left( \int_\sigma^{t^*} \sum_{j=1}^m |g_j(t,c)|^q)ds \right)^{\frac{-1}{p}}$. This is the time-optimal control, and as observed before, it is uniquely determined when the system is normal. The proof is complete. $\qquad\square$


**Proof of Theorem 7.6.2**    Because of the Euclidean controllability and the stability assumptions, there is indeed a $t_1$ such that the solution $x$ of (7.5.1) satisfies

$$x_\sigma(\cdot, \sigma, \phi, u) = \phi, \qquad x(t_1, \sigma, \phi, u) = 0.$$

We observe that in this case $y(t)$ defined in (7.5.19), $y(t) = x_1(t) - x(t, \sigma, \phi, 0) = -x(t_1, \sigma, \phi, 0) \equiv y_1 \neq 0$ and $y(t_1) \in \mathbb{R}(t_1)$. Observe that $u^*$ in (7.6.7) is a boundary control in the sense that if

$$z(t_1, c) = \int_0^{t_1} U(t_1, t)B(t)u^*(t,c)dt,$$

then $z(t_1, c)$ is on the boundary of the reachable set $\mathbb{R}(t)$, so that

$$c^T z(t_1, c) = \sum_{j=1}^m \int_\sigma^{t_1} |g_j(t,c)|dt = F(c, t_1) = \|S_{t_1}^*(c^T)\| ,$$

and

$$c^T z(t_1, c) > c^T y, \qquad \forall\, y \in \mathbb{R}(t_1), \qquad y \neq z(t_1, c).$$

With $y(t_1) \equiv y_1 \in \mathbb{R}(t_1)$, we can extend this to reach the boundary as follows:
    Let

$$\alpha = \max\{\beta : \beta y(t_1) \in \mathbb{R}(t_1)\}. \qquad (7.6.13)$$

Since $y(t_1) \neq 0$, $\alpha$ can be assumed to be positive, and obviously $\alpha y(t_1)$ is a boundary point of $\mathbb{R}(t_1)$. This means that $\alpha y(t_1) = z(t_1, c)$ for some $c$, where

$$c^T y(t_1) = 1 = -c^T x(t_1, \sigma, \phi, 0). \tag{7.6.14}$$

It is easy to verify that the control

$$\overline{u}(t) = \frac{u^*(t, c)}{\alpha} \tag{7.6.15}$$

steers $\phi$ to 0 in time $t_1$ while minimizing $E_1(u(t_1))$. Also

$$\frac{1}{\alpha} = \frac{1}{M(t_1)} = \min E_1(u(t_1)), \tag{7.6.16}$$

where

$$\alpha = \min\{F_1(t, c) : c \in P = \{c \in E^n : c^T y_1 = 1\}\}. \tag{7.6.17}$$

Details of the verification are the same as in the case of ordinary linear differential equations (see Neustadt [38]). We conclude that

$$\overline{u}(t) = \frac{u^*(t, c^*)}{M(t_1)} = \frac{u^*(t)}{\alpha}, \qquad 0 \le t \le t_1 \tag{7.6.18}$$

is the minimum energy control, where $c^* \in E^n$ is vector in $P$ on which the minimum in (7.6.17) is attained. Because $u^*(t) = \mathrm{sgn}(g(t, c))$, we deduce that

$$\overline{u}(t) = \frac{\mathrm{sgn}\{g(t, c^*)\}}{M(t_1)}, \tag{7.6.19}$$

where $c^*$ is the minimizing vector in (7.6.17). Normality ensures that $\overline{u}$ is uniquely determined. If $y(t_1) = y_1 = 0$, then the choice $\overline{u} \equiv 0$ is appropriate. This completes the proof of Theorem 7.6.2. $\qquad\square$

**Proof of Theorem 7.6.3**  In this case, $U_p$ in (7.5.7) is used to define the reachable set

$$\mathbb{R}(t_1) = \left\{ \int_\sigma^t U(t_1, t) B(t) u(t) dt : u : \|u\|_p \le 1 \right\}$$

which, as we have noted earlier, is closed. Just as in the proof of Theorem 7.5.2, the time-optimal control $u^*(t, c)$ in (7.6.8) is a boundary control that is uniquely defined. Thus if

$$z(t_1, c) = \int_0^{t_1} U(t_1, t) B(t) u^*(t, c) dt, \qquad z(t_1, c) \in \partial\mathbb{R}(t_1).$$

With $\alpha$ as defined in (7.6.13), it is easy to verify that the minimum-effort control is given by

$$\overline{u}(t, c^*) = \frac{u^*(t, c^*)}{\alpha}, \tag{7.6.20}$$

where

$$\frac{1}{\alpha} = \frac{1}{M(t_1)} = \min E_2(u(t_1)),\qquad(7.6.21)$$

and

$$\alpha = \min_{c \in P = \{c \in E^n : c^T y_1 = 1\}} F(t, c),\qquad(7.6.22)$$

and $F_2(t_1 c)$ is as given in (7.5.26):

$$F_2(t_1, c) = \left(\sum_{j=1}^{m} \int_{\sigma}^{t_1} |g_j(t, c)|^q dt\right)^{\frac{1}{q}}.$$

The vector $c^*$ is that which minimizes the function in (7.6.22). Because $u^*(t, c^*)$ in (7.6.20) is given by (7.6.8), the minimum-effort strategy is given by

$$\bar{u}_j(t, c^*) = \frac{k|g_j(t, c^*)^{\frac{q}{p}} \operatorname{sgn}(g_j(t, c^*)}{\alpha},\qquad(7.6.23)$$

where

$$k = \left(\int_{\sigma}^{t_1} \sum_{j=1}^{m} |g_j(t, c^*)|^q dt\right)^{-\frac{1}{p}}.$$

Through Theorem 7.6.2 and the general ideas of Hájek and Krabs [36], we can link up the time-optimal controls and the minimum-effort controls. In some situations we see that they are the same. We observe that $\|S_t^*(c^T)\|$ is given by (7.6.11) and it is to be minimized in (7.6.22) and (7.6.17) over the hyperplane $P$. We are led to the following definition: For each $t \in [0, t_1]$ let

$$\alpha(t) = \inf\{\|S_t^*(c^T)\| : c \in P\} \quad \text{where} \quad P = \{c \in E^n : c^T y_1 = 1\}.\qquad(7.6.24)$$

$\square$

**Theorem 7.6.5**  *Let the prevailing Assumptions I–V hold for (7.5.1). Then for each $t \in (0, t_1]$, we have that if $t^*$ is the minimum time, then*

$$\frac{1}{\alpha(t)} \left\{\begin{matrix}>\\=\\<\end{matrix}\right\} 1 \leftrightarrow t \left\{\begin{matrix}<\\=\\>\end{matrix}\right\} t^*.\qquad(7.6.25)$$

**Remark 7.6.2**  Note that 1 is the norm bound of the control set $U$, and $t_1$ is defined in the minimum-effort problem in (7.6.2).

**Proof**  Quite rapidly from (7.6.24), one has that

$$c^T y_1 \le \frac{1}{\alpha(t)} \|S_t^*(c^T)\|,\qquad \forall\, c \in E^n.\qquad(7.6.26)$$

From an elementary approximation theory argument we can verify that for each $t \in (0, t_1]$ there exists some $c(t) \in P$ such that

$$\|S_t^*(c^T(t))\| = \alpha(t) > 0. \tag{7.6.27}$$

Also there exists a $u_t \in L$ such that $S_t(u_t) = y_1$ and $\|u_t\| = \frac{1}{\alpha(t)}$. From (7.6.27) we argue that there exists a $c(t) \in E^n$ such that

$$c(t)y_1 = 1 = \frac{1}{\alpha(t)}\|S_t^*(c^T(t))\|.$$

Invoke Theorem 7.6.1 and Theorem 7.6.2 to validate (7.6.25). $\qquad\square$

**Theorem 7.6.6** *Let the prevailing Assumptions I–VI hold for System (7.5.1), in which (7.5.11) is complete. The optimal control $u^*$ that steers $\phi \in C$ to 0 in minimum time $t^*$ while minimizing $E_1(u(t^*))$ is given by*

$$u^*(t) = \mathrm{sgn}[g(t, c^*)], \qquad \sigma \leq t \leq t^*, \tag{7.6.28}$$

*where $c^*$ is the vector in $P$ that minimizes (7.6.17), i.e.,*

$$\alpha = \min_{c \in P} F(t^*, c) \tag{7.6.29}$$

*where*

$$F(t^*, c) = \sum_{j=1}^{m} \int_{\sigma}^{t^*} |g_j(t, c)| dt, \tag{7.6.30}$$

$$g(t, c) = c^T U(t^*, t) B(t). \tag{7.6.31}$$

*The optimal strategy that is time-optimal and minimizes $E_2(u(t^*))$ is $u^*$ with component*

$$u_j^*(t) = k|g_j(t, c^*)|^{\frac{q}{p}} \mathrm{sgn}\, g_j(t, c^*), \tag{7.6.32}$$

*where*

$$k = \left( \int_{\sigma}^{t^*} \sum_{j=1}^{m} |g_j(t, c^*)|^q dt \right)^{-\frac{1}{p}} \tag{7.6.33}$$

*with $c^*$ the minimizing vector in*

$$\alpha = \min_{c \in P} F_1(t^*, c), \tag{7.6.34}$$

$$F(t^*, c) = \left( \sum_{j=1}^{m} \int_{\sigma}^{t^*} |g_j(t, c)|^1 dt \right)^{\frac{1}{q}}. \tag{7.6.35}$$

**Proof**    By assumption, the feasible $t_1$ of (7.5.2) in the minimum-effort problem is the minimum time $t^*$. But then

$$\alpha(t^*) = \inf_{c \in P} \{\|S_{t^*}^*(c^T)\|\} = 1$$

by (7.6.25) in Theorem 7.6.5. Because the minimum fuel controls are the functions

$$\bar{u}(t) = \frac{u^*(t, c^*)}{\alpha(t^*)} = u^*(t, c^*)$$

by (7.6.18) or (7.6.20), the corresponding expressions in (7.6.7) and (7.5.8) for time-optimal controls of Theorem 7.5.4 prove that (7.5.8) and (7.5.32) are correct. The following fundamental principle is now obvious:

*In our dynamics (7.5.1), the time-optimal control is also the control that minimizes the effort function.*    $\square$


## 7.7    Optimal Absolute Fuel Function

The solutions of the minimum time/minimum fuel problems contained in Theorems 7.5.1–7.5.3 and 7.6.4 depend on the closure of the reachable sets. Since the reachable sets are also convex, optimal controls are boundary controls in the sense that they generate points on the boundary of $\mathbb{R}(t_1)$. But if the effort function is defined by $E_3(u(t_1))$ in (7.5.8), as the absolute fuel function

$$E_3(u(t_1)) = \|u\|_1 = \int_0^{t_1} \sum_{j=1}^m |u_j(t)| dt\,, \tag{7.7.1}$$

and if $U$ is defined by

$$U = \{u \text{ measurable } \|u\|_1 \le 1\}\,, \tag{7.7.2}$$

then the reachable set

$$\mathbb{R}(t_1) = \left\{ \int_\sigma^{t_1} U(t_1, t) B(t) u(t) dt : u \in U \right\} \tag{7.7.3}$$

is not closed, but open. Because the reachable set is open, its boundary points can only be "reached" by convex combination of delta functions. Optimal controls that yield boundary points must therefore be delta or dirac functions, which are impulsive in nature. If we admit such functions as controls in absolute fuel minimization problems, optimal controls exist. If we rule them out and $\mathbb{R}(t)$ is open, there is no integrable optimal control. We consider

$$\dot{x}(t) = A_0 x(t) + A_1 x(t - h) + B u(t)\,. \tag{7.7.4}$$

**Lemma 7.7.1**    *Suppose (7.7.4) is metanormal, i.e., for each* $j = 1, \dots, m$,

$$\text{rank}[Q_{1j}(s), \dots, Q_{nj}(s),\ s \in [0, t_1]] = n\,, \tag{7.7.5}$$

*where $Q_{kj}$ is as defined in (7.5.43). Then the reachable set $\mathbb{R}(t)$ of (7.7.4) is open, bounded, convex, and symmetric.*

**Proof** The symmetry, boundedness, and convexity of $\mathbb{R}(t)$ are easy to verify. We prove that $\mathbb{R}(t)$ is open. Assume on the contrary that $\mathbb{R}(t_1)$ is not open and that

$$y_1 = \int_\sigma^{t_1} U(t_1, t) B u_0(t) dt$$

is a boundary point. Let $c \neq 0$ be an outer normal to (the closure of) $\mathbb{R}(t)$ at $y_1$. With this $c$, define the index of (7.7.4) $g(t, c) = c^T U(t_1, t) B$. Obviously, $u_0$ is a control that maximizes

$$\int_\sigma^{t_1} g(t, c) u(t) dt \leq \int_\sigma^{t_1} g(t, c) u_0(s) ds \quad \text{whenever} \quad \|u\|_1 \leq 1. \tag{7.7.6}$$

But the mapping $u(\ ) \to \int_\sigma^{t_1} g(t, c) u(t) dt$ is a linear functional on $L_1([\sigma, t_1])$ with norm

$$\|g\|_\infty = \max_{1 \leq j \leq m} \max_{t \in [\sigma, t_1]} |g_j(t, c)|.$$

The inequality on (7.7.6) implies that the value $\|g\|_\infty$ is attained at the element $u_0$ of the unit ball in $L_1$. Therefore

$$\|g\|_\infty = \int_\sigma^{t_1} g(t, c) u_0(t) dt$$

$$\leq \int_\sigma^{t_1} |g_j| |u_{0j}|$$

$$\leq \int_\sigma^{t_1} \sum_{j=1}^m \|g\|_\infty |u_{0j}|$$

$$= \|g\|_\infty \|u_0\|_1 \leq \|g\|_\infty.$$

This shows that we have equality throughout, so that

$$\int_\sigma^{t_1} \sum_{j=1}^M (\|g\|_\infty - |g_j(t, c)|) |u_{0j}(t)| dt = 0.$$

It follows that

$$(\|g\|_\infty - |g_j(t, c)|) \cdot |u_{0j}(t)| = 0 \text{ a.e. for each } j = 1, \ldots, m, \quad t \in [\sigma, t_1].$$

Because the system is metanormal, $|g_j|$ is constant $(= \|g\|_\infty)$ only on a set of measure zero. Hence, $u_0 = 0$ a.e. on $[\sigma, t_1]$. But with $u_0 = 0$, our boundary point is $y_1 = 0$. This contradicts the following containment:

$$\mathbb{R}_\alpha(t_1) \subset \text{Int } \mathbb{R}_\beta(t_1), \qquad \text{Int}(1/\alpha)\mathbb{R}_\alpha(t_1) \supset \left(\frac{1}{\beta}\right)\mathbb{R}_\beta, \tag{7.7.7}$$

whenever $0 < \alpha < \beta \leq mt_1$, which is an easy consequence of metanormality and completeness as can be proved by the methods of Hájek [35, p. 432]. Recall that (in Hájek's notation)

$$\|u\|_1 = \int_\sigma^{t_1} \sum_{j=1}^m |u_j(t)| dt\,, \qquad \|u\|_\infty = \max_{1 \leq j \leq 1} \text{ess sup} |u_j(t)|\,,$$

and

$$\mathbb{R}_\alpha(t) = \left\{ \int_\sigma^{t_1} U(t_1,t)Bu(t)dt : \|u\|_\infty \leq 1,\ \|u\|_1 \leq \alpha \right\}. \tag{7.7.8}$$

Thus the reachable set $\mathbb{R}(t_1)$ is open. $\qquad\qquad\square$

**Proof of Theorem 7.5.4**    Because $\mathbb{R}(t_1)$, the reachable set using $L_1$ controls with bound $\|u\|_1 \leq 1$, is open, then

$$\mathbb{R}(t_1) = \bigcup_{\alpha > 0} \frac{1}{\alpha} \mathbb{R}_\alpha = \left\{ \int_\sigma^{t_1} U(t_1,t)Bv(t)dt : \|v\| \leq 1\,, \|v\|_\infty < \infty \right\}, \tag{7.7.9}$$

where $\frac{1}{\alpha}u = v$ is an open set.

We conclude that if the controls $u$ are measurable with $\|u\| \leq k$, then the reachable set is $k\mathbb{R}(t_1)$ for $k > 0$. Since this is open, the minimal $k$ can never be attained unless $y_1 = -x(t_1, \sigma, \phi, 0) = 0$. Continuing, if $g$ is the index, i.e., $g(t) = g(t,c) = c^T U(t_1,t)B$, then

$$c^T y_1 = c^T y(t_1) = \int_\sigma^{t_1} g(t,c)u(t)dt\,. \tag{7.7.10}$$

The control that realizes a boundary point of $\mathbb{R}(t_1)$ definitely maximizes

$$\int_\sigma^{t_1} g(s)u(s) \leq \int_\sigma^{t_1} g(s)u_0(s) \tag{7.7.11}$$

whenever $\|u\|_1 \leq 1$, i.e., over $u \in U$. We have that

$$c^T y(t_1) = \int_\sigma^{t_1} \sum_{j=1}^m g_j(t)u_j(t)dt$$

$$\leq \int_\sigma^{t_1} \sum_{j=1}^m |g_j|\,|u_j|d\,,$$

$$\leq \max_{1 \leq j \leq m} \max_{\sigma \leq t \leq t_1} |g_j(t,c)| = \|g\|_\infty\,. \tag{7.7.12}$$

If $t^*$ is a minimum time for the time-optimal control problem with $u \in U$, then

$$c^T y(t^*) = \|S_{t^*}^*(c^T)\|_\infty \tag{7.7.13}$$

where

$$\|S_{t^*}^*(c^T)\|_\infty = \max_{1\le j\le m} \max_{\sigma\le t\le t^*} |g_j(c,t)| .\tag{7.7.14}$$

If equality holds in (7.7.12), then impulse functions applied at the points where $g_j$ is largest maximizes (7.7.11). The maximum values in (7.7.14) occur at multiple $j$ and at multiple instances $\tau_{ji}$, $i = 1, 2, \ldots N_j$, where $N_j$ is taken to be zero if $g_j$ does not contain the maximum. Because of these, the time-optimal controls (which are "boundary" controls) are $u^*$,

$$u_j^*(t,c) = \sum_{i=1}^{N_j} \mathrm{sgn}(g_j(\tau_{ji},c))\delta(t - \tau_{ji}) \bigg/ \sum_{j=1}^{m} N_j , \quad 1 \le j \le m .\tag{7.7.15}$$

For the minimum-fuel problem we deduce as before that the optimal control is

$$\overline{u}(t) = \frac{u^*(t,c^*)}{\alpha} ,\tag{7.7.16}$$

where

$$\frac{1}{\alpha} = \frac{1}{M(t_1)} = \min E_3(u(t_1))\tag{7.7.17}$$

and

$$\alpha(t_1) = \min_{c\in P} F(t_1,c) , \qquad P = \{c \in E^n : c^T y(t_1) = 1\} ,\tag{7.7.18}$$

$$F(t_1,c) = \max_{1\le j\le m} \sup_{\sigma\le t\le t_1} |g_j(t,c)| = \|g\|_\infty .\tag{7.7.19}$$

In (7.7.16) $c^*$ is the minimizing vector in (7.7.17). The proof is complete. $\square$

**Remark 7.7.1** If $U$ in (7.7.2) is replaced by

$$U = \left\{u \text{ measurable } u(t) \in E^n : \int_\sigma^{t_1} |u_j(t)|dt \le 1, \; j = 1,\ldots,m\right\} ,\tag{7.7.20}$$

then

$$c^T y(t_1) = \int_\sigma^{t_1} \sum_{j=1}^{m} g_j(t,c)u_j(t)dt \le \sum_{j=1}^{m} \max |g_j(t,c)|, \quad \sigma \le t \le t_1 .\tag{7.7.21}$$

The maximum in (7.7.21) occurs as before at multiple instances of time $\tau_{ji}$, $i = 1, 2, \ldots, M_j$, where $M_j \ge 1$. The maximizing impulsive controls that approximate reachable points on the boundary of $\mathbb{R}(t)$ are

$$u_j^*(t,c) = \sum_{i=1}^{M_j} \frac{1}{M_j}\mathrm{sgn}(g_j(\tau_{ji},c))\delta(t - \tau_{ji}) , \qquad i \le j \le m .\tag{7.7.22}$$

Therefore, with

$$F(t_1, c) = \sum_{j=1}^{M} \max_{1 \le t \le t_1} |g_j(t, c)|, \qquad (7.7.23a)$$

$$\alpha(t_1) = \min_{c \in P} F(t_1, c) = F(c^*), \qquad (7.7.23b)$$

the optimal control that minimizes the absolute fuel is

$$\bar{u} = \frac{u^*(t, c^*)}{\alpha(t_1)}, \qquad (7.7.24)$$

where $c^*$ and $\alpha$ are determined by (7.7.23b).

**Proof of Theorem 7.5.5**    Recall the following definitions:

$$\mathbb{R}(t_1) = \left\{ \int_{\sigma}^{t_1} U(t_1, t) B u(t) dt : \|u\|_\infty \le 1 \right\}, \qquad (7.7.25)$$

$$\mathbb{R}_\alpha = \left( \mathbb{R}_\alpha(t_1) = \int_{\sigma}^{t_1} U(t_1, t) B u(t) dt : \|u\|_\infty \le 1, \|u\|_1 \le \alpha \right\}, \quad (7.7.26)$$

where $\alpha \ge 0$. These sets are nonvoid, convex, and symmetric about 0. Clearly $\mathbb{R}_\alpha \subset \mathbb{R}(t_1)$ and equality holds if $\alpha \ge mt_1$. Indeed, if $\alpha \ge mt_1$ and $y \in \mathbb{R}(t_1)$, then $y \in \mathbb{R}(t_1)$ corresponds to some admissible control $u$ such that

$$y = \int_{\sigma}^{t_1} U(t_1, t) B u(t) dt, \qquad \|u\|_\infty \le 1.$$

But then

$$\|u\|_1 = \int_0^{t_1} \sum_{j=1}^{m} |u_j(t_1)| dt \le \int_{\sigma}^{t_1} \sum_{j=1}^{m} = mt_1.$$

This shows that $y \in \mathbb{R}_{mt_1} \subset \mathbb{R}_\alpha$. We also observe that for $0 < \alpha \le \beta$,

$$\mathbb{R}_\alpha \subset \mathbb{R}_\beta, \qquad \alpha^{-1} \mathbb{R}_\alpha \supset \beta^- \mathbb{R}_\beta. \qquad (7.7.27)$$

The first is obvious from the definition. The second follows from the following containment:

$$\lambda \mathbb{R}_\gamma + (1 - \lambda) \mathbb{R}_\delta \subset \mathbb{R}_{\lambda\gamma + (1-\lambda)\delta}$$

for all $\gamma, \delta \ge 0$, $0 \le \lambda \le 1$. If $\delta \ge 0$, $\gamma = \beta$, and $\lambda = \alpha/\beta$, then the second containment follows. From (7.7.27) we deduce that

$$\mathbb{R}_\alpha \supset (\alpha/mt_1) \mathbb{R}(t_1), \qquad 0 \le \alpha \le mt_1.$$

If (7.5.1) is controllable, then $\mathbb{R}(t_1)$ has a nonvoid interior, and this forces $\mathbb{R}_\alpha$ to have nonvoid interior for each $\alpha > 0$. From the usual weak-star compactness argument we can prove that $\mathbb{R}(t_1)$ is compact. $\qquad\square$

Other properties of $\mathbb{R}(t_1)$ are contained in the next lemma.

**Lemma 7.7.2**  *For each $\phi \in C$ such that $x(t_1, \sigma, \phi, 0) \in \mathbb{R}(t_1)$ (and for no other points) there is an admissible control $u$ that steers $\phi$ to 0 at time $t_1$ while minimizing $E_3(u(t_1)) = \|u\|_1$. Also $x(t_1, \sigma, \phi, 0) \in \partial\mathbb{R}_\theta$ for $\theta = \|u\|$.*

**Proof**  By definition, $\phi$ can be steered to zero at time $t_1$ by a control if and only if

$$x(t_1, \sigma, \phi, 0) \in \mathbb{R}(t_1).$$

Let

$$\theta = \inf\{\alpha \geq 0 : x(t_1, \sigma, \phi, 0) \in \mathbb{R}\}, \qquad (7.7.28)$$

where

$$\mathbb{R}(t_1) = \bigcup_{\alpha \geq 0} \mathbb{R}_\alpha. \qquad (7.7.29)$$

It is clear that there is a sequence of admissible controls $u(\ )$, each steering $\phi$ to 0 at time $t_1$ with $\|u\|_1 \to \theta^+$. The usual weak compactness argument yields an optimal control $\overline{u}$ with $\theta = \|\overline{u}\|_1 = E_3(\overline{u}(t_1))$ and

$$E_3(\overline{u}(t_1)) \leq E_3(u(t_1)), \qquad \forall\, u.$$

To see this, interpret admissible controls as points in $L_2$-space, i.e., $L_2([\sigma, t_1], E^m)$. Thus the conditions

$$\|u\|_\infty \leq 1, \qquad \|u\|_1 \leq \theta + \epsilon,$$

determine a convex set that is bounded in $L_2$ norm. It is weakly closed since if a sequence converges weakly, then a sequence of convex combination converges strongly in $L_2$, and as a result a subsequence converges pointwise a.e. Thus $x(t_1, \sigma, \phi, 0) \in \mathbb{R}_\theta$. We now verify that $x(t_1, \sigma, \phi, 0) \in \partial\mathbb{R}_\theta$. We assume this is false and aim at a contradiction: $x(t_1, \sigma, \phi, 0) \in \text{Int } \mathbb{R}_\theta$. This implies that for sufficiently small $\lambda - 1 > 0$, $\lambda x(t_1, \sigma, \phi, 0) \in \mathbb{R}_\theta$. But if $0 < \alpha \leq \theta$, then by what was proved before

$$\mathbb{R}_\theta \subset (\theta/\alpha)\mathbb{R}(t_1).$$

Let $\alpha = \theta/\lambda$, then

$$\lambda x(t_1, \sigma, \phi, 0) \in \lambda\mathbb{R}_\alpha, \qquad x(t_1, \sigma, \phi, 0) \in \mathbb{R}_\alpha,$$

even though $\alpha < \theta$ and $\theta$ is minimal. If $\theta = 0$, $\mathbb{R}_0 = 0$. The proof is complete. $\qquad\square$

The next lemma is the Maximum Protoprinciple of Hájek, which corresponds to our System (7.7.4).

**Lemma 7.7.3** *If $\phi \in C$ is such that $x(t_1, \sigma, \phi, 0) \in \partial \mathbb{R}_\alpha$, and $c$ is an exterior normal to $\mathbb{R}_\alpha$ at $x(t_1, \sigma, \phi, 0)$, and if $u_0(\cdot)$ is any admissible control steering $\phi$ to $0$ at $t_1$, then*

$$\int_\sigma^{t_1} g(t)u(t)dt \leq \int_0^{t_1} g(t)u_0(t)dt, \qquad (7.7.30)$$

*for any $u(\cdot)$ with $\|u\|_\infty \leq 1$, $\|u\|_1 \leq \alpha$, where $g$ is the index of the control system:*
*$g(t) = c^T U(t_1, t)B$.*

*It is not necessary that $\|u_0\|_1 \leq \alpha$. Conversely, if $c \neq 0$, and $\|u_0\|_\infty \leq 1$, $\|u_0\|_1 \leq \alpha$, and if (7.7.30) holds for all controls described there, then $u_0(\cdot)$ reaches the point $x(t_1, \sigma, \phi, 0) \in \partial \mathbb{R}_\alpha$ for some $\phi \in C$, at which $c$ is an external normal.*

**Proof**   Let $\phi \in C$ be a point for which $x(t_1, \sigma, \phi, 0)$ is on the boundary of $\mathbb{R}_\alpha$ with $c$ an external normal. If $c \neq 0$, then this is equivalent to $c^T y(t_1) \leq c^T x(t_1, \sigma, \phi, 0)$, $\forall\, y(t_1) \in \mathbb{R}_\alpha$. But then

$$x(t_1, \sigma, \phi, 0) = \int_\sigma^{t_1} g(t_1, c)u_0(t)dt,$$

and

$$y(t_1) = \int_\sigma^{t_1} g(t, c)u(t)dt.$$

The proof is complete.        $\square$

**Corollary 7.7.1**   *Assume that the boundary control $u_0$ satisfies*

$$\|u_0\|_1 \leq \alpha.$$

*Then each coordinate $u_{0j}$ of $u_0$ satisfies*

$$g_j(t, c)u_{0j}(t) \geq 0 \qquad a.e.\ [\sigma, t_1]. \qquad (7.7.31)$$

**Proof**   Define a control $u$ with coordinates

$$u_j = (\text{sign } g_j) \cdot |u_{0j}|.$$

This control satisfies the assumptions. Since (7.7.30) is valid,

$$\int_\sigma^{t_1} gu_0 = \int_\sigma^{t_1} \sum_{j=1}^m g_j u_{0j} \geq \int_\sigma^{t_1} gu = \int_\sigma^{t_1} \sum_{j=1}^m |g_j| \cdot |u_{0j}|.$$

Because of this,

$$\int_\sigma^{t_1} \sum_{j=1}^m (g_j u_{0j} - |g_j u_{0j}|) \geq 0,$$

so that

$$g_j u_{0j} = |g_j u_{0j}| \geq 0$$

a.e., since the summands are nonpositive.

We know from the lemma that optimal controls exist as boundary controls of $\mathbb{R}_\alpha$. We obtain more information by applying the extension of Pontryagin's maximum principle reported by Manitius [37, p. 98]. Define the Hamiltonian

$$H = y(t)\left[A_0 x(t) + \sum_{j=1}^{N} A_j x(t - h_j) + Bu(t) - \eta|\sum_{j=1}^{m} |u_j|\right] \qquad (7.7.32)$$

for some constant $\eta \geq 0$, where the adjoint equation may be written as

$$\dot{y}(t) = -y(t)A_0(t) - \sum_{j=1}^{N} y(t+h_j)A_j(t+h_j) \text{ a.e. in } [\sigma, t_1]\,,\ y(t) = 0\,,\ t > t_1\,. \qquad (7.7.33)$$

The optimal control that maximizes $H$ is the $u(t)$, which maximizes the expression

$$y(t)Bu(t) - \eta|\sum_{j=1}^{M} |u_j|\,. \qquad (7.7.34)$$

But $y(t) = c^T U(t_1, t)$ is the solution of the adjoint equation [8, pp. 147–149] with $y(t_1) = c^T$.

Thus with the index of the control system

$$g(t, c) \equiv y(t)B = c^T U(t_1, t)B\,,$$

we desire that $u$ that maximizes

$$\sum_{j=1}^{M} (g_j(t, c)u_j - \eta|u_j|) \qquad (7.7.35)$$

over all $m$-vectors $u$ whose coordinates satisfy $|u_j| \leq 1$. If $\eta = 0$, then the optimal control would coincide with that for the time-optimal problem, and we should therefore have to assume $\eta > 0$. In this case we maximize each summand in (7.7.35) separately:

$$u_j = \begin{cases} \text{sign } g_j(t, c)\,, & \text{if } |g_j(t, c)/\eta \geq 1\,, \\ 0\,, & \text{otherwise}\,, \end{cases} \qquad (7.7.36)$$

provided that $g_j \not\equiv \eta$. The coordinates of $u$ have values $\pm 1$ and $0$ only. As $t$ varies, there is no switch from $1$ to $-1$, nor vice versa. Just as in ordinary differential systems studied by Hájek, optimal controls either always use all the available capacity or are completely dormant. The expression (7.7.36) uniquely determines optimal controls provided $g_j \not\equiv \eta$ on set of positive measure, i.e., provided (7.5.1) is metanormal, which is assumed in the theorem. The expression for optimal control depends on a careful choice of $\eta$. The following construction follows very closely those of Hájek for ordinary linear systems. First, choose any $\eta \geq 0$, such that (7.7.36) holds and hunt for some $\overline{u} : \|\overline{u}\|_\infty \leq 1$. But

$$\|u\|_1 = \int_\sigma^{t_1} \sum_{j=1}^m |u_j| = \sum_{j=1}^m \text{meas}\{|y_j| \geq \eta\}.$$

We select some suitable $\eta$ for which $\|u\|_1 = \alpha$. Indeed, since (7.5.1) is metanormal, $g_j(t,c)$ is nonconstant a.e., and the mapping $m(\eta)$:

$$\eta \to \text{meas}\{t \in [\sigma, t_1] : |g_j(t,c)| \geq \eta\}$$

is continuous, strictly decreasing, $0 \leq m(\eta) \leq t_1$. Because $0 \leq \alpha \leq mt_1$, there exists a unique value of $\eta \geq 0$ such that

$$\sum_{j=1}^m \text{meas}\{|g_j| \geq \eta\} = \alpha.$$

With this $\eta$ set

$$E_j = \{t \in [\sigma, t_1] : |g_j(t,c)| \geq \eta\}, \qquad F_j = [\sigma, t_1]/E_j,$$

so that

$$\sum_{j=1}^m \text{meas}\, E_j = \alpha \quad |g_j(t,c)| \geq \eta \text{ on } E_j, \quad |g_j(t,c)| \leq \eta \text{ on } F_j.$$

With this, define the control $\overline{u}(\ )$ as follows

$$\overline{u}_j(t) = \begin{cases} \text{sign } g_j(t,c) & \text{on } E_j, \\ 0 & \text{on } F_j. \end{cases}$$

This $\overline{u}$ has the following properties:

$$\|\overline{u}\|_\infty \leq 1, \qquad \|\overline{u}\|_1 = \alpha.$$

It is easy to see that

$$\int_\sigma^{t_1} g(t,c)u(s)ds \leq \int_\sigma^{t_1} g(t,c)\overline{u}(t)dt,$$

for all controls $u(\cdot)$ with

$$\|u\|_\infty \leq 1, \qquad \|u\|_1 \leq \alpha.$$

Because (7.7.4) is metanormal, $\overline{u}$ is unique a.e.

The Theorem is completely proved.      $\square$

## 7.8 Control of Nonlinear Functional Differential Systems with Finite Delay: Existence, Uniqueness, and Continuity of Solutions

In this section we give some known result on the existence, uniqueness, and continuous dependence on initial data and parameter of the system

$$\dot{x}(t) = f(t, x_t), \qquad t \geq \sigma,$$
$$x_\sigma = \phi, \qquad x_t(s) = x(t+s), \qquad -h \leq s \leq 0. \tag{7.8.1}$$

Also considered is the control system monitored by

$$\dot{x}(t) = F(t, x_t, u(t)), \qquad x_\sigma = \phi. \tag{7.8.2}$$

As usual, $C$ is space of continuous functions from $[-h, 0] \to E^n$ with the sup norm. If $O$ is a subset of $E^n$, and $f : E \times C([-h, 0], O) \to E^n$, we say $x$ is a solution of (7.8.1) on $[\sigma - h, \ \sigma + a)$ if there is an $a > 0$ such that $x : [\sigma - h, \ \sigma + a) \to E^n$ is continuous and $x(t)$ satisfies (7.8.1) for all $t \in [\sigma, \sigma + a)$. We assume $\dot{x}(\sigma) = f(t, \phi)$.

For a subset $D$ of $E \times C \times E^m$ given by $D = E \times C([-h, 0] \ O) \times E^m$ and $F : D \to E^n$, a solution $x$ of (7.8.2) is defined similarly. For the system (7.8.1) we have the following:

**Theorem 7.8.1** *Suppose $O \subset E^n$ is open, $\mathcal{O} = E \times C([-h, 0], O)$, and $f : \mathcal{O} \to E^n$ is continuous and satisfies a local Lipschitz condition in the second variable. If $(\sigma, \phi) \in \mathcal{O}$, then there exists a unique solution for (7.8.1) on the interval $[\sigma, \sigma + a)$ for some $a > 0$. This solution varies continuously with initial data, and satisfies the integral equation*

$$x_\sigma = \phi x(t) = \phi(0) + \int_\sigma^t f(s, x_s) ds. \tag{7.8.3}$$

The solution $x$ given by Theorem 7.8.1 satisfies $x_t \in C$. There are various function spaces where the solution of (7.8.2) can be considered to lie. It depends on the choice of the control space from which $u$ is selected. For the existence problems, we select the control space to be the Banach space $C([\sigma, \tau], E^m)$ of continuous functions mapping $[\sigma, \tau]$, $(\sigma \leq \tau < \infty)$ into $E^m$ with the sup norm. The results we shall state are valid for controls in $L_p$ spaces $(2 \leq p \leq \infty)$ treated in Rudin [39]. Thus we can replace $C([\sigma, \tau], E^m)$ by $L_\infty([\sigma, \tau], E^m)$, whose members are essentially bounded measurable functions defined on $[\sigma, \tau]$ with values in $E^m$. We adopt the following notation in the theorem: the Fréchet derivative of $g$ at $\phi$ is denoted by $Dg(\phi)$ where $\phi \in C$, a Banach space. If $C$ is the product of several Banach spaces, then the partial derivative with respect to the $i$th variable will be denoted by $D_i g(\phi)$.

**Theorem 7.8.2** [40] *Consider System (7.8.2), where*

$$F : [\sigma, \tau] \times C([-h, 0], O) \times E^m \to E^n$$

*satisfies the following conditions:*

(i) $F(t, \cdot, \cdot)$ *is continuously differentiable for each* $t$.

(ii) $F(\cdot, \phi, \omega)$ *is measurable for all* $\phi$ *and* $\omega$.

(iii) *For each compact* $k \subset O$ *there exists an integrable function* $M_1 : [\sigma, \tau] \to [0, \infty)$ *and square integrable functions* $M_i : [\sigma, \tau] \to [0, \infty)$, $i = 2, 3$, *such that*

$$\|D_2 F(t, \phi, \omega)\| \leq M_1(t) + M_2(t)|\omega|, \quad \|D_3 F(t, \phi, \omega)\| \leq M_3(t),$$

*for all* $t$ *and* $\omega$ *and all* $\phi \in C([-h, 0], O)$.

(iv) $F(t, 0, 0) = 0$ *for all* $t$. *Let the function* $x(\sigma, \phi, u) \in C([\sigma - h, \tau], E^n)$ *be the solution of (7.8.2) for any pair of* $(\phi, u)$ *such that the solution exists.*

*Then there exist open neighborhoods* $N_1$ *and* $N_2$ *of the origin in* $C([-\sigma, 0], E^n)$ *and* $C([\sigma, \tau], E^m)$ *such that, for each* $\phi \in N_1$, $u \in N_2$, *(7.8.2) has a unique solution* $x = x(\sigma, \phi, u)$ *and such that* $x$ *is continuously differentiable on* $N_1 \times N_2$.

**Remark 7.8.1**    If a measure theoretic analysis is undesirable, we can formulate the following version of Theorem 7.8.2:

**Theorem 7.8.3**    *In (7.8.2), assume that:*

(i) $F(t, \cdot, \cdot)$ *is continuously differentiable for each* $t$.

(ii) $F(\cdot, \phi, \omega)$ *is continuous for all* $\phi$ *and* $\omega$.

(iii) *For each compact* $k \subset 0$ *there exists continuous functions* $M_i : [\sigma, \tau] \to [0, \infty)$ $i = 1, 2, 3$, *such that*

$$\|D_2 F(t, \phi, \omega)\| \leq M_1(t) + M_2(t)|\omega|,$$

$$\|D_3 F(t, \phi, \omega)\| \leq M_3(t), \qquad \forall\, t, \omega \quad and \quad \phi \in C([-h, 0], 0).$$

(iv) $F(t, 0, 0) = 0$ *for all* $t$. *Let the function* $x(\sigma, \phi, u)$ *be the solution of (7.8.2) for any pair of* $(\phi, u)$ *such that the solution exists.*

*Then there exist open neighborhoods* $N_1$ *and* $N_2$ *of the origin in* $C([-h, 0], E^n)$ *and* $C([\sigma - h, \tau], E^m)$ *such that for each* $\phi \in N_1$, $u \in N_2$, *(7.8.2) has a unique solution* $x = x(\sigma, \phi, u)$, *and* $x$ *is continuously differentiable on* $N_1 \times N_2$.

**Remark 7.8.2**    Theorem 7.8.2 is valid with $u \in L_p([\sigma, \tau], E^m)$, though this seems to be more useful when $p = \infty$, since difficulties are encountered when differentiating on $L_p$ $(2 \leq p \in \infty)$.

**Proof of Theorem 7.8.2**    We first observe that if $x$ is any continuous map from $[\sigma - h, \tau]$ to $O$ and $u \in L_\infty([\sigma, \tau], E^m)$, there exists a compact convex set $k \subset O$ containing the origin such that $x(t) \in k$, $\sigma - h \leq t \leq \tau$. For this $k$ let $M_1$ and $M_3$ be as in hypothesis (iii) of Theorem 7.8.2. It follows that for all $t \in [\sigma, \tau]$, $\phi \in C([-h, 0], k)$, $\omega \in E^m$ we have

$$|F(t, \phi, \omega)| \leq M_1(t)\|\phi\| + M_3(t)|\omega|$$

so that $F(t, x_t, u(t))$ is an integrable function of $t$.

Next we note that there is at most one solution $x$ of (7.8.2) that satisfies $x_\sigma = \phi$. The Gronwall's type argument of Hale [8, p. 42] is applicable for its proof. We express (7.8.2) in integrated form as follows: Define the operator $T : C \to C([\sigma - h, \tau]), E^n)$ by

$$(T\phi)(t) = \begin{cases} \phi(t - \sigma), & \text{if } \sigma - h \leq t < \sigma, \\ \phi(0), & \text{if } \sigma \leq t \leq \tau. \end{cases} \qquad (7.8.4)$$

Then $x$ is a solution of (7.8.2) on $[\sigma - h, \tau]$ if and only if

$$x(t) = (T\phi)(t) + \int_\sigma^{\alpha(t)} F(s, x_s, u(s))ds, \qquad \sigma - h \leq t \leq \tau,$$

where $\alpha(t) = \max(\sigma, t)$. If we define $H : C([\sigma - h, \tau], N) \times L_\infty([\sigma, \tau], E^m) \to C([\sigma - h, \tau], E^n)$ by

$$H(x, u)(t) = \int_\sigma^{\alpha(t)} F(s, x_s, u(s))ds, \qquad (7.8.5)$$

then $H$ is continuously differentiable, and

$$(D_1 H(\overline{x}, \overline{u})x)(t) = \int_\sigma^{\alpha(t)} D_2 F(s, \overline{x}_s, \overline{u}(s))x_s ds,$$

$$(D_2 H(\overline{x}, \overline{u})u)(t) = \int_\sigma^{\alpha(t)} D_3 F(s, \overline{x}_s, \overline{u}(s))u(s)ds,$$

where $t \in [\sigma - h, \tau]$,

$$\overline{x} \in C([\sigma - h, \tau], N), \quad x \in C([\sigma - h, \tau], E^n), \quad u, \overline{u} \in L_\infty([\sigma, \tau], E^m).$$

Thus

$$(D_1 H(0, 0)x)(t) = \int_\sigma^{\alpha(t)} D_2 F(s, 0, 0)x_s ds,$$

$$(D_2 H(0, 0)u)(t) = \int_\sigma^{\alpha(t)} D_3 F(s, 0, 0)u(s)ds,$$

for all $x \in C([\sigma - h, \tau], E^n)$, $u \in L_\infty([\sigma, \tau], E^m)$, and $t \in [\sigma - h, \tau]$.

Detailed proof of these formulae require the use of the Dominated Convergence Theorem, which is permissible because of condition (iii) of Theorem 7.8.2. To prove continuity of $D_i H \ \ i = 1, 2$, we show that

$$\lim_{(x,u) \to (\overline{x}, \overline{u})} D_i H(x, u) = D_i H(\overline{x}, \overline{u}),$$

by proving for example that for $i = 1$,

$$\lim_{(x,y)\to(\overline{x},\overline{u})} \int_\sigma \|D_2F(s,x_s,u(s)) - D_2F(s,\overline{x}_s,\overline{u}(s))\|ds = 0\,.$$

The argument involves extraction of subsequences, and the use of dominated convergence theorem. With $H$ defined in (7.8.5), the solution $x$ of (7.8.2) can be rewritten as $x = H(x,u) + T\phi$, where $H(0,0) = 0$, since condition (iv) is valid.

Now define

$$G(x,u,\phi) = x - H(x,u) - T\phi\,. \tag{7.8.6}$$

Clearly $G(0,0,0) = 0$, and $G$ is continuously differentiable on

$$C([\sigma - h, \tau], N) \times L_\infty([\sigma, \tau], E^m) \times C\,,$$

and $D_1G(0,0,0) = I - D_1H(0,0)$. Carefully note that if $D_2F(t,0,0)\phi = A(t,\phi)$ and conditions (i) and (iii) hold, then the operator $K$, defined by

$$(Kx)(t) = \int_\sigma^{\alpha(t)} A(s,x_s)ds\,, \qquad \sigma - h \le t \le \tau\,,$$

is such that $(I-K)^{-1}$ exists. This is true since $\{Kx : x \in C([\sigma-h,\tau], E^n)\ \|x\| \le 1\}$ is equicontinuous, and therefore $K$ is compact. Furthermore, since

$$\dot{x}(t) = A(t,x_t)\,, \qquad x_\sigma = \phi\,,$$

has a unique solution on $[\sigma - h, \tau]$, the unique solution of $x = Kx$ is $x = 0$. The assertion that $(I-K)^{-1}$ exists follows readily from Fredholm's alternative [40]. As a consequence $(I-D_1H(0,0))^{-1} = (I-G(0,0,0))^{-1}$ exists. We are now assured by the Implicit Function Theorem that there exists a continuously differentiable function $\overline{x}$ defined on an open neighborhood $\overline{N}$ of the origin in $C([-h,0], E^n) \times L_\infty([\sigma,\tau], E^m)$, such that

$$G(\overline{x}(\phi,u),u,\phi) = 0\,, \quad \forall\, (\phi,u) \in \overline{N}\,.$$

By choosing $\overline{N}$ appropriately small, the proof is complete.                    $\square$


## 7.9    Sufficient Conditions for the Existence of a Time-Optimal Control

Consider the general delay differential control system

$$\dot{x}(t) = f(t,x_t,u(t))\,,$$
$$x_\sigma = \phi\,, \tag{7.9.1}$$

whose time-optimal control problem is formulated in Sec. 1.9. The initial state $\phi$ is given in the space $C$ of continuous functions from $[-h,0]$ into $E^n$ with the sup norm. The state space is either the $n$-dimensional Euclidean space $E^n$, the space

$C$, or the Sobolev space $W_p^{(1)}$, $1 \le p \le \infty$, of absolutely continuous functions $x : [-h, 0] \to E^n$ whose derivatives are $L_p$-integrable. If $E^n$, $C$, or $W_\infty^{(1)}$ is the state space, admissible controls are measurable functions $u : [0, \infty) \to E^m$ whose values $u(t) \in U(t, x_t)$, where $U(t, x_t)$ are compact, convex, set-valued function with $U(t, x_t) \subset E^m$. Thus if $2^{E^m}$ is the collection of nonempty, compact, convex subsets of $E^m$, the set map $U : E \times C \to 2^{E^m}$ is the control set. In particular, we shall consider the fixed control set $U$ that is the $m$-dimensional unit cube, $C^m$, of $E^m$, that is

$$C^m = \{u : u \in E^m, \ |u_j| \le 1, \ j = 1, \ldots, m\}.$$

When controls are taken as measurable functions with values in $U$, we say the controls are constrained. Thus we shall study constrained controls for $E^n$, $C$, and $W_\infty^{(1)}$ state spaces. If we restrict our attention to the state space $W_2^{(1)}$, we use the control space $L_2([\sigma, \tau], E^m)$ of square integrable functions $u : [\sigma, \tau] \to E^m$.

The problem resolved in this section is to show the existence of a constrained control $u^*$ such that the solution $x(\sigma, \phi, u^*)$ of (7.9.1) reaches a continuously moving target $z_t \in C([-h, 0], E^n) = C$ in minimum time $t^* \ge \sigma$. Thus we shall investigate the existence of optimal control in $C$.

We assume that $f$ is smooth enough to guarantee global existence and uniqueness of a solution passing through each $(\sigma, \phi) \in [0, \infty) \times C$ for each $u \in L_\infty([\sigma, \infty), E^m)$. We also assume that $f$ satisfies the inequality

$$x^T(t) f(t, x_t, u(t)) \le K(1 + \|x_t\|^2) \tag{7.9.2}$$

for some $K > 0$, all $t \in E$, $x_t \in C$, and $u(t) \in U(t, x_t)$. This second assumption on $f$ is given to avoid finite escape times. We also assume for any bounded set $N \subset C$ that there exists an almost-everywhere bounded measurable function $\xi_N : E \to E^n$ such that

$$|f(t, x_t, u(t))| \le |\xi_N(t)| \quad \text{a.e. in } t, \tag{7.9.3}$$

for all $x_t \in N$, $u(t) \in U(t, x_t)$. Associated with (7.9.1) is the contingent equation,

$$\dot{x}(t) \in F(t, x_t), \quad x_\sigma = \phi, \tag{7.9.4}$$

where

$$F(t, \phi) = \{f(t, \phi, u) : u \in U(t, \phi)\}. \tag{7.9.5}$$

Since $f$ is continuous and $U(t, \phi)$ nonempty and compact, $F(t, \phi)$ is nonempty and compact. When $U$ is not fixed, we assume it is upper semicontinuous with respect to inclusion in $t, \phi$ in the following sense:

**Definition 7.9.1** The set map, $F(\cdot, \cdot) : E \times C \to 2^{E^n}$ is said to be upper semicontinuous with respect to inclusion in $t, x_t$, if for any $\epsilon > 0$ there is a $\delta > 0$ such that

$F(t_1, x_{t_1}) \subset F_\epsilon(t, x_t)$ whenever $|t_1 - t| < \delta$, $\|x_{t_1} - x_t\| \le \delta$. Here $F_\epsilon$ represents the $\epsilon$-neighborhood of $F$ in $E^n$. A solution $x$ of (7.9.4) is an element $x \in C([\sigma - h, \tau], E^n)$ such that

(i) $x$ is absolutely continuous on $[\sigma, \tau]$,
(ii) $x_\sigma = \phi$, and
(iii) $\dot{x}(t) \in F(t, x_t)$.

We deduce from McShane and Warfield's generalization [23] of an important result of Filippov [41] that every solution of (7.9.4) can be viewed as a solution of (7.9.1). Indeed, we have:

**Lemma 7.9.1** *Let $F(t, \phi)$ be convex for each $(t, \phi) \in E \times C$. A function $x$ is a solution of (7.9.4) if and only if $x$ is a solution of (7.9.1) for some admissible control $u$.*

**Definition 7.9.2** A subset $A$ of $C$ is called the attainable set if it is given by $A(t, \sigma) = \{x_t : x \text{ is a solution of (7.9.1) for some } u(t) \in U(t, x_t)\}$. This is the same as $A(t, \sigma) = \{x_t : \dot{x}(t) \in F(t, x_t) : t \ge \sigma,\ x_\sigma = \phi\}$. The closure of $\mathcal{A}(t, \sigma)$ in $C$ is needed to prove the existence of an optimal control. Indeed the following is true:

**Theorem 7.9.1** *Suppose*

(i) *$f : E \times C \times E^m \to E^n$ is continuous and is continuously differentiable in the last arguments.*
(ii) *$f$ verifies the inequalities (7.9.2) and (7.9.3).*
(iii) *Let $F(t, \phi)$, as defined in (7.9.5), be convex for each $t, \phi$.*
(iv) *Suppose $U : E \times C \to 2^{E^m}$ is upper semicontinuous with respect to inclusion in the two arguments, and is compact valued.*
 *Then $\mathcal{A}(t, \sigma)$, $t \ge \sigma$, is compact.*

**Proof** Because $f : E \times C \times E^m \to E^n$ is continuous and $U$ upper semicontinuous with respect to inclusion, $F(\cdot, \cdot)$ is upper semicontinuous with respect to inclusion. Because $U$ is compact, $F$ is closed valued. Let $\{x_T^k\}$ be a sequence of points in $\mathcal{A}(T, \sigma)$ with $x_\sigma^k = \phi$. Then $\dot{x}^k$ is a solution of (7.9.4):

$$\dot{x}^k(T) \in F(T, x_T), \qquad T \ge \sigma, \qquad x_\sigma^k = \phi.$$

We now prove that $\{x^k\}$ is equicontinuous and equibounded, so that $\{x_T^k\}$ is equicontinuous and equibounded. Since (7.9.2) and (7.9.3) are valid,

$$(\dot{x}^k(t), \dot{x}^k(t)) \le K(1 + \|x_t^k\|^2),$$

and

$$\|x^k(t)\|^2 \le (1 + \|x_\sigma^k\|^2)(\exp[2K(t - \sigma)] - 1),$$

for $t \in [\sigma, T]$, so that

$$\|x^k(t)\|^2 \le (1 + \|\phi\|^2)\exp[2K(t - \sigma)] - 1.$$

We have proved that the function $\{x^k\}$ is equibounded. Because $\{x^k\}$ is equibounded, (7.9.3) guarantees that there exists an almost-everywhere bounded function $\xi$ such that $|\dot{x}^k(t)| \le |\xi(t)| \le M$ a.e. in $t$. It follows that $\{|\dot{x}^k(t)|\}$ is equibounded a.e. As a result, $\{x^k\}$ is equicontinuous and therefore compact. By the Ascoli Theorem there is a subsequence (and we can take it to be the original sequence) $x^k$ converging uniformly as $k \to \infty$ on $[\sigma, T]$ to some $x$. Note that $x_\sigma = \phi$ and $|\dot{x}(t)| \le M$, and $x$ is absolutely continuous. We now prove that $x$ is a solution of (7.9.4). Suppose at the point $t_0 \in [\sigma, T]$, $\dot{x}(t_0)$ exists. Then

$$\frac{x(t) - x(t_0)}{t - t_0} = \lim_{k \to \infty} \frac{x^k(t) - x^k(t_0)}{t - t_0}$$

$$= \lim_{k \to \infty} \frac{1}{t - t_0} \int_{t_0}^{t} \dot{x}^k(\tau)d\tau$$

$$= \lim_{k \to \infty} \int_0^1 \dot{x}^k((t - t_0)s)ds.$$

For a given $\epsilon > 0$ we can choose $\delta > 0$ such that

$$\left|\frac{x(t) - x(t_0)}{t - t_0} - \dot{x}(t_0)\right| \le \epsilon \quad \text{and} \quad F(t, x_t) \subset F_\epsilon(t_0, x_{t_0}), \tag{7.9.6}$$

whenever $|t - t_0| < \delta$, $\|x_t - x_{t_0}\| < \delta$. The inclusion in (7.9.6) follows from the upper semicontinuity of $F(t, x_t)$. Thus for almost all $\tau \in [\sigma, t]$, $\dot{x}^t(\tau) \in F(\tau, x^k)$. Since $x^k$ converges uniformly to $x$ so that for $N \ge N_0$ (say), $\|x_t - x^k\| \le \delta/2$. Therefore $\dot{x}^k(\tau) \in F_\epsilon(t_0, x_{t_0})$ for almost all $\tau \in [t_0, t]$. Thus $\dot{x}^k(t_0 + (t - t_0)s) \in F_\epsilon(t_0, x_{t_0})$ a.e. $s \in (0, 1)$. Because $F_\epsilon(t_0, x_{t_0})$ is convex, the Mean Value Theorem yields that

$$\int_0^1 x^k(t_0 + (t - t_0)s)ds \in F_\epsilon(t_0, x_{t_0})$$

for $N \ge N_0$, say. Thus

$$\frac{x(t) - x(t_0)}{t - t_0} \in F_\epsilon(t_0, x_{t_0}).$$

Therefore, $\dot{x}(t_0) \in F_{2\epsilon}(t_0, x_{t_0})$, for arbitrary $\epsilon > 0$. Since $F(t, x_t)$ is closed,

$$\dot{x}(t_0) \in F(t_0, x_{t_0}) = \bigcap_{\epsilon > 0} F_{2\epsilon}(t_0, x_{t_0}).$$

Since this holds a.e. on $[\sigma, T]$, we have indeed verified that $x$ satisfies (7.9.4). Thus $x_T \in \mathcal{A}(T, \sigma)$, and $\mathcal{A}(T, \sigma)$ is closed. Summing, we have shown that $x_T^k$ is an equibounded sequence which converges to a point of $\mathcal{A}(T, \sigma)$. Hence $\mathcal{A}(T, \sigma)$ is compact. $\qquad\square$

**Definition 7.9.3**    Let $z_t \in C$ be a target function that is time varying and for each $t$ is an element of the function space $C$. Suppose $z_T \in \mathcal{A}(T, \sigma)$, $T \geq \sigma$, then the system (7.9.1) is controllable to the target. In other words, for each $\phi \in C$ there exists a $T \geq \sigma$ and an admissible control $u, u(t) \in U(t, x_t)$, $t \in [\sigma, T]$ such that the solution of (7.9.1) satisfies

$$x_\sigma(\sigma, \phi, u) = \phi, \qquad x_T(\sigma, \phi, u) = z_T .$$

In this case we also say that (7.9.1) is function-space $z$-controllable on $[\sigma, T]$ with constraints.

System (7.9.1) is null controllable with constraints if for each $\phi \in C$, there exists a $t_1 < \infty$ and an admissible control $u \in U$ such that the solution $x(\cdot)$ of (7.9.1) satisfies $x_\sigma(\sigma, \phi, u) = \phi, x_{t_1}(\sigma, \phi, u) = 0$. In what follows it will be very convenient to work in the space $E^n$, $C$, or $W_\infty^{(1)}$ as the state space with controls $L_\infty$ that are constrained to lie in the unit cube $U$ in $E^m$. Controllability concepts can also be defined without assuming constraints.

**Definition 7.9.4**    System (7.9.1) is controllable on $[\sigma, t_1]$, $t_1 > \sigma + h$, if for each $\phi, \psi \in C$ there is a control $u \in L_\infty([\sigma, t_1], E^n)$ such that the solution $x(\sigma, \phi, u)$ of (7.9.1) satisfies $x_\sigma(\sigma, \phi, u) = \phi$, $x_{t_1}(\sigma, \phi, u) = \psi$. It is null controllable on $[\sigma, t_1]$, $t_1 > \sigma + h$, if for each $\phi \in C$ there is a control $u \in L_\infty([\sigma, t_1], E^m)$ such that the solution of (7.9.1) satisfies $x_\sigma(\sigma, \phi, u) = \phi$, $x_{t_1}(\sigma, \phi, u) = 0$. The qualifying phrase "on the interval $[\sigma, t_1]$" is dropped if they hold on every interval with $t_1 > \sigma + h$.

In the above definition, we consider Euclidean controllability by replacing $\psi$ by $x_1 \in E^n$.

We now investigate the existence of optimal control in the state space $C$.

**Theorem 7.9.2**    (*Existence of a Time-Optimal Control*)

   (i) *Assume that in (7.9.1), conditions (i)–(iv) of Theorem 7.9.2 hold.*
  (ii) *For some $T \geq \sigma$, System (7.9.1) is function-space $z$-controllable on $[\sigma, T]$ with constraints.*

*Then there exists an optimal control.*

**Proof**    Let

$$t^* = \inf\{t : T \geq t \geq \sigma, \ z_t \in \mathcal{A}(T, \sigma)\} .$$

Because of (ii) there is at least one $T$ such that $z_T \in \mathcal{A}(T, \sigma)$; as a consequence $t^*$ is well defined as a finite number. We now prove that $z_{t^*} \in \mathcal{A}(t^*, \sigma)$. Let $\{t_n\}$ be a sequence of times that converge to $t^*$ such that $z_{t_n} \in \mathcal{A}(t_n, \sigma)$. For each $n$, let $x^n$ be a solution of the contingent equation (7.9.4) with $z_{t_n} = x_{t_n}^n$. Then

$$\|x_{t^*}^n - z_{t^*}\| \leq \|x_{t^*}^n - x_{t_n}^n\| + \|x_{t_n}^n - z_{t^*}\| ,$$

$$\leq \|x_{t^*}^n - x_{t_n}^n\| + \|z_{t_n} - z_{t^*}\| ,$$

$$\leq \|z_{t_n} - z_{t^*}\| + \int_{t^*}^{t_n} \xi(\tau)d\tau \,,$$

where, by condition (7.9.3), $\xi$ is bounded almost everywhere. Since the target $z_t$ is continuous in $t$, $z_{t_n} \to z_{t^*}$ as $t_n \to t^*$. Also $\int_{t^*}^{t_n} \xi(\tau)d\tau \to 0$ as $t_n \to t^*$, since $\xi$ is bounded a.e. and is measurable. It follows that $x_{t^*}^n \to z_{t^*}$ as $n \to \infty$. Because $x_{t^*}^n \in \mathcal{A}(t^*, \sigma)$, and this is a closed set, $z_t \in \mathcal{A}(t^*, \sigma)$. This completes the proof. $\square$

**Corollary 7.9.1**  *Consider the linear system*

$$\dot{x}(t) = L(t, x_t) + B(t)u(t) \,, \qquad t \geq \sigma \,,$$
$$x_\sigma = \phi \,, \tag{7.9.7}$$

*where*

$$L(t, x_t) = \sum_{k=1}^{N} A_k(t)x(t - w_k) + \int_{-h}^{0} A(t, s)x(t + s)ds \,, \tag{7.9.8}$$

*with $0 = w_0 \leq w_1 \leq \cdots < w_N \leq h$, $A_k$ is a continuous $n \times n$ matrix, $B(t)$ is a continuous $n \times m$ matrix function, $A(t, s)$ is an $n \times n$ matrix function continuous in $t$ and integrable in $s$ for each $t$ with the following property. There is a locally essentially bounded measurable function $a(t) \in (-\infty, \infty) \to E$ such that*

$$\left|\int_{-h}^{0} A(t, s)x(t + s)ds\right| \leq a(t)\|x_t\| \tag{7.9.9}$$

*for $t \in E$, $x_t \in C$. The admissible controls $u \in U$ are measurable $L_\infty$ functions with values in the unit cube*

$$C^m = \{u \in E^m, \; |u_j| \leq 1, \; j = 1, \ldots, m\} \,. \tag{7.9.10}$$

*Suppose that in (7.9.7) where (7.9.8), (7.9.9), and (7.9.10) hold, there is some $t_1 \geq \sigma$ such that (7.9.7) is function-space $z$-controllable with constraints on $[\sigma, t_1]$, $t_1 > \sigma + h$, where $z_t$ is continuous in $t$ and is in $C$.*

*Then there exists time-optimal control.*

**Proof**

$$f(t, x_t, u(t)) = L(t, x_t) + B(t)u(t)$$

is a map $f : E \times C \times E^m \to E^n$ is continuous and is continuously differentiable in the last two arguments. Now, $F(t, x_t) = L(t, x_t) + B(t)U$ is convex for each $t, x_t$, because it is linear in $u$. Since $U$ is compact, and constant, it is upper semicontinuous, and conditions (i), (iii), and (iv) of Theorem 7.9.1 are satisfied. That (ii) holds as well follows from the following argument:

$$|f(t, x_t, u(t))| = |L(t, x_t) + B(t)u(t)| \leq (c(t) + a(t))\|x_t\| + \beta(t)$$

for some continuous functions $c(t), \beta(t)$, and a locally integrable $a(t)$. If $\|x_t\|$ is bounded,

$$(c(t) + a(t))\|x_t\| + \beta(t) \le \xi(t) \,,$$

where $\xi(t)$ is essentially bounded measurable function on $(-\infty, \infty)$, and therefore satisfies (7.9.3). Condition (7.9.2) is needed to prevent finite escape times for solutions, i.e., to ensure that the solution of (7.9.7) is bounded on every finite time $T$. From Proposition 2.2.1 any solution $x(\sigma, \phi)$ of (7.9.7) is defined and continuous on $[\sigma - h, \infty)$. It is therefore uniformly bounded on any finite interval $[\sigma, T]$ $T \ge \sigma$. It follows now that Theorem 7.9.3 can be invoked for the system (7.9.7), and the existence of a time-optimal control can be inferred.

Theorem 7.9.2 is of fundamental importance in time-optimal control theory. It asserts that one must make the controllability assumption and have conditions that guarantee constrained controllability before the time-optimality questions can be resolved. Controllability results with controls constrained to lie in $C^m$ are therefore very useful in the resolution of the time-optimal control problems of delay equations. This is tackled in Chaps. 6 and 8. Questions of controllability with unlimited power are next taken up.  $\square$

## 7.10  Optimal Control of Nonlinear Delay Systems with Target in $E^n$

Consider the following problem:
Minimize

$$J(u) = \int_0^T f^0(x(t), x(t - h), u(t)) \tag{7.10.1}$$

subject to

$$\dot{x}(t) = f(x(t), x(t - h), u(t)), \qquad t \in [0, T], \tag{7.10.2}$$

$$x(t) = \phi(t) \in E^n, \qquad t \in [-h, 0], \tag{7.10.3}$$

$$x(T) = x_1 \in E^n, \tag{7.10.4}$$

where $f(x, y, u)$ and $f^0(x, y, u)$ are continuously differentiable on $E^n \times E^n \times E^m$. $f^0 : E^n \times E^n \times E^m \to E$, while $f : E^n \times E^n \times E^m \to E^n$. The class of admissible controls are measurable functions $u$ with values $u(t) \in U$, where $U$ is the unit $m$-dimensional cube.

For each $(x, y, u, p) \in E^n \times E^n \times E^m \times E^n$, and $p_0 \in E$, define the Hamiltonian

$$H(x, y, u, p) = p_0 f^0(x, y, u) + p(t) f(x, y, u) \,, \tag{7.10.5}$$

where $p(t)$ is a row-vector function.

**Theorem 7.10.1** (cf. [46]) (**Maximum Principle**) *Let $u^*(t)$ be an optimal control and $x^*(t, u^*)$ the corresponding optimal trajectory. For this optimal pair $(u^*, x^*)$ there exists a nonzero function $p(t)$, and a $p_0^*$ such that*

(i)

$$H(x^*(t), x^*(t - h), u^*(t), p^*(t)) = \sup_{u \in U} H(x^*, x^*(t - h), u, p^*) \qquad (7.10.6)$$

*holds for all $t \in [0, T]$.*

(ii) *At $T$, the relation*

$$H(x^*(T), x^*(T - h), u^*(T), p^*(T)) = 0 \qquad (7.10.7)$$

*is satisfied. Also the vector $p^*(T)$ is orthogonal to the tangent plane to the manifold (which defines the target) at the point $x^*(t_1)$.*

(iii) *$p_0^* \leq 0$, where $p_0 \equiv$ constant.*

*Furthermore,*

$$\dot{x}(t) = \frac{\partial H}{\partial p} = f(x(t), x(t - h),\ u(t)), \qquad (7.10.8)$$

$$\dot{p}(t) = -p(t)A_0(t) - p(t + h)A_1(t + h) - p_0(t)f_x^0 \ \text{a.e. on } [0, T], \qquad (7.10.9)$$

$$p(t) \equiv 0, \quad \forall\, t > T, \qquad p(T) \ \text{satisfies (ii)}.$$

*Here $A_i(t)$, $i = 0, 1$, are given by*

$$\begin{aligned}
A_0(t) &= f_x(x^*(t), x^*(t - h), u^*(t)), \\
A_1(t) &= f_y(x^*(t), x^*(t - h), u^*(t)).
\end{aligned} \qquad (7.10.10)$$

**Remark 7.10.1** Because $p(t) \equiv 0$, $\forall\, t > T$, the advanced term in the adjoint equation vanishes on $[T - h, T]$. The time-optimal control problem is a special case of the situation in Theorem 7.10.1. Indeed, let $f^0(x(t),\ x(t - h), u(t)) \equiv 1$.

In Theorem 7.10.1 one assumes the existence of optimal control. This existence was proved for the time-optimal situation in Theorem 5.2.3. If the state space is $C$, the resulting general cost functional situation was treated by Angell in [44] and Angell and Kirsch [45]. It is treated in Sec. 7.3.

An important problem to be resolved is to find when the maximum principle is a necessary and sufficient condition for optimality. The following general result for linear systems is available.

Consider

$$\dot{x}(t) = A_0(t)x(t) + A_1 x(t - h) + B(t)u(t), \qquad t \geq 0, \qquad (7.10.11)$$

$$x(t) = \phi(t), \qquad t \in [-h, 0], \qquad (7.10.12)$$

where $h > 0$ constant $\phi$ is a continuous function, $u$ is measurable with $u(t) \in C^m = \{u \in E^m : |u_j| \leq 1, j = 1, \ldots, m\}$, and $A_i := 0, 1$ $B(\cdot)$ are continuous matrices.

Let $G \subset E^n$ be a closed target set. The problem is to minimize

$$J(u) = g^0(x(T)) + \int_0^T [f^0(x(t), t) + h^0(t, u(t))]dt \qquad (7.10.13)$$

subject to $x(\cdot)$ satisfies (7.10.11), (7.10.12), and the terminal condition

$$x(T_1, \phi, u) \in G. \qquad (7.10.14)$$

In (7.10.13), assume that $x \to g^0(x)$ is continuously differentiable and convex, and $f^0(\cdot, \cdot)$ and $h^0(\cdot, \cdot)$ are nonnegative continuous functions that are convex with respect to $x$ for all $t \in [0, T]$. We assume $x \to f^0(x, t)$ is continuously differentiable for all $t \in [0, T]$. The following theorems are valid.

**Theorem 7.10.2**  **(Existence)** *If there exists an admissible control $u \in U$ that transfers the response $x(t)$ in (7.10.11) from the initial function $\phi$ to the target $G$ at time $T$, then there exists an optimal controller that steers the response to $G$ at time $T$ and minimizes $J(u)$. The optimal controller is given as follows:*

*There exists a nontrivial function $t \to (p_0(t), p(t))$ satisfying the equation*

(i)  $p_0(t) \equiv constant < 0$,

(ii)  $\overset{\circ}{p}(t) = -p(t)A_0(t) - p(t+h)A_1(t) - p_0(t)f_x^0(x(t), t)$ *a.e. on* $[0, T]$, *such that*

$$p_0 h^0(u^*(t), t) + p(t)B(t)u^*(t) = \max_{u \in U}(p_0 h^0(u, t) + p(t)B(t)u).$$

*Conversely, if $u^*$, an admissible controller, is given as above then it is an optimal controller.*

**Remark 7.10.2**  The proof is similar to the reachable set approach of Theorem 7.1.1, and is contained in Manitius [37, pp. 104–108].


## 7.11  Optimal Control of Delay Systems in Function Space

For $h > 0$ let $C = C([-h, 0), E^n)$ be the Banach space of continuous functions defined on $[-h, 0]$ with values in $E^n$ equipped with the sup norm. If $x$ is a function on $[-h, T]$, define a Banach space $C([-h, T], E^n)$, analogously, where $T > h$, and for each $t \in [0, T]$ let $x_t \in C$ be defined by $x_t(s) = x(t + s)$, $-h \leq s \leq 0$.

Consider the following problem:

Minimize

$$J(u) = \int_0^T f^0(t, x_t, u(t))dt, \qquad (7.11.1)$$

subject to the constraints

$$\dot{x}(t) = (t, x_t, u(t)) \quad \text{a.e. on } [0, T], \tag{7.11.2}$$

$$x_0 = \phi \in W_\infty^{(1)}([-h, 0], E^n),$$

$$x_T = \psi \in C^{(1)}([-h, 0], E^n), \tag{7.11.3}$$

$$u(t) \in C^m \quad \text{a.e. on } [0, T], \tag{7.11.4}$$

where

$$C^m = \{u \in E^m : |u_j| \leq 1, \; j = 1, \ldots, m\},$$

$$f_i^{(1)}(t, x_t) \leq 0, \quad i = 1, \ldots, p, \quad t \in [0, T], \tag{7.11.5}$$

and

$$x \in W_\infty^1([-h, T], E^n), \; u \in L_\infty([0, T], E^m). \tag{7.11.6}$$

For this problem we assume that $W_\infty^{(1)}([-h, T], E^n)$ is the state space of absolutely continuous $n$-vector valued functions on $[-h, T]$ with essentially bounded derivatives, and $L_\infty([0, T], E^m)$ is the set of functions that are essentially bounded $m$-vector valued functions. Also, $\phi \in C^1([-h, 0], E^n)$ if $\phi$ is continuously differentiable. The functions $f^0, f, f^1$ are mappings

$$f^0 : [0, T] \times C \times E^m \to E,$$

$$f : [0, T] \times C \times E^m \to E^n,$$

$$f^1 : [0, T] \times C \to E^p,$$

which we assume to satisfy the following basic conditions: $f, f^0$ are continuous, Fréchet-differentiable with respect to their second argument, and continuously differentiable with respect to their third argument, the derivative being assumed continuous with respect to all arguments. Also $f^1 : [0, T] \times C \to E^p$ is continuous and continuously Fréchet-differentiable with respect to its second argument. Denote by $NBV_{k([a,b])}$ and $NBV_{k \times p(([a,b])}$ respectively the $k$-vector and the $k \times p$-matrix valued functions whose components are of bounded variation left continuous at each point of $[a, b]$ and normalized to zero at $t = b$. Associated with (7.11.2) is the linear equation

$$\dot{x}(t) = L(t, x_t) + B(t)u(t), \qquad 0 \leq t \leq T, \tag{7.11.7}$$

$$x_0 = 0, \tag{7.11.8}$$

where $u \in U$, $x \in X$, with these subspaces $U, X$ defined as follows:

$$X = \{x \in W_\infty^{(1)}([-h, T], E^n) : x_T \in C^1([-h, 0], E^n)\},$$

$$U = \{u \in L_\infty([0, T], E^m) : u_T \in C([-h, 0], E^m)\}.$$

In (7.11.7),

$$L(t, x_t) = [f_\phi(t, x_t^*, u^*(t))]x_t, \qquad B(t) = f_u(t, x_t^*, u^*(t))$$

are the Fréchet derivatives with respect to $\phi$ and $u$ respectively. Thus there exists $\eta(t, \cdot) \in NBV_{n \times n}([-h, 0])$ such that $L(t, x_t) = \int_{-h}^0 d_\theta \eta(t, \theta) x(t + \theta)$. Angell and Kirsch [45] have proved the following fundamental result:

**Theorem 7.11.1**

   (i) *Let $(x^*, u^*)$ be the optimal solution pair of the problem (7.11.1)–(7.11.6).*
   (ii) *Assume that all the smoothness conditions are satisfied.*
   (iii) *Suppose the map $t \to B^+(t)$ is continuous on $[T - h, T]$, where $B^+(t)$ is the generalized inverse of $B(t)$, and $\mathrm{rank}\, B(t) = n$, $\forall\, t \in [T - h, T]$.*

   *Then there exists*

$$(\lambda, a, \rho, v, q) \in E \times L_\infty([0, T], E^n) \times NBV_p[0, T] \times NBV_n[0, T] \times E^n,$$

$$(\lambda, a, \rho, v, q) \not\equiv (0, 0, 0, 0, 0),$$

*such that $\lambda \geq 0$, $\rho = (\rho_1, \ldots, \rho)$, $\rho_j$ is nondecreasing on $[0, T]$, $\rho_j$ is constant on every interval where $f_j^{(1)}(t, x_t^*) < 0$,*

$$a(t) = \lambda \int_t^T \eta_0(s, t - s)ds + \int_t^T \eta_1(s, t - s)^T d\rho(s)$$

$$- q - \int_t^T \eta(s, t - s)^T a(s)ds + \int_t^T \eta(s, t - s)^T dv(s - T), \quad \forall\, t \in [0, T],$$

$$\tag{7.11.9}$$

*and*

$$\lambda \int_0^T B_0^T(t)u(t)dt + \int_0^T a^T(t)B(t)u(t)dt - \int_{T-h}^T dv(t - h)^T B(t)u(t)$$

$$\leq \lambda \int_0^T B_0^T(t)u^*(t)dt + \int_0^T a^T(t)B(t)u^*(t)dt$$

$$- \int_{T-h}^T dv^T(t - h)B(t)u^*(t) \tag{7.11.10}$$

*for all $u \in U_{ad}$. The multiplier $\lambda$ can be taken to be 1. We now state a Kuhn–Tucker type necessary optimality theorem on which the maximum principle is based. The notations are first stated.*

**Definitions**   Let $X$ be a real Banach space.

(i) $Y \subset X$ is a linear subspace if $x, y \in Y$, $\lambda, \mu \in E$ implies $\lambda x + \mu y \in Y$.

(ii) $Z \subset X$ is an affine manifold if $x, y \in Z$, $\lambda \in E$ implies $(1 - \lambda)x + \lambda y \in Z$.

(iii) $H \subset X$ is convex if $x, y \in H$, $\lambda \in [0, 1]$ implies $(1 - \lambda)x + \lambda y \in H$.

(iv) $M \subset X$ is a cone (with vertex at 0) if $x \in M$, $\lambda \geq 0$ implies $\lambda x \in M$.

**Definition 7.11.1**   Let $X$ be a real Banach space and $S \subset X$ a subset. The

(i) linear span of $S$, denoted by span $(S)$, is the smallest linear subspace of $X$ that contains $S$.

(ii) the affine hull (or affine span) of $S$ is the smallest affine manifold containing $S$.

**Definition 7.11.2**   Let $X$ be a real Banach space. A transformation $f : X \to E$ is a functional. The space of all bounded linear functionals $f : X \to E$ is called the dual of $X$ and is denoted by $X^*$. We recall that $X^*$ is a Banach space with norm

$$\|f\| = \sup_{\|x\| \leq 1} |f(x)|.$$

Let $X$ be a real Banach space with $X^*$ as dual, and let $Y \subset X$. The interior of $Y$ relative to $X$ will be denoted as $Y^0$, while its interior relative to its closed affine hull will be denoted by $Y^{00}$. Let $Z \subset X$ be a convex subset of $X$. Define

$$Z^+ = \{\ell \in X^* : \ell(x) \geq 0, \ \forall\, x \in X\}.$$

Let $f : X \to Y$ be a map of the Banach space $X$ into the Banach space $Y$. The Fréchet derivative of $f$ at $x^* \in X$ is denoted by $Df(x^*)$. We use $D_i f(x)$ to denote the partial Fréchet derivative of $f$ with respect to the $i$th variable.

**Theorem 7.11.2**   *Let $\mathbb{R}$, $Z_1$, and $Z_2$ be Banach spaces, $V \subset \mathbb{R}$ a closed convex set with nonempty interior relative to its closed affine hull $W$, and let $Y \subset Z_1$ be a closed convex cone with vertex at the origin and having a nonempty interior in $Z_1$. Let*

$$g^0 : \mathbb{R} \to E, \quad g^1 : \mathbb{R} \to Z_1, \quad g^2 : \mathbb{R} \to Z_2,$$

*and let*

$$M = \{x \in V : g^1(x) \in -Y, \ g^2(x) = 0\}.$$

*Suppose that $g^0$ takes on a local minimum at $x^* \in M$, and that $g^1$, $g^1$, and $g^2$ are continuously Fréchet differentiable at $x^*$. Suppose that $Dg^2(x^*)W$ is closed in $Z_2$, but either of the following two conditions fails:*

(i) *There exists an $x^0 \in V^{00}$ with $Dg^2(x^*)x^0 = 0$ and*

(ii) *$g^1(x^*) + Dg^1(x^*)x^0 \in -Y^0$.*

*Also*

(iii) $Dg^2(x^*)(W) = Z_2$.

*Then there exists a nontrivial triple $(\lambda, \ell_1, \ell_2) \in E \times Z_1^* \times Z_2^*$ such that*

(i) $\lambda \geq 0$, $\ell_1 \in Y^+$,

(ii) $\ell_0 \circ g^1(x^*) = 0$,

(iii) $\lambda Dg^0(x^*)x + \ell_1 Dg^1(x^*)x + \ell_2 Dg^2(x^*)x \geq 0$, $\forall\, x \in V - x^*$.

To apply the multiplier rule to the optimal problem, we define $Z_1$, $Z_2$, as follows:

$$Z_1 = C([0,T], E^p),$$

$$Z_2 = \{(z, v, \beta) \in L_\infty([0,T], E^n) \times C([-h, 0], E^n) \times E^n : z_T \in C\}.$$

Also

$$\mathbb{R} = X \times U, \quad W = \{(x, u) \in \mathbb{R} \mid x_0 = \phi\}, \quad V = \{(x, u) \in W \mid u \in U\}.$$

Consider the three mappings $g^0 : X \times U \to E$, $g^1 : X \times U \to Z_1$, and $g^2 : X \times U \to Z_2$ defined respectively by

$$g^0(x, u) = \int_0^T f^0(t, x_t, u(t))dt, \tag{7.11.11}$$

$$g^1(x, u) = f^1(\cdot, x_{(\cdot)}), \tag{7.11.12}$$

$$g^2(x, u) = (\dot{x}() - f(\cdot, x_{(\cdot)}, u(\cdot)),\ \dot{x}_T - \dot{\psi},\ x(T) - \psi(T)). \tag{7.11.13}$$

Under the basic prevailing assumptions on $f^0$, $f^1$, $f$, the mappings $g^0$, $g^1$, $g^2$ are continuously Fréchet differentiable at each point $(x^*, u^*) \in X \times U$, and the derivatives are of the forms:

$$D_x g^i(x^*, u^*)x = \int_0^T L_i(t)x_t dt, \qquad x \in X, \qquad i = 0, 1,$$

where

$$L_i(t)x_t = [f_\phi^i(t, x_t^*, u^*(t))]x_t, \qquad x_t \in C,$$

$$D_u g^i(x^*, u^*)u = \int_0^T B_i(t)u(t)dt, \qquad u \in U,$$

where $B_i(t) = [f_u^i(t, x_t^*, u^*(t))$. Here

$$D_u g^1(x^*, u^*) = 0.$$

Also

$$D_x g^2(x^*, u^*)x = (\dot{x}(\cdot) - L^*(\cdot)x_{(\cdot)}, \dot{x}_T, x(T)), \qquad x \in X,$$

where

$$L^*(t)y = [f_\phi(t, x_t^*, u^*(t))]y\,, \quad y \in C\,,$$

$$D_u g^2(x^*, u^*)u = (-B(\cdot)u(\cdot), 0, 0)\,, \quad u \in U\,;$$

where

$$B(t) = f_u(t, x_t^*, u^*(t))\,.$$

Now set

$$\mathbb{R} = X \times U\,, \qquad W = \{(x, u) \in \mathbb{R} \,|\, x_0 = \phi\}\,,$$

$$V = \{(x, u) \in W \,|\, u \in U_{ad}\}\,.$$

We note that $g^i$, $i = 0, 1, 2$, satisfy the conditions of the multiplier rule of Theorem 7.11.2 relative to the spaces $X, U, Z_1$, and $Z_2$. Denote by $L_\infty^c = \{x \in L_\infty([0, T], E^n) : x_T \in C\}$ and observe that by Theorem 7.11.2 there exist $\ell \in [L_\infty^c]^*$, the dual of $L_\infty^c$, $v \in NBV_n([-h, 0])$, $q \in E^n$, $\rho \in NBV_p([0, T]$, and $\lambda \geq 0$, $(\lambda, \ell, v, q, \rho) \neq (0, 0, 0, 0, 0)$ such that $\rho$ is nondecreasing and

$$\int_0^T d\rho(t)(t)^T f^1(t, x^*(t)) = 0\,, \tag{7.11.14}$$

$$\lambda \int_0^T L_0^*(t)x_t dt + \int_0^T dp^T(t)L_1^*(t)x_t + \ell(\dot{x}(\cdot) - L_2^*(\cdot)x_{(\cdot)}]$$

$$+ \int_{T-h}^T dv(t-T)^T \dot{x}(t) + q^T x(t) = 0\,, \tag{7.11.15}$$

$\forall\, x \in X$ with $x_0 = 0$, and

$$\lambda \int_0^T B_0^*(t)^T u(t)dt + \ell[-B_2^*(\cdot)u(\cdot)] \geq 0\,, \qquad \forall\, u \in U_{ad} - u^*\,. \tag{7.11.16}$$

That the map $D_x g^2(x^*, u^*)$, $(x, u) \to (\dot{x}(\cdot) - L_2^* \cdot x(\cdot) - B_2^*(\cdot)u(\cdot)\dot{x}_T, x(T))$ is a surjection is a consequence of condition (iii), which is the criteria for the controllability of (7.11.7) in the space $C^1([-h, 0], E^n)$. The proof is essentially contained in Theorem 6.2.2. We now affirm that (7.11.14)–(7.11.16) hold. The function $v$ is defined on $[-h, 0]$. Outside this interval, we let $v(\theta) = \begin{cases} v(-h)\,, & 0 < -h \\ 0 \end{cases}$. We claim that the functional $\ell$ is of the form

$$\ell(z) = \int_0^T a^T(t)z(t)dt - \int_{T-h}^T dv(t-T)^T z(t)\,,$$

for $z \in L_\infty^c$, $a \in L_\infty([0,T])$. Indeed, consider (7.11.15). Let $z \in L_\infty^c$. The solution of

$$\dot{x}(t) - L_2(t)x_t = z(t) \text{ in } [0,T], \qquad x_0 = 0 \tag{7.11.17}$$

is given by

$$x(t,z) = \int_0^t X(t,s)z(s)ds, \qquad t \in [0,T],$$

where $X$ is the fundamental matrix solution of

$$\dot{x}(t) = L_2(t)x_t.$$

Thus the system (7.11.15) becomes

$$\lambda \int_0^T L_0(t)x_t(\cdot,z)dt + \int_0^T d\rho^T(t)L_1(t)x_t(\cdot,z) + \ell(z) + \int_{T-h}^T dv(t-T)^T z(t)$$

$$+ \int_{T-h}^T dv(t-T)^T L_2(t)x_t(\cdot,z) + q^T x(T,z) = 0.$$

If we use the representation of $L_i$ and of $x(\cdot,z)$ and change the order of integration, then we deduce that

$$\ell(z) = \int_0^T a(t)^T z(t)dt - \int_{T-h}^T dv(t-T)^T z(t),$$

for all $z \in L_\infty^c$, where $a(t)$ depends on $\lambda$, $q$, $X$, $\rho$, $v$, $\eta_0$, $\eta_1$, $\eta_2$. With this we rewrite (7.11.15) as follows:

$$\lambda \int_0^T L_0(t)x_t(\cdot,z)dt + \int_0^T d\rho(t)^T L_1(t)x_t + \int_0^T a^T(t)[\dot{x}(t) - L_2(t)x_t]dt$$

$$- \int_{T-h}^T dv(t-T)[\dot{x}(t) - L_2(t)x_t] + \int_{T-h}^T dv(t-T)^T[\dot{x}(t)]$$

$$+ q^T x(t) = 0,$$

for all $x \in X$ with $x_0 = 0$. This is the same as

$$\lambda \int_0^T \int_{t-h}^t d_\theta\eta_0(t,\theta-t)^T x(\theta)dt + \int_0^T \int_{t-h}^t d\rho(t)^T d_\theta\eta_1(t,\theta-t)x(\theta)$$

$$+ \int_0^T a(t)^T[\dot{x}(t) - \int_{t-h}^t d\theta\eta_2(t,\theta-t)x(\theta)]dt$$

$$+ \int_{T-h}^T \int_{t-h}^t dv(t-T)^T d_\theta\eta_2(t,\theta-t)x(\theta)$$

$$+ q^T x(T) = 0,$$

for all $x \in X$ with $x_0 = 0$. We now integrate by parts to obtain

$$-\lambda \int_0^T \eta_0(t, -h)^T x(t - h) - \lambda \int_0^T \int_{t-h}^t \eta_0(t, s - t)\dot{x}(s)ds dt$$

$$- \int_0^T d\rho(t)^T \eta_1(t, -h)x(t - h) - \int_0^T \int_{t-h}^t d\rho(t)^T \eta_1(t, s - t)\dot{x}(s)ds$$

$$+ \int_0^T a(t)^T [\dot{x}(t) + \eta_2(t, s - t)\dot{x}(s)ds]dt + \int_0^T a(t)^T \eta_2(t, -h)x(t - h)dt$$

$$+ \int_{T-h}^T dv(t - T)^T \eta_2(t, -h)x(t - h) - \int_{T-h}^T \int_{t-h}^t dv(t - T)^T \eta_2(t, s - t)\dot{x}(s)ds.$$

Now define a function $\mu$ as follows:

$$\mu(t) = \begin{cases} -\displaystyle\int_t^T a(s)ds, & 0 \le t \le T - h, \\ -v(t - T) - \displaystyle\int_t^T a(s)ds, & T - h \le t \le T, \end{cases}$$

and for any $y \in L_\infty^c$ extend $y$ by zero and then define

$$x(t) = \int_0^t y(s)ds \quad \text{for} \quad t \in E.$$

Use these definitions and change the order of integration to deduce

$$-\lambda \int_0^T \int_s^{T-h} \eta_0(\theta + h, -h)^T d\theta y(s)ds - \lambda \int_0^T \int_s^{s+h} \eta_0(\theta, s - \theta)^T d\theta y(s)ds$$

$$- \int_0^T \int_s^{T-h} d\rho(\theta)^T \eta_1(\theta + h, -h)y(s)ds$$

$$- \int_0^T \int_s^{s+h} d\rho(\theta)^T \eta_1(\theta, s - \theta)^T d\theta y(s)ds$$

$$+ \int_0^T a(t)^T y(t)dt + \int_0^T \int_s^{s+h} d\mu(\theta)^T \eta_2(\theta, s - \theta)y(s)ds$$

$$+ \int_0^T \int_s^{T-h} d\mu(\theta + h)^T \eta_2(\theta + h, -h)y(s)ds + q \int_0^T y(s)ds = 0, \qquad (7.11.18)$$

for all $y \in L^c_\infty([0,T])$. Because the equality in (7.11.18) holds for all $y$, the integrand has to vanish pointwise. As a consequence we have

$$-\lambda \int_s^{T-t} \eta_0(\theta + h, -h)^T ds - \lambda \int_0^{s+h} \eta_0(\theta, s - \theta)^T d\theta$$

$$- \int_s^{T-h} d\rho(\theta)^T \eta_1(\theta + h, -h) - \int_0^{s+h} d\rho(\theta)^T \eta_1(\theta, s - \theta)$$

$$+ a(s)^T + \int_s^{s+h} d\mu(\theta)^T \eta_2(\theta, s - \theta)$$

$$+ \int_s^{T-h} d\mu(\theta + h)^T \eta_2(\theta + h, -h) + q^T = 0 \,,$$

which yields

$$-\lambda \int_s^T \eta_0(\theta, s - \theta) d\theta - \int_s^T d\rho(\theta)^T \eta_1(\theta, s - \theta) + a(s)^T$$

$$+ \int_s^T a(\theta)^T \eta_2(\theta, s - \theta) d\theta$$

$$- \int_s^T dv(\theta)^T \eta_2(\theta, s - \theta) + q^T = 0 \,.$$

This proves (7.11.9). To show that (7.11.10) is also valid, we substitute the form of $\ell$ into (7.11.16).

We observe that $\rho = (\rho_1, \ldots, \rho_\ell)$ $\rho_j$ is nondecreasing on $[0,T]$. Because $(\lambda, \ell, v, q, \rho) = (0, 0, 0, 0, 0)$ is impossible, $(\lambda, a, p, v, q)$ cannot vanish simultaneously. The proof is complete.

As an immediate consequence of this, we state in the next section the solution of the time-optimal control problem.


## 7.12    The Time-Optimal Problem in Function Space

**Theorem 7.12.1    (The Time-Optimal Problem)** *Consider the following problem: Minimize $T$ subject to the constraints*

$$\dot{x}(t) = f(t, x_t, u(t)) \quad \text{a.e. on } [0,T] \,,$$

$$x_0 = \phi \in W^1_\infty([-h, 0], E^n) \,,$$

$$x_T = \psi \in C^1([-h, 0], E^n) \,, \tag{7.12.1}$$

$$u(t) \in C^m \quad \text{a.e. on } [0,T] \,,$$

*where*

$$C^m = \{u \in E^m : |u_j| \le 1, \ j = 1, \ldots, m\}\,.$$

(i) *We assume that $f$ is continuous, and is Fréchet differentiable with respect to its second argument and continuously differentiable with respect to its third argument, the derivative being assumed continuous with respect to all arguments. Associated with (7.12.1) is the linear equation*

$$\dot{x}(t) = L(t, x_t) + B(t)u(t)\,, \quad 0 \le t \le T\,, \quad x_0 = 0\,, \tag{7.12.2}$$

*where $u \in U$, $x \in X$ where these subspaces are defined as follows:*

$$X = \{x \in W_\infty^{(1)}([-h, T], E^n),\ x_T \in C^1([-h, 0], E^n)\}\,,$$

$$U = \{u \in L_\infty([0, T], E^m),\ u_T \in C([-h, 0], E^m)\}\,.$$

*In (7.12.2),*

$$L(t, x_t) = [f_\phi(t, x_t^*, u^*(t))]x_t\,, \qquad B(t) = f_u(t, x_t^*, u^*)\,,$$

*are the Fréchet derivatives with respect to $\phi$ and $u$ respectively. As usual,*

$$L(t, x_t) = \int_{-h}^{0} d_\theta \eta(t, \theta) x(t + \theta)\,.$$

(ii) *Suppose the map $t \to B^+(t)$ is continuous on $[T - h, T]$, where $B^+(t)$ is the generalized inverse of $B(t)$, and*

$$\operatorname{rank} B(t) = n\,, \qquad \forall\, t \in [T - h, T]\,.$$

*Then there exists*

$$(a, v, q) \in L_\infty([0, T], E^n) \times NBV_n([0, T] \times E^n\,, \quad (a, v, q) \ne (0, 0, 0)\,,$$

*such that*

$$a(t) = -q - \int_t^T \eta(s, t - s)^T a(s)\,ds + \int_t^T \eta(s, t - s)^T dv(s - T), \forall\, t \in [0, T]\,,$$

*and*

$$\int_0^T a^T(t)B(t)u(t)\,dt - \int_{T-h}^T dv(t - T)^T B(t)u(t)$$

$$\le \int_0^T a^T(t)B(t)u^*(t) - \int_{T-h}^T dv(t - T)^T B(t)u^*(t)\,,$$

*for all $u \in U_{ad}$.*

## References


1. N. Dunford and J. T. Schwartz, *Linear Operators Part I*, Interscience, New York, 1957.
2. H. O. Fattorini, "The Time Optimal Control Problem in Banach spaces", *Applied Mathematics and Optimization* **1** (1974) 163–168.
3. R. Gabasov and F. Kirillova, *The Qualitative Theory of Optimal Processes*, Marcel Dekker, New York, 1976.
4. H. Hermes and J. P. LaSalle, *Functional Analysis and Time Optimal Control*, Academic Press, New York, 1969.
5. E. B. Lee and L. Markus, *Foundations of Optimal Control Theory*, John Wiley and Sons, New York, 1967.
6. R. B. Zmood and N. H. McClamrock, "On the Pointwise Completeness of Differential-Difference Equations," *J. Differential Equations* **12** (1972) 474–486.
7. C. Castillo-Chavez, K. Cooke, W. Huang, and S. A. Levin, "The Role of Long Periods of Infectiousness in the Dynamics of Acquired Immunodeficiency Syndrome," in C. Castillo-Chavez, S. A. Levin, C. Shoemaker (eds.) "Mathematical approaches to Resource Management and Epidemiology (*Lect. Notes Biomath*), Springer-Verlag, Berlin, Heidelberg, New York, 1989.
8. J. Hale, *Theory of Functional Differential Equations*, Springer-Verlag, New York, NY, 1977.
9. E. N. Chukwu, *The Time Optimal Control of Nonlinear Delay Systems in Operator Methods for Optimal Control Problems*, Edited by S. J. Lee, Marcel Dekker, Inc., New York, 1988.
10. O. Hájek, "Geometric Theory of Time Optimal Control," *SIAM J. Control* **9** (1971) 338–350.
11. G. Tadmor, "Functional Differential Equations of Retarded and Neutral Type: Analytic Solutions and Piecewise Continuous Controls," *J. Differential Equations* **51** (1984) 151–181.
12. S. Nakagiri, "On the Fundamental Solution of Delay-Differential Equations in Banach Spaces," *Journal of Differential Equations* **41** (1981) 349–368.
13. E. N. Chukwu, *The Time Optimal Control Theory of Functional Differential Equations Applicable to Mathematical Ecology*, Proceedings of the Conference on Mathematical Ecology–International Center for Theoretical Physics, Trieste, Italy, edited by T. Hallam and L. Gross.
14. E. N. Chukwu, "Global Behaviour of Retarded Linear Functional Differential Equations", *J. Mathematical Analysis and Applications* **162** (1991) 277–293.
15. E. N. Chukwu, "Function space null-controllability of linear delay systems with limited power," *J. Mathematical Analysis and Applications* **121** (1987) 293–304.
16. A. Manitius and H. Tran, "Numerical Simulation of a Nonlinear Feedback Controller for a Wind Tunnel Model Involving a Time Delay," *Optimal Control Applications and Methods* **7** (1986) 19–39.
17. R. Bellman, I. Glicksberg and O. Gross, "On the 'Bang-Bang' Control Problems," *Quarterly of Applied Mathematics* **14** (1956) 11–18.
18. H. T. Banks and M. Q. Jacobs, "The Optimization of Trajectories of Linear Functional Differential Equations," *SIAM J. Control* **8** (1970) 461–488.
19. H. J. Sussman, "Small-Time Local Controllability and Continuity of the Optimal Time Function of Linear Systems," *J. Optimization Theory and Applications* **53** (1987) 281–296.
20. E. N. Chukwu and O. Hájek, "Disconjugacy and Optimal Control," *J. Optimization*

*Theory and Applications* **27** (1979) 333–356.

21. H. T. Banks and G. A. Kent, "Control of Functional Differential Equations of Retarded and Neutral Type to Target Sets in Function Space," *SIAM J. Control* **10** (1972) 567–594.

22. H. T. Banks, M. Q. Jacobs, and C. E. Langenhop, "Characterization of the Controlled States in $W_2^{(1)}$ of Linear Heredity Systems," *SIAM J. Control* **13** (1975) 611–649.

23. E. J. McShane and R. B. Warfield, "On Filippov's Implicit Function Lemma," *Proc. Amer. Math. Soc.* **18** (1967) 41–47.

24. O. Hájek, "On Differentiability of the Minimal Time Functions," *Funkcialaj Ekvacioj* **20** (1976) 97–114.

25. G. Tadmor, "Functional Differential Equations of Retarded and Neutral Type: Analytic Solutions and Piecewise Continuous Controls," *J. Differential Equations* **51** (1984) 151–181.

26. L. E. El'sgol'ts and S. B. Norkin, *Introduction to the Theory and Application of Differential Equations with Deviating Arguments*, Academic Press, New York, 1973.

27. G. Stépan, *Retarded Dynamical Systems: Stability and Characteristic Functions*, Longman Scientific and Technical, New York, 1991.

28. H. T. Banks and J. A. Burns, "Hereditary Control Problems: Numerical Methods on Averaging Approximations," *SIAM J. Control* **16** (1978) 169–208.

29. H. T. Banks and K. Ito, "A Numerical Algorithm for Optimal Feedback Gains in High Dimensional Linear Quadratic Regulator Problems," *SIAM J. Control and Optimization*, **29** (1991) 499–511.

30. E. N. Chukwu, "Time Optimal Control of Delay Differential System in Euclidean Space," Preprint.

31. O. Hájek, "Terminal Manifold and Switching Locus," *Mathematical Systems Theory* **6** (1973) 289–301.

32. D. S. Yeung, "Synthesis of Time-Optimal Controls," Case Western Reserve University, Ph.D Thesis, 1974.

33. D. S. Yeung, "Time-Optimal Feedback Control," *J. Optimization Theory and Applications* **21** (1977) 71–82.

34. H. A. Antosiewicz, "Linear Control Systems," *Arch. Rat. Mech. Anal.* **12** (1963) 313–324.

35. O. Hájek, "$L_1$-Optimization in Linear Systems with Bounded Controls," *J. Optimization Theory and Applications* **29**(3) (1979) 409–432.

36. O. Hájek and W. Krabs, "On a General Method for Solving Time-Optimal Linear Control Problems," Preprint No. 579, Technische Hochschule Darmstadt, Fachbereich Mathematik, Jan. 1981.

37. A. Manitius, "Optimal Control of Hereditary Systems," in *Control Theory and Topics in Functional Analysis, Vol. III*, International Center for Theoretical Physics, Trieste International Atomic Energy Agency, Vienna, 1976.

38. L. W. Neustadt, "Minimum Effort Control Systems," *SIAM J. Control* **1** (1962) 16–31.

39. W. Rudin, *Real and Complex Analysis*, McGraw-Hill, New York, 1974.

40. R. G. Underwood and D. F. Young, "Null Controllability of Nonlinear Functional Differential Equations," *SIAM Journal of Control and Optimization* **17** (1979) 753–768.

41. A. F. Filippov, "On a Certain Question in the Theory of Optimal Control," *SIAM J. Control* **1** (1962) 76–84.

42. O. Hájek, "Control Theory in the Plane," Springer-Verlag Lecture Notes in Control and Information Sciences, New York, 1991.

43. T. S. Angell, "Existence Theorems for a Class of Optimal Problems with Delay,"

University of Michigan, Ann Arbor, Michigan, Doctoral Dissertation, 1969.

44. T. S. Angell, "Existence Theorems for Optimal Control Problems Involving Functional Differential Equations," *J. Optimization Theory and Applications* **7** (1971) 149–169.

45. T. S. Angell and A. Kirsch, "On the Necessary Conditions for Optimal Control of Retarded Systems," *Appl. Math. Optim.* **22** (1990) 117–145.

46. H. T. Banks, "Optimal Control Problems with Delay," Purdue, Doctoral Dissertation, 1967.

# Chapter 8

# Controllable Nonlinear Delay Systems

## 8.1  Controllability of Ordinary Nonlinear Systems

Consider the autonomous linear control system

$$\dot{x}(t) = Ax(t) + Bu(t), \quad x(0) = x_0, \tag{8.1.1}$$

with controls-measurable functions whose values $u(t)$ lie on the $m$-dimensional cube

$$C^m = \{u \in E^m : |u_j| \le 1, \ j = 1, \ldots, m\}. \tag{8.1.2}$$

The solution of (8.1.1) is given by

$$x(t, x_0; u) = e^{At}\left[x_0 + \int_0^t e^{-As}Bu(s)ds\right]. \tag{8.1.3}$$

**Definition 8.1.1**  The attainable set of (8.1.1) is the subset $\mathcal{A}(t, x_0) = \{x(t, x_0, u) : u$ measurable $u(t) \in C^m$, $x$ is a solution of (8.1.1)$\}$ of $E^n$. If $x_0 = 0$, $\mathcal{A}(t, x_0) \equiv \mathcal{A}(t)$.

**Definition 8.1.2**  The domain $\mathcal{C}$ of (Euclidean) null controllability of (8.1.1) is the set of all initial points $x_0 \in E^n$, each of which can be steered to 0 in the same finite $t_1$ with $u$ measurable $u(t) \in C^m$, $t \in [0, t_1]$:

$$\mathcal{C} = \{x_0 \in E^n : x(t_1, x_0, u) = 0 \text{ for some } t_1 \text{ some } u \in C^m\}.$$

**Definition 8.1.3**  If

$$0 \in \text{Int } \mathcal{C}, \tag{8.1.4}$$

then (8.1.1) is locally (Euclidean) null controllable with constraints.

**Lemma 8.1.1**  *Assume (8.1.1) is controllable, and this holds if and only if*

$$\text{rank}[B, AB, \ldots, A^{n-1}B] = n. \tag{8.1.5}$$

289

*Then* $0 \in \text{Int } C$.

**Proof**    Consider the mapping

$$u \to x(t,0,u) = e^{At} \int_0^t e^{-As} Bu(s)ds \,,$$

$$T \ : \ L_\infty([0,t], E^m) \to E^n \,, \quad Tu = x(t,0,u)\,.$$

Equation (8.1.1) is controllable if and only if

$$T(L_\infty([0,t], E^m)) = E^n \,.$$

But $T$ is a bounded linear map since $u \to Tu$ is continuous. By the open mapping [5, pp. 99] $T$ is an open map. Therefore if $U$ is an open ball such that $U \subset L_\infty([0,t], C^m)$, then $T(U)$ is open and $T(U) \subset T(L_\infty([0,t], C^m)) = \mathcal{A}(t)$. But $0 \in T(U) \subset \mathcal{A}(t)$. As a consequence,

$$0 \in \text{Int } \mathcal{A}(t)\,. \tag{8.1.6}$$

It readily follows from (8.1.6) that (8.1.4) is valid. Indeed, assume that 0 is not contained in Int $C$. Then there is a sequence $\{x_{0n}\}_1^\infty$, $x_{0n} \in E^n$, $x_{0n} \to 0$ as $n \to \infty$, and no $x_{0n}$ is in $C$. The trivial solution is a solution of (8.1.1), so that $0 \in C$. Hence $x_{0n} \neq 0$ for any $n$. We also have that $0 \neq x(t, x_{0n}, u)$ for any $t > 0$ and any $u \in L_\infty([0,t], C^m)$. Thus

$$y_n \equiv -e^{At}x_{0n} \neq e^{At} \int_0^t e^{-As} Bu(s)ds$$

for any $n$, $t > 0$ and any $u \in L_\infty([0,t], C^m)$. It follows that $y_n \notin$ in $\mathcal{A}(t)$ for any $n$. But $y_n \to 0$ as $n \to \infty$. Thus the sequence $\{y_n\}_1^\infty$ has the property: $y_n \to 0$ as $n \to \infty$, $y_n \notin \mathcal{A}(t)$ for any $t > 0$, $y_n \neq 0$ for any $n$. This means $0 \notin \text{Int } \mathcal{A}(t)$, a contradiction.      $\square$

**Remark 8.1.1**   The necessity and sufficiency of the rank condition (8.1.5) for controllability is old and due to Kalman, (see [2, p. 74]).

The statement in Lemma 8.1.1 can be generalized to nonlinear systems

$$\dot{x}(t) = f(t, x(t), u(t))\,, \tag{8.1.7}$$

where $f : E \times E^n \times E^m \to E^n$ is continuous and in the second and third argument is continuously differentiable. We assume all solutions $x(t, x, u)$ of (8.1.7) exist and $(t, x_0, u) \to x(t, x_0, u)$ are continuous. The assumptions on $f$ ensure this.

In addition to (8.1.7), consider the linearized system

$$\dot{x}(t) = A(t)x(t) + B(t)u(t)\,, \tag{8.1.8}$$

where

$$D_2 f(t,0,0) = A(t)\,, \quad D_3 f(t,0,0) = B(t)\,. \tag{8.1.9}$$

**Theorem 8.1.1**  *Consider* (8.1.7) *in which* $f : E \times E^n \times E^m \to E^n$

(i)  *is continuous, and continuously differentiable in the second and third arguments.*

(ii)  $f(t, 0, 0) = 0, \ \forall \ t \geq 0.$

(iii)  *System* (8.1.8) *with* $A(t)$ *and* $B(t)$ *given by* (8.1.9) *is Euclidean controllable on* $[0, t_1]$.

*Then the domain* $\mathcal{C}$ *of null controllability of* (8.1.7) *has* $0 \in \text{Int } \mathcal{C}$.

**Proof**  The solution of

$$\dot{x}(t) = f(t, x(t), u(t)), \quad x(0) = 0 \tag{8.1.10}$$

is given by the integral equation

$$x(t, u) = \int_0^t f(s, x(s), u(s))ds \,.$$

Consider the mapping

$$T : L_\infty([0, t_1], E^m) \to E^n$$

defined by

$$Tu = x(t, u) \,.$$

Because of the smoothness assumptions on $f$, the partial derivative of $x(t, u)$ with respect to $u$ is given by

$$D_u x(t, u) = \int_0^t D_2 f(s, x(s, u), u(s)) D_u x(s, u) ds$$

$$+ \int_0^t D_3 f(s, x(s, u), u(s)) ds \,, \quad t \geq 0 \,.$$

On differentiating this with respect to $t$, we have

$$\frac{d}{dt} D_u x(t, u) v = D_2 f(t, x(t, u), u(t)) D_u x(t, u) v$$

$$+ D_3 f(t, x(t, u), u(t)) v \,. \tag{8.1.11}$$

Because (ii) holds, the function $x(t, 0, 0) = 0$ is a solution of (8.1.10). We deduce from these calculations that $T'(0)v = D_u x(t, 0)v = z(t, v)$ is a solution of (8.1.8). Therefore $T'(0) : L_\infty([0, t_1], E^m) \to E^n$ is a surjection if and only if the system (8.1.8) with $z(0, v) = 0$ is controllable on $[0, t_1]$. It follows from Graves' Theorem [3, p. 193] that $T$ is locally open: There is an open ball radius $\rho$, $\mathbb{B}_\rho \subset L_\infty([0, t_1], C^m)$ center 0, and an open ball radius $r$ center 0, $\mathbb{B}_r \subset E^n$, such that

$$B_r \subset T(\mathbb{B}_\rho) \subset T(L_\infty([0, t_1], C^m)) \,,$$

since

$$T(L_\infty([\sigma, t_1], C^m)) = \mathcal{A}(t),$$

where

$$\mathcal{A}(t) = \{x(t, u) : \ u \in L_\infty([0, t_1], C^m) \ x \text{ is a solution of } (8.1.10)$$

$$\text{with } x(0, u) = 0\}.$$

We have proved that $0 \in \text{Int } \mathcal{A}(t)$. With this result one proves, as in the linear case, that $0 \in \text{Int } C$. The proof is complete. $\qquad\square$

Theorem 8.1.1 is a statement on the constrained local controllability of (8.1.7). It is interesting to explore conditions that will yield the global result of $C = E^n$. Such a statement is contained in the next theorem.

**Theorem 8.1.2**   *For* (8.1.7),

   (i)  *Assume that hypotheses* (i)–(iii) *of Theorem 8.1.1 are valid.*
   (ii) *The solution $x(t)$ of*

$$\dot{x}(t) = f(t, x(t), 0), \quad x(0) = x_0, \tag{8.1.12}$$

   *tends to $x_1 = 0$, as $t \to \infty$.*

*Then $C = E^n$, that is, (8.1.7) is globally (Euclidean) controllable with constraints.*

**Proof**   From Theorem (8.1.1) there is a neighborhood $\mathcal{O}$ of zero in $E^n$ that is contained in $C$. Because of (ii), each solution $x(t, x_0)$ for any $x_0 \in E^n$ glides into $\mathcal{O}$ in time $t_0$: $x(t_0, x_0) \in \mathcal{O} \subset C$. This point $x(t_0, x_0)$ can be brought to the precise zero in time $t_1$. Thus the control $u \in L_\infty([0, t_1], C^m)$,

$$u = \begin{cases} 0 & \text{on} \quad [0, t_0], \\ v \in L_\infty([t_0, t_1], C^m) \end{cases}$$

drives $x_0$ into 0 in time $t_1$. This completes the proof. $\qquad\square$

**Remark 8.1.2**   Conditions are available for the behavior

$$x(t, x_0) \to 0 \quad \text{as} \quad t \to \infty \tag{8.1.13}$$

needed for (ii).

**Proposition 8.1.1**   *In* (8.1.12), *assume:*

   (i)  *There exists a symmetric positive definite $n \times n$ constant matrix $A$ such that the eigenvalues $\lambda_k(x, t)$, $k = 1, 2, \ldots, n$ of the matrix*

$$\frac{1}{2}(AJ + J^T A)$$

*satisfy*

$$\lambda_k \leq -\delta < 0, \quad k = 1, 2, \ldots, n,$$

$\forall \, (x, t) \in E^{n+1}$, *where* $\delta$ *is a constant and*

$$J = \frac{\partial f(t, x, 0)}{\partial x}.$$

(ii) *There are constants* $r > 0$, $\rho$, $1 \leq \rho \leq 2$, *such that*

$$\int_t^{t+r} |f(\tau, 0, 0)|^\rho d\tau \to 0 \quad as \quad t \to \infty.$$

*Then every solution of* (8.1.12) *satisfies* (8.1.13).

The proof is contained in [1].

**Example 8.1.1**   Consider the model of a mass spring system

$$\ddot{x} + a\dot{x} + bx = g(u), \tag{8.1.14}$$

where

$$a > 0, \ b > 0, \quad u : E \to [-1, 1],$$

$g : E \to E$ with

$$g(0) = 0, \ g(-b) < 0 < g(b), \quad \int_0^\infty g(u(s))ds < \infty.$$

Let $\underline{x} = (x_1, x_2)$ where $x_1 = x$, $x_2 = \dot{x}$, so that the mass spring equation becomes

$$\underline{\dot{x}} = \begin{bmatrix} \dot{x}_1 \\ \dot{x}_2 \end{bmatrix} = \begin{bmatrix} x_2 \\ -bx_1 - ax_2 + g(u) \end{bmatrix} = f(\underline{x}, u).$$

Here

$$A(t) = A = \begin{bmatrix} 0 & 1 \\ -b & -a \end{bmatrix}.$$

We may write (8.1.8) as $\dot{x} = Ax + Bu$ where $B = \begin{bmatrix} 0 \\ 1 \end{bmatrix}$, $v$ is in the unbounded closed convex cone of $g([-1, 1])$. The rank of $[B, AB] = \begin{bmatrix} 0 & 1 \\ 1 & -a \end{bmatrix}$ is 2. In (i), $A$ has its characteristic values negative. All the hypotheses of Theorem 8.1.2 are seen to be satisfied. The system (8.1.14) is (globally) Euclidean null controllable with constraints.

## 8.2   Controllability of Nonlinear Delay Systems

In this section we resolve the problem of controllability of general nonlinear delay systems in function space on which the solution of the optimal problem had rested. In [11], for example, the problem of minimizing a cost functional in nonlinear delay systems with $W_2^{(1)}$ boundary conditions was explored under the condition of controllability. The time-optimal control theory of nonlinear systems requires controllability (see Theorem 7.9.3). As has been argued, if the controls are $L_p$ functions, the natural state space is $W_p^{(1)}$. For general nonlinear systems, a natural way of investigating this problem requires the Fréchet differentiability of an operator $F : W_p^{(1)} \times L_p \to W_p^{(1)}$, which is very difficult to realize for nonlinear systems. To overcome this difficulty we use weaker concepts of differentiability in $W_p^{(1)}$ and more powerful open-mapping theorems. The results obtained are analogous to those of ordinary differential systems. We shall first treat the problem in $C$, the space of continuous functions.

Consider the general nonlinear system

$$\dot{x}(t) = f(t, x_t, u(t)), \quad t \geq \sigma,$$
$$x_\sigma = \phi \in C,$$

$$(8.2.1)$$

where $f : E \times C \times E^m \to E^n$ is continuous and continuously Fréchet differentiable in the second and third argument. We assume also that there exist integrable functions $N_i : E \to [0, \infty)$, $i = 1, 2, 3$, such that the partial derivatives $D_i f(t, \phi, w)$ satisfy

$$\|D_2 f(t, \phi, w)\| \leq N_i(t) + N_2(t)|w|, \quad \|D_3 f(t, \phi, w)\| \leq N_3(t),$$

for all $t \in E$, $w \in E^m$, $\phi \in C$. Under these assumptions we are guaranteed existence and uniqueness of a solution through each $(\sigma, \phi) \in E \times C$ for each $u \in L_\infty([\sigma, \infty), E^m)$, (see Underwood and Young [15], Chukwu [6], and Sec. 7.8). Thus for each $(\sigma, \phi) \in E \times C$ and for each $u \in L_\infty([\sigma, \infty), E^m)$ there exists a unique response $x(\sigma, \phi, u) : E^t = [\sigma, \infty) \to C$ with initial data $x_\sigma(\sigma, \phi, u) = \phi$ corresponding to $u$. The mapping $x(\sigma, \phi, u) : E^+ \to C$ defined by $t \to x_t(\sigma, \phi, u)$ represents a point of $C$. We now study the mapping $u \to x_t(\sigma, \phi, u) : x_t : L_\infty([\sigma, \infty), E^m) \to C$.

**Lemma 8.2.1**   *For each $v \in L_\infty([\sigma, T], E^m)$, $T > \sigma$, we have that the partial Fréchet derivative of $x_t(\sigma_0, \phi_0, u_0)$ with respect to $u$,*

$$D_u x_t(\sigma_0, \phi_0, u_0)(v) = z_t(\sigma_0, \phi_0, u_0, v)$$

*where the mapping $t \to z(t, \sigma_0, \phi_0, u_0, v)$ is the unique absolutely continuous solution of the linear differential equation*

$$\dot{z}(t) = D_2 f(t, x_t(\sigma_0, \phi_0, u_0), u_0(t))z_t + D_3 f(t, x_t(\sigma_0, \phi_0, u_0), u_0(t))v(t),$$

$$(8.2.2)$$

$$z_{\sigma_0}(\sigma_0, \phi_0, u_0, v) = 0.$$

**Proof**  The solution $x$ of (8.2.1) is a solution of the integral equation

$$x_\sigma = \phi \quad \text{in} \quad [-h, 0]\,,$$

$$x(t, \sigma, \phi, u) = \phi(0) + \int_\sigma^t f(s, x_s(\sigma, \phi, u), u(s))ds, \quad t \geq \sigma\,.$$

Let $D_u$ denote the partial derivative of $x_t(t, \sigma, \phi, u)$ with respect to $u$. Then $D_u x(t, \sigma, \phi, u) = 0$ in $[-h, 0]$, and

$$D_u x(t, \sigma, \phi, u) = \int_\sigma^t D_2 f(s, x_s(\sigma, \phi, u), u(s))D_u x_s(\sigma, \phi, u)ds$$

$$+ \int_\sigma^t D_3 f(s, x_s(\sigma, \phi, u), u(s))ds, \quad t \geq \sigma\,.$$

($D_i f$ denotes the Fréchet derivative with respect to the $i$th argument.)  These differentiation formulas follow from results in Dieudonne [9, pp. 107, 163].  On differentiating with respect to $t$ we have

$$\frac{d}{dt}[D_u x(t, \sigma, \phi, u)](v) = D_2 f(t, x_t(\sigma, \phi, u), u(t)) \cdot D_u x_t(\sigma, \phi, u)v$$

$$+ D_3 f(t, x_t(\sigma, \phi, u), u(t))v\,.$$

This proves the lemma. $\qquad\qquad\square$

**Lemma 8.2.2**   *For $\phi \in C$, $u \in L_\infty([\sigma, t_1], E^m)$, $t_1 > \sigma + h$, let $x(\sigma, \phi, u)$ be a solution of (8.2.1) with $x_\sigma(\sigma, \phi, u) = \phi$. Consider the mapping $u \to x_{t_1}(\sigma, \phi, u)$, $F : L_\infty([\sigma, t_1], E^m) \to C([-h, 0], E^n) = C$, $F(u) = x_{t_1}(\sigma, \phi, u)$. Then the Fréchet derivative*

$$DF(u) = \frac{d}{du}F(u) : L_\infty([\sigma, t_1], E^m) \to C$$

*has a continuous local right inverse and is a surjective linear mapping if and only if, the variational control system (8.2.2) along the response $t \to x_t(\sigma, \phi, u)$, namely*

$$\dot{z}(t) = D_2 f(t, x_t, u(t))z_t + D_3 f(t, x_t, u(t))v(t)\,, \quad z_\sigma(\sigma, \phi, u, v) = 0\,,$$

*is controllable on $[\sigma, t_1]$ $t_1 > \sigma + h$.*

**Proof**   System (8.2.2) is controllable on $[\sigma, t_1]$ if and only if the mapping $u \to z_t(\sigma, 0, \phi, u, v)$ $t \in [\sigma, t_1]$ is surjective. The lemma follows the observation in Lemma 8.2.1 that

$$z_{t_1}(\sigma, \phi, u, v) = D_u x_{t_1}(\sigma, \phi, u)(v) = DF(u)v\,.$$

In what follows, we assume in (8.2.1) that

$$f(t, 0, 0) = 0\,. \tag{8.2.3}$$

As a consequence, if $\phi \equiv 0$ in (8.2.1), we have a unique trivial solution when $u = 0$.

$\square$

**Definition 8.2.1** The $C$-attainable set of (8.2.1) is a subset of $C([-h, 0], E^n)$ given by

$$\mathcal{A}(t, \phi, \sigma) = \{x_t(\sigma, \phi, u) : u \in L_\infty([\sigma, t], E^m), \quad x_\sigma = \phi,$$

$$x \text{ is a solution of (8.2.1)}\}.$$

If $\phi \equiv 0$, we write $\mathcal{A}(t, 0, \sigma) \equiv \mathcal{A}(t, \sigma)$. Let $C^m$ denote the unit cube

$$C^m = \{u \in E^m : |u_j| \le 1, \ j = 1, \ldots, m\}. \tag{8.2.4}$$

The constrained $C$-attainable set is the subset of $C$ defined by

$$a(t, \phi, \sigma) = \{x_t(\sigma, \phi, u) : u \in L_\infty([\sigma, t], C^m), \quad x \text{ is a solution of (8.2.1)}$$

$$\text{with } x_\sigma = \phi\}.$$

Whenever $\phi \equiv 0$, we simply write

$$a(t, 0, \sigma) = a(t, \sigma).$$

**Definition 8.2.2** System (8.2.1) is proper on $[\sigma, t]$, $t > \sigma + h$, if

$$0 \in \text{Int } a(t, \sigma). \tag{8.2.5}$$

**Proposition 8.2.1** *System (8.2.1) is proper on $[\sigma, t_1]$, $t_1 > \sigma + h$, whenever*

$$\dot{z}(t) = D_2 f(t, 0, 0) z_t + D_3 f(t, 0, 0) v(t) \tag{8.2.6}$$

*is controllable on $[\sigma, t_1]$.*

**Proof** Consider the response $x(\sigma, \phi, u)$ to $u \in L_\infty([\sigma, t_1], E^m)$ for (8.2.1), and the associated map $u \to x_{t_1}(\sigma, 0, u)$ given by $Fu = x_{t_1}(\sigma, 0, u)$, where $F : L_\infty([\sigma, t_1], E^m) \to C$. Evidently $F(L_\infty([\sigma, t_1], C^m)) = a(t, \sigma)$. It follows from Lemma 8.2.1 and Lemma 8.2.2 that

$$DF(0) = \frac{d}{du}(F(u))|_{u=0} = D_3 x_{t_1}(\sigma, 0, u)|_{u=0}$$

is a surjective mapping of $L_\infty([\sigma, t_1], E^n) \to C$. Therefore, (Lang [3, p. 193]), $F$ is locally open: There is an open ball $\mathbb{B}_\rho \subset L_\infty([\sigma, t_1], E^m)$ containing zero, radius $\rho$, and an open ball $\mathbb{B}_r \subset C$ containing zero, radius $r$ such that $\mathbb{B}_r \subset F(\mathbb{B}_\rho)$. Because $L_\infty([\sigma, t_1], C^m)$ contains an open ball containing zero, $r > 0$, $\rho > 0$, can easily be chosen such that

$$\mathbb{B}_r \subset F(\mathbb{B}_\rho \cap (L_\infty([\sigma, t_1], C^m))).$$

Thus

$$\mathbb{B}_r \subset F(L_\infty([\sigma, t_1], C^m)) = a(t, \sigma),$$

so that $0 \in \text{Int } a(t, \sigma)$. This completes the proof. $\square$

Denote by $\mathcal{U}$ the set of admissible controls

$$\mathcal{U} = L_\infty([\sigma, t_1], C^m), \tag{8.2.7}$$

where $C^m$ is as defined in (8.2.4).

**Definition 8.2.3** Consider (8.2.1). The domain $\mathcal{D}$ of null controllability of (8.2.1) is the set of initial functions $\phi \in C$ such that the solution $x(\sigma, \phi, u)$ of (8.2.1) for some $t_1 < \infty$, and some $u \in L_\infty([\sigma, t_1], C^m)$, satisfies $x_\sigma(\sigma, \phi, u) = \phi$, $x_{t_1}(\sigma, \phi, u) = 0$. If $\mathcal{D}$ contains an open neighborhood of $x_{t_1} = 0$, then (8.2.1) is said to be locally null controllable with constraints when the initial data is restricted to a certain small neighborhood of zero in $W_2^{(1)}$.

**Proposition 8.2.2** *In (8.2.1), assume that*

(i)

$$f : E \times C \times E^m \to E^n$$

*is continuous and continuously differentiable in the second and third argument; and there exists integrable functions $N_i : E \to [0, \infty)$, $i = 1, 2, 3$ such that*

$$\|D_2 f(t, \phi, w)\| \le N_1(t) + N_2(t)\,|w|, \quad \|D_3 f(t, \phi, w)\| \le N_3(t),$$

*for all $t \in E$, $w \in E^m$, $\phi \in C$.*

(ii) $f(t, 0, 0) = 0, \ \forall \, t \ge \sigma$.
(iii) *System (8.2.6) is controllable in $[\sigma, t_1]$, $t_1 > \sigma + h$.*

*Then the domain of null controllability $\mathcal{D}$ of (8.2.1) contains zero in its interior, that is, $0 \in \text{Int } \mathcal{D}$, so that (8.2.1) is locally null controllable with constraints.*

**Proof** Assume that $0$ is not contained in the interior of $\mathcal{D}$, and aim at a contradiction. By this assumption there is a sequence $\{\phi_n\}^\infty$, $\phi_n \in C$, $\phi_n \to 0$ as $n \to \infty$, and so $\phi_n$ is in $\mathcal{D}$. Because the trivial solution is a solution of (8.2.1) an account of (ii), $0 \in \mathcal{D}$, therefore $\phi_n \ne 0$ for any $n$. Also $x_t(\sigma, \phi_n, u) \ne 0$ for any $t > \sigma + h$, and any $u \in L_\infty([\sigma, t], C^m) = \mathcal{U}$. Thus on setting $\xi_n = x_t(\sigma, \phi_n, u)$, and noting that for $u = 0$, as $n \to \infty$, $\xi_n = x_t(\sigma, \phi_n, 0) \to x_t(\sigma, 0, 0)$ (from continuity and uniqueness of solutions), and because $x_t(\sigma, 0, 0) = 0$ is the trivial solution of (8.2.1) that is contained in $a(t, \sigma)$, $\xi_n \notin a(t, \sigma)$ for any $t \ge \sigma + h$. We have constructed a sequence $\{\xi_n\}_1^\infty \in C$ that has the following properties: $\xi_n \to 0$ as $n \to \infty$, $\xi_n \ne 0$ for any $n$, and $\xi_n \notin a(t, \sigma)$ for any $t \ge \sigma + h$. This proves that $0 \notin \text{Int } a(t, \sigma)$ for any $t > \sigma + h$. This contradicts the implication of Proposition (8.2.1) that (8.2.1) is proper, since (8.2.6), by (iii), is controllable. The proof is complete: $0 \in \text{Int } \mathcal{D}$. $\square$

Recall that tests for the controllability of the linear system

$$\dot{x}(t) = L(t, x_t) + B(t)u(t) \tag{8.2.8}$$

in the space $C$ are available in Theorem 6.2.2.

Thus, for (8.2.8) the domain of null controllability is open if $B$ is continuous and rank $B(t) = n$ on $[t_1 - h, t_1]$, $t_1 > \sigma + h$.

Proposition 8.2.2 is a local result. To obtain a global result in which $\mathcal{D} = C$, we need a global stability result for the system

$$\dot{x}(t) = f(t, x_t, 0). \tag{8.2.9}$$

It is contained in the next Theorem.

**Theorem 8.2.1**   *In (8.2.1), assume that:*

(i)  *Conditions (i)–(iii) of Proposition 8.2.2 are valid.*

(ii)  *System (8.2.9) is uniformly exponentially stable, that is, each solution of (8.2.9) satisfies*

$$\|x_t(\sigma, \phi)\| \le k\|\phi\| \exp[-\alpha(t - \sigma)], \ t \ge \sigma, \tag{8.2.10}$$

*for some $k > 1$, $\alpha > 0$.*

*Then (8.2.1) has its domain of null controllability $\mathcal{D} = C$.*

**Proof**   By Proposition 8.2.2, $0 \in \mathrm{Int} \ \mathcal{D}$, so that whenever (8.2.10) is valid, every solution of (8.2.1) with $0 = u \in \mathcal{U}$, that is every solution of (8.2.9), satisfies $x_t(\sigma, \phi, 0) \to 0$ as $t \to \infty$. Thus there exists a finite $t_0 < \infty$, such that $x_{t_0}(\sigma, \phi, 0) \equiv \psi \in \mathcal{O} \subset \mathcal{D}$, where $\mathcal{O}$ is an open ball contained in $\mathcal{D}$ with zero as center. With this $t_0$ as initial time now and $\psi$ as initial function in $\mathcal{O}$, there exists some control $v \in \mathcal{U}$ such that the solution $x(t_0, \psi, v)$ of (8.2.1) satisfies $x_{t_0}(t_0, \psi, v) = \psi, x_{t_1}(t_0, \psi, v) = 0$. Using

$$w(s) = \begin{cases} 0, & s \in [\sigma, t_0], \\ v(s), & s \in [t_0, t_1], \end{cases}$$

we have $w \in L_\infty([\sigma, t_1], C^m)$, and the solution $x = x(\sigma, \phi, w)$ satisfies $x_\sigma = \phi$, $x_{t_1} = 0$. The proof is complete.     $\square$

We now restrict our attention to the state space $W_2^{(1)}$ and use the control space $L_2([\sigma, t_1], E^m)$ for some $t_1 > \sigma + h$. We have:

**Theorem 8.2.2**   *In (8.2.1), assume that:*

(i)  *$f : E \times C \times E^m \to E^n$ is continuous and continuously differentiable in the second and third arguments; and there exist integrable $N_i : E \to [0, \infty)$,*

$i = 1, 2, 3$, *such that*

$$\|D_2 f(t, \phi, w)\| \leq N_1(t) + N_2(t)|w|,$$

$$\|D_3 f(t, \phi, w)\| \leq N_3(t), \quad \forall\, t \in E, \ w \in E^m, \ \phi \in C.$$

(ii)  $f(t, 0, 0) = 0, \ \forall\, t \geq \sigma.$
(iii)  *The system*

$$\dot{z}(t) = D_2 f(t, 0, 0)z_t + D_3 f(t, 0, 0)v(t), \qquad (8.2.6)$$

*with $v \in L_2([\sigma, t_1], E^m)$, is controllable in the state space $W_2^{(1)}([-h, 0], E^n)$.*

*Then the system*

$$\dot{x}(t) = f(t, x_t, u(t)) \qquad (8.2.1)$$

*is locally null controllable with constraints; that is, there exists an open ball $\mathcal{O}$ center zero in $W_2^{(1)}$ such that all initial functions in $\mathcal{O}$ can be driven to zero with controls in*

$$V = \{u \in L_2([\sigma, t_1], E^m) : \|u\|_2 \leq 1\}.$$

**Proof**   The proof parallels that of the state space $C$ except that here the map $F$ may not be Fréchet differentiable.  More precisely, if $u \in L_2([0, t_1], E^n)$, then the corresponding solution of (8.2.1) $x(u)$ is an absolutely continuous function with derivative in $L_2([0, t_1], E^n)$, so that $x(u) \in W_2^{(1)}([0, t_1], E^n)$.  Consider the mapping

$$F : L_2([\sigma, t_1], E^n) \to W_2^{(1)}([-h, 0], E^n), \quad \text{defined by } Fu = x_t(u).$$

To proceed as in the previous case, we need $Fu$ to be Fréchet differentiable. Because of the norms of $L_2$ and $W_2^{(1)}$, this requirement will place very stringent conditions on $f$. Unless $u$ appears in an affine linear fashion in $f$, Fréchet differentiability is impossible because of Vainberg [19]. We use Gateaux derivative, which does not require such stringent conditions, and we shall see that this will suffice. For this, let $F'(u)$ denote (formally) the Gateaux derivative of $x_t(u) \in W_2^{(1)}$, with respect to $u$. Then we have $F'(u) : L_2 \to W_2^{(1)}$ exists and is given by

$$F'(u)v = D_u x(t, u)(v) = z(t, u, v),$$

where the mapping $t \to z(t, u, v)$ of $E$ into $E^n$ is the unique solution of (8.2.2). This assertion follows from the following argument: The solution $x(u) = x(\sigma, \phi, u)$ of (8.2.1) is given as the integral

$$x(t) = \phi(t), \quad t \in [-h, 0],$$

$$x(t) = \int_\sigma^t f(s, x_s(u), u(s))ds, \quad t \geq \sigma. \qquad (8.2.11)$$

Now consider $u \to F(u) = x(t, u)$ given as

$$(Fu)(t) = \int_{\sigma}^{t} f(s, x_s(u), u(s))ds,$$

which is the map $F : L_2 \to W_2^{(1)}$. Let $u_0, h \in L_2([\sigma, t_1], E^m)$. Then

$$F(u_0 + \tau h) - F(u_0) = \int_{\sigma}^{t} f(s, x_s(u_0 + \tau h), u(s) + \tau h(s))$$

$$- f(s, x_s(u_0), u_0(s))ds$$

$$\equiv z(t).$$

We note that

$$f(s, x_s(u_0 + \tau h), u_0(s) + \tau h(s)) - f(s, x_s(u_0), u_0(s))/\tau$$

$$\to \delta f(u_0(s), h(s)) \quad \text{as} \quad \tau \to 0,$$

where $\delta f$ denotes the Gateaux differential of $f$. Now

$$\|T(u_0 + \tau h) - T(u_0)/\tau\|_{W_2}^2$$

$$= \int_0^T (\dot{z}(t), \dot{z}(t))/\tau \; dt$$

$$= \int_0^T (f(t, x_t(u_0 + \tau h), u_0(t) + \tau h(t)) - f(t, x_t(u_0), u_0(t))/\tau)^2$$

$$\to \int_0^T [\delta f(u_0(t), h(t))]^2 dt \quad \text{as} \quad \tau \to 0,$$

where $\delta f$ denotes the Gateaux differential of $f$. Since $f : E \times C \times E^m$ is continuously Fréchet differentiable

$$\delta f(u_0(t), h(t)) = D_u f(t, x_t(u_0), u_0(t))h$$

where $D_u f$ denotes the Fréchet derivative with respect to $u(t)$ ([14, Lemma 6.3]). Thus we obtain

$$\delta T(u_0, h) = \lim_{\tau \to 0} (T(u_0 + \tau h) - T(u_0))/\tau$$

$$= \int_0^T D_u f(t, x_t(u_0), u_0(t))h(t)dt,$$

since condition (i) of Theorem 8.2.2 holds and the Lebesgue Dominated Convergence Theorem is available.

We now prove that $\delta T(\cdot, \cdot)$ is continuous at $(u_0, 0) \in L_2([0, \tau]) \times L_2([0, \tau])$, so that by Problem 6.61 of [14], $\delta T(u_0, h) = T'(u_0)h$, $T'(u_0) \in B(L_2, W_2^{(1)}) : T'(u_0) :$

$L_2 \to W_2^{(1)}$ is a bounded linear map, from $L_2$ into $W_2^{(1)}$. Suppose $(u_k, h_k) \to (u_0, 0)$. Then

$$\lim_{k \to \infty} \delta T(u_k, h_k) = \lim_{k \to \infty} \int_0^\tau \| D_u f(s, x_s(u_k), u_k(s)) h_k(s)$$

$$- D_u f(s, x_s(u_0)) \| ds$$

$$= 0,$$

since $D_u f(s, x_s(u_k), u_k(s)) h_k(s) \to 0$ as $k \to \infty$ and assumption (i) of Theorem 8.2.2 and the Lebesgue Dominated Theorem is valid. $\qquad\square$

We note that

$$\| T'(u) - T'(u_0) \| \le \int_0^\tau \| D_u f(s, x_s(u), u(s)) - D_u f(s, x_s(u_0), u_0(s)) \| ds,$$

so that from the continuity of $D_u f$ in $u$ we have: for every $r > 0$ and $u \in B_r(u_0) = \{u : \|u - u_0\|_p \le r\}$, there exists some finite constant $L < \infty$ such that

$$\| T'(u) - T'(u_0) \| \le L. \tag{8.2.12}$$

We now observe that the solution $x(u)$ of (8.2.11) is Fréchet differentiable with respect to $u(t) \in E^m$. Indeed, if $D_u$ denotes this partial derivative, then by Dieudonne [9]

$$D_u x(u) = \int_0^t D_2 f(s, x_s(u), u(s)) D_u x_s(u) ds + \int_0^t D_3 f(s, x_s(u), u(s)) ds.$$

We now differentiate with respect to $t$ to obtain

$$\frac{d}{dt} [D_u x(u) v] = D_2 f(t, x_t(u), u(t)) D_u x_t(u) v + D_3 f(t, x_t(u), u(t)) v.$$

Using the assertions previously proved,

$$\delta T(u_0, v) = T'(u_0) v = D_u x(u) v,$$

and

$$\frac{d}{dt} [T'(u_0) v] = D_2 f(t, x_t(u), u(t)) D_u x_t(u) v + D_3 f(t, x_t(u), u(t)) v.$$

If $u \in L_2([0, T], E^m)$, $T > h$ and $u \to x_t(u)$ are the mapping

$$F : L_2([0, T], E^m) \to W_2^{(1)}([-h, 0] E^n)$$

given by $Fu = x_t(u)$ where $x(u)$ is the solution of (8.2.1), then by what we have proved

$$F'(u) v = D_u x_t(u) v = y_t(u, v),$$

where $y$ is a solution of the variational equation (8.2.2) and

$$F'(u) : L_2([0,T], E^m) \to W_2^{(1)}[-h, 0].$$

Evidently $F'(0)$ is a bounded linear surjection if and only if the control system (8.2.2) is controllable on $[\sigma, t_1]$. To sum up, consider the mapping

$$F : L_2([\sigma, t_1], E^m) \to W_2^{(1)}, \quad F'(0) : L_2([\sigma, t_1], E^m) \to W_2^{(1)},$$

is a surjection by condition (iii), so that $F$ satisfies all the requirements of Corollary 15.2 [10, p. 155], an open-mapping theorem. Thus for $u_0 \equiv 0 \in L_2 ([\sigma, t_1], E^m)$, $F(u_0) = F(0) = 0 \in W_2^{(1)}$, to every $r > 0$ and open ball $\mathbb{B}(0, r) \subset L_2([\sigma, t_1], E^m)$ center $0 \in L_2$ and radius $r$, there is an open ball $\mathbb{B}(0, \rho) \subset W_2^{(1)}$ center $0$ of radius $\rho$ such that

$$\mathbb{B}(0, \rho) = \mathbb{B}(F(u_0), \rho) \subset F(\mathbb{B}(u_0, r)) = F(\mathbb{B}(0, v)).$$

Thus

$$0 \in \mathbb{B}(0, \rho) \subset F(\mathbb{B}(0, r)) \subset F(L_2([\sigma, t_1], E^m)).$$

It follows from this that

$$0 \in \text{Int } \overline{\mathcal{A}}(t, \sigma) \subset W_2^{(1)},$$

where

$$\overline{\mathcal{A}}(t_1, \sigma) = \{x_{t_1}(\sigma, 0, u) : u \in L_2([\sigma, t_1], E^m), \quad \|u\|_2 \leq r,$$

$$x(u) \text{ a solution of } (8.2.1)\}. \tag{8.2.13}$$

We have proved that for any $r > 0$, $0 \in \text{Int } \overline{\mathcal{A}}(t_1, \sigma)$ where the attainable set is as defined in (8.2.13) above with controls in

$$U_{ad} = \{u \in L_2([0, T], E^m) : \|u\|_2 \leq r, \ r \text{ arbitrary}\}.$$

Since $r$ is arbitrary, we can select it to be 1. What we have proved implies that $0 \in \text{Int } \mathcal{D}$, where $\mathcal{D}$ is the domain of null controllability, i.e., the set of all initial functions $\phi$ such that the solution $x(\sigma, \phi, u)$ of (8.2.1) with $u \in U_{ad}$ satisfies

$$x_\sigma(\sigma, \phi, u) = \phi, \quad x_{t_1}(\sigma, \phi, u) = 0.$$

Indeed, suppose not. Then there is a sequence

$$\{\phi_n\}_1^\infty, \ \phi_n : [-h, 0] \to E^n, \ \phi_n \to 0,$$

as $n \to \infty$ and no $\phi_n$ is in $\mathcal{D}$. Since $x(\sigma, 0, 0) = 0$ is a solution of (8.2.1), $0 \in \mathcal{D}$. We can therefore assume that $\phi_n \neq 0 \ \forall n$. Thus, $x_{t_1}(\sigma, \phi_n, u) \neq 0$, for any $u \in U_{ad}$, and any $t_1 > h$. If $\xi_n \equiv x_{t_1}(\sigma, \phi_n, u)$, $\xi_n = x_{t_1}(\sigma, \phi_n, 0) \to x_{t_1}(\sigma, 0, 0)$ (from continuity and uniqueness of solution). Because $x_{t_1}(\sigma, 0, 0) = 0 \in \mathcal{A}(t_1, \sigma)$, $\xi_n \notin \mathcal{A}(t_1, \sigma)$ for any $t_1 > \sigma$. We now have a sequence $\{\xi_n\} \in W_2^{(1)}$ that has the following property:

$$\xi_n \to 0 \quad \text{as} \quad n \to \infty, \quad \xi_n \neq 0 \text{ for any } n, \quad \xi_n \notin \mathcal{A}(t_1, \sigma) \text{ for any } t_1 > h.$$

We conclude that $0 \notin \operatorname{Int} \mathcal{A}(t_1, \sigma)$ for any $t_1 > h$, a contradiction. This concludes the proof that (8.2.1) is locally null controllable with constraints.

**Remark 8.2.1**  If $D_3 f(t, 0, 0) \equiv B(t)$ is continuous, a necessary and sufficient condition for $W_2^{(1)}$ controllability on $[\sigma, t_1]$, $t_1 > \sigma + h$ is rank $B(t) = n$ on $[t_1 - h, t_1]$. See Theorem 6.2.1. This condition is strong. It is interesting to know whether the controllability condition (iii) of Theorem 8.2.2 can be weakened. Indeed, it can be replaced by the closure of the attainable set of (8.2.6). One such sharp result is obtained for the simple variant of (8.2.6), namely

$$\dot{x}(t) = A_0(t)x(t) + A_1(t)x(t - h) + B(t)u(t), \tag{8.2.14}$$

where

$$D_2 f(t, 0, 0)x_t = A_0(t)x(t) + A_1(t)x(t - h),$$

$$D_3 f(t, 0, 0)v = B(t)v.$$

**Theorem 8.2.3**  *In* (8.2.1),

(i) *assume conditions* (i) *and* (ii) *of Theorem* 8.2.2.

(ii) *Also assume that its linearized system is* (8.2.14) *where* $t \to A_0(t)$, $A_1(t)$, $B(t)$ *are analytic on* $[0, t_1]$.

(iii) *The rank of* $B(t)$ *is constant on* $[t_1 - h, t_1]$.

(iv) $\operatorname{Im} A_1(t)\Gamma^i(t)B(t) \subset \operatorname{Im} B(t)$, $i = 0, \ldots, n-1$, *for all but isolated points in* $[\sigma, t_1]$, *where the operator* $\Gamma$ *is defined by*

$$\Gamma(t) = -A_0(t) + \frac{d}{dt},$$

*and* $\operatorname{Im} H$ *denotes the image of* $H$.

*Then* (8.2.1) *is locally null controllable with constraints when the initial data is restricted to a certain subspace* $Y$ *of* $W_2^{(1)}$.

**Proof**  The proof is as before, but because (iii) and (iv) replaces the controllability condition, the mapping $F : L_2([\sigma, t_1], E^m) \to W_2^{(1)}$ does not have its Gateaux derivative $F'$ a surjection. Instead, $F'(L_2([\sigma, t_1], E^m) \equiv Y$ is closed as guaranteed by (iii) and (iv) and Theorem 2 of [13]. As a consequence, $Y$ is a subspace of $W_2^{(1)}([-h, 0], E^n)$. Consider the mapping $F : L_2([\sigma, t_1], E^m) \to Y$ of one Banach space into another. We have that $F' : L_2([\sigma, t_1], E^m) \to Y$ is a surjection and satisfies all the requirements of Corollary 15.2 of [10, p. 155]. The proof is concluded as in the previous proof of Theorem 8.2.2. $\square$

The controllability assumption of Theorem 8.2.2 was relaxed in Theorem 8.2.3 by requiring the closure of the attainable set of the linearized equation. The theory of ordinary differential control systems suggests a further relaxation. In this theory, the linear system

$$\dot{x}(t) = Ax(t) + Bu(t)$$

is locally null controllable with constraints, and the domain of null controllability is open if and only if $\mathrm{rank}[B, AB, \ldots, A^{n-1}B] = n$. Here the controls are taken to be in the unit cube $C^m$. This rank condition, which is equivalent to controllability with $L_2$ unrestrained controls, is also equivalent to null controllability. But in functional differential systems, controllability is not equivalent to null controllability. It is interesting to assume the weaker condition of null controllability. Thus the problem posed for (8.2.1) can be stated as follows: Suppose (8.2.6) is null controllable. Does the same version of Theorem 8.2.2 hold? The next result in the space $C$ states further conditions that will guarantee this.

In (8.2.6) we let

$$D_2 f(t,0,0)z_t = \int_{-h}^{0} d_s \eta(t,s) z(t+s), \qquad (8.2.15)$$

where $\eta(t, \cdot)$ is of bounded variation on $[-h, 0]$, left continuous on $(-h, 0)$, and $\eta(t, 0) = 0$. We need the following notation: Let the integer $j \in [1, n]$ be fixed. Let $x \in E^n$. Let $\pi_1 x$ be the projection of $x$ onto its first $j$ components, and let $\pi_2 x$ denote the projection of $x$ onto its last $n - j$ components. We write $x = (x^1, x^2)$ where $x^1 = \pi_1 x$, $x^2 = \pi_2 x$. Define

$$\bar{\pi}_1 : x \to C([-h, 0], E^j), \quad \bar{\pi}_2 : C([-h, 0], E^n) \to C([-h, 0], E^{n-j})$$

$$\text{by } (\bar{\pi}_i \phi)(s) = \pi_i \phi(s), \quad i = 1, 2.$$

The following result is extracted from the statement and proof of Theorem 2.1 of Underwood and Young [15]. It is the main contribution of Chukwu [7].

**Theorem 8.2.4**    *Consider the system*

$$\dot{x}(t) = f(t, x_t, u(t)) \qquad (8.2.1)$$

*and its linearization* (8.2.6) :

$$\dot{x}(t) = D_2 f(t,0,0)x + B(t)u(t). \qquad (8.2.6)$$

*Assume:*

   (i)  *that conditions* (i) *and* (ii) *of Proposition 8.2.2 hold.*
   (ii) *For each $t$ and $s$, the range of $f$ is contained in the null space of $\eta(t, s)$, the function defined in (8.2.5).*

(iii) *For any $u \in L_2([\sigma, t_1], E^m)$ satisfying $u(t) = 0$ for $\bar{t} \leq t \leq t_1$, and any $y \in C([\sigma - h, t_1], E^n)$ satisfying $y(t) = 0$ for $\bar{t} - h \leq t \leq t_1$ there exists no solution $z$ of $\dot{z}(t) = f(t, y_t + z_t, u(t)) - D_2 f(t, 0, 0) y_t - B(t) u(t)$ on $[\sigma - h, t_1]$ that satisfies both $z(t_1) = 0$ and $z_{t_1} \neq 0$.*

(iv) *The only solution $z \in C([\sigma - h, t_1], E^n)$ of $\dot{z}(t) = D_2 f(t, 0, 0) z_t$, $\sigma \leq t \leq t_1$ are $z(t_1) = 0$ that is constant on $[\sigma - h, \sigma]$ is $z = 0$.*

(v) *Suppose that $t_1 - t_0 > 3h$, and suppose there exist functions*

$$f_1 : E \times E^j \times C([-h, 0], E^{n-j}) \times E^m \to E^j,$$

$$f_2 : E \times E^{n-j} \times E^m \to E^{n-j},$$

*such that*

$$f(t, \phi, w) = (f_1(t, \pi_1 \phi(0), \ \overline{\pi}_2 \phi, w), \ f_2(t, \pi_2 \phi(0), w)), \quad \forall \, t, \phi, w$$

*in the domain of $f$.*

(vi) *Let (8.2.6) be null controllable on $[t_0, t_1 - 2h]$. Then (8.2.1) is locally null controllable with restraints on $[t_0, t_1]$, i.e., with controls in a unit closed sphere of $L_2([t_0, t_1], E^m)$ with center the origin. Furthermore, if*

(vii) *the trivial solution of*

$$\dot{x} = f(t, x_t, 0) \tag{8.2.16}$$

*is globally, uniformly, exponentially stable so that for some $M \geq 1$, $\alpha > 0$, the solution of (8.2.15) satisfies*

$$\|x_{t_0}(\sigma, \phi)\| \leq M \|\phi\| e^{-\alpha(t_0 - \sigma)}, \quad t_0 \geq \sigma.$$

(viii) *In (i)–(iii) above, $t_0$ is sufficiently large so that $t_1 - t_0 > 3h$.*

*Then (8.2.1) is globally null controllable with constraints.*

**Remark 8.2.2**  If (8.2.6) is null controllable on every interval $[t_0, t_1]$, $t_1 > t_0 + h$, then condition (iv) of Theorem 8.2.4 prevails if $t_1 - t_0 > 3h$. Conditions are available in Theorem 6.2.3 and its corollaries. Thus if we assume that (8.2.6) is null controllable, we need not assume $t_0$ large as in (viii).

There are various criteria for the stability requirement of condition (viii). They are contained in [8]. They are needed for the global result. In Theorem 8.2.2, a global constrained null controllability theorem can be deduced by imposing the required global stability hypothesis.

## 8.3   Controllability of Nonlinear Systems with Controls Appearing Linearly

We now consider the special case of (8.2.1) of the form

$$\dot{x}(t) = g(t, x_t, u(t)) + B(t, x_t)u(t) \,. \tag{8.3.1}$$

Here $g : E \times C \times C^m \to E^n$ is a nonlinear function. $B(\cdot, \cdot) : E \times C \to E^{n \times m}$ is an $n \times m$ matrix function. We assume that

$$f(t, x_t, u(t)) \equiv g(t, x_t, u(t)) + B(t, x_t)u(t)$$

satisfies all the hypotheses of Sec. 8.2 for the existence of a unique solution. It is summed up in the following statement:

**Proposition 8.3.1**   *In* (8.3.1), *assume that*

(i)   $B : E \times C \to E^{n \times m}$ *is continuous*; $B(t, \cdot) : C \to E^{n \times m}$ *is continuously differentiable.*

(ii)   *There exist integrable functions* $N_i : E \to [0, \infty)$ $i = 1, 2$ *such that*

$$\|D_2 B(t, \phi)\| \le N_1(t), \quad \|B(t, \phi)\| \le N_2(t), \quad \forall t \in E, \quad and$$

$$\phi \in C([-h, 0], E^n) \,.$$

(iii)   $g(t, \cdot, \cdot)$ *is continuously differentiable for each* $t$.

(iv)   $g(\cdot, \phi, w)$ *is measurable for each* $\phi$ *and* $w$.

(v)   *For each compact set* $K \subset E^n$, *there exists an integrable function* $M_i : E \to E^+$ *and square integrable functions* $M_i : E \to [0, \infty)$, $i = 2, 3$ *such that*

$$\|D_2 g(t, \phi, w)\| \le M_1(t) + M_2(t)|w|, \quad \|D_3 g(t, \phi, w)\| \le M_3(t) \,,$$

$$\forall t \in E, \quad w \in E^m, \quad \phi \in C([-h, 0], E^n) \,.$$

*Under assumptions* (i)–(v), *for each* $u \in L_2$, $\phi \in C$, *there exists a unique solution* $x = x(\sigma, \phi, u)$ *of* (8.3.1); *that is, an absolutely continuous function* $x : [\sigma - h, \infty) \to E^n$ *such that* (8.3.1) *holds almost everywhere and* $x_\sigma = \phi$.

The proof is given in Underwood and Young [15]. They also show [15, p. 761] that $(\phi, u) \to x_t(\sigma, \phi, u) \in C$ is continuously differentiable.

Let the matrix $H$ be defined as follows:

$$H = \int_\sigma^{t_1} B(s, \phi)B^*(s, \phi)ds, \quad t_1 > \sigma \,, \tag{8.3.2}$$

where $B^*$ is the transpose of $B$, and $\phi \in C([-h, 0], E^n)$. The full rank of $H$ is needed to prove that (8.3.1) is Euclidean controllable.

**Theorem 8.3.1**   *In* (8.3.1), *assume the following*:

(i) *Conditions* (i)–(v) *of Proposition 8.3.1 on $f$ and $B$ are valid, and there is a continuous function $N_1^*(t)$ such that*

$$\|B^*(t,\phi)\| \le N_1^*(t), \quad \forall t \in E, \ \phi \in C.$$

(ii) *The matrix $H$ in (8.3.2) has a bounded inverse.*

(iii) *There exist continuous functions $G_j : C \times E^m \to E^+$ and integrable functions $\alpha_j : E \to E^+$, $j = 1, \ldots, q$, such that*

$$|g(t,\phi,u(t))| \le \sum_{j=1}^{q} \alpha_j(t) G_j(\phi, u(t)),$$

*for all $(t,\phi,u(t)) \in E \times C \times E^m$, where the following growth condition is satisfied:*

$$\limsup_{r \to \infty} \left( r - \sum_{j=1}^{q} c_i \, \sup\{G_j(\phi,u) : \|(\phi,u)\| \le r\} \right) = +\infty.$$

*Then (8.3.1) is Euclidean controllable on $[\sigma, t_1]$, $t_1 > \sigma$.*

**Remark 8.3.1**  If $g$ is uniformly bounded, then condition (iii) is met.

**Proof**  Let $\phi \in W_2^{(1)}$, $x_1 \in E^n$. Then the solution of (8.3.1) is the solution of the integral equation

$$x(t+\sigma) = \phi(t), \quad t \in [-h, 0],$$

$$x(t) = \phi(0) + \int_{\sigma}^{t} g(s, x_s, u(s))ds + \int_{\sigma}^{t} B(s, x_s)u(s)ds, \quad t \ge \sigma. \quad (8.3.3)$$

A control that does the transfer is defined as follows:

$$u(t) = B^*(t, x_t) H^{-1} \left[ x_1 - \phi(0) - \int_{\sigma}^{t_1} g(s, x_s, u(s))ds \right], \quad (8.3.4)$$

where $x(\cdot)$ is a solution of (8.3.1) corresponding to $u$ and $\phi$. $\qquad\square$

We now prove that such a $u$ exists. It is obviously an $L_2$ function, since $t \to B(t, \phi)$ is continuous. We need to prove that such a $u$ exists as a solution of the integral equation (8.3.4).

**Proof**  Introduce the following spaces:

$$X = C([-h, t_1], E^n) \times L_2([\sigma, t_1], E^m),$$

with norm $\|(\phi, u)\| = \|\phi\| + \|u\|_2$, where

$$\|\phi\| = \sup_{s \in [-h, q]} |\phi(s)|, \quad \|u\|_2 = \left( \int_{\sigma}^{t_1} |u(s)|^2 ds \right)^{\frac{1}{2}}.$$

We show the existence of a positive constant, $r_0$, and a subset $A(r_0)$ of $X$ such that

$$A(r_0) = A_1(t_1, r_0) \times A_2(t_1, r_0),$$

where

$$A_1(t_1, r_0) = \{\xi : [-h, t_1] \to E^n \text{ continuous } \xi_\sigma = \phi, \quad \|\xi_t\| \le r_0, t \in [\sigma, t_1]\},$$

$$A_2(t_1, r_0) = \{u \in L_2((0, t_1], E^m), (i), |u(t)| \le r_0 \quad \text{a.e. in } t \in [\sigma, t_1],$$

and $\int_\sigma^{t_1} |u(t+s) - u(t)|^2 dt \to 0$ as $s \to 0$ uniformly with respect to $u \in A_1(t_1, r_0)\}$. It is obvious that the two conditions for $A_2$ ensure that $A_2$ is a compact convex subset of the Banach space $L_2$ ([12, p. 297]). Define the operator $T$ on $X$ as follows:

$$T(x, u) = (z, v),$$

where

$$z(t + \sigma) = \phi(t), \quad t \in [-h, 0],$$

$$z(t) = \phi(0) + \int_\sigma^t g(s, x_s, u(s))ds + \int_\sigma^t B(s, x_s)v(s)ds, \quad t \ge \sigma, \quad (8.3.5)$$

$$v(t) = B^*(t, x_t)H^{-1}\left[x_1 - \phi(0) - \int_\sigma^{t_1} g(s, x_s, u(s))ds\right]. \quad (8.3.6)$$

Obviously the solutions $x(\cdot)$ and $u(\cdot)$ of (8.3.3) and (8.3.4) are fixed points of $T$, i.e., $T(x, u) = (x, u)$. Using Schauder's fixed-point theorem we shall prove the existence of such fixed points in $A$. Let

$$F_i(r) = \sup\{F_i(\phi, u) : \|(\phi, u)\| \le r\}.$$

Because the growth condition of (iii) is valid there exists a constant $r_0 > 0$ such that $r_0 - \sum_{i=1}^q c_i F_i(r_0) \ge d$ or $\sum_{i=1}^q c_i F_i(v_0) + d \le r_0$. See a recent paper by Do [18, p. 44]. With this $r_0$, define $A(r_0)$ as described above. To simplify our argument we introduce the following notation:

$$\beta = \max\{\|B(t, \phi)\| : \sigma \le t \le t_1\},$$

$$k = \max\{\beta(t_1 - \sigma), 1\},$$

$$\lambda = \max\{\|B^*(t, \phi)\| \cdot \|H^{-1}\| : t \in [\sigma, t_1]\},$$

$$\|\alpha_i\| = \int_\sigma^{t_1} |\alpha_i(s)|ds, \quad a_i = 3k\lambda\|\alpha_i\|, \quad b_i = 3\|\alpha_i\|,$$

$$c_i = \max\{a_i, b_i\}, \quad d_1 = 3k\lambda[|x_1| + |\phi(0)|],$$

$$d_2 = 3|\phi(0)|, \quad d = \max\{d_1, d_2\}.$$

If $(x, u) \in A(r_0)$, from (8.3.5) and (8.3.6), we have

$$|v(t)| \leq \lambda \left[ |x_1| + |\phi(0)| + \int_\sigma^{t_1} \sum_{i=1}^q \alpha_i(s) G_i(x_s, u(s)) ds \right]$$

$$\leq \lambda \left[ |x_1| + |\phi(0)| + \sum_{i=1}^q \int_\sigma^{t_1} \alpha_i(s) G_i(r_0) ds \right]$$

$$\leq \frac{1}{3k} \left( d + \sum_{i=1}^q c_i G_i(r_0) \right)$$

$$\leq \left( \frac{1}{3k} \right) r_0 \leq \frac{r_0}{3}.$$

Also

$$\|z\| \leq |\phi(0)| + \beta(t_1 - \sigma) r_0 / 3k$$

$$+ \int_\sigma^{t_1} \sum_{i=1}^q \alpha_i(s) G_i(x_s, u(s)) ds$$

$$\leq \frac{d}{3} + \frac{r_0}{3} + \sum_{i=1}^q \int_\sigma^{t_1} \alpha_i(s) G_i(r_0) ds$$

$$\leq \frac{d}{3} + \frac{r_0}{3} + \sum_{i=1}^q \frac{c_i}{3} G_i(r_0)$$

$$\leq \frac{r_0}{3} + \frac{r_0}{3} = \frac{2r_0}{3}.$$

We now verify that

$$\int_\sigma^{t_1} |v(t+s) - v(t)|^2 dt \to 0 \quad \text{as} \quad s \to 0$$

uniformly with respect to $v \in A_2(t_1, r_0)$. Indeed,

$$\int_\sigma^{t_1} |v(t+s) - v(t)|^2 dt \leq \int_\sigma^{t_1} |[B^*(t+s, x_{t+s}) - B^*(t, x_t)] H^{-1} \xi|^2 dt$$

$$\leq \|H^{-1}\xi\|^2 \int_\sigma^{t_1} \|B(t+s, \ x_{t+s}) - B^*(t, x_t)\|^2 dt,$$

where

$$\xi = x_1 - \phi(0) - \int_\sigma^{t_1} g(s, x_s, u(s)) ds.$$

Because $t \to B^*(t, x_t)$ and $t \to x_t$ are continuous, we assert that indeed

$$\int_\sigma^{t_1} |v(t+s) - v(t)|^2 dt \to 0 \quad \text{as} \quad s \to 0. \qquad \square$$

This proves that $v \in A_2$, and we have completed the proof that $T$ maps $A(r_0)$ into itself. We next prove that $T$ is a continuous operator. This is obvious if

$$(t, \phi) \to B(t, \phi), \ (t, \phi, u) \to g(t, \phi, u)$$

are continuous since $u \to x(\cdot, u)$ is continuous. To prove continuity in the general situation, we argue as follows: Let $(x, u), (x', u') \in A(r_0)$, and

$$T(x, u) = (z(t), u(t)), \quad T(x', u') = (z'(t), v'(t)),$$

$$(z(t), v(t)) = T(x, u), \quad (z'(t), v'(t)) = T(x', u').$$

Then

$$|v(t) - v'(t)|^2 = \left| B^*(t, x_t) \left[ x_1 - \phi(0) - \int_\sigma^{t_1} g(s, x_s, u(s)) ds \right] \right.$$

$$\left. - B^*(t, x_t') \left[ x_1 - \phi(0) - \int_\sigma^{t_1} g(s, x_s', u'(s)) ds \right] \right|^2$$

$$\leq \|B^*(t, x_t) - B^*(t, x_t')\|^2 |x_1 - \phi(0)|^2$$

$$+ \|B^*(t, x_t') - B^*(t, x_t)\|^2 \int_\sigma^{t_1} |g(s, x_s', u'(s))|^2 ds$$

$$+ \|B^*(t, x_t)\| \int_\sigma^{t} |g(s, x_s', u'(s)) - g(s, x_s, u(s))|^2 ds. \qquad (8.3.7)$$

Because $u \to B^*(t, x_t^u)$ and $u(t) \to g(t, x_t, u(t))$ are continuous, given any $\epsilon > 0$ there exists an $\eta > 0$ such that if $|u(t) - u(t)|$, then

$$|B^*(t, x_t^u) - B^*(t, x_t^{u'})| < \epsilon, \quad |g(t, x_t^u, u) - g(t, x_t^{u'}, u)| < \epsilon, \quad \forall t \in [\sigma, t_1].$$

Divide $[\sigma, t_1]$ into two sets $e_1$ and $e_2$; put the points at which $|u(t) - u'(t)| < \eta$ to be $e_1$ and the remainder $e_2$. If we write $\|u - u'\|_2 = \gamma$. Then

$$\gamma^2 = \int_\sigma^{t_1} |u(t) - u'(t)|^2 dt \geq \int_{e_2} |u(t) - u'(t)|^2 dt \geq \eta^2 \ \text{mes} \ e_2,$$

so that mes $e_2 \leq \gamma^2 / \eta^2$. Consider the integral

$$I = \int_\sigma^{t_1} |g(s, x_s^{u'}, u'(s)) - g(s, x_s^u, u(s))|^2 ds.$$

Then

$$I = \int_{e_1} + \int_{e_2} |g(s, x'_s, u'(s)) - g(s, x_s, u(s))|^2 ds$$

$$\leq \epsilon^2 \text{ mes } e_1 + \frac{4\gamma^2}{\eta^2} \{\sup g(\cdot)\}^2$$

$$\leq \epsilon^2 \text{ mes } e_1 + \frac{4\gamma^2}{\eta^2} R^2 \,,$$

for some $R$ since $(x, u), (x', u') \in A(r_0)$. If we insert this last estimate in (8.3.7), we deduce that if $|u(t) - u'(t)| < \eta$, then

$$|v(t) - v'(t)|^2 \leq \epsilon^2 |x_1 - \phi(0)|^2 + \epsilon^2 (t_1 - \sigma)R^2 + N_2^*(t)[\epsilon^2 (t_1 - \sigma) + 4\gamma^2 R^2/\eta^2]\,.$$

Thus

$$\|v - v'\|^2 = \int_\sigma^{t_1} |v(s) - v'(s)|^2 ds$$

$$\leq (t_1 - \sigma)[\epsilon^2 \{|x_1| + |\phi(0)|\}^2 + \epsilon^2 (t_1 - \sigma)R^2$$

$$+ \epsilon^2 (t_1 - \sigma) + \frac{4\gamma R^2}{\eta^2} \int_\sigma^{t_1} N_2^*(s)ds\,.$$

Since $\gamma^2 = \|u - u'\|^2$ and $N_2^*$ is integrable, $v, v'$ can be made as close as possible if $u, u'$ are sufficiently close. We next consider the term $|z(t) - z'(t)|$.

$$|z(t) - z'(t)| \leq \int_\sigma^t |g(s, x_s, u(s)) - g(s, x'_s, u'(s))|ds$$

$$+ \int_\sigma^t |B(s, x_s)v(s) - B(s, x'_s)v'(s)|ds$$

$$\leq \int_\sigma^t \|B(s, x_s) - B(s, x'_s)\| |v(s)|ds$$

$$+ \int_\sigma^t \|B(s, x'_s)\| |v(s) - v'(s)|ds$$

$$+ \int_\sigma^t |g(s, x_s, u(s)) - g(s, x'_s, u'(s))|ds\,.$$

Because of this inequality and an argument similar to the above, $z, z'$ can be made as close as possible in $A_1$ if $u, u'$ are sufficiently close. We have proved that $T$ is continuous in $u$. It is easy to see that $T$ is continuous in $x$, the first argument, and thus, by a little reasoning based on the continuity hypothesis on $g$ and $B$, that $T(x, u)$ is continuous on both arguments.

To be able to use Schauder's fixed-point theorem we need to verify that $T(A(r_0))$ is compact. Since $A_2(t_1, r_0)$ is compact, we need only verify that if $(x, u) \in A(r_0)$

and $(z, v) = T(x, u)$, then $z$ as defined in (8.3.5) is equicontinuous for each $r_0$. To see this we observe that for each $(x, u) \in A(r_0)$ and $s_1, s_2 \in [\sigma, t_1]$, $s_1 < s_2$, we have

$$|z(s_2) - z(s_1)| \leq \int_{s_1}^{s_2} |g(s, x_s, u(s))| ds + \int_{s_1}^{s_2} \|B(s, x_s)\| \|v(s)\| ds$$

$$\leq \beta \frac{r_0}{3} |s_2 - s_1| + \int_{s_1}^{s_2} \sum_{i=1}^{q} \alpha_i(s) G_i(r_0) ds$$

$$\leq \beta \frac{r_0}{3} |s_2 - s_1| + \sum_{i=1}^{q} \|\alpha_i\| G_i(r_0) |s_2 - s_1| . \qquad (8.3.8)$$

In the above estimate we have used the fact that

$$\beta = \max_{s \in [\sigma, t_1]} \|B(s, \phi)\| ,$$

and

$$|v(t)| \leq \lambda[|x_1| + |\phi(0)|] + \sum_{i=1}^{q} |(\alpha_i)| G_i(r_0) \leq r_0/3 .$$

It now follows that the right hand of this last inequality does not depend on particular choices of $(x, u)$. Hence, the set of the first components of $T(A(r_0))$ is relatively compact. Thus $T(A(r_0))$ is compact, which by an earlier remark proves that $T$ is a compact operator.

Gathering results, we have proved that $T : A(r_0) \to A(r_0)$ is a continuous compact operator from a closed convex subset into itself. By Schauder's fixed point theorem, there exists a fixed-point $(x, u) = T(x, u)$, given by (8.3.3) and (8.3.4),

$$x_\sigma = \phi ,$$

$$x(t_1) = \phi(0) + \int_{\sigma}^{t_1} g(s, x_s, u(s)) ds$$

$$+ \int_{\sigma}^{t_1} B(s, x_s) B^*(s, x_s) ds \, H^{-1} \xi$$

$$= x_1 .$$

Euclidean controllability is proved.

Our next effort is to obtain criteria for controllability in $W_2^{(1)}$ for System (8.3.1).

**Theorem 8.3.2**    *In* (8.3.1), *assume that*

    (i) *Conditions* (i) *and* (iii) *of Theorem* 8.3.1 *are valid.*
    (ii) $\mathrm{rank}[B(t, \xi)] = n$ *on* $[t_1 - h, t_1]$ *for each* $\xi \in C$, $t \in [t_1 - h, t_1]$.
    (iii) $\xi \to B^+(t, \xi)$ *is continuous for each* $t$ *where* $B^+$ *is the generalized inverse.*

*Then* (8.3.1) *is controllable on* $[\sigma, t_1]$ *with* $t_1 > \sigma + h$.

**Proof**  The first step is to show that (8.3.1) is Euclidean controllable on $[\sigma, t_1 - h]$. Indeed, let $\phi \in W_2^{(1)}$, $x_1 \in E^n$. Then the solution $x$ of (8.3.1) is given by (8.3.3). Because of the rank condition (ii), $B(t, \xi)B^*(t, \xi)$ has rank $n$ on $[t_1 - h, t_1]$, and therefore $B(t_1 - h, x_{t_1 - h})B^*(t_1 - h, x_{t_1 - h})$ has rank $n$. Since $t \to B(t, x_t)$ is continuous, there exists some $\epsilon > 0$ such that for $0 \le s < \epsilon$, $B(t_1 - h - s, x_{t_1 - h - s})B^*(t_1 - h - s, x_{t_1 - h - s})$ has rank $n$. Because of this,

$$H(t_1 - h) = \int_\sigma^{t_1 - h} B(s, x_s)B^*(s, x_s)ds$$

$$= \int_\sigma^{t_1 - h - \epsilon} B(s, x_s)B^*(s, x_s)ds + \int_{t_1 - h - \epsilon}^{t_1 - h} B(s, x_s)B^*(s, x_s)ds$$

has rank $n$, since the last integral is positive definite and $H(t_1 - h)$ is positive semidefinite. By Theorem 8.3.1, System 8.3.1 is Euclidean controllable on $[\sigma, t_1 - h]$, $t_1 > \sigma + h$, so that for given any $\phi, \psi \in W_2^{(1)}$ there exists a $u \in L_2([\sigma, t_1 - h], E^m)$ such that the solution of (8.3.1) satisfies $x_\sigma = \phi$, $x(t_1 - h, \sigma, \phi, u) = \psi(-h)$. To conclude the proof we extend $u$ and $x(\cdot, \sigma, \phi, u) = x(\cdot)$ to the interval $[\sigma, t_1]$, $t_1 > \sigma + h$, so that

$$\dot{\psi}(t - t_1) = g(t, x_t, u(t)) + B(t, x_t)u(t) \tag{8.3.9}$$

a.e. on $[t_1 - h, t_1]$ where $x(\tau) = \psi(\tau - t_1)$, $t_1 - h \le \tau \le t_1$. Since (ii) holds, a control $u$ can be defined as follows:

$$u(t) = B^+(t, x_t^u)\left[\frac{d}{dt}[\psi(t - t_1)] - g(t, x_t^u, u(t))\right] \tag{8.3.10}$$

for $t_1 - h \le t \le t_1$. That such a $u$ exists can be proved as follows:

We define the following set:

$$A_1(r_0) = \left\{ u \in L_2([t_1 - h, t_1], E^m) : \|u(t)\|_{L_2} \le r_0 \quad \text{and with} \right.$$

$$\left. \times \int_{t_1 - h}^{t_1} |u(t + s) - u(t)|dt \to 0 \text{ as } s \to 0 \text{ uniformly with } u \in A_1(r_0) \right\}.$$

It follows from [12, p. 297] that $A = A_1(r_0)$ is compact.

Let $T$ be a map on $A$ defined as follows:

$$T(u)(t) = (v(t)),$$

where

$$v(t) = B^+(t, x_t^u)[\dot{\psi}(t - t_1) - g(t, x_t^u, u(t))].$$

We shall prove that there is a constant $r_0$ such that with $A = A_1(r_0)$, $T : A \to A$, where $T$ is continuous. Because of [12, p. 297] and [12, p. 645], $T$ is guaranteed a fixed point, that is

$$T(u) = (u) \in A,$$

which implies that (8.3.9) and (8.3.10) hold. Observe that $A$ is a compact and convex subset of the Banach space $L_2$. Because of a result of Campbell and Meyer [17, p. 225] and hypotheses (i) and (ii) of the theorem, the generalized inverse $t \to B^+(t, \xi)$ is continuous and therefore uniformly bounded on $[t_1 - h, t_1]$. Since the growth condition (iii) is valid, there exists an $r_0 > 0$ such that

$$\sum_{j=1}^{q} c_j F_j(r_0) + d \le r_0$$

for some $d$. With this $r_0$, define $A = A_1(r_0)$. Now introduce the following notations:

$$B = \max\{\|B(t, \phi)\|,\ \sigma \le t \le t_1\},$$

$$B^+ = \max\{\|B^+(t, \phi)\|,\ \sigma \le t \le t_1\},$$

$$b = \max[\beta, \beta^+],$$

$$k = \max\{\beta(t_1 - \sigma), 1, b\},$$

$$\lambda = \max\{\beta^+, \beta^+\|H^{-1}\|\},$$

$$\|\alpha_i\| = \max\left\{\int_\sigma^{t_1} |\alpha_i(s)|ds, \|\alpha_i\|_2, \max_{\sigma \le t \le t_1} |\alpha_j(t)|\right\},$$

$$a_i = 3k\lambda\|\alpha_i\|,\quad b_i = 3\|\alpha_i\|,\quad c_i = \max\{a_i, b_i\},$$

$$d_1 = \max 3k\lambda[|x_1| + |\phi(0)|,\ \|\dot\psi\|_{L_2}, \sup_{\sigma \le t \le t_1} \dot\psi(t - t_1)],$$

$$d_2 = 3|\phi(0)|,\quad d = \max[d_1, d_2].$$

Let $(u) \in A$. Then

$$|T(u)(t)| \le |v(t)|,$$

where

$$|v(t)| \le |B^+(t, x_t^u)|[|\dot\psi(t - t_1)| - g(t, x_t^u, u(t))]$$

$$\le \beta^+[|\dot\psi(t - t_1)| + |g(t, x_t, u(t))|]$$

$$\le \beta^+\left[|\dot\psi(t - t_1)| + \sum_{j=1}^{q} \alpha_j(t)G_j(x_t, u(t))\right]$$

$$\le \beta^+\left[|\dot\psi(t - t_1)| + \sum_{j=1}^{q} \alpha_j(t)G_j(r_0)\right]$$

$$\le \frac{d_1}{3k} + \sum_{j=1}^{q} \frac{\|\alpha_j\|}{3k}G_j(r_0)$$

$$\leq \frac{1}{3k}\left[d_1 + \sum_{j=1}^{q} G_j(r_0)\right] \, .$$

Therefore

$$\|v\| \leq \beta^+[\|\dot{\psi}\|_{L_2} + \sum_{j=1}^{q} \|\alpha_j\|G_j(r_0)]$$

$$\leq \lambda\|\dot{\psi}\|_{L_2} + \sum_{j=1}^{q} \|\alpha_j\|G_j(r_0)$$

$$\leq \lambda\|\dot{\psi}\|_2 + \sum_{j=1}^{q} \frac{a_j}{3k}G_j(r_0)$$

$$\leq \frac{d_i}{3k} + \sum_{j=1}^{q} \frac{a_j}{3k}G_j(r_0)$$

$$\leq \frac{1}{3k}\left[d + \sum_{j=1}^{q} c_jG_j(r_0)\right]$$

$$\leq \frac{1}{3k}r_0 \, .$$

Therefore

$$\|T(u)\| \leq r_0 \, .$$

Hence $T : A \to A$, if we can verify the second condition. Now

$$\int_{t_1-h}^{t_1} |v(t+s) - v(t)|^2 dt = \int_{t_1-h}^{t_1} |B^+(t+s, x_{t+s})\xi(t+s) - B^+(t, x_t)\xi(t)|^2 dt \, ,$$

where

$$\xi(t) = \dot{\psi}(t - t_1) - g(t, x_t, u(t)) \, .$$

The function $u(t) \equiv B^+(t, x_t)\xi(t)$ is measurable in $t$ and is $L_2$. We can therefore choose a sequence $\{k_n(t)\}$ of continuous functions such that

$$\int_{t_1-h}^{t_1} |k(t) - k_n(t)|^2 dt \to 0 \, ,$$

as $n \to \infty$. Therefore

$$\left[\int_{t_1-h}^{t_1} |k(t+s) - k(t)|^2 dt\right]^{\frac{1}{2}} \leq \left[\int_{t_1-h}^{t_1} |k(t+s) - k_n(t+s)|^2 dt\right]^{\frac{1}{2}}$$

$$+ \left[ \int_{t_1-h}^{t_1} |k_n(t+s) - k_n(t)|^2 dt \right]^{\frac{1}{2}}$$

$$+ \left[ \int_{t_1-h}^{t_1} |k_n(t) - k(t)|^2 dt \right]^{\frac{1}{2}}.$$

We choose $n$ large so that the last and first integral on the right-hand side of this inequality are less than an arbitrary $\epsilon > 0$. Also $s$ can be made small enough for the second integral to be less than $\epsilon > 0$. This completes the verification of the first part that $T : U \to U$.

We now turn to the problem of continuity. Let $(u)$, $(u') \in A(r_0)$, $T(u) = (v)$, $T(u') = (v')$. Then

$$|v(t) - v'(t)| \leq B^+(t, x_t^u)[\dot{\psi}(t - t_1) - g(t, x_t^u, u(t))]$$

$$- B^+(t, x_t^{u'})[\dot{\psi}(t - t_1) - g(t, x_t^{u'}, u'(t))]$$

$$\leq \|B^+(t, x_t) - B^+(t, x_t')\| \, |\dot{\psi}(t - t_1)|$$

$$+ \|B^+(t, x_t') - B^+(t, x_t)\| \, |g(t, x_t', u'(t))|$$

$$+ \|B^+(t, x_t)\| [|g(t, x_t', u'(t)) - g(t, x_t, u(t)|].$$

Since $u \to B^+(t, x_t^u)$ and $u(t) \to g(t, x_t, u(t)$ are continuous, given $\epsilon > 0$ there exists an $\eta > 0$ such that if $|u(t) - u'(t)| < \eta$, then

$$|B^+(t, x_t^u) - B^+(t, x_t^{u'})| < \epsilon,$$

$$|g(t, x_t^u, u(t)) - g(t, x_t^{u'}, u'(t))| < \epsilon, \quad \forall t \in [t_1 - h, t_1] \equiv I.$$

Divide $I$ into two sets $e_1$ and $e_2$, and put the points at which $|u(t) - u'(t)| < \eta$ to be $e_1$ and the other to be $e_2$. If we set $\|u - u'\|_2 = \gamma$, then

$$\gamma^2 = \int_{t_1-h}^{t_1} |u(t) - u'(t)|^2 dt$$

$$\geq \int_{e_2} |u(t) - u'(t)|^2 dt$$

$$\geq \eta^2 \text{ mes } e_2,$$

so that mes $e_2 \leq \gamma^2/\eta^2$. A simple analysis shows that

$$\|v - v'\|^2 \leq \epsilon^2 \text{ mes } e_1 \|\dot{\psi}\|_2 + 2\beta^+ \frac{\gamma^2}{\eta^2} \|\dot{\psi}\|_2$$

$$+ \epsilon^2 \text{ mes } e_2 \{\sup\}\{|g(\cdot)|\}^2$$

$$+ 4\beta^{+2} \frac{\gamma^2}{\eta^2} \sup\{|g(\cdot)|\}^2$$

$$+ \beta^{+2}\epsilon^2 \text{ mes } e_1 + (\beta^{+2})\frac{4\gamma^2}{\eta^2}\{\sup|g(\cdot)|\}^2\,.$$

It follows from these estimates that $\|v - v'\|^2$ can be made arbitrarily small if $\|u-u'\|^2$ is small. This proves that $T : A \to A$ is a continuous mapping of a compact convex subset of $L_2$ into itself. By Schauder's fixed-point theorem [12, p. 645], $T$ has a fixed point. We invoke Schauder's fixed-point theorem to conclude that

$$u = Tu = B^+(t, x_t^u)[\dot\psi(t - t_1) - g(t, x_t^u, u(t)]\,,$$

proving that $u$ is well defined. With this $u$ (8.3.6) is satisfied.

As a result of Theorem 8.3.2 and Theorem 8.3.1, the following conclusion is valid: $\qquad\qquad\qquad\qquad\qquad\qquad\qquad\qquad\qquad\qquad\qquad\qquad\quad\square$

**Theorem 8.3.3**　*In* (8.3.1), *assume that*

- (i)  $g(t, 0, 0) = 0,\ \forall\, t \geq 0.$
- (ii)  *Conditions* (i)–(iii) *of Theorem 8.3.1 hold.*
- (iii)  *The equation*

$$\dot x(t) = g(t, x_t, 0) \tag{8.3.11}$$

*is such that each solution satisfies*

$$x_t \to 0 \quad as \quad t \to \infty\,. \tag{8.3.12}$$

*Then* (8.3.1) *is locally null Euclidean controllable with constraints, i.e., with controls in*

$$U_{ad} = \{u \in E^m : u \in L_2([\sigma, t_1], E^m) : \|u\|_2 \leq 1\}\,.$$

**Proof**　From the proof of Theorem 8.3.1, for each $\phi \in W_2^{(1)}$, $x_1 \in E^n$, there is a control that transfers $\phi$ to $x_1$ in time $t_1$. The control that does the transfer is given in (8.3.4):

$$u = B^*(t, x_t)H^{-1}\xi \equiv H(\phi)\,.$$

Since $\phi \to x_t(\phi, u)$ is continuous, $u$ is a continuous function of $\phi : H$:

$$H : W_2^{(1)} \to L_2([\sigma, t_1], E^n)$$

is continuous. It is continuous at zero. For each neighborhood $V$ of zero in $L_2([\sigma, t_1], E^m)$ there is a neighborhood $\mathcal{O}$ of zero in $W_2^{(1)}$ such that $H(\mathcal{O}) \subset V$. Let $U_{ad} = \{u \in L_2([\sigma, t_1],\ E^m) : \|u\|_2 \leq 1\}$. Choose an open set $V$, a neighborhood of zero in $L_2$ that is a subset of $U_{ad}$ such that $H(\mathcal{O}) \subset V$. Every $\phi \in \mathcal{O}$ can be steered to zero by the control

$$u = H(\phi) = B^*(t, x_t)H^{-1}\xi\,,$$

where

$$\xi = \left[ -\phi(0) - \int_{\sigma}^{t_1} g(s, x_s, u(s)) ds \right].$$

Hence there is a neighborhood $\mathcal{O}$ of zero in $W_2^{(1)}$ such that a control $u \in U_{ad}$ drives every point of $\phi \in \mathcal{O}$ to zero in some time $t_1$. Local null controllability with constraints is established. Now use the control $u = 0 \in U_{ad}$ in (8.3.1). Then every solution of (8.3.11) is driven into $\mathcal{O}$ in some finite time $t_0$. With this point $x_{t_0} \in \mathcal{O}$ as an initial point of $W_2^{(1)}$, one reaches zero in $E^n$ in time $t_1$.       $\square$

**Remark 8.3.2** Since Theorem 8.3.2 is also valid, a theorem analogue to Theorem 8.3.3 can be formulated and proved as above. We note that

$$u = B^+(t, x_t^u) \left[ \frac{d}{dt} \psi(t - t_1) - g(t, x_t^u, u(t)) \right]$$

is implicitly a function of $\phi$; $\phi \to H\phi = u$ is continuous and one argues as before.

## References


1. E. N. Chukwu, "Finite Time Controllability of Nonlinear Control Process," *SIAM J. Control* **13** (1975) 807–816.
2. H. Hermes and J. P. LaSalle, *Functional Analysis and Time Optimal Control*, Academic Press, New York, 1969.
3. S. Lang, *Analysis II*, Addison-Wesley, Reading, MA, 1969.
4. E. B. Lee and L. Markus, *Foundations of Optimal Control Theory*, John Wiley, New York, 1967.
5. W. Rudin, *Real and Complex Analysis*, McGraw-Hill, New York, 1974.
6. E. N. Chukwu, *The Time Optimal Control of Nonlinear Delay Equations in Operator Methods for Optimal Control Problems*, edited by Sung J. Lee, Marcel Dekker, 1988.
7. E. N. Chukwu, "Global Null Controllability of Nonlinear Delay Equations with Controls in a Compact Set," *J. Optimization Theory and Applications* **53** (1987) 43–57.
8. E. N. Chukwu, "Global Behavior of Retarded Functional Differential Equations," in *Differential Equations and Applications, Vol. I*, edited by A. R. Affabizadeh, Ohio University Press, Athens, 1989.
9. J. Dieudonne, *Foundations of Modern Analysis, Vol. I*, Academic Press, New York, 1969.
10. K. Deimling, *Nonlinear Functional Analysis*, Springer-Verlag, New York, 1985.
11. M. Q. Jacobs and T.-J. Kao, "An Optimum Settling Problem for Time Lag Systems," *J. Math. Anal. Appl.* **40** (1972) 687–707.
12. L. V. Kantorovich and G. P. Akilov, *Functional Analysis in Normed Spaces*, Pergamon Press Book, Macmillan, New York, 1964.
13. S. Kurcyusz and A. W. Olbrot, "On the Closure in $W_1^q$ of Attainable Subspaces of Linear Time Lag Systems," *J. Differential Equations* **24** (1977) 29–50.
14. G. E. Ladas and V. Lakshmikantham, *Differential Equations in Abstract Spaces*, Academic Press, New York, 1972.
15. R. Underwood and D. Young, "Null controllability of nonlinear functional differential equations," *SIAM J. Control Optim.* **17** (1979) 753–772.

16. J. P. Dauer, "Nonlinear Perturbations of Quasilinear Control Systems," *J. Math. Anal. Appl.* **54** (1976) 717–725.

17. S. L. Campbell and C. D. Meyer, Jr., *Generalized Inverses of Linear Transformations*, Pitman Publishing, London, 1979.

18. V. N. Do, "Controllability of semilinear systems," *J. Optimization Theory and Applications* **65** (1990) 41–52.

19. M. M. Vainberg, "Some problems in differential calculus in linear spaces," *Uspehi Mat. Nauk* **7**(4), 55–102 (in Russian).

Chapter 9

# Control of Interconnected Nonlinear Delay Differential Equations in $W_2^{(1)}$

## 9.1 Introduction

Our main interest in this section is the resolution of the problem of controllability of interconnected nonlinear delay systems in function space, from which the existence of an optimal control law can be later deduced. We insist that each subsystem is controlled by its own variables while taking into account the interacting effects. This is the recent basic insight of Lu, Lu, Gao, and Lee [5] for ordinary differential systems. Controllability is deduced for the overall system from the assumption of controllability of each free subsystem and a growth condition of the interconnecting structure. Two applications are presented. In one, the insight it provides for the growth of global economy has important policy implications.

For linear-free systems, criteria for $W_2^{(1)}$ controllability have been provided and is reported in Sec. 6.2. For nonlinear cases, a similar investigation was recently carried out in [2], and presented in Sec. 8.3. Recently Sinha [7] treated controllability in Euclidean space of composite systems in which the base is linear. We extend the scope of the treatment in [7] by treating interconnected systems with delay when the state space is $W_2^{(1)}$ and the base system is not necessarily linear. We state criteria for controllability of the free subsystem by defining an $L_2$ control that does the steering both in the linear and nonlinear case. We prove that such a control exists as a solution of an integral equation in a Banach space. For this we use Schauder's fixed-point theorem. Assuming the free subsystem is controllable and the interaction function has a certain growth condition, we prove the controllability of the overall interconnected system.

We begin with a simple system. A linear state equation of the $i$th subsystem of a large-scale control system can be described by

$$\dot{x}^i(t) = A_{1i}x^i(t) + A_{2i}x^i(t-h) + A_{3i}y^i(t) + A_{4i}y^i(t-h) + B_iu^i(t), \qquad (9.1.1)$$

where $x^i(t) \in E^{ni}$ is the $n_i$-dimensional Euclidean state vector of the $i$th subsystem, $u_i \in E^{m_i}$ is the control vector, and $A_{1i}\,A_{2i}\,B_i\,A_{3i}\,A_{4i}$ are time-invariant matrices of appropriate dimensions. Also, $y^i(t)$ is the supplementary variable of the

321

$i$th subsystem and is a function of its own Euclidean state vector $x^i(t)$ and other subsystem state vector $x^j(t)$, $j = 1, \ldots, \ell$. We express this as follows:

$$y^i(t) = M_{ii}x^{(i)}(t) + \sum_{\substack{j=1 \\ j \neq i}}^{\ell} M_{ij}x^j(t) \,, \qquad (9.1.2)$$

where $M_{ii}$, $M_{ij}$ $(j = 1, 2, \ldots, \ell, \, j \neq i)$ are constant matrices.

By substituting (9.1.2) into (9.1.1) we obtain the state equation of the overall interconnected system

$$\dot{x}^i(t) = H_i x^i(t) + G_i x^i(t - h) + B_i u^i(t) + m_i(t) + e_i(t - h) \,, \qquad (9.1.3)$$

where

$$H_i = A_{1i} + A_{3i} M_{ii} \,, \qquad\qquad G_i = A_{2i} + A_{4i} M_{ii} \,,$$

$$m_i(t) = \sum_{\substack{j=1 \\ j \neq i}}^{\ell} M_{3i} M_{ij} x^j(t) \,, \quad e_i(t - h) = \sum_{\substack{j=1 \\ j \neq i}}^{\ell} M_{ij} x^j(t - h) \,.$$

In (9.1.3), $m_i(t)$ and $e_i(t - h)$ describe the interaction, the effects of other subsystems on the $i$th subsystem. This can be measured locally. The decomposed system (9.1.3) can be viewed as an interconnection of $\ell$-isolated subsystems

$$\dot{x}^i(t) = H_i x^i(t) + G_i x^i(t - h) + B_i u^i(t) \,, \qquad (S_i)$$

with interconnection structure characterized by

$$g^i(t, t - h) = m_i(t) + e_i(t - h) \,,$$

which does not depend on the state variables $x^i(t)$.

We now consider the more general free linear subsystem

$$\dot{x}^i(t) = L_i(t, x_t^i) + B_i u^i(t) \,,$$
$$x_\sigma^{(i)} = \phi^i \,, \qquad\qquad\qquad\qquad (L_i)$$

and the interconnected system

$$\dot{x}^i(t) = L_i(t, x_t^i) + B_i u^i(t) + g_i(t) \,,$$
$$x_\sigma^{(i)} = \phi^i \,, \quad i = 1, \ldots, \ell \,, \qquad\qquad (I_i)$$

where

$$L_i(t, x_t^i) = \int_{-h}^{0} d_\theta \eta_i(t, \theta) x^i(t + \theta) \,,$$

$$g_i(t) = \sum_{\substack{j=1 \\ j \neq i}}^{\ell} \int_{-h}^{0} d_\theta \eta_{ij}(t, \theta) x^j(t + \theta) \,.$$

We assume $B_i(t)$ is an $n_i \times m_i$ *continuous* matrix. The linear operator $\phi \to L_i(t, \phi)$ is described by the integral in the Lebesgue–Stieltjes sense, where $(t, \theta) \to \eta_i(t, \theta)$ is an $n_i \times n_i$ matrix function. It is assumed that $t \to \eta_i(t, \theta)$, $t \in E$, is continuous for each fixed $\theta \in [-h, 0]$, and $\theta \to \eta_i(t, \theta)$ is of bounded variation on $[-h, 0]$ for each fixed $t \in E$. Also, $\eta_i(t, \theta) = 0$, $\theta \geq 0$, $\eta_i(t, \theta) = \eta_i(t, -h)$, $\theta \geq -h$, and $\theta \to \eta_i(t, \theta)$ is left continuous on $(-h, 0)$. It is assumed that

$$\operatorname*{var}_{\theta \in E} \eta_i(t, \theta) \leq \rho_i(t) , \quad t \in E ,$$

where $\rho_i(t)$ is locally integrable. These conditions also hold for $\eta_{ij}$.

An easy adaptation of the argument in [1] yields the following result on the system $(I_i)$:

**Theorem 9.1.1** *Consider the interconnected decomposed system $(I_i)$ in which $B_i^+(t)$ is essentially bounded. Suppose*

$$\operatorname{rank} B_i(t) = n_i \quad \text{on} \quad [t_1 - h, t_1] ,$$

*so that $B_i^+$ is continuous.*

*Then $(I_i)$ is controllable on $[\sigma, t_1]$, $t_1 > \sigma + h$.*

**Proof** Let $X_i(t, s)$ be the fundamental matrix solution of

$$\dot{x}^i(t) = L_i(t, x_t^i) .$$

Then

$$G_i(\sigma, t_1) = \int_\sigma^{t_1 - h} X_i(t_1 - h, s) B_i(s) B_i(s)^* X_1^*(t_1 - h, s) ds$$

has rank $n_i$, so that $(I_i)$ is Euclidean controllable. Here $B^*$ is the algebraic adjoint of B. This is proved by letting $\phi^i \in W_2^{(1)} \equiv W_2^{(1)}([-h, 0], E^{ni}), x_1^i \in E^{n_i}$, and by defining a control

$$u^i(t) = [B_i^*(s) X_i^*(t_1 - h, t)] G^{-1}(\sigma, t_1 - \sigma) \bigg[ x_1^i - x^i(t_1, \sigma, \phi, 0)$$

$$- \int_\sigma^{t_1 - h} X_i(t_1 - h, s) g_i(s, s - h) ds \bigg] ,$$

where $x^i(t, \sigma, \phi, 0)$ is the solution of $(L_i)$ with $u^i \equiv 0$. Using the variation of parameter, one verifies that $u^i$ indeed transfers $\phi^i$ to $x_1^i$ in time $t_1 - h$. Thus there is a control $u^i \in L_2([\sigma, t_1 - h], E^{m_i})$ such that $x^i(t_1 - h, \sigma, \phi, u^i) = \psi^i(-h)$. We extend $u^i$ and $x^i$ to the interval $[\sigma, t_1]$, so that

$$\psi^i(t - t_1) = x^i(t), \quad t_1 - h \leq t \leq t_1 ,$$

and

$$\dot{\psi}^i(t - t_1) = L_i(t, x_t^i) + B_i(t) u^i(t) + g_i(t)$$

a.e. on $[t_1 - h, t_1]$. To do this note that

$$L_i(t, \phi^i) = -\eta_i(t, -h)\phi^i(-h) - \int_{t-h}^{t_1-h} \eta_i(t_1, \alpha - t)\dot{\phi}^i(\alpha)d\alpha$$

$$- \int_{t_1-h}^{t} \eta_i(t, \alpha - t)\dot{\phi}^i(\alpha)d\alpha, \quad t_{1-h} \leq t \leq t_1,$$

and

$$g_i(t) = \sum_{\substack{j=1 \\ j \neq i}}^{\ell} \int_{-h}^{0} d_\theta \eta_{ij}(t, \theta)x^j(t + \theta)$$

$$= -\sum_{\substack{j=1 \\ j \neq i}}^{\ell} \left\{ \eta_{ij}(t, -h)x^j(t - h) + \int_{t-h}^{t_1-h} \eta_{ij}(t, \alpha - t)\dot{x}^j(\alpha)d\alpha \right.$$

$$\left. + \int_{t_1-h}^{t} \eta_{ij}(t, \alpha - t)\dot{x}^j(\alpha)d\alpha \right\}.$$

Now define

$$u^i(t) = B_i^+(t) \left\{ \dot{\psi}^i(t - t_1) + \eta_i(t, -h)x^i(t - h) + \int_{t-h}^{t_1-h} \eta_i(t, \alpha - t)\dot{x}^i(\alpha)d\alpha \right.$$

$$+ \int_{t_1-h}^{t} \eta_i(t, \alpha - t)\dot{\psi}^i(\alpha - t)d\alpha$$

$$+ \sum_{\substack{j=1 \\ j \neq i}}^{\ell} \left[ \eta_{ij}(t, -h)x^j(t - h) + \int_{t-h}^{t_1-h} \eta_{ij}(t, \alpha - t)\dot{x}^j(\alpha)d\alpha \right.$$

$$\left. \left. + \int_{t_1-h}^{t} \eta_{ij}(t, \alpha - t)\dot{x}^j(\alpha)d\alpha \right] \right\},$$

for $t_1 - h \leq t \leq t_1$. Because of the smoothness properties of $x^i, x^j$, and $\psi^i, u^i$ is indeed appropriate. Thus the controllability of the large-scale system can be deduced from that of the subsystems so long as the interconnection is as proposed. We now turn our attention to the nonlinear situation.     $\square$

## 9.2   Nonlinear Systems

Consider the general nonlinear large-scale system,

$$\dot{x}(t) = f(t, x_t, u(t)) + B(t, x_t)u(t) + g(t, x_t, u(t)), \tag{9.2.1}$$

where $f : E \times C \times E^m \to E^n$ is a nonlinear function, $g : E \times C \times E^m \to E^n$ is a nonlinear interconnection, and the $n \times m$ matrix function $B : E \times C \to E^{n \times m}$

is possibly nonlinear. Conditions for the existence of a unique solution $x(\cdot, \sigma, \phi, u)$, when $u \in L_2, \phi \in C([-h, 0], E^n)$, are given in Proposition 8.3.1. It is shown there that $(\phi, u) \to x_t(\cdot, \sigma, u) \in C$ is continuously differentiable. These conditions are assumed to prevail here.

**Remark 9.2.1**   Note that conditions of Proposition 8.3.1 imply that for all

$$t \in [\sigma, t_1], \quad \phi \in C([-h, 0], K), \quad \omega \in E^m,$$

we have

$$|f(t, \phi, \omega)| \le M_1(t)\|\phi\| + M_2(t)|\omega|.$$

System (9.2.1) may be decomposed as

$$\dot{x}^i(t) = f_i(t, x_t^i, u^i(t)) + B_i(t, x_t^i)u^i(t) + \sum_{\substack{j=1 \\ j \ne i}}^{\ell} g_{ij}(t, x_t^j, v^i(t)), \quad i = 1, \dots, \ell,$$

$$(9.2.2)$$

where

$$f_i : E \times C([-h, 0], E^{ni}) \times E^{mi} \to E^{ni},$$

$$g_{ij} : E \times C([-h, 0], E^{ni}) \times E^{mi} \to E^{ni},$$

$$B_i : E \times C([-h, 0], E^{ni}) \to E^{ni \times mi}.$$

Let

$$\sum_{i=1}^{\ell} n_i = n, \quad \sum_{\ell=1}^{\ell} m_i = m,$$

$$x^T = [(x^1)^T, \dots, (x^\ell)^T] \in E^n,$$

$$u^T = [(u^1)^T, \dots, (u^\ell)^T] \in E^m,$$

$$[f(t, x_t, u)]^T = [(f_1(t, x_t^1, u^1))^T \cdots (f_\ell(t, x^\ell, u^\ell))^T],$$

$$g_i(t, x_t, u^i(t)) = \sum_{\substack{j=1 \\ j \ne i}}^{\ell} g_{ij}(t, x_t^j, u^i(t)),$$

$$g(t, x_t, u^i(t)) = [g_1(t, x_t, u^i(t))^T, \dots, g_\ell(t, x_t, u^i(t))^T],$$

$$B(t, x_t) = \text{diag}[B_1(t, x^1), \dots, B_\ell(t, x^\ell)].$$

Then we can view (9.2.1) with decomposition 9.2.2 as an interconnection of $\ell$ isolated subsystems $(S_i)$ described by the equation

$$\dot{x}^i(t) = f_i(t, x_t^i, u^i(t)) + B_i(t, x_t^i)u^i(t), \qquad (S_i)$$

with interconnecting structure characterized by

$$g_i(t, x_t, v^i(t)) = \sum_{\substack{j=1 \\ j \neq i}}^{\ell} g_{ij}(t, x_t^j, v^i(t)) \equiv g_i(t, x_t, v^i(t)).$$

Conditions for the existence and uniqueness of solutions are assumed. In particular, $t \to g_i(t, x_t, u^i(t))$ is assumed integrable.

Conditions for controllability of each isolated subsystem $(S_i)$ are stated in Theorems 8.3.1 and 8.3.2. We defined a matrix $H_i$:

$$H_i = \int_\sigma^{t_1} B_i(s, \phi^i) B_i^*(s, \phi^i) ds, \quad t_1 > \sigma, \tag{9.2.3}$$

for each $\phi^i \in C([-h, 0], E^{ni}) \equiv C^{ni}$.

**Theorem 9.2.1** *In $(S_i)$, assume that:*

(i) *There is a continuous function $N_{2i}^*(t)$ such that $\|B_i^*(t, \phi^i)\| \leq N_{2i}^*(t)$, $\forall \phi^i \in C^{ni}$.*

(ii) *$H_i$ in (9.2.3) has a bounded inverse.*

(iii) *There exist continuous functions $F_{ij} : C \times E^{mi} \to E^+$ and integrable functions $\alpha_j : E \to E^+$, $j = 1, \ldots, q$, such that*

$$|f_i(t, \phi^i, u^i(t))| \leq \sum_{j=1}^{q} \alpha_j(t) F_{ij}(\phi^i, u^i(t))$$

*for all $(t, \phi^i, u^i(t)) \in E \times C \times E^{mi}$ where the following growth conditions are satisfied:*

$$\limsup_{r \to \infty} \left( r - \sum_{j=1}^{q} c_j \sup\{G_j(\phi^i, u^i) : \|(\phi, u)\| \leq r\} \right) = +\infty.$$

*Then $(S_i)$ is Euclidean controllable on $[\sigma, t_1]$.*

We now consider criteria for controllability in $W_2^{(1)}$ for System $(S_i)$.

**Theorem 9.2.2** *In $(S_i)$, assume:*

(i) *Conditions (i) and (iii) of Theorem 9.2.1.*

(ii) *$\text{rank}[B_i(t, \xi)] = n_i$ on $[t_1 - h, t_1]$ for each $\xi \in C([-h, 0], E^{n_i})$, $t \in [t_1 - h, t_1]$.*

(iii) *The Moore Penrose generalized inverse of $B_i$, $B_i^+(t, \xi)$ is essentially uniformly bounded on $[t_1 - h, t_1]$, for each $\xi \in C$, $t \in [t_1 - h, t_1]$.*

(iv) *$\xi \to B_i^+(t, \xi)$ is continuous.*

*Then $(S_i)$ is controllable on $[\sigma, t_1]$, with $t_1 > \sigma + h$.*

In Theorem 9.2.2 we have stated conditions that guarantee the controllability of each isolated free subsystem $(S_i)$. Next we assume these conditions and give additional conditions to the interconnection $g_i$ that will ensure that the composite system (9.2.2) is controllable. It should be carefully noted that

$$g_i(t, x_t, v^i(t)) = \sum_{\substack{j=1 \\ j \neq i}}^{\ell} g_{ij}(t, x_t^j, v^i(t))$$

is independent of $x^i$ the state of the $i$th subsystem, though it is measured locally in the $(S_i)$ system.

**Theorem 9.2.3**  *Consider (9.2.1) and its decomposition (9.2.2). Assume that:*

(i) *Conditions (i)–(iii) of Theorem 9.2.1 are valid. Thus each isolated subsystem is Euclidean controllable on $[\sigma, t_1]$.*

(ii) *For each $i, j = 1, \ldots, \ell, \ i \neq j$,*

$$g_i(t, x_t^i, v^i) = \sum_{\substack{j=1 \\ j \neq i}}^{\ell} g_{ij}(t, x_t^j, v^i(t))$$

*satisfies the following conditions: There are continuous functions*

$$C_{ij} : E^{n_i} \times E^{m_i} \to E^+$$

*and $L'$ functions $\beta_j : E \to E^+ \ j = 1, \ldots, q$, such that*

$$|g_{ij}(t, \psi^i, v^i)| \leq \sum_{j=1}^{q} \beta_j C_{ij}(\psi^i, u^i)$$

*for all $(t, \psi^i, u^i)$, $\beta_i < \alpha_i$, where*

$$\limsup_{r \to \infty} \left( r - \sum_{j=1}^{q} C_j \sup\{C_{ij}(\psi^i, u^i) : \|(\psi^i, u^i)\| \leq r\} \right).$$

*Then (9.2.1) is Euclidean controllable on $[\sigma, t_1]$.*

The proof is contained in [14] and is similar to the proof of Theorem 9.2.1.

**Theorem 9.2.4**  *Consider (9.2.1) and its decomposition (9.2.2). Assume that:*

(i) *Conditions (i)–(iv) of Theorem 9.2.2 hold.*

(ii) *Condition (ii) of Theorem 9.2.3 holds.*

*Then (9.2.1) is controllable on $[\sigma, t_1], \ t_1 > \sigma + h$.*

**Proof**   Just as in the proof of Theorem 9.2.2, define

$$H(t_1 - h) = \int_\sigma^{t_1 - h} B(s, x_t) B^*(s, x_t) ds$$

in place of $H_i(t_1 - h)$, and conclude Euclidean controllability on $[\sigma, t_1 - h]$. Define $u$ as

$$u(t) = B^+(t, x_t^u)[\dot{\psi}(t - t_1) - f(t, x_t^u, u(t)) - g(t, x_t, u(t))]$$

for $t_1 - h \leq t \leq t_1$ in place of $u^i(t)$, and deduce controllability. The growth condition (ii) of $g$ comes in handy in the argument to conclude the proof as before.     $\square$

## 9.3   General Nonlinear Systems

In (9.2.1) it is very important that the system is of the form in which some term is linear in $u$. Here we consider the more general situation

$$\dot{x}(t) = f(t, x_t, u(t)), \tag{9.3.1}$$

where $f : E \times C \times E^m \to E^n$ is continuously differentiable in the second and third arguments, and is continuous, and also satisfies all the conditions of Proposition 8.3.1. Details of the proof of the following are contained in Proposition 8.2.1.

**Theorem 9.3.1**   *In* (9.3.1), *assume that*:

(i)   $f(t, 0, 0) = 0, \ \forall \ t \geq \sigma.$
(ii)   *The system*

$$\dot{z}(t) = L(t, z_t) + B(t)v(t) \tag{9.3.2}$$

*is controllable on* $[\sigma, t_1]$, *where* $t_1 \geq \sigma + h$, *and where*

$$D_2 f(t, 0, 0) z_t = L(t, z_t), \quad D_3 f(t, 0, 0)v = B(t)v.$$

*Then*

$$0 \in \text{Int } \mathcal{A}(t, \sigma), \tag{9.3.3}$$

*where*

$$\mathcal{A}(t, \sigma) = \{x_t(\sigma, 0, u) : u \in L_2([\sigma, t], E^m) \|u\|_{L_2} \leq 1 \quad x(u)$$

$$\textit{is a solution of } (9.3.1) \textit{ with } x_\sigma = 0\} \tag{9.3.4}$$

*is the attainable set associated with* (9.3.1).

We shall now investigate the large-scale system

$$\dot{x}^i(t) = f_i(t, x_t^i, u^i(t)) + \sum_{\substack{j=1 \\ i \neq j}}^{\ell} g_{ij}(t, x_t^j, u^j(t)), \tag{9.3.5}$$

where $f_i$ and $g_{ij}$ are as defined following (9.2.2). Thus if

$$g_i(t, x_t, v(t)) = \sum_{\substack{j=1 \\ j \neq i}}^{\ell} g_{ij}(t, x_t^j, u^j(t)),$$

and $g(t, x_t, v(t)) = [g_1(t, x_t, v(t))^T, \ldots, g_\ell(t, x_t, v(t))^T]$, by considering (9.3.5) we are investigating

$$\dot{x}(t) = f(t, x_t, u(t)) + g(t, x_t, v(t)), \tag{9.3.6}$$

where $f$ is identified following (9.2.2). We state the following result:

**Theorem 9.3.2**  *Consider the large-scale system (9.3.6) with its decomposition (9.3.5), where*

(i)  $f_i(t, 0, 0) = 0,\ g_{ij}(t, \phi, 0) = 0.$
(ii)  *$f_i, g_i$ satisfy all the requirements of Proposition 8.3.1.*
(iii)  *Assume that the linear variational system*

$$\dot{z}^i(t) = L_i(t, z_t^i) + B^i(t)v(t), \tag{9.3.7}$$

*of*

$$\dot{x}^i(t) = f_i(t, x_t^i, u^i(t)), \tag{9.3.8}$$

*where*

$$D_2 f_i(t, 0, 0)z_t^i = L_i(t, z_t^i), \qquad D_3 f_i(t, 0, 0)v = B^i(t)v$$

*is controllable on $[\sigma, t_1]$, $t_1 > \sigma + h$.*

*Then the interconnected large-scale system (9.3.5) is locally null controllable with constraints.*

**Corollary 9.3.1**  *Assume:*

(i)  *Conditions (i)–(iii) of Theorem 9.3.2.*
(ii)  *The system*

$$\dot{x}^i(t) = f_i(t, x_t^i, 0) \tag{9.3.9}$$

*is globally exponentially stable.*

*Then the interconnected system is (globally) null controllable, with controls in*

$$U_i = \{u^i \in L_2([\sigma, t_1], E^{m_i}), \ \|u^i\|_{L_2} \le 1\}.$$

**Proof of Theorem 9.3.2**　By Theorem 9.3.1,

$$0 \in \operatorname{Int} \mathcal{A}_i(t, \sigma) \quad \text{for} \quad t > \sigma + h, \tag{9.3.10}$$

where $\mathcal{A}_i$ is the attainable set associated with (9.3.8). Let $x^i$ be the solution of (9.3.5), with $x^i_\sigma = 0$. Then

$$x^i(\sigma, 0, u^i, u^j)(t) = \int_\sigma^t f_i(s, x^i_s, u^i(s))ds + \int_\sigma^t \sum_{\substack{j=1 \\ i \ne j}}^\ell g_{ij}(s, x^j_s, u^j(s))ds.$$

Thus, if we define the set

$$H_i(t_1, \sigma) = \{x^i(\sigma, 0, u^i, u^j) \in W_2^{(1)}([0, t_1], E^{n_i}) : u^i \in U_i, \ u^j \in U_j\},$$

we deduce that

$$\mathcal{A}_i(t, \sigma) \subset H_i(t_1, \sigma).$$

Because $f_i(t, 0, 0) = g_{ij}(t, \phi, 0) = 0$, and because $x^i(t, 0, 0) = 0$ is a solution of (9.3.5), $0 \in H_i(t_1, \sigma)$. As a result of this and (9.3.10), we deduce

$$0 \in \operatorname{Int} \mathcal{A}_i(t_1, \sigma) \subset H_i(t_1, \sigma). \tag{9.3.11}$$

There is an open ball $\mathbb{B}(0, r)$, center zero, radius $r$, such that

$$0 \in \mathbb{B}(0, r) \subset \mathbb{R}_i(t_1, \sigma) \subset H_i(t_1, \sigma).$$

The conclusion

$$0 \in \operatorname{Int} H_i(t_1, \sigma) \tag{9.3.12}$$

follows at once. Using this, one deduces readily that $0 \in \operatorname{Int} \mathcal{D}$, the interior of the domain of null controllability of (9.3.5), proving local null controllability with constraints. $\square$

**Proof of Corollary 9.3.1**　One uses the control $u^i = 0 \in U_i$, $i = 1, \dots, \ell$ to glide along System (9.3.5) and approach an arbitrary neighborhood $\mathcal{O}$ of the origin in $W_2^{(t)}([-h, 0], E^{n_i})$. Note that

$$0 \in U_i = \{u^i \in L_2([0, \sigma], E^{m_i}), \|\|u^i\|_{L_2} \le 1\}.$$

Because of stability in hypothesis (ii) of (9.3.9), every solution with $u^i = 0$ is entrapped in $\mathcal{O}$ in time $\sigma \ge 0$. Since (i) guarantees that all initial states in this neighborhood $\mathcal{O}$ can be driven to zero in finite time, the proof is complete. $\square$

**Remark**　Conditions for global stability of hypothesis (ii) are available in Chukwu [10, Theorem 4.2].

**Remark 9.3.1**  The condition

$$0 \in \text{Int } \mathcal{A}_i(t, \sigma) \subset H_i(t, \sigma),\tag{9.3.11}$$

from which we deduced that

$$0 \in \text{Int } H_i(t, \sigma),\tag{9.3.12}$$

is of fundamental importance. If the condition

$$0 \in \text{Int } \mathcal{A}_i(t, \sigma)\tag{9.3.13}$$

fails, the isolated system is "not well behaved" and cannot be controlled. Condition (9.3.12) may still prevail and the composite system will be locally controllable. To have this situation, we require

$$0 \in \text{Int } G_i(t, \sigma),\tag{9.3.14}$$

where

$$G_i(t, \sigma) = \left\{ \int_\sigma^t \sum_{\substack{j=1 \\ i \neq j}}^{\ell} g_{ij}(s, x_s^j, u^j(s))ds \ : \ u^j \in U_i \right\}.$$

In words, we require a sufficient amount of control impact (i.e., (9.3.14)) to be brought to bear on $(S_i)$ that is not an integral part of $S_i$. Thus knowing the limitations of the control $u^i \in U_i$, a sufficient signal $\sum_{\substack{j=1 \\ j \neq i}}^{\ell} g_{ij}(t, x_t^j, u^j) = g_i(t, x_t, u)$ is dispatched to make (9.3.14) hold. And (9.3.12) will follow.

**Remark 9.3.2**  The same type of reasoning yields a result similar to Theorem 9.3.2 if we consider the system

$$\dot{x}_i(t) = f_i(t, x_t^i, u^i(t)) + g_i(t, x_t^1, \ldots, x_t^\ell, u^1(t), \ldots, u^\ell(t)),\tag{9.3.15}$$

where

(i)  $f(t, 0, 0) = 0.$
(ii)

$$g_i(t, x_t^1, \ldots, x_t^{i-1}, 0, x_t^{i+1}, \ldots, x_t^\ell, u^1(t), \ldots, u^{i-1}(t), 0, u^{i+1}(t), \ldots, u^\ell(t)) = 0,$$

$$g_i(t, x_t, \ldots, x_t^\ell, 0, 0, \ldots, 0) = 0.$$

Also conditions (ii) and (iii) of Theorem 9.3.2 are satisfied.
If we consider

$$\dot{x}^i(t) = f_i(t, x_t^i, u^i(t)) + \sum_{\substack{j=1 \\ i \neq j}}^{\ell} g_{ij}(t, x_t^j, u^i(t))\tag{9.3.16}$$

instead of (9.3.5), we can obtain the following result:

**Theorem 9.3.3**     *In (9.3.16), assume* (i)–(iii) *of Theorem 9.3.2. But in (9.3.7),* $L_i(t, z_t^i)$ *and* $B^i(t)$ *are defined as*

$$D_2 f_i(t, 0, 0) z_t^i = L_i(t, z_t^i), \quad D_3(f_i(t, 0, 0)) + g_i(t, x_t, 0)) = B^i(t).$$

*Then* (9.3.16) *is locally null controllable with constraints.*

The proof is essentially the same as that of Theorem 9.3.1. We note that the essential requirement for (9.3.16) to be locally null controllable is the controllability of

$$\dot{x}(t) = L(t, x_t) + (B_1(t) + B_2(t))u(t), \tag{9.3.17}$$

where

$$B_1(t) = D_3 f(t, 0, 0),$$

$$B_2(t) = D_3 g(t, x_t, 0).$$

If the isolated system (9.3.8) is not "proper" (and this may happen when $B_1(t)$ does not have full rank on $[\sigma, t_1]$, $t > \sigma + h$), the solidarity function $g_i$ can be brought to bear to force the full rank of $B = B_1 + B_2$, from which (9.3.16) will be proper because (9.3.17) is controllable. Even if $B_1$ has full rank and (9.3.8) is proper, the interconnected system need not be locally null controllable. The function has to be so nice that $B_2 + B_1$ has full rank. An adequate proper amount of regulation is needed in the form of a solidarity function $g_i$.

In applications it is important to know something about $g_i$ and to decide its adequacy. It is possible to consider $g_i$ as a control, and view

$$\dot{x}_i(t) = f_i(t, x_t^i, u^i(t)) + g_i(t)$$

as a differential game. Considered in this way, we must describe how big the control set for $g_i$ must be. In the linear case see Chukwu [13]. The linear case is treated in Sec. 9.5.

## 9.4    Examples

**Example 9.4.1     (Fluctuations of Current)** The flow of current in a simple circuit in Fig. 9.4.1 is described by the equation

$$L\frac{d^2 x(t)}{dt^2} + (R + R_1)\frac{dx(t)}{dt} + \frac{1}{c}x(t) = 0.$$

The voltage across $R_1$ is applied to a nonlinear amplifier $A$. The output is provided a special phase shifting network $P$. This introduces a constant time lag between the input and output of $P$. The voltage across $R$ in series with the output of $P$ is

$$e(t) = qg(\dot{x}(t - h)),$$

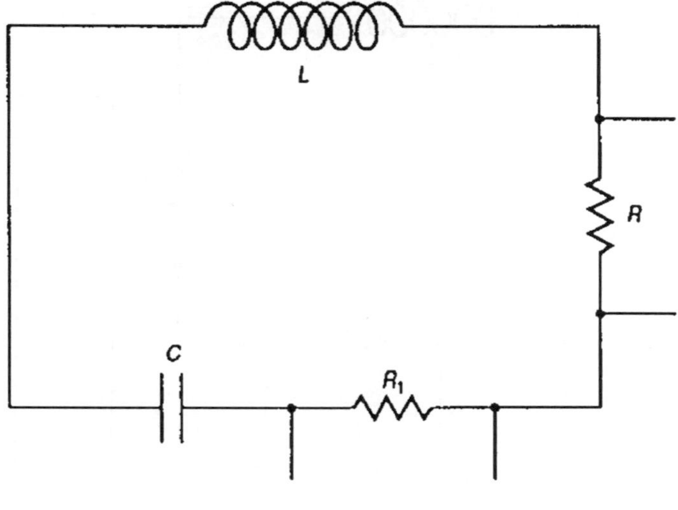

Fig. 9.3.1

$q$ is the gain of the amplifier to $R$ measured through the network. The equation becomes

$$L\frac{d^2x(t)}{dt^2} + R\dot{x}(t) + qg(\dot{x}(t-h)) + \frac{1}{c}x(t) = 0.$$

Finally, a control device is introduced to help stabilize the fluctuations of the current. If $\dot{x}(t) = y(t)$, the "controlled" system is now given by

$$\dot{x}(t) = y(t) + u_1(t),$$
$$\dot{y}(t) = -\frac{R}{L}y(t) - \frac{q}{L}g(y(t-h)) - \frac{1}{cL}x(t) + u_2(t). \qquad (9.4.1)$$

The question of interest is the following: With the control $u = (u_1, u_2)$, which is "created" by the stabilizer, is it possible to bring any "wild fluctuations" of the current (any initial position) to a normal equilibrium position in finite time? Is (9.4.1) controllable?

**Example 9.4.2** **(Fluctuations of Current in a Network)** A six-loop electric network is connected up as in Figs. 9.4.1 and 9.4.2. Each loop is a simple circuit as in Fig. 9.4.1, and is described by the equation

$$L_i\frac{d^2x^i(t)}{dt^2} + (R_i + R_{1i})\frac{dx^i}{dt}(t) + \frac{1}{c}x^i(t) = 0.$$

The voltage across $R_{1i}$ is applied to a nonlinear amplifier $A_i$. The output is provided to a special shifting network $P_i$. This introduces a constant time lag between the input and the output of $P_i$. The voltage across $R_i$ in series with the output of $P_i$

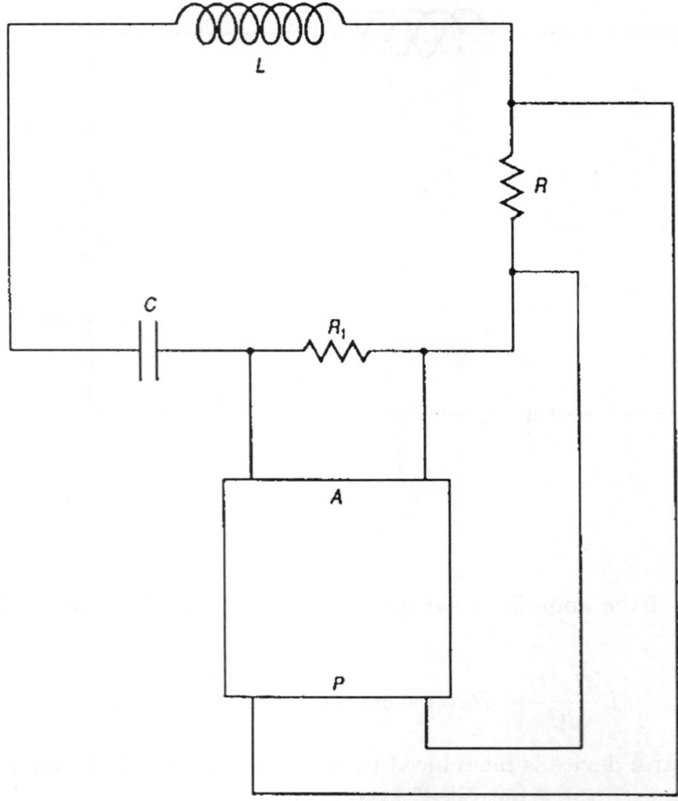


Fig. 9.4.2

is $e_i(t) = q_i g(\dot{x}^i(t - h))$, $q_i$ is the gain of the amplifier to $R_i$ measured through the network. The equation becomes

$$L_i \frac{d^2 x^i(t)}{dt^2} + R_i \dot{x}^i(t) + q_i g_i(\dot{x}^i(t - h)) + \frac{1}{c_i} x^i(t) = 0 \,.$$

For this $i$th loop, a control device is introduced to help stabilize the fluctuations of current. If $\dot{x}^i(t) = y^i(t)$, the "controlled" $i$-subsystem is

$$\dot{x}^{(i)}(t) = y^i(t) + u_1^i(t) \,,$$

$$\dot{y}^i(t) = -\frac{R_i}{L_i} y^i(t) - \frac{q_i}{L_i} g_i(y^i(t - h)) - \frac{1}{c_i L_i} x^i(t) + u_2^{(i)}(t) \,.$$

The question of interest is the following: With a control $u^i = (u_1^i, u_2^i)$, which is "created" by the stabilizer, is it possible to bring any "wild fluctuations" of current (any initial position) to a normal equilibrium position of the $i$th subsystem in finite time? Is the $i$th-subsystem controllable? Furthermore, we study the overall large-scale system in Fig. 9.4.3, which we consider as a decentralized control problem,

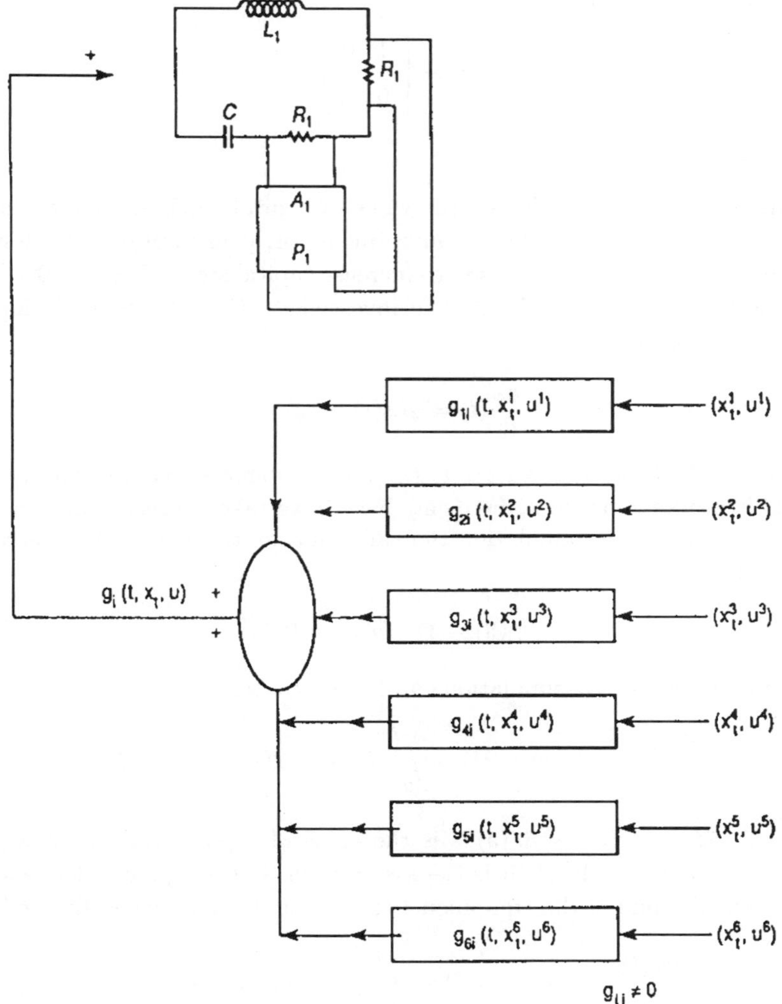



Fig. 9.4.3

where each subsystem is influenced by interaction terms $g_i(t, x) = \sum_{\substack{j=1 \\ j \neq i}} g_{ij}(t, x^j, u^i)$ that are monitored (measured) locally. Is the large-scale system controllable?

**Proposition 9.4.1**   *Assume*

(i)   $R_i$ $L_i, q_i$ *and* $C_i$ *are positive and* $g_i(\phi)$ *is continuous;*
(ii)  $g_{ij}(t, \phi^j, u^i)$ *satisfies the growth condition of Theorem 9.2.4.*

*Then the interconnected system is controllable, and each subsystem is controllable.*

**Proof**   Note that the matrix

$$B_i = \begin{bmatrix} 1 & 0 \\ 0 & 1 \end{bmatrix}$$

has full rank. $\qquad\qquad\qquad\qquad\qquad\qquad\qquad\qquad\qquad\qquad\qquad\qquad\quad \square$

**Example 9.4.3**   Let $x(t)$ denote the value of capital stock at time $t$. Suppose $x$ can be used in one of two ways: (i) investment; (ii) consumption. We assume the value allocated to investment is used to increase capital stock. Let $u(t)$ $0 \le u(t) \le 1$ denote the fraction of $x(t)$ allocated to investment. Capital accumulation can be said to satisfy the equation

$$\frac{dx}{dt}(t) = g(x(t), u(t))\,.$$

For example, $g(x(t), u(t)) \equiv ku(t)x(t)$ ($k$ constant) shows that the rate of increase of capital is proportional to investment. Now if we take depreciation into account, and if at time $a$ after accumulation the value of $a$ unit of capital has decreased by a factor $p(a)$,

$$p(0) = 1, \quad p(L) = 0\,,$$

the equation of capital accumulation can be represented by

$$\frac{dx(t)}{dt} = g(x(t), u(t)) - \int_0^L f(t - s, x(t - s))p'(s)ds\,,$$

where $-\int_0^L f(t - s, x(t - s))p'(s)ds$ is the value of capital that has "evaporated" per unit time at time $t$. Implicit is the assumption that equipment depreciates over a time $L$ (the lifetime of the equipment) to a value 0. More generally, we have

$$\frac{dx(t)}{dt} = g(x(t), u(t)) - \int_0^L f(t - s, x(t - s))dp(s)\,, \qquad (9.4.2)$$

where we use a Riemann–Stieltjes integral.

Equation (9.4.2) is a special case of (9.3.1). Along with (9.4.2) are an initial capital endowment function $\phi \in W_2^{(1)}$ and $\psi \in W_2^{(1)}$, its long run "desired stock of capital function". In time-optimal control theory, one minimizes $T$ (time) subject to (9.4.2) and

$$x_0 = \phi, \quad x_T = \psi\,. \qquad (9.4.3)$$

One can also minimize consumption

$$\int_0^T (1 - u(t))x(t)dt \qquad (9.4.4)$$

subject to (9.4.2) and (9.4.3). System (9.4.2) is an isolated one. It may form part of a large-scale system

$$\frac{dx^i}{dt}(t) = g_i(x^i(t), u^i(t)) - \int_0^L f_i(t - s, x^i(t - s))dP_i(s)$$

$$+ \sum_{\substack{j=1 \\ i \neq j}}^{\ell} M_{ij}(x^j(t - h), u^i(t)),\qquad(9.4.5)$$

and the problem of controllability can be investigated.

From our analysis we have the following insight: If the isolated system is controllable and the interconnection "nice", then the large-scale system is also controllable. Detailed investigations of (9.4.5) will be pursued in Sec. 9.5, where comments on global economy will be made. See [11] and [12, p. 688].

It is seen from Remark 9.3.1 that a sufficient amount of control regulation is needed from outside the subsystem to make the large-scale system controllable. When this is lacking, and the subsystem is not well behaved, the controllability of the large-scale system is not assured. This has grave policy implications for global economy. Deregulation may not be a cure. It may lead to global depression, or lack of growth of the economy.

## 9.5   Control of Global Economic Growth

### 9.5.1   *Introduction*

In Sec. 1.8 we formulated a nonlinear general model of the dynamics of capital accumulation of a group of $n$ firms whose destiny are linked with $\ell$ other systems to form a large-scale organization. As argued by Takayama [12], time lags are incorporated into the control procedure in a distributed fashion: The state and the control history are taken into account. Our aim is to control the growth of capital stock from an initial endowment to some target point, which can be taken to be the origin. From the theory, some qualitative laws are formulated for the control of the growth of capital stock. They help validate the correctness of the new economic policies of movement away from centralization of firms of the East; they raise questions on the wisdom of deregulations in the West. The principles cast doubt on the popular myth that to grow, the firms of the Third World should dismantle the "solidarity function".

In Sec. 1.8 we designated $x(t)$ to be the value of capital stock at time $t$, and then postulated that the net capital formation $\dot{x}(t)$ is given by

$$\dot{x}(t) = ku(t)x(t) - \delta x(t),\qquad(9.5.1)$$

where $k$ and $\delta$ are positive constants and $-1 \leq u(t) \leq 1$, $u$ being the fraction of stock that is used for payment of taxes or for investment. A more general version of (9.5.1) is the system

$$\dot{x}(t) = L(t, x_t, u_t)x_t + B(t, x_t, u_t)u_t \,. \tag{9.5.2}$$

Later we denote $x(t) = (x_1(t), \ldots, x_n(t))$ to be the value of $n$ capital stocks at time $t$, with investment and tax strategy $u = (u_1, \ldots, u_n)$, where $-1 \leq u_j(t) \leq 1$. We therefore consider (9.5.2) as the equation of the net capital function for $n$ stocks in a region that is isolated. They are linked to $\ell$ other such systems in the globe, and the interconnection or solidarity function is given by

$$g_i(x_{1t}, \ldots, x_{\ell t},\ u_{1t}, \ldots, u_{\ell t}) \,.$$

This function describes the effects of other subsystems on the $i$th subsystem as measured locally at the $i$th location. Thus,

$$\dot{x}_i(t) = L_i(t, x_{it}, u_{it})x_{it} + B_i(t, x_{it}, u_{it})u_{it}$$

$$+ \, g_i(t, x_{1t}, \ldots, x_{\ell t}, u_{1t}, \ldots, u_{\ell t}), i = 1, \ldots, \ell \,, \tag{9.5.3}$$

is the decomposed interconnected large-scale system whose free subsystem is

$$\dot{x}_i(t) = L_i(t, x_{it}, u_{it})x_{it} + B_i(t, x_{it}, u_{it})u_{it} \,. \tag{9.5 Si}$$

We now introduce the following notation:
    Let

$$\sum_{i=1}^{\ell} n_i = n, \ \sum_{i=1}^{\ell} m_i = m \,.$$

$$\phi = [\phi_1, \ldots, \phi_\ell] \in E^n, \quad x = [x_1, \ldots, x_\ell] \in E^n, \quad u = [u_1, \ldots, u_\ell] \in E^m \,.$$

$$L(t, x_t, u_t) = [L_1(t, x_{1t}, u_1), \ldots, L_\ell(t, x_{\ell t}, u_{\ell t})] \,,$$

$$g(t, x_t, u_t) = [g_1(t, x_t, u_t), \ldots, g_\ell(t, x_t, u_t)] \,,$$

$$B(t, x_t, u_t) = \mathrm{diag}[B_1(t, x_{1t}, u_{1t}), \ldots, B_\ell(t, x_{\ell t} u_\ell(t)] \,.$$

Then (9.5.3) is given as

$$\dot{x}(t) = L(t, x_t, u_t) + B(t, x_t, u_t)u_t + g(t, x_t, u_t) \,. \tag{9.5 S}$$

### 9.5.2   *Preliminaries*

The problem of controllability of (9.5 S) will be explored in this section. Conditions are stated for the controllability of the isolated system (9.5.2). Assuming that the subsystem (9.5 Si) is controllable, we shall deduce conditions for (9.5 S) to be controllable when the solidarity function is "nice".

    The following basic assumptions will be maintained: In (9.5.2),

$$L(t, \phi, \psi)x_t = \int_{-h}^{0} d\eta(t, s, \phi, \psi)x(t+s) \,,$$

where $\eta(t, s, \phi, \psi)$ is measurable in $(t, s) \in E \times E$ normalized so that

$$\eta(t, s, \phi, \psi) = 0, \quad s \geq 0$$

$$\eta(t, s, \phi, \psi) = \eta(t, -h, \phi, \psi), \quad s \leq -h.$$

$\eta(t, s, \phi, \psi)$ is continuous from the left in $s$ on $(-h, 0)$, and $\eta(t, s, \phi, \psi)$ has bounded variation in $s$ on $[-h, 0]$ for each $t, \phi, \psi$, and there is an integrable $m$ such that

$$|L(t, s, \phi, \psi)x_t| \leq m(t)\|x_t\|$$

for all $t \in (-\infty, \infty), \phi, \psi, \ x_t \in C$. We assume $L(t, \cdot, \cdot)$ is continuous. The assumption on $B(t, \phi, \psi)$ in (9.5.2) is continuity on all the variables. Also continuous is the function $g : E \times C \times E^m \rightarrow E^n$. Enough smoothness conditions on $L$ and $g$ are imposed to ensure the existence of solutions of (9.5 S) and the continuous dependence of solutions on initial data. See [19] and [8], and Sec. 8.3.

### 9.5.3 *Main results*

To solve the Euclidean controllability problem for the system (9.5.2), we consider the simpler system

$$\dot{x}(t) = L(t, z, v)x_t + B(t, z, v)v, \tag{9.5.4}$$

where the arguments $x_t$ and $u_t$ of $L$ and $B$ have been replaced by specified functions $z \in C([-h, 0], E^n) \equiv C, \ v \in L_\infty([-h, 0], E^m)$. Here $C$ is the space of continuous $E^n$-valued functions on $[-h, 0]$, and $L_\infty$ is the space of essentially bounded measurable $E^m$-valued functions on $[-h, 0]$. For each $(z, v) \in C \times L_\infty$, the system (9.5.4) is linear and one can deduce its variation of parameter by the ideas of Klamka [23]. Let $X(t, s) = X(t, s, z, v)$ be the transition matrix for the system

$$\dot{x}(t) = L(t, z, v)x_t, \tag{9.5.5}$$

so that

$$\frac{\partial X(t, s)}{\partial t} = L(t, \cdot, z, v)X_t(\cdot, s), \tag{9.5.6}$$

where

$$X(t, s) = \begin{cases} 0, & s - h \leq t < s, \\ I, & t = s \ (\text{I identity}), \end{cases}$$

and where $X_t(\cdot, s)(\theta) = X(t + \theta, s), \ -h \leq \theta \leq 0$. Then the solution of (9.5.4) is given by

$$x(t) = x(t, \sigma, \phi, 0) + \int_\sigma^t X(t, x) \left[ \int_{-h}^0 d_\theta H(s, z(\theta), v(\theta))u(s + \theta) \right] ds, \quad \sigma \leq t \leq t_1,$$

$$x(t) = \phi(t), \quad t \in [\sigma - h, \sigma], \quad u(t) = \eta(t), \quad t \in [\sigma - h, \sigma]. \tag{9.5.7}$$

Here $x(t, \sigma, \phi, 0)$ is the solution of (9.5.5). The last term in (9.5.7) contains values of $u$ on $[\sigma - h, \sigma]$. As usual in such matters [7], [16], [20], and [23], we use Fubini's Theorem to write (9.5.7) as

$$x(t) = x(t, \sigma, \phi, 0) + \int_{\sigma+s}^{\sigma} \left\{ \int_{-h}^{0} X(t, s - \theta) d_\theta [H(s - \theta, z(\theta), v(\theta)] \eta(\theta) ds \right\}$$

$$+ \int_{\sigma}^{t} \left( \int_{-h}^{0} X(t, s - \theta) d_\theta [H(s - \theta, z(\theta), u(\theta))] \right) u(s) ds. \qquad (9.5.8)$$

Define $q$ as

$$q = q(t, \eta, z, v) = \int_{\sigma+s}^{\sigma} \left\{ \int_{-h}^{0} X(t, s - \theta) d_\theta [H(s - \theta, z(\theta), v(\theta))] \eta(\theta) \right\} ds, \qquad (9.5.9)$$

and

$$S(t, s, z, v) = \int_{-h}^{0} X(t, s - \theta) d_\theta H(s - \theta, z(\theta), v(\theta)). \qquad (9.5.10)$$

Thus (9.5.8) can be written as

$$x(t) = x(t, \sigma, \phi, 0) + q + \int_{\sigma}^{t} S(t, s, z, v) u(s) ds. \qquad (9.5.11)$$

The controllability map of (9.5.4) at time $t$ is

$$W(t) = W(\sigma, t, z, v) = \int_{\sigma}^{t} S(t, s, z, v) S^*(t, s, z, v) ds, \qquad (9.5.12)$$

with $W(t_1) = W$, where the star denotes the matrix transpose. Define the reachable set $\mathbb{R}(t)$ of (9.5.4) as follows:

$$\mathbb{R}(t) = \left\{ \int_{\sigma}^{t} S(t, s, z, v) u(s) ds : u \in U_{\mathrm{ad}} \right\},$$

where

$$U_{\mathrm{ad}} = \{ u \text{ measurable } u \in E^m | u_j(t) | \leq 1, \ j = 1, \ldots, m \}.$$

Thus the reachable set is all solution $x(u)$ of (9.5.4) with $\phi = 0$, $\eta = 0$ as initial data and with controls $u \in L_\infty([\sigma, t_1], \ C^m)$, where $C^m$ is the unit $m$-dimensional cube. The Euclidean attainable set of (9.5 S) is the subset of $E^n$ defined as follows:

$$\mathcal{A}(t) = \{ x(u)(t) : u \in U_{\mathrm{ad}} : x \text{ is a solution of (S) with } x_\sigma = 0, \eta = 0 \}.$$

The domain $\mathcal{D}$ of null controllability is the set of initial $\phi$, such that there exists a $t_1$ and and a $u \in L([\sigma, t_1], C^m)$ such that the solution $x$ of (9.5 S) satisfies

$$x_\sigma(u) = \phi, \quad x(t_1, u) = 0.$$

The following lemma is fundamental in the sequel:

**Lemma 9.5.1** *In* (9.5.4)

    (i) *Maintain the basic assumptions of $L$ and $B$.*
    (ii) *There exists a constant $c > 0$ such that $\det W(\sigma, t_1, v) \geq c > 0$ for all $(z, v) \in C \times L_\infty$.*

*Then*

$$0 \in \text{Int } \mathcal{A}(t). \tag{9.5.13}$$

*Also for* (9.5.2),

$$0 \in \text{Int } \mathcal{D}. \tag{9.5.14}$$

**Proof**  From standard arguments, the nonsingularity of $W = W(t_1) \equiv W(\sigma, t_1, z, v)$ is equivalent to the Euclidean controllability of (9.5.4) [22, p. 92] and of (9.5.4) being "proper" [22, p. 77]. As usual, this is equivalent to

$$0 \subset \text{Int } \mathbb{R}(t, z, v). \tag{9.5.15}$$

Because (ii) is uniform for all $(\phi, v) \subset C \times L_\infty$, $0 \in \text{Int } \mathbb{R}(t, z, v)$ for all $(\phi, v) \in C \times L_\infty$. That (9.5.15) holds follows from the fact that the map

$$T : L_\infty \to E^n ,$$
$$(Tu)(t) = \int_\sigma^t X(t, \tau, z, v) B(\tau, z, v) u(\tau) d\tau \tag{9.5.16}$$

is a continuous linear surjection for each $(z, v) \in C \times L_\infty$, which is therefore an open map for each $(z, v)$. The transfer of any arbitrary $\phi \in C$ to any $x_1 \in E^n$ is accomplished by the control

$$u(t) = S^*(t_1, t, z, v) W^{-1}[x_1 - x(t, \phi, 0) - q(t, \eta, z, v)], \quad t \in [\sigma, t_1],$$
$$u(t) = \eta(t), \quad t \in [\sigma - h, t_1], \tag{9.5.17}$$

which when inserted into (9.5.11) gives an expression of the solution (9.5.4):

$$x(t) = x(t, \sigma, \phi, 0) + q(t, \eta, z, v)$$
$$+ W(\sigma, t, z, v) W^{-1}(\sigma, t_1, z, v)[x_1 - x(t_1, t_0, \phi, 0) - q(t_1, \eta, z, v)]. \tag{9.5.18}$$

With initial data $w(\sigma) = (x(\sigma), \phi, \eta)$ and $x(t_1) = x_1$, controllability of (9.5.4) is established. We now show controllability for (9.5.2). Suppose $(x, u)$ solves the set of integral equations

$$u(t) = S^*(t_1, t, x, u) W^{-1}[x_1 - x(t_1, \sigma, \phi, 0) - q(t_1, \eta, z, u)] \quad \text{for} \quad t \in [\sigma, t_1],$$
$$u(t) = \eta(t), \quad t \in [\sigma - h, \sigma], \tag{9.5.19}$$

and

$$x(t) = x(t, \sigma, \phi, 0) + q(t, u, x, u) + \int_\sigma^t S(t, s, x, u)u(s)ds, \quad t \geq \sigma,$$

$$x(t) = \phi(t), \quad t \in [\sigma - h, \sigma].$$

(9.5.20)

Then $u$ is $L_\infty([\sigma - h, t_1])$ and $x$ solves (9.5.2) with this $u$ and with initial data $w(\sigma) = (x(\sigma), \phi, \eta)$, where $u_\sigma = \eta$. Also $x(t_1) = x_1$. Next we prove that a solution pair indeed exists for (9.5.19) and (9.5.20). Let $\mathbb{B}$ be the Banach space of all functions $(x, u) : [\sigma - h, t_1] \times [\sigma - h, t_1] \to E^n \times E^m$ where $x \in C([\sigma - h, t_1], E^n)$ $u \in L_\infty([\sigma - h, t_1], E^m)$, with norms defined as $\|(x, u)\| = \|x\| + \|u\|_\infty$, where

$$\|x\| = \max_{s \in [\sigma - h, t_1]} |x(s)|, \quad \|u\|_\infty = \max_{1 \leq j \leq m} \sup_{\sigma - h \leq s \leq 0 \leq s \leq t_1} |u_j(s)|.$$

Define the operator $P : \mathbb{B} \to \mathbb{B}$ by $P(x, u) = (y, v)$, where

$$v(t) = S^*(t_1, t, x, u)W^{-1}[x_1 - x(t_1, \sigma, \phi, 0) - q(t_1, \eta, x, u)] \quad \text{for} \quad t \in [\sigma, t_1], \quad (9.5.21)$$

and

$$u(t) = \eta(t) \quad \text{for} \quad t \in [\sigma - h, \sigma].$$

$$y(t) = x(t, \sigma, \phi, 0) + q(t, \eta, x, v) + \int_\sigma^t S(t, s, x, u)v(s)ds \quad \text{for} \quad t \in [\sigma, t_1],$$

(9.5.22)

and $y(t) \equiv \phi(t) \quad \text{for} \quad t \in [\sigma - h, \sigma]$.

By a tedious but standard analysis (see Sinha [7], and Balachandran and Dauer [15, 16], we can identify a closed bounded and convex subset $Q(\lambda)$ of $\mathbb{B}$ where

$$Q(\lambda) = \{(x, u) \in \mathbb{B} : \|x\| \leq \lambda, \ \|u\|_\infty \leq \lambda\},$$

and $P : Q(\lambda) \to Q(\lambda)$, and $P$ is a relatively compact mapping. We are then assured by Schauder's fixed-point theorem that $P$ has a fixed-point: $P(x, u) = (x, u)$. Indeed, let

$$a_1 = \sup\{\|S^*(t_1, t, x_{t_1}, u_{t_1})\|, t \in [\sigma, t_1]\},$$

$$a_2 = \|W^{-1}(\sigma, t_1, x_{t_1}, u_{t_1})\|,$$

$$a_3 = \sup\{|x(t, \sigma, \phi, 0, 0)| + |g(t, \eta, x_t, u_t)| + |x_1|, t \in [\sigma, t_1]\},$$

$$a_4 = \sup\{\|S(t, s, x_t, u_t)\|, (t, s) \in [\sigma, t_1] \times [\sigma, t_1]\},$$

$$b = \max\{(t_1 - \sigma)a_4, 1\}.$$

Then

$$|v(t)| \leq a_1 a_2 a_3, \quad t \in [\sigma, t_1],$$

$$|v(t)| = |\eta(t)| \leq \sup_{t\in[\sigma-h,\sigma]} |\eta(t)| = a_5 \,.$$

If

$$r_0 = \max[a_1 a_2 a_3, a_5]\,,$$

then

$$|v(t)| \leq r_0, \quad t \in [\sigma - h, t_1], \quad \|v\|_\infty \leq r_0\,.$$

Also

$$|y(t)| \leq a_3 + a_4(t_1 - \sigma)a_1 a_2 a_3 \leq a_3 + ba_1 a_2 a_3, \quad t \in [\sigma, t_1]\,,$$

$$|y(t)| = |\phi(t)| \leq a_6, \quad t \in [\sigma - h, \sigma]\,.$$

If

$$r_1 = \max[a_6, a_3 + ba_1 a_2 a_3]\,,$$

then $|y(t)| \leq r_1, t \in [\sigma - h, t_1]$.

Thus $\|y\| \leq r_1$. If $\max[r_0, r_1] = \lambda$, then $\|v\|_\infty \leq \lambda$, $\|y\| \leq \lambda$.
Therefore, if we let

$$\mathbb{Q}(\lambda) = \{(x, u) \in C([\sigma - h, t_1], E^n) \times L_\infty([\sigma - h, t_1], E^m) \equiv \mathbb{B} :$$

$$\|(x, u)\| = \|x\| + \|u\|_\infty \leq 2\lambda\}\,,$$

we have proved that $T(\mathbb{Q}(\lambda)) \subset \mathbb{Q}(\lambda)$. It is clear that $\mathbb{Q}(\lambda)$ is closed, bounded, and convex. Also $P$ is continuous since $(t, \phi, v) \to B(t, \phi, v)$ is continuous. To prove that the set of $y(t)$ defined in (9.5.22) is equicontinuous is routine. Observe that $v$ defined in (9.5.21) is relatively compact since all such $v$ are uniformly bounded in $L_\infty$ norm, and $\|v_s - v\|_\infty \to 0$ as $s \to 0$ uniformly, where

$$\|v_s - v\| = \max_{1\leq j\leq m} \sup|v_j(t + s) - v_j(s)|$$

[4, pp. 296, 645]. Indeed, since $v(t)$ is measurable in $t$ and in $L_\infty$, there exists a sequence $\{v_n(t)\}$ of continuous functions such that $\|v - v_n\|_\infty \to 0$ as $n \to \infty$. Hence

$$\|v_s - v\|_\infty \leq \|v_s - v_{ns}\| + \|v_{ns} - v_n\|_\infty + \|v_n - v\|_\infty \,.$$

The last and first terms can be made arbitrarily less than $\epsilon$ by selecting $n$ very large. The second term can be made less than $\epsilon$ if $s$ is selected small enough. This proves that $\|v_s - v\|_\infty \to 0$ as $s \to 0$. All this analysis verifies that $P(Q(\lambda))$ is relatively compact. Invoke Schauder's fixed-point theorem to conclude that there is a fixed point $(x, u) \in P(Q(\lambda))$ such that $P(x, u) = (x, u)$. With this fixed point established, we verify (9.5.13) by arguing as follows: Since $T$ is an open map for each $(z, v) \in C \times L_\infty$, and in particular for $(z, v) = (x, u)$, we have that $T$ defined by

$$Tu = \int_\sigma^{t_1} S(t, s, x, u)u(s)ds, \quad T : L_\infty([t_\sigma, t_1], E^m) \to E^n$$

is open. Because of this, $0 \in \text{Int } T(L_\infty(\sigma, t_1), C^m) = \mathcal{A}(t_1)$, where $\mathcal{A}(t_1)$ is the reachable set of (9.5.2), so that

$$0 \in \text{Int } \mathcal{A}(t_1) \tag{9.5.23}$$

and

$$0 \in \text{Int } \mathcal{D}. \tag{9.5.24}$$

Details of the needed arguments are contained in [19, p. 111].      $\square$

**Theorem 9.5.1**    *Consider the large-scale system (9.5 S) with its decomposition (9.5.3), and assume that:*

(i)

$$g_i(t, x_{1t}, x_{i-1t}, 0, x_{i+1t}, \ldots, x_{\ell(t)}, u_{1t}, \ldots, u_{i-1t}, 0, u_{i+1t}, \ldots, u_{\ell t}) = 0,$$

$$g_i(t, x_{1t}, \ldots, x_{\ell t}, 0, \ldots, 0) = 0.$$

(ii) *The system*

$$\dot{x}_i(t) = L_i(t, \psi, v)x_t + B_i(t, \psi, v)u(t) \tag{9.5 Si}$$

*satisfies all the conditions of Lemma 9.5.1 for each $\psi, v \in C \times L_\infty([\sigma, t_1], E^m)$.*

(iii) *$g_i : E \times E^n \times E^m \to E^n$ is continuous.*

*Then (9.5 S) is locally null controllable on $[\sigma, t_1]$ with constraints. If, in addition, the system*

$$\dot{x}_i(t) = L_i(t, x_{it})x_{it} \tag{9.5.25}$$

*is globally exponentially stable, then (9.5 S) is (globally) null controllable with controls in $U_i = \{u_i \in L_\infty([\sigma, t_1], E^{mi})$ for some $t_1, \|u_i\|_\infty \leq 1, j = 1, \ldots, \ell\}$.*

**Proof**    Because of Lemma 9.5.1,

$$0 \in \text{Int } \mathcal{A}_i(t_1) \quad \text{for} \quad t_1 > \sigma + h, \tag{9.5.26}$$

where $\mathcal{A}_i$ is the attainable set associated with (9.5 Si). Let $x_i$ be the solution of (9.5.3) with $x_\sigma^i = 0$, and $u_\sigma^i = \eta = 0$. Then

$$x_i(\sigma, 0, u^i, u)(t_1) = x_i(u^i, u)(t_1)$$

$$= \int_\sigma^{t_1} S(t, s, x_{is}, u_i) u_i(s) ds$$

$$+ \int_\sigma^{t_1} S(t, s, x_{is}, u_i)[g_i(s, x_s, u_s)] ds. \qquad (9.5.27)$$

If we define the set

$$H_i(t_1) = \{x_i(u^i, u)(t_1) \in E^{ni} : u^i \in U_i, u^j \in U_i, j \neq i, j = 1, \ldots, \ell\},$$

then we deduce that

$$\mathcal{A}_i(t_1) \subset H_i(t_1).$$

Because $g_i(t, x_{1t}, \ldots, 0, \ldots, x_{\ell t}, u_1, \ldots, 0, \ldots, u_{i\ell}) = 0$, and because $x_i(\sigma, 0, 0)(t) = 0$ is a solution of (9.5.3), $0 \in H_i(t_1)$. Because (9.5.26) holds, $0 \in \text{Int } \mathcal{A}_i(t_1) \subset H_i(t_1)$ so that

$$0 \in \text{Int } H_i(t_1). \qquad (9.5.28)$$

Using this, one deduces readily that

$$0 \in \text{Int } \mathcal{D}. \qquad (9.5.29)$$

The rest of the proof is routine.

Appropriate the control $u_i \in U_i$, $u_i \equiv 0$, $i = 1, \ldots, \ell$, and use this in (9.5.3); then the solution of (9.5.3) is that of (9.5.25). There is a time $\sigma < \infty$ such that $\psi \equiv x(\sigma, \phi, 0) \in \mathcal{O}$ where $\mathcal{O} \subset \mathcal{D}$, and $\mathcal{O}$ is an open ball containing zero. With $\psi$ as initial state, there is a time $t_1 > \sigma$ and an $n$-tuple $u = (u_1, \ldots, u_\ell)$ control $u_i \in L_\infty([\sigma, t_1], C^{mi}) = U_i$ such that the solution of (9.5.3) satisfies

$$x_\sigma(\cdot, \phi, u_i) = \phi, \quad x(t_1, \sigma, \phi, u_i) = 0.$$

The proof is complete. $\qquad\qquad\qquad\qquad\qquad\qquad\qquad\qquad\qquad\qquad\qquad\square$

Note that Theorem 9.5.1 generalizes the main results in [7, 15, 18]. It removes the growth conditions of the nonlinear functions present in the equations. All that is needed here are the usual smoothness conditions required for the existence of solutions, and bigger "independent" control sets.

**Remark 9.5.1** Our analysis describes Euclidean null controllability of large-scale systems. The results hold also for arbitrary point targets in $E^n$. Indeed, if $x_1$ is a nontrivial arbitrary target and $y(t, \phi, u^0) = y(t)$ is any solution of

$$\dot{x}(t) = f(t, x_t, u(t)), \qquad (9.5.30)$$

with $y_\sigma = \phi$ and with $u^0$ admissible, and $y(t_1) = x_1$, one can equivalently study the null controllability of the system

$$\dot{z}(t) = f(t, z_t + y_t, u_t) - f(t, y_t, u(t)), \qquad (9.5.31)$$

where $z_\sigma = 0$ so that $x_\sigma = y_\sigma = \phi$ and $x(t) = z(t) + y(t)$. Indeed, if we integrate (9.5.31) we obtain

$$z(t, \sigma, \phi, u^0, u) = -y(t) + \int_\sigma^t f(s, x_s, u(s)) ds \, .$$

Note that

$$z(t, \sigma, \phi, u^0, u^0) = 0, \quad \forall \, t \geq \sigma \, .$$

If we can show that there is a neighborhood $\mathcal{O}$ of the origin in $z$ space such that all initial $\phi \in \mathcal{O}$ can be brought to $z(t_1, u, \phi, u^0) = 0$ by some admissible control $u$ at time $t_1$, then

$$z(t_1, u, \phi, u^0) = x(t_1, \phi, u) - y(t_1) = 0 \, ,$$

that is, $x(t_1) = y(t_1) = x_1$.


### 9.5.4    *Universal laws for the control of global economic growth*

The main theorem and some assertions in its proof provide very useful broad policy prescription for the control of a large-scale economic system's growth. It describes qualitatively when it is possible to control the growth of capital stock from an initial endowment to the long-run desired stock of capital, and when this control is impossible. Indeed, from the proof we deduce that if the isolated system (9.5 Si) is "proper", and this is valid if (9.5.26) holds, then the large-scale system (9.5.3) will also be proper, i.e., (9.5.28) will hold provided the "solidarity function" $g_i$ is nice in the sense of being integrable, etc. From (9.5.28) one proves that the domain $\mathcal{D}$ of null controllability of the large-scale system is such that

$$0 \in \text{Int } \mathcal{D} \, . \tag{9.5.32}$$

The condition

$$0 \in \text{Int } \mathcal{A}_i(t) \, , \tag{9.5.33}$$

which implies

$$0 \in \text{Int } H_i(t) \tag{9.5.34}$$

when the interconnection is nice, yields the following obvious but fundamental principles for the control of large-scale organizations.


**Fundamental Principle 1**    *If the isolated system is "proper", which means that it is locally controllable with constraints, and if the interconnection that we will call a "solidarity function" is nice and measured locally at the ith subsystem and is a function of the state of all subsystems and their controls, then the composite system is also proper and therefore locally null controllable with constraints.*

Some isolated system may fail to have property (9.5.33), and is not well behaved. Condition (9.5.34) on which (9.5.32) is deduced can still be salvaged. Obviously what is needed is a condition that ensures that

$$0 \in \text{Int } G_i(t_1) \tag{9.5.35}$$

where

$$G_i(t_1) = \left\{ \int_0^{t_1} g_i(s, x_{1s}, \ldots, x_{\ell s}, u_1(s), \ldots, u_\ell(s)) : u_i(s) \in U_i, \ i = 1, \ldots, \ell \right\}.$$

With (9.5.35), one easily deduces (9.5.34) since $0 \in \mathcal{A}_i(t)$. A criteria such as in (9.5.35) can be deduced from a controllability of a linearization of the nonlinear control system

$$\dot{x}_i(t) = g_i(t, x_{1t}, \ldots, x_{it}, \ldots, x_{\ell t}(t), u_1(t), \ldots, u_i(t), \ldots, u_\ell(t)).$$

In other words, we require a sufficient amount of control impact from external sources $g_i$ that is monitored locally at the $i$-subsystem to be "forced" on (9.5 Si). It works as follows: The external source of power senses the control $u_i$ available to (9.5 Si) and constructs a sufficient controlling signal $g_i$ that is a function of all the states of the subsystems and their controls including $u_i$. With this the right behavior is enforced. It is formalized in the following principle:

**Fundamental Principle 2** *If some subsystem is not proper and is not locally controllable with constraints, the large-scale system can be proper and be made locally controllable provided there is an external source of sufficient power available to the subsystem to enforce controllability and proper behavior. There is no theorem at the present time that ensures that the large-scale system is proper when some system is not proper, and there is no compensating proper behavior from an external source of power that is monitored locally.*

We observe that the solidarity function does not ignore the subsystems controls $u_i$ or initiative. Were it to be ignored, for example, and

$$g_i(t, x_1, \ldots, x_\ell)$$

to be still operative and nontrivial, the condition $0 \in H_i(t)$ and subsequently (9.5.34) will fail. The large-scale system will not be proper and locally null controllable.

**Fundamental Principle 3** *If subsystems' control initiatives are ignored in the construction of a nontrivial solidarity function that is monitored and applied locally, the interconnected system will fail to be proper and locally null controllable.*

Failure to acknowledge individual subsystem's control initiative in highly centralized economic systems that are described by our dynamics may lead to lack of growth from initial endowment $\phi$ to the desired economic target (which is an equilibrium in this case) in centralized systems. On the other hand, failure to enforce regulation bringing to bear the right force $g_i$ when some subsystems misbehave and are locally uncontrollable in individualistic systems may make the large-scale system locally uncontrollable and may trigger an economic depression. These observations seem to have a bearing as an explanation of what is happening in the East and what may happen in the West as a consequence of the 1980–89 U.S. dismantling of regulations (see [17, pp. 264] and [26]) on the economy. In the East, economic targets cannot be reached because of too much centralization: inflexible $g_i$: to make firms controllable, market forces are being advocated. In the Third World, the popular policy prescription of dismantling the "solidarity function", which rides the wave of deregulations of the 1980s in the West, is at work with fury. Aside from its hardships, it does not seem to have a solid theoretical foundation. A certain amount of solidarity function is effective for economic growth. In other words, carefully timed government interventions (or solidarity functions) are crucial for economic growth. This can be provided by central governments that cherish individual initiatives. A qualitative description of the solidarity function will be explored in Sec. 9.6.

## 9.6    Effective Solidarity Functions

In (9.5.3) the dynamics of the $i$th subsystem are described by an equation of the form

$$\dot{x}(t) = L(t, x_t, u_t)x_t + B(t, x_t, u_t)u(t) + g(t, z_t, v_t), \qquad (9.6.1)$$

where $g(t, z_t, v_t)$ is the interconnection or the solidarity function. We now study simpler versions of (9.6.1) and in the process isolate the basic properties of effective solidarity functions. We consider them "controls", "disturbances", "quarry controls", and appropriate an earlier formulation from [13].

The system we study is the linear one,

$$\dot{x}(t) = L(t, x_t) + B(t)u(t) + g(t, v(t)), \qquad (9.6.2)$$

which can be written as

$$\dot{x}(t) = L(t, x_t) - p(t) + q(t), \qquad (9.6.3)$$

where

$$B(t)u(t) = -p(t), \quad q(t) = g(t, v(t)). \qquad (9.6.4)$$

We assume $p \in L_\infty([\sigma, t_1], P)$, $P \subset E^n$, $q \in L_\infty([\sigma, t_1], Q)$, $Q \subset E^n$, and where for each $t \in E$, the linear operator $\phi \to L(t, \phi)$, $\phi \in C$, has the form

$$L(t, \phi) = \int_{-h}^{0} d_\theta \eta(t, s)\phi(s).$$

Here the integral is in the Lebesgue–Stieltjes sense, and $\eta(t, s)$ is an $n \times n$ matrix. Basic assumptions on $\eta$ are contained in Chukwu [13, p. 327]. Viewed in our setting, $Q$ in [13] is now called a solidarity set, and $P$ is a control set. Define the set

$$U : U = P \underline{*} Q = \{u : u + Q \subset P\} \tag{9.6.4}$$

(the Pontryagin difference of $P$ and $Q$). Associate with (9.6.3) the control system

$$\dot{x}(t) = L(t, x_t) - u(t), \quad u(t) \in U(t). \tag{9.6.5}$$

Let $X(t, s)$ be the fundamental matrix of

$$\dot{x}(t) = L(t, x_t), \tag{9.6.6}$$

i.e., the solution of

$$\frac{\partial}{\partial t} X(t, s) = L(t, X_t(\cdot, s)), \quad t \geq s \text{ a.e. in } s, t,$$

$$X(t, s) = \begin{cases} 0, & s - h \leq t < s, \\ I \text{ (Identity)}, & t = s, \end{cases}$$

where $X_t( , s)(\theta) = X(t + \theta, s), -h \leq \theta \leq 0$. The statement and proof of the following theorem is available in [13].

**Theorem 9.6.1**  *Assume* $0 \in Q$ *and* $P$ *compact. System* (9.6.3) *is Euclidean controllable at time* $t_1$ *(on the interval* $[\sigma, t_1]$*) if and only if the associated linear control system* (9.6.5) *with* $u(t) \in U(t)$,

$$U(t) \equiv (P + \ker X(t_1, t)) \underline{*} Q, \tag{9.6.7}$$

*is Euclidean controllable on* $[\sigma, t_1]$. *Furthermore,*

$$p(t) \equiv \sigma(q, t) \equiv u(t) + q \quad modulo \ker X(t_1, t). \tag{9.6.8}$$

*Then for all* $q \in Q$, $t \in [\sigma, t_1]$ *can be used to determine a suitable strategy from* $u \in L_\infty([\sigma, t_1], U)$ *in* (9.6.5) *and vice versa.*

It is important to describe how the control $p(t)$ that drives into a target $x_1 \in E^n$ is achieved. From the knowledge of each initial endowment $\phi$, the subsystem constructs a vector-valued function $t \to u(t)$. Next, an unpredictable but observable regulation or solidarity function $t \to q(t)$ is measured locally by the subsystem. The controller of the subsystem responds simultaneously via a stroboscopic strategy using the solidarity function by taking

$$p(t) = u(t) + q(t) \quad modulo \ker X(t_1, t) \equiv \tau(q(t), t).$$

One might object to this instantaneous observation and response. But if the subsystem is allowed to observe and retain the past history of its control $p(t)$ and of the state variable $x$ in (9.6.3), then as was argued by Hájek [25, p. 56], it already has available to it almost everywhere the current value $q(t)$ of the "regulation". The problem then is that of the controllability of (9.6.5). This problem is solved in [20, 21]. The applied problem of economic growth of the large-scale economy is reduced to the solution of the controllability questions of (9.6.5). The desirable solidarity function $q(t)$ is constrained to lie in $Q$. Further properties of $Q$ can be deduced from Hájek's studies [25, p. 61, pp. 110–148]. It is possible to take into account delays in the transmission of information [25, p. 95]. Note also that the minimum-time problems for (9.6.3) and (9.6.5) coincide if $P$ and $Q$ are "nice".

**The Limitations of $Q$**   It is very important to determine how "big" the solidarity set $Q$ should be. To determine it, we need the following definition and corollary to Theorem 9.6.1:

**Definition 9.6.1**   Consider a mapping $\tau : Q \times E \to P$, ($\tau(q,t)$ is a point in $P$ whenever $q$ is a point of $Q$ and $t \in E$). Let $\sigma$ be an induced mapping on the collection of $Q$ of solidarity functions $q(\cdot)$, which is defined by $\sigma[q](t) = \tau(q(t),t)$. The function $\sigma$ is called stroboscopic if it preserves measurability.

Suppose $t \to u(t)$ is given in advance. Then

$$\sigma(q)(t) = q(t) + u(t)$$

is a stroboscopic strategy, if $u$ is measurable and $Q + u(t) \subset P$ for all $t$. We have:

**Corollary 9.6.1**   *In (9.6.3), assume that $0 \in Q$, $P$ is compact, and $P, Q$ are nonvoid, convex, and symmetric. Then*

$$Q \subset P + \ker X(t_1,t), \quad \sigma \le t \le t_1$$

*is necessary and sufficient for the presence of initial $\phi$ that can be transferred to zero in time $t_1$ when the subsystem uses stroboscopic strategy.*

**Proof**   Theorem 9.6.1 asserts that the set of functions $\phi$ that can be forced to zero at time $t_1$ via (9.6.3) and the set of initial functions that can be steered to zero within the control system 9.6.5 are the same. But

$$U(t) = (P + \ker X(t_1,t)) \underline{*} Q$$

is convex and symmetric. It is nonvoid if and only if $0 \in U(t)$, i.e.,

$$Q \subset P + \ker X(t_1,t).$$

Conversely, if $Q \subset P + \ker X(t_1,t)$, then $0 \in (P + \ker X(t_1,t)) \underline{*} Q$ is an admissible control steering 0 to 0. $\qquad\qquad\square$

**Corollary 9.6.2**   *The condition*

$$Q \subset \mathrm{Int}(P + \ker X(t_1,t)), \quad \sigma \le t \le t_1$$

*is necessary and sufficient for the domain of null controllability of (9.6.3) to have zero in its interior. The given condition is valid if and only if*

$$0 \in \mathrm{Int}\, U = \mathrm{Int}(P + \ker X(t_1,t) \underline{*} Q).$$

*But $0 \in \text{Int } U$ is needed for (9.6.5) to have zero in the interior of the Euclidean reachable set, and therefore in the interior of the domain of null controllability of (9.6.5) that coincides with that of (9.6.3). Note that the Euclidean reachable set is*

$$\mathbb{R}(t_1) = \left\{ \int_\sigma^{t_1} X(t_1, s)u(s)ds : \ u(s) \in U(s) \right\}.$$

*These two corollaries give the last rule of thumb and fourth universal principle for the control of economic growth of interconnected systems.*

**Fundamental Principle 4**  *No initial endowment can be controlled to the origin in our dynamics unless*

$$Q \subset P + \ker X(t_1, t), \quad \sigma \le t \le t_1.$$

*To guarantee control to the equilibrium of each initial endowment, one must require that*

$$Q \subset \text{Int}(P + \ker X(t_1, t)).$$

*In words, to guarantee economic growth from any initial endowment to a target it is necessary that subsystem's control set, together with the system's internal power for waste, dominate the solidarity set or external power. Loosely translated, the subsystem's power should dominate the effects of government "power". Power may be understood to be investment and or consumption/taxation.*

# References


1. H. T. Banks, M. C. Jacobs and C. E. Langenhop, "Characterization of the Controlled States in $W_2^{(1)}$ of Linear Hereditary Systems," *SIAM J. Control* **13**(3) (1975) 611–649.
2. E. N. Chukwu, "Control of Nonlinear Delay Differential Equations in $W_2^{(1)}$," Preprint.
3. J. P. Dauer, "Nonlinear Perturbations of Quasilinear Control Systems," *J. Math. Anal. Appl.* **54**(3) (1976).
4. L. V. Kantorovich and G. P. Akilov, *Functional Analysis in Normal Spaces*, Macmillan, New York, 1964.
5. Qiung Lu, J. N. Lu, Jixans Gao and Gordon K.F. Lee "Decentralized Control for Multimachine Power Systems," *Optimal Control Application and Methods* **10** (1989) 53–64.
6. N. B. Mirza and B. F. Womack, "On the Controllability of Nonlinear Time Delay Systems," *IEEE Trans. Automatic Control* **AC-17**(16) (1972) 812–814.
7. A. S. C. Sinha, "Controllability of Large-Scale Nonlinear Systems with Distributed Delays," in "Control and States" *Proceedings of the 1988 International Conference on Advances in Communications and Control Systems*, **II**, October 19–21, 1988, Baton Rouge, LA, edited by W. A. Porter and S. C. Kak.
8. R. G. Underwood and D. F. Young, "Null Controllability of Nonlinear Functional Differential Equations," *SIAM J. Control and Optimization* **17** (1979) 753–772.
9. K. Deimling, *Nonlinear Functional Analysis*, Springer-Verlag, New York, 1985.

10. E. N. Chukwu, "Global Behaviour of Retarded Functional Differential Equations," in *Differential Equations and Applications*, edited by A. R. Affabizadeh, **I**, Ohio University Press, Athens, 1989.

11. K. L. Cooke and J. A. Yorke, "Equations Modelling Population Growth, Economic Growth and Gonorrhea Epidemiology," in *Ordinary Differential Equations*, edited by L. Weiss, Academic Press, New York, 1972.

12. A. Takayama, *Mathematical Economics*, Cambridge University Press, 1974.

13. E. N. Chukwu, "Capture in Linear Functional Games of Pursuit," *J. Math. Anal. Appl.* **70** (1979) 326–336.

14. E. N. Chukwu, "Control of Interconnected Nonlinear Delay Differential Equations in $W_2^{(1)}$," Preprint.

15. K. Balachandran and J. P. Dauer, "Relative Perturbations of Nonlinear Systems," *J. Optimization Theory and Applications* **63**(1) (1989) 51–56.

16. K. Balachandran and J. P. Dauer, "Null Controllability of Nonlinear Infinite Delay Systems with Distributed Delays in Control," *J. Math. Anal. Appl.* **145** (1990) 274–281.

17. R. Batra, *Surviving the Great Depression of 1990*, Dell Publishing, New York, 1988.

18. E. N. Chukwu, "Global Null Controllability of Nonlinear Delay Equations with Controls in a Compact Set," *J. Optimization Theory and Applications* **53** (1987).

19. E. N. Chukwu, "The Time Optimal Control of Nonlinear Delay Equations," in *Operator Methods for Optimal Control Problems*, edited by Sung J. Lee, Marcel Dekker, New York, 1988.

20. E. N. Chukwu, "Euclidean Controllability of Linear Delay Systems with Limited Controls," *IEEE Trans. Automatic Control* **AC-24** (1979).

21. E. N. Chukwu, "Function Space Null Controllability of Linear Delay Systems with Limited Power," *J. Math. Anal. Appl.* **124** (1987) 293–304.

22. H. Hermes and J. P. LaSalle, *Functional Analysis and Time Optimal Control*, Academic Press, New York, 1969.

23. J. Klamka, "Controllability of Nonlinear Systems with Distributed Delays in Control," *International J. Control* **31** (1980) 811–819.

24. A. Manitius, "Optimal Control of Hereditary Systems," *Control Theory and Topics in Functional Analysis, Vol. I*, International Atomic Energy Agency, Vienna, 1976.

25. O. Hájek, *Pursuit Games*, Academic Press, New York, 1975.

26. K. Phillips, *The Politics of Rich and Poor*, Random House, New York, 1990.

27. E. N. Chukwu, "Control of Global Economic Growth: Will the Center Hold?," in *Ordinary Delay Differential Equations*, Ed., J. Wiener and J. K. Hale, Pitman Research Notes in Mathematics Series **272**, Logman group UK. 1993.

# Chapter 10

# The Time-Optimal Control of Linear Differential Equations of Neutral Type

In this chapter we study the time-optimal control problem of various forms of (1.9.1). Controllability questions are answered. For linear systems, the existence and forms of an optimal control are established in both Euclidean and in function space. The theory developed here was reported in [13] and [14].

## 10.1   The Time-Optimal Control Problem of Linear Neutral Functional Systems in $E^n$: Introduction

In this section, the time-optimal problem is explored for the linear neutral system of the form

$$\frac{d}{dt}D(t, x_t) = L(x, x_t) + B(t)u(t), \quad t \geq \sigma, \quad x_\sigma = \phi, \tag{10.1.1}$$

where the control set is a unit $m$-dimensional cube and the target is a continuous set function in an $n$-dimensional Euclidean space. Necessary and sufficient conditions for the existence and uniqueness of optimal controls are stated and proved. The bang-bang form of optimal control is given. For zero target, easily computable conditions are stated for the system to be proper or normal. Also proved is a sufficient condition for (10.1.1) to be Euclidean null controllable. As an application we consider a differential pursuit game described by the equation

$$\frac{d}{dt}D(t, x_t) = L(t, x_t) - p(t) + q(t), \quad t \geq \sigma, \quad x_\sigma = \phi,$$
$$q \in L_\infty^{\mathrm{loc}}([\sigma, \infty), Q), \ p \in L_\infty^{\mathrm{loc}}([\sigma, \infty), P), \quad P, Q \subseteq E^n. \tag{10.1.1$'$}$$

We show that the differential game is equivalent to a control system of the form (10.1.1) where the control set $U$ is a Pontryagin difference of sets.

In (10.1.1), $B \in L_1([\sigma, v], E^{n \times n})$, $\sigma \leq v$, and

$$D(t, \phi) = \phi(0) - g(t, \phi), g(t, \phi), L(t, \cdot)$$

353

are bounded linear operators from $C$ into $E^n$ for each fixed $t$ in $[0, \infty)$, $g(t, \phi)$ is continuous for $(t, \phi) \in [0, \infty) \times C$.

$$g(t, \phi) = \int_{-h}^{0} d_s \mu(t, s) \phi(s) , \quad L(t, \phi) = \int_{-h}^{0} d_s \eta(t, s) \phi(s) ,$$

$$|g(t, \phi)| \leq k \|\phi\| , \quad |\ell(t, \phi)| \leq L(t) \|\phi\| , \quad (t, \phi) \in [0, \infty) \times C ,$$

for some nonnegative constant $k$, continuous nonnegative $\ell$ and $\mu(t, \cdot)$, $\eta(t, \cdot)$ are $n \times n$ matrix functions of bounded variation on $[-h, 0]$. We also assume that $g$ is uniformly nonatomic at zero, that is, there exists a continuous, nonnegative, nondecreasing function $\ell(s)$ for $s \in [0, h]$ such that $\ell(0) = 0$,

$$\left| \int_{-s}^{0} d, \mu(t, \tau) \phi(\tau) \right| \leq \ell(s) \|\phi\| .$$

We shall consider the set of admissible controls that are measurable functions $u :$ $[\tau, t] \to C^m$, $t \geq \tau$, where $C^m$ is the unit cube in $E^m$ given by

$$C^m = \{ u \in E^m : |u_j| \leq 1 \} .$$

Such controls will simply be denoted by $u \in C^m$.

Finally we assume that

$$g(t) \equiv B(t) u(t)$$

is locally integrable from $[\sigma, \infty)$ into $E^n$, where $u \in L_p([\sigma, \infty), E^n)$.

Under the above hypotheses, there exists a continuous function $x : [\sigma - h, \infty) \to E^n$ such that $x_\sigma = \phi$, which is a solution of system (10.1.1) for all $t \geq \sigma$. If we denote the solution $x = x(\sigma, \phi, u)$, then for each $t \geq \sigma$, $\phi \in C$, $u \in L_p([\sigma, t], E^n)$, $x_t(\sigma, \phi, u) = x_t(\sigma, \phi, 0) + x_t(\sigma, 0, u)$. Define the operators

$$T(t, \sigma) : C \to C, \quad K(t, \sigma) : L_p([\sigma, t], E^m) \to C$$

by

$$x_t(\sigma, \phi, u) = T(t, \sigma) \phi + K(t, \sigma) u . \tag{10.1.2}$$

Then $T, K$ are bounded linear operators.

The solution of System (10.1.1) is also given by

$$x(t, \sigma, \phi, u) = Y(\sigma, t) D(\sigma, \phi) + \int_{\sigma-h}^{\sigma-} d_s \nu(t, s) \phi(s) + \int_{\sigma}^{t} Y(s, t) B(s) u(s) ds , \tag{10.1.3a}$$

where

$$\nu(t, s) \equiv - \int_{\sigma}^{t+} d_\alpha(\alpha, t) \mu(\alpha, s) + \int_{\sigma}^{t} Y(\alpha, s) \eta(\alpha, s) d\alpha$$

and $Y$ is an $n \times n$ matrix defined by:

$$Y(\sigma, t) = I + \int_\sigma^{t+} d_\alpha Y(\alpha, t)\mu(\alpha, \sigma) - \int_\sigma^t Y(\alpha, t)\eta(\alpha, \sigma)d\alpha\,, \tag{10.1.3b}$$

$$Y(t, t) = I, \quad Y(\sigma, t) = 0 \quad \text{for} \quad \sigma > t\,.$$

All integrals are understood to be Lebesgue–Stieltjes integrals. The function $Y(\sigma, t)$ is Borel measurable subject to the conditions given in System (10.1.1). It is left continuous on its first argument and

$$\|Y(\sigma, t)\| \leq \beta\,, \quad \mathrm{Var}[\sigma, t](Y(\cdot, t)) \leq \beta \quad \text{for} \quad (\sigma, t) \in [\sigma, t_1] \times [\sigma, t_1]\,,$$

where $\beta < \infty$ and is independent of $(\sigma, t)$. (See Henry [16] or Banks and Kent [3].) Because of these properties of $Y$, we can easily study the adjoint of the operator $T(t, \sigma)$ in (10.1.2).

We need some properties of the adjoint of the operator $T(t, \sigma)$, which is defined as follows: Let $B_0$ denote the Banach space of functions of bounded variation $\psi :$ $[-h, 0] \to E^{n*}$ (the row vectors) that are continuous from the left on $(-h, 0)$ and vanish at 0; we may use the norm $\|\psi\| = \mathrm{Var}_{[-h, 0]}\psi$. We identify $B_0$ with the conjugate space $C$ with the pairing

$$(\psi, \phi) = \int_{-h}^0 d\psi(s)\phi(s), \quad \psi \in B_0\,, \quad \phi \in C\,. \tag{10.1.4}$$

The adjoint of $T(t, \sigma)$, $T^*(\sigma, t) : B_0 \to B_0$ with domain in $B_0$ is defined by

$$(T^*(\sigma, t)\psi, \phi) = (\sigma, T(t, \sigma)\phi)\,,$$

whenever $\sigma \leq t$, $\psi \in B_0$, $\phi \in C$. We now state the following lemma:

**Lemma 10.1.1** *Let $\psi \in B_0]$, $\psi \neq 0$. Define*

$$-y(s, t) = [T^*(s, t)\psi](0^-)\,, \quad 0 \leq s \leq t\,. \tag{10.1.5}$$

*Then $s \to y(s)$ satisfies the adjoint equation*

$$\frac{d}{ds}\left\{y(s) - \int_s^{t+} dy(\alpha)\mu(\alpha, s - \alpha) + \int_s^t y(\alpha)\eta(\alpha, s - \alpha)d\alpha\right\} = 0, \quad s \leq t\,. \tag{10.1.6}$$

This follows from Henry [16].

**Remark** $T^*(s, t)$ is characterized in [16, Theorem 3] in a way similar to [9, Theorem 4.1]. We now designate the strongly continuous semigroup of linear transformation defined by

$$\frac{d}{dt}[D(t)x_t] = L(t, x_t)\,, \tag{10.1.7}$$

by $T(t, \sigma)$, $t \geq \sigma$, so that

$$T(t, \sigma)\phi = x_t(\sigma, \phi, 0)\,. \tag{10.1.8}$$

Let the function $U(t,s) \in L_\infty([\sigma,t], E^{n^2})$ for each $t$ be defined by

$$U(t,s) = \frac{\partial \underline{W}(t,s)}{\partial s} \quad \text{a.e. in } s\,,$$

where $\underline{W}(t,s)$ is the unique solution of

(a) $\underline{W}_s(\cdot,s) = 0$,
(b) $\underline{W}(t,s) = g(t,\underline{W}_t(\cdot,s)) + \int_s^t L(\lambda,\underline{W}_\lambda(\cdot,s))d\lambda - (t-s)I$, $0 \le s \le t$.

Then the variation of constant formula for a more general version of System (10.1.1), namely,

$$\frac{d}{dt}[D(t)x_t - G(t)x_t] = L(t,x_t) + B(t)u(t)\,, \tag{10.1.9}$$

is given by

$$x_t(\sigma,\phi,u) = T(t,\sigma)\phi + \int_\sigma^t d_s U_t(\cdot,s)\{G(\sigma)(\phi) - G(s)(x_x)\}$$

$$+ \int_\sigma^t U_t(\cdot,s)B(s)u(s)ds\,. \tag{10.1.10a}$$

Here $G : E \times C \to E^n$ is a linear operator. If we set

$$X_0(\theta) = \left\{ \begin{array}{ll} 0, & -h \le \theta < 0\,, \\ t, & \theta = 0, \ I \text{ identity}\,, \end{array} \right.$$

we are justified in writing

$$T(t,\sigma)X_0 = U_t(\cdot,s)\,. \tag{10.1.10b}$$

Here $U(t,t) = I$, the identity matrix.

If we consider the system in $E^n$, we have

$$x(\sigma,\phi,u)(t) = T(t,\sigma)\phi(0) + \int_\sigma^t d_s U(t,s)[G(\sigma)(\phi) - G(s)(t)]$$

$$\times \int_\sigma^t U(t,s)B(s)u(s)ds\,. \tag{10.1.10c}$$

The Euclidean constrained reachable set is given by

$$R(t,\sigma) = \left\{ \int_\sigma^t U(t,s)B(s)u(s)ds : u \in L_\infty([\sigma,t], C^m) \right\}, \tag{10.1.11}$$

and the constrained reachable set in $C$ is given by

$$\alpha(t,\sigma) = \left\{ \int_\sigma^t U_t(\cdot,s)B(s)u(s)ds : u \in L_\infty([\sigma,t], C^m) \right\}. \tag{10.1.12}$$

We summarize the basic properties of these sets in the following proposition.

**Proposition 10.1.1a**   *Assume that $D$ and $L$ satisfy the basic assumptions in the introduction, and furthermore there exist $t > 0$, $L > 0$ such that*

$$\int_{t-\ell}^{t} |d_\theta\{\mu(t,\theta) - \mu(s,\theta)\}| \le L|t-s|, \quad s \le t. \tag{10.1.13}$$

*Then*

    (i)  $0 \in R(t,\sigma)$ *for each* $t \ge \sigma$.

    (ii)  $W(t,s)R(s,\sigma) \subseteq R(t,\sigma)$, $\sigma \le s \le t$, *where* $W(t,s) : C \to E^n$ *is the operator,* $W(t,s)\phi = x(t,\sigma,\phi,0)$ *is a solution of* (10.1.7).

    (iii)  $R(t,\sigma)$ *is compact in* $E^n$, *and convex.*

*Also,*

    (iv)  $0 \in \alpha(t,\sigma)$ *for each* $t \ge \sigma$.

    (v)  $T(t,s)\alpha(s,\sigma) \subseteq \alpha(t,\sigma)$, $\sigma \le s \le t$.

    (vi)  $\alpha(t,\sigma)$ *is compact in* $C$, *and convex.*

    (vii)  *Also* $\mathcal{A}(t)$ *is compact and convex where* $\mathcal{A}(t)$ *is defined as follows:*

*Let $S$ be a compact and convex subset of $C$. The attainable set $\mathcal{A}(t)$ of* (10.1.1) *at time $t$ is defined by*

$$\mathcal{A}(t) = \{x(t) : \phi \in S,\ u \in C^m\} \subset E^n.$$

**Proof**   Convexity of $\mathcal{A}(t)$ follows trivially from that of $S$ and $C^m$. That of $\mathbb{R}(t)$ follows from the convexity of $C^m$. Because $S$ is compact and $x(t,\sigma,\cdot,0)$ continuous, $x(t,\sigma,S,0)$ is bounded. Since $U(t,s)B(s)$ is integrable and $u(s) \in C^m$, $\mathcal{A}(t)$ is bounded in $E^n$. Also $\mathbb{R}(t)$ is bounded. We use a weak compactness argument and the compactness of $S$ to deduce that $\mathcal{A}(t)$ is closed in $E^n$. The same weak compactness argument proves the closedness of $\mathbb{R}(t)$. It now remains to prove (10.1.7). Let $r \in \mathbb{R}(s)$; then for some $u \in C^m$, $r = \int_\sigma^s U(s,\tau)B(\tau)u(\tau)d\tau$. Define

$$u^*(\tau) = \begin{cases} u(\tau), & \sigma \le \tau \le s, \\ 0, & s < \tau \le t. \end{cases}$$

Then $u^*(\tau) \in C^m$.

Consider the point

$$q = W(t,s)r = \int_\sigma^s W(t,s)U(s,\tau)B(\tau)u(\tau)d\tau$$

$$= \int_\sigma^s W(t,s)U(s,\tau)B(\tau)u(\tau)d\tau + 0$$

$$= \int_\sigma^s U(t,\tau)B(\tau)u(\tau)d\tau + \int_s^t U(t,\tau)B(\tau)\circ d\tau$$

$$= \int_\sigma^t U(t,\tau)B(\tau)u^*(\tau)d\tau \in \mathbb{R}(t).$$

Hence

$$W(t,s)\mathbb{R}(s) \subseteq \mathbb{R}(t), \quad \sigma \leq s \leq t.$$

$\square$

**Theorem 10.1.1**   *Let $C_0^m = \{u \in C^m : |u_j| = 1\}$. If*

$$\mathcal{A}^0(t) = \{x(t) : \phi \in S, \ u^0 \in C_0^m\},$$

$$\mathbb{R}^0(t) = \left\{\int_\sigma^t U(t,s)B(s)u^0(s)ds : u^0 \in C_0^m\right\}.$$

*Then $\mathbb{R}^0 = \mathbb{R}(t)$ and $\mathcal{A}^0(t) = \mathcal{A}(t)$ for each $t \geq \sigma$.*

**Proof**   Because $U(t,\cdot) \in L_\infty([\sigma,t], E^{n^2})$ and $B(\cdot) \in L_1([\sigma,t], E^{n \times m})$, we have that

$$U(t,s)B(s) \in L_1([\sigma,t], E^{n \times m}).$$

It follows from Lemma 2 of LaSalle [2] that

$$\mathbb{R}^0(t) = \mathbb{R}(t) \quad \text{for each } t.$$

Hence $\mathcal{A}(t) = \mathcal{A}^0(t)$.      $\square$

**Remark 10.1.2**   The bang-bang principle of Theorem 10.1.2 is not valid for $\alpha(t,\sigma)$ in function space as shown by the following example of Banks and Kent [3]. Consider the neutral system

$$\dot{x}(t) = \dot{x}(t-1) + u(t), \quad t \in [0,2],$$

$$U = \{v \in E : |v| \leq 1\},$$

$$\xi(t) = 2 - t, \quad t \in [1,2].$$

The assumption that $u$ is bang-bang and $\xi$ is attained leads to the conclusion

$$|\dot{\phi}(t)| = 0 \quad \text{or} \quad \dot{\phi}(t) = -3,$$

where $\phi$ is the initial function. However, beginning with initial function $\phi \equiv 1$ one can attain $\xi$ using an admissible control $u$. Thus, there are initial functions $\phi$ for which $T(t,\sigma)\phi + \alpha(t,\sigma)$ using bang-bang controls is a proper subset of the set attained using all admissible controls.

**Definition 10.1.1**   Let $r^n$ denote the metric space of all nonempty compact subsets of $E^n$ with the metric $\rho$ defined as follows: The distance of a point $x$ from $\mathcal{A}(t_1) = d(x, \mathcal{A}(t_1)) = \inf\{|x - \alpha| : \alpha \in \mathcal{A}(t)\}$,

$$N(\mathcal{A}(t_1), \epsilon) = \{x \in E^n : d(x, \mathcal{A}(t_1)) \leq \epsilon\},$$

$\rho(\mathcal{A}(t_1), \mathcal{A}(t_2)) = \inf\{\epsilon : \mathcal{A}(t_1) \subseteq N(\mathcal{A}(t_2), \epsilon) \text{ and } \mathcal{A}(t_2) \subseteq N(\mathcal{A}(t_1), \epsilon)\}$. The target in our system (10.1.1) is the continuous set function: $G : [\tau, \infty) \to r^n$. The problem of reaching $G$ in minimum-time will be called the general problem. For this we want at some $t \geq \sigma$ to have $x(t, \sigma, \phi, u) \in G(t)$, that is, to have

$$w(t) = \int_\sigma^t U(t,s)B(s)u(s)ds, \qquad (10.1.14)$$

where $w(t) = z(t) - x(t, \sigma, \phi, 0)$, for some $z(t)$ with $z(t) \in G(t)$. This is equivalent to requiring

$$\mathcal{A}(t) \cap G(t) \neq \emptyset$$

for some $t \geq \sigma$.

**Remark 10.1.3** Theorem 10.1.2 states the bang-bang principle of LaSalle is valid for (10.1.1): that if of all bang-bang steering functions there is an optimal one relative to $C_0^m$, then it is optimal (relative to $C^m$). Also, if there is an optimal steering function, there is always a bang-bang steering function that is optimal.

**Proposition 10.1.1b** *The attainable set* $\mathcal{A}(\cdot) : [\sigma, \infty) \to r^n$ *is continuous. Also* $t \to \mathbb{R}(t, \sigma)$ *and* $t \to \alpha(t, \sigma)$ *are continuous.*

**Remark 10.1.4** The proof of this proposition is essentially the same as the assertion which is contained in Theorem 4.1 of Banks and Kent [3]. One proves that there exists an $M > 0$ such that $\|x, (\sigma, \phi, u)\| \leq M$ for $t \in [\sigma, t_1], \phi \in S, u \in C^m$. With this, one then proves that

$$\mathcal{A} = \{x(\sigma, \phi, u) : \phi \in S, u \in C^m\}$$

is an equicontinuous subset of $C([\sigma - h, t_1], E^n)$. As a consequence, $t \to x(t, \sigma, \phi, u)$ is continuous uniformly in $\phi, u$. The continuity of $t \to \mathcal{A}(t)$ follows readily from this. As pointed out in [3], because the fundamental matrix $t \to U(t, s)$ is not continuous one cannot prove $\mathcal{A}$ continuous directly using the variation of parameter formula (10.1.3) as is done for the example in [4].

**Theorem 10.1.2** *If for the general problem there exists a pair* $\phi \in S, u \in C^m$ *such that* $x(t, \phi, u) \in G(t)$ *for some $t$, then there is an optimal pair* $\phi^* \in S, u^* \in C^m$.

**Proof** By assumption there is some $t \geq \sigma, \phi \in S$ such that

$$\mathcal{A}(t) \cap G(t) \neq \emptyset.$$

If we now define the minimal-time function

$$t^*(S) = \inf\{t \geq \sigma : \mathcal{A}(t) \cap G(t) \neq \emptyset\},$$

where $\mathcal{A}(t) = \mathcal{A}(t, S)$, we can prove first as is done by Strauss [5, p. 63] that $\mathcal{A}(t^*) \cap G(t^*) \neq \emptyset$. We use the compactness of $G(t), \mathcal{A}(t)$, and their continuity. $\square$

**Definition 10.1.2** Let

$$D(t, \phi) = \int_{-h}^0 d_s\eta(t, s)\phi(s),$$

and suppose

$$A(t, \beta) = \eta(t, \beta^+) - \eta(t, \beta^-).$$

Then $D(t, \phi)$ is atomic at $\beta$ on $E \times C$ if $\det A(t, \beta) \neq 0$ for all $t \in E$. In particular, if $\beta \neq 0$, $\beta \in [-h, 0]$, and $D(t, \phi) = \phi(0) + M(t)\phi(\beta)$, then $M(t) = A(t, \beta)$ and $D(t, \phi)$ is atomic at $\beta$ on $E \times C$ if $\det M(t) \neq 0$ for all $t \in E$. See Hale [9, p. 50]. For the linear system (10.1.1), a fundamental theorem of Hale shows that the solution operator

$$T(t, \sigma) : C \to C,$$

defined by $T(t, \sigma)\phi = x_t(\sigma, \phi)$, where $x$ is a solution of system (10.1.1), is a homeomorphism, provided $D$ is atomic at $0$ and at $-h$ [9, p. 279]. As asserted by Hale, since

$$L(t, x_t) = f(t, x_t)$$

is linear and therefore continuously differentiable in the second argument, the solution operator is a diffeomorphism. It follows from an argument similar to Hale [9] that the solution operator

$$T(t, \sigma) : W_p^{(1)} \to W_p^{(1)}, \quad 1 \leq p \leq \infty,$$

is also a homeomorphism.

**Definition 10.1.3**    The control $u \in C^m$ is an extremal control on $[\sigma, t_1]$ if for some $\phi \in C$ and each $t \in [\sigma, t_1]$ the solution $x(\sigma, \phi, u)$ of (10.1.1) through $\sigma, \phi$ belongs to the boundary $\partial \mathcal{A}(t)$ of $\mathcal{A}(t)$.

**Theorem 10.1.3**    *Assume that conditions of Proposition (10.1.1a) hold. Assume the solution of (10.1.7) is pointwise complete. Let $u^*$ be optimal on $[\sigma, t^*]$. Then there is a nonzero $n$-dimensional row vector $c^*$ depending on $\phi^*$ and $t^*$ such that*

$$\{u^*(t)\} = \text{sgn}\{c^* U(t^*, t) B(t)\}_j, \quad \sigma \leq t \leq t^*, \tag{10.1.15}$$

*for each $1 \leq j \leq m$ for which*

$$\{c^* U(t^*, t) B(t)\}_j \neq 0.$$

**Proof**    To prove that any optimal control is extremal, one uses the method of Strauss and the following proposition which is proved exactly as in [5].          $\square$

**Proposition 10.1.2**    *If $\alpha$ is an interior point of $\mathcal{A}(t)$, then there is some $\epsilon > 0$ such that the $\epsilon$-neighborhood $N(\alpha, \epsilon)$ of $\alpha$ satisfies*

$$N(\alpha, \epsilon) \subseteq \mathcal{A}(s)$$

*for all $s \in [t - \epsilon, t]$.*

To prove the last assertion, we note that $x(t^*, \sigma, \phi^*, u^*) \in \partial \mathcal{A}(t^*)$. Because $\mathcal{A}(t^*)$ is convex and closed, there exists a support plane $\pi$ through $x(t^*, \sigma, \phi^*, u^*) = x^*$ such that $\mathcal{A}(t)$ lies on one side of $\pi$. Let $v$ be a unit normal to $\pi$ directed away from $\mathcal{A}(t^*)$. Clearly, for each

$$u \in C^m, \quad x = x(t^*, \phi^*, u) \in \mathcal{A}(t^*).$$

Hence $(v, x^* - x) \leq 0$. From the variation of parameter (10.1.3a), this is equivalent to

$$\left( v, \int_\sigma^{t^*} U(t^*, s) B(s) u(s) ds \right) \leq \left( v, \int_\sigma^{t^*} Y(t^*, s) u * (s) ds \right)$$

for all $u \in C^m$. Let $\lambda(s) = v^T U(t^*, s) B(s)$. Then

$$\int_\sigma^{t^*} v, U(t^*, s) B(s) u^*(s) ds \leq \sum_{j=1}^m \int_\sigma^{t^*} \lambda_j(s) \ \text{sgn} \ \lambda_j(s) ds.$$

Hence we see that, on any interval of positive length where $\lambda(s) \neq 0$, it must be that $u_j^*(s) = \text{sgn} \ \lambda_j(s)$. We have proved the Theorem.

**Definition 10.1.4**   System (10.1.1) is said to be normal on $[\sigma, t]$ if no component of $c^T Y(t, s)$, $c \neq 0$, vanishes on a subinterval of $[\sigma, t]$ of positive length. This definition is equivalent to the following: Let $y_j(t, s)$ be the $j$th column vector of $Y(t, s)$. For each $j = 1, \ldots, m$ the functions $y_{j_\ell}(t, s), \ldots, y_{j_n}(t, s)$ are linearly independent on each subinterval of $[\sigma, t]$ of positive length.

**Corollary 10.1.1**   *If System* (10.1.1) *is normal on* $[\sigma, t]$, *then* $u^*(t)$, *the optimal control, is uniquely determined by* (10.1.9) *and is bang-bang.*

**Definition 10.1.5**   Let $S \subseteq E^n$ be a set. $S$ is strictly convex if for every $x_1$ and $x_2$ in $S$, $x_1 \neq x_2$, the open line segment

$$\{x_1 + (1 - \lambda)x_2 : 0 < \lambda < 1\}$$

is in the interior of $S$.

Following the methods of Strauss [5, p. 70] we can prove the following:

**Theorem 10.1.4**   *Assume conditions of Theorem* 10.1.3. *Suppose* (10.1.1) *is normal and* $G : [\tau, \infty) \to r^n$ *is continuous and strictly convex valued. Then if there exists a pair* $\phi \in C$, $u \in C^m$ *such that* $x(t, \sigma, \phi, u) \in G(t)$ *for some* $t \geq \sigma$, *then there is exactly one (time) optimal control and it is bang-bang.*


## 10.2   Forcing to Zero

In this section, we consider System (10.1.1) with a fixed target $G(t) \equiv \{0\}$, where the aim is to start at any initial state $\phi \in C$ at time $\sigma \in [\tau, \infty)$ and to reach the origin.

The next definition will enable us to derive a computational criterion for solving our problem.

**Definition 10.2.1**  System (10.1.1) is proper on an interval $[\sigma, t^*]$, if $c^T U(t^*, s) B(s) \equiv 0$ almost everywhere on $[\sigma, t^*]$ implies $c = 0$. If (10.1.1) is proper on $[\sigma, \sigma+\delta]$ for each $\delta > 0$, we say that (10.1.1) is proper at time $\sigma$. If (10.1.1) is proper on each interval $[\sigma, t]$, $t > \sigma$, we say the system is proper.

**Definition 10.2.2**  System (10.1.1) is Euclidean null controllable if given any initial state $\phi \in C = C([-h, 0], E^n)$ there exists a $t_1$ and an admissible control $u \in L_\infty([\sigma, t_1], E^m)$ such that the solution $x(\sigma, \phi, u)$ of (10.1.1) satisfies $x_\sigma(\sigma, \phi, u) = \phi, x(t_1, \sigma, \phi, u) = 0$.

**Definition 10.2.3**  The domain $\mathbf{N}$ of null controllability of (10.1.1) is the set of all initial functions $\phi \in C$ for which the solution $x = x(\sigma, \phi, u)$ of (10.1.1) with $x_\sigma = \phi$ satisfies $x(t_1) = 0$ at some $t_1$ using an admissible control $u \in L_\infty([\sigma, t_1], C^m)$.

**Theorem 10.2.1**  *The domain $\mathbf{N}$ of null controllability contains zero in its interior whenever (10.1.1) is proper.*

To prove Theorem 10.2.1 we need the following Proposition:

**Proposition 10.2.1**  *System (10.1.1) is proper on $[\sigma, t]$ if and only if the origin is an interior point of $\mathbf{R}(t)$.*

**Proof of Proposition 10.2.1**  Observe that 0 is always in $\mathbf{R}(t)$ for each $t \geq \sigma$. Suppose (10.1.1) is proper but zero is not in the interior of $\mathbf{R}(t)$ for some $t \geq \sigma$. Because $\mathbf{R}(t)$ is a closed and convex set and because the constraint set $C^m$ is the unit cube, 0 being on the boundary of $\mathbb{R}(t)$ is equivalent to

$$c^T \int_\sigma^t U(t, s) B(s) u(s) ds = 0$$

for some nonzero $c \in E^n$ and for $u \in L_1([\sigma, t], C^m)$ given by $u(s) = \text{sgn}(c^T U(t, s) B(s))$ where $c$ is the outward normal to the support plane to $\mathbb{R}(t)$ at 0. Thus

$$c^T U(t, s) B(s) \equiv 0, \quad \text{almost everywhere } s \in [\sigma, t],$$

where $c \neq 0$, which contradicts the fact that (10.1.1) is proper on $[\sigma, t]$. We now reverse the argument to complete the proof. $\square$

**Proof of Theorem 10.2.1**  With $0 = u \in C^m$, $x(t) = 0$ is a solution of (10.1.1) so that $0 \in \mathbf{N}$. Suppose 0 is in the interior of $\mathbf{R}(t)$ for each $t$, but 0 is not in the interior of $\mathbf{N}$. Then there is a sequence $\{\phi_m\}_1^\infty \subseteq C$ such that $\phi_m \to 0$ as $m \to \infty$ (the convergence is in the sup norm of $C([-h, 0], E^n)$) and no $\phi_m$ is in $\mathbf{N}$ (so $\phi_m \neq 0$). From the variation of constant formula we deduce that

$$0 \neq x(t_1, \sigma, \phi_m, u) = T(t_1, \sigma)\phi(0) + \int_\sigma^{t_1} U(t_1, s) B(s) u(s) ds$$

for any $t_1 > \sigma$ and any $u \in C^m$.

Hence $y_m = T(t_1, \sigma)\phi_m(0)$ is not in $\mathbf{R}(t_1)$ for any $t_1$. We therefore obtain a sequence $\{y_m\}_1^\infty \subseteq E^n$, $y_m \notin \mathbf{R}(t_1)$ for any $t_1, y_m \neq 0$ such that $y_m \to 0$, as $m \to \infty$. We conclude that 0 is not in the interior of $\mathbb{R}(t_1)$ for any $t_1$, a contradiction. Hence $0 \in \mathrm{Int}\,\mathbf{N}$. $\qquad\square$

**Theorem 10.2.2** *If (10.1.1) is proper and (10.1.2) is uniformly asymptotically stable, then (10.1.1) is Euclidean null controllable.*

**Proof** Let $\mathbf{N}$ be the domain of null controllability of (10.1.1). Since (10.1.1) is proper we have from Theorem 10.2.1 that $0 \in \mathrm{Int}\,\mathbf{N}$. In (10.1.1) consider the trivial admissible control $u = 0$ on $[\sigma, \infty)$. Then the solution $x(\sigma, \phi, 0)$ of (10.1.1) with $u = 0$ (that is the solution of (10.1.2)) satisfies

$$\|x_t(\sigma, \phi, 0)\| \leq ke^{-c(t-\sigma)}\|\phi\|, \quad t \geq \sigma, \quad \phi \in C,$$

for some $k > 1$ and $c > 0$. This behavior of (10.1.2) follows from Cruz and Hale [1]. Hence $x_t(\sigma, \phi, 0) \to \phi_0 = 0 \in \mathrm{Int}\,\mathbf{N}$, as $t \to \infty$. Therefore there exists a $t_1 > \sigma$ such that $x_{t_1}(\sigma, \phi, 0) \in \mathbf{B}$ where $\mathbf{B}$ is a ball in $\mathbf{N}$ with the origin as center and radius sufficiently small. With $x_{t_1}(\sigma, \phi, 0) \in \mathbf{N}$ as an initial point, there exists some $t_2$ and some $u \in L_1([t_1, t_2], C^m)$ such that the solution $x(t_1, x_{t_1}, u)$ of (10.1.2) satisfies $x(t_1, x_{t_1}, u) = 0$. This proves the theorem. $\qquad\square$

We conclude this section by considering a more general class of controls, in which we only require them to be integrable on finite intervals. If a control $u$ satisfies this condition, we simply write $u \in C^*$.

**Definition 10.2.4** System (10.1.1) is Euclidean controllable on $[\sigma, t_1]$ if for each $\phi \in C$ and each $x_1 \in E^n$ there exists a control $u \in C^*$ such that the solution $x(\sigma, \phi, u)$ of (10.1.1) satisfies $x_\sigma(\sigma, \phi, u) = \phi$ and $x(t_1, \sigma, \phi, u) = x_1$.

**Theorem 10.2.3** *System (10.1.1) is proper on $[\sigma, t_1]$ if and only if (10.1.1) is Euclidean controllable on $[\sigma, t_1]$.*

**Remark 10.2.1** The ideas behind the proof are standard in the literature. See, for example, Zmood [7, Theorem 2], and Lemma 6.1.1 here.

## 10.3 Normal and Proper Autonomous Systems

In this section an autonomous special case of (10.1.1) will be considered. Sharp necessary and sufficient conditions for normality and for the system to be proper are deduced. The method of Gabasov and Kirillova [12] is used to obtain results that generalize the recent normality conditions of Kloch [8], and that specialize to well-known conditions for autonomous linear ordinary differential equations.

Consider the system

$$\frac{d}{dt}(x(t) - A_{-1}x(t-h)) = A_0 x(t) + A_1 x(t-h) + Bu(t), \quad t \geq 0, \tag{10.3.1}$$

$$x(t) = \phi(t) \quad \text{for} \quad t \in [-h, 0],$$

where $A_j$ are $n \times n$ constant matrices and $B$ is an $n \times m$ constant matrix. Here $u \in C^m$ and $\phi \in C([-h, 0], E^n) \equiv C$.

It is well known that for the above assumptions, for each admissible control $u \in L_1^{\mathrm{loc}}([0, \infty), C^m)$, there exists a unique solution to (10.3.1) on $[-h, \infty)$ through $\phi$. Furthermore, if $A_{-1} \neq 0$, this solution exists on $(-\infty, \infty)$ and is unique. (See Hale [9, p. 26].) The fundamental matrix of

$$\frac{d}{dt}(x(t) - A_{-1}x(t-h)) = A_0 x(t) + A_1 x(t-h) \tag{10.3.2}$$

is a solution of (10.3.2) with initial data

$$X_0(t) = \begin{cases} 0, & t < 0, \\ I, & t = 0 \ (\text{Identity}), \end{cases} \tag{10.3.3}$$

for which $U(t) - A_{-1}U(t-h)$ is continuous and satisfies (10.3.2) for $t \geq 0$ except at $kh$, $k = 0, 1, 2, \ldots$. Indeed, $U(t)$ has a continuous first derivative on each interval $(kh, (k+1)h)$, $k = 0, 1, 2, \ldots$, the right- and left-hand limits of $U(t)$ exist at each $kh$, $k = 0, 1, 2, \ldots$, so that $U(t)$ is of bounded variation on each compact interval and satisfies

$$U(t) - A_{-1}U(t-h) = A_0 U(t) + A_1 U(t-h), \quad t \neq kh, \quad k = 0, 1, 2, \ldots. \tag{10.3.4}$$

Also if $U(t)$ is the fundamental matrix solution of (10.3.2) described above, then the solution $x(\phi, u)$ of (10.3.1) is given by

$$x(t, \phi, u) = x(t, \phi, 0) + \int_0^t U(t-s)Bu(s)ds, \tag{10.3.5}$$

where

$$x(t, \phi, 0) = U(t)(\phi(0) - A_{-1}\phi(-h)) + A_1 \int_{-h}^0 U(t-s-h)\phi(s)ds$$

$$- A_{-1} \int_{-h}^0 dU(t-s-h)\phi(s).$$

In order to introduce computational criteria to check when (10.3.1) is proper, or normal, we introduce the following notation by defining

$$Q_k(s) = A_0 Q_{k-1}(s) + A_1 Q_{k-1}(s-h) + A_{-1}Q_k(s-h),$$

$$k = 0, 1, 2, \ldots, \quad s = 0, h, 2h, \ldots,$$

$$Q_0(0) = I \text{ identity matrix}, \quad Q_0(s) \equiv 0 \quad \text{if} \quad s < 0.$$

**Theorem 10.3.1** *A necessary and sufficient condition for System (10.3.1) to be proper on the interval $[0, T]$ is that the matrix*

$$\prod(T) = \{Q_k(s)B, \ k = 0, 1, \ldots, n-1, \ s \in [0, T]\}$$

*has* rank *$n$.*

**Proof** The proof is exactly as in [12, pp. 51–60]. $\qquad\qquad\square$

**Corollary 10.3.1** *In (10.3.1), assume that $A_{-1} = 0$. Then a necessary and sufficient condition that (10.3.1) is proper on $[0, T]$ is that*

$$\prod(T) = \{Q_k(s), \ k = 0, 1, \ldots, n-1, \ s \in [0, T]\}$$

*has* rank *$n$ where*

$$Q_k(s) = A_0 Q_{k-1}(s) + A_1 Q_{k-1}(s - h),$$

$$Q_0(0) = B, \quad Q_0(s) \equiv s \neq 0.$$

This is Theorem 6.1.1, located in this book.

Corollary 10.3.1 is the algebraic criterion for complete controllability given by Gabasov and Kirillova for the delay system

$$\dot{x}(t) = A_0 x(t) + A_1 x(t - h) + Bu(t), \tag{10.3.6}$$

when the controls are not restrained to lie on a compact set but are only required to be integrable on compact intervals. We note that an algebraic criterion for the delay equation (10.3.6) to be proper is given by Corollary 10.3.1. This is a generalization of the fundamental result of LaSalle in Hermes and LaSalle [6, p. 74] on the autonomous system

$$\dot{x} = Ax + Bu(t). \tag{10.3.7}$$

Recall that (10.3.1) is normal on $[0, T]$, $T > 0$, if for any $r = 1, \ldots, m$,

$$\eta^T U(T - s)B_r = 0 \quad \text{almost everywhere} \quad s \in [0, T],$$

implies $\eta = 0$. If we follow the idea of Theorem 10.3.1, we deduce the following theorem:

**Theorem 10.3.2** *A necessary and sufficient condition for (10.3.1) to be normal on the interval $[0, T]$ is that for each $r = 1, 2, \ldots, m$, the matrix*

$$\prod_r(T) = \{Q_k(s)B_r, \ k = 0, 1, \ldots, n-1, \ s \in [0, T]\}$$

*has* rank *$n$.*

We now apply our result to the general $n$th order scalar autonomous neutral equations of the form

$$x^n(t) = \sum_{i=0}^{n} b_i x^{(i)}(t-1) + \sum_{i=0}^{n-1} a_i x^{(i)}(t) + u(t), \qquad (10.3.8)$$

where $|u| \leq 1$. Define

$$B = \begin{bmatrix} 0 \\ 0 \\ \vdots \\ 0 \\ 1 \end{bmatrix}, \qquad A_{-1} = \begin{bmatrix} 0 & 0 & 0 & \cdots & 0 \\ \cdot & \cdot & \cdot & \cdots & \cdot \\ \cdot & \cdot & \cdot & \cdots & \cdot \\ \cdot & \cdot & \cdot & \cdots & \cdot \\ 0 & 0 & & \cdots & b_n \end{bmatrix},$$

$$A_0 = \begin{bmatrix} 0 & 1 & 0 & \cdots & & 0 \\ 0 & 0 & 1 & & & 0 \\ \cdot & \cdot & \cdot & & & \cdot \\ 0 & \cdot & 0 & & & \cdot \\ a_0 & a_1 & a_n & \cdots & & a_{n-1} \end{bmatrix}, \qquad A_1 = \begin{bmatrix} 0 & 0 & \cdots & & 0 \\ 0 & 0 & & & 0 \\ \cdot & \cdot & \cdots & & \cdot \\ \cdot & \cdot & \cdots & & \cdot \\ 0 & 0 & & & 0 \\ b_0 & b_1 & \cdots & & b_{n-1} \end{bmatrix}.$$

$$(10.3.9)$$

Then System (10.3.8) is equivalent to (10.3.1) with $A_j, B$ so defined.

**Corollary 10.3.2**    *The scalar control system* (10.3.8) *is proper for every $T > 0$.*

**Proof**    The result follows immediately from (10.3.9) and Theorem 10.3.1. It is fairly obvious from Theorem 10.3.1 that if (10.3.1) is proper on $[0, T_1]$, then it is proper on $[0, T]$ for $T \geq T_1$.      $\square$

**Theorem 10.3.3**    *Consider the pointwise complete system* (10.3.1). *Suppose*

(i) *for each $T \geq 0$ and for each $r = 1, 2, \ldots, m$, the matrix*

$$\prod_r(T) = \{Q_k(s)B_r, \; k = 0, 1, \ldots, n-1, \; s \in [0, T]\} \qquad (10.3.10)$$

     *has* rank $n$, *and*

(ii) $a = \sup\{R \in \lambda : \det \Delta(\lambda) = 0\} < 0$ *where*

$$\Delta(\lambda) = \lambda(I - A_{-1}e^{-\lambda h}) - A_0 - A_1 e^{-\lambda h}.$$

*Then there is precisely one time-optimal control that drives any $\phi \in C([-1,0),\ E^n)$
to the origin in minimum-time $t^*$. It is given by*

$$u_j^*(t) = \operatorname{sgn}(c^T X(t^* - t)B)_j, \quad j = 1,\ldots,m,\ 0 \le t \le t^*. \qquad (10.3.11)$$

**Proof**   Because of (i), (10.3.1) is normal and a fortiori proper. Because of (ii),
(10.3.2) is uniformly asymptotically stable. Hence (10.3.1) is Euclidean null controllable; see Theorem 10.2.2. Because (10.3.1) is null controllable, Theorem 10.1.3
guarantees there is an optimal control that is extremal by Theorem 10.1.4 and
uniquely determined by (10.1.9) because of Corollary 10.1.1.                     $\square$

## 10.4   Pursuit Games and Time-Optimal Control Theory

In this section we show that a class of differential pursuit games is equivalent to some
time-optimal control problems. We can therefore appropriate the earlier theory that
we have developed for System (10.1.1) to solve this later problem. The emphasis
in this section is exactly as in Hájek [10]: the emphasis is on the reduction rather
than on the consequences.

We consider the linear neutral differential game described by

$$\frac{d}{dt}D(t, x_t) = L(t, x_t) - p(t) + q(t), \quad t \ge s,$$
$$x_\sigma = \phi \in C, \qquad (10.4.1)$$

where $p(t) \in P$, $q(t) \in Q$ with $P \subseteq E^n$, $Q \subseteq E^n$ is the pursuer control and
quarry control constraints. The functions $p \in L_\infty([\sigma, t], P)$, $q \in L_\infty([\sigma, t], Q)$ are
called pursuer and quarry controls. They are said to steer the initial function
$\phi \in C([-h, 0], E^n)$ to the origin in $E^n$ in time $t_1$ if the solution $x(\sigma, \phi, p, q)$ of
(10.4.1) with $x_\sigma(\sigma, \phi, p, q) = \phi$ satisfies $x(t_1, \sigma, \phi, p, q) = 0$.

The information pattern of our game can be described as follows:

For any quarry control $q$,

(i)   there exists (that is, "the pursuer can choose") a pursue control $p$ such that
      for each $s \in [\sigma, t]$ the value of $p(s)$ depends only on $q(s)$ (and of course on
      $\phi, D, L$).

(ii)  The pair of controls $p, q$ steer $\phi$ to $0 \in E^n$.

(iii) This is done in minimum-time.

Associated with the game (10.4.1) is a linear control system

$$\frac{d}{dt}D(t, x_t) = L(t, x_1) - v(t), \quad t \ge \sigma,$$
$$x_\sigma = \phi, \qquad (10.4.2)$$

where $v(t) \in V(t)$. Here $D$ and $L$ are as given in (10.4.2), but the control constraint set is defined by

$$V = P \underline{*} Q = \{x : x + Q \subseteq P\} \tag{10.4.3}$$

where $\underline{*}$ denotes the Pontryagin difference of $P$ and $Q$.

It is important to observe that we can define

$$p(t) = B(t)u(t), \quad q(t) = C(t)v(t),$$

where $B$ is an $n \times m$ matrix function and $C$ is an $n \times r$ matrix function, provided we assume $u(t) \in E^m$, $v(t) \in E^r$ with the constraint sets $P_1 \subseteq E^m$ and $Q_1 \subseteq E^r$. Viewed in this way there is nothing in (10.4.1) that suggests that the control functions have the same dimensions as the state space. These same comments are valid for (10.4.2).

**Definition 10.4.1**    A point $\phi \in C$ is said to be in position to win in time $t_1 > \sigma$ if, for any quarry control $q$, there is a pursuer control subject to the information pattern (i)–(iii) in the introduction.

**Theorem 10.4.1**    *Assume $0 \in Q$ and $P$ is compact. The game (10.4.1) is equivalent to the associated time-optimal control problem (10.4.2), where $V$ is given by*

$$V(t) = (P + \ker U(t_1, t)) \underline{*} Q, \quad \text{for some } t_1.$$

*Furthermore,*

$$f(q, t) = u(t) + q \text{ modulo } \ker U(t_1, t) \tag{10.4.4}$$

*(for all $q \in Q$, $t \in [\sigma, t_1]$) can be used to obtain a suitable strategy from $u \in L_\infty([\sigma, t_1], V)$.*

*In detail, $\phi$ is in position to win in time $t_1$ if and only if $\phi$ can be steered to 0 in time $t_1$ within the control system (10.4.2), and the corresponding minimum-times coincide.*

**Proof**    Assume $\phi \in C$ is in a position to win at time $t_1$. Then from the definition, given any quarry control $q : [\sigma, t_1] \to Q$, the mapping $f : E^n \times [\sigma, t_1] \to E^n$ exists with values $p(t) = f(q(t), t)$, $t \to p(t)$ is integrable, $p(t) \in P$, and

$$0 = x(t_1, \phi, 0, 0) - \int_\sigma^{t_1} U(t_1, s)(f(q(s), s) - q(s))ds$$

or

$$x(t_1, \phi, 0, 0) = \int_\sigma^{t_1} U(t_1, s)(f(q(s), s) - q(s))ds, \tag{10.4.5}$$

where $x(t_1, \phi, 0, 0)$ is the solution of the homogeneous equation (10.1.7). Because $0 \in Q$, we can consider the quarry control to be 0. Then

$$x(t_1, \phi, 0, 0) = \int_\sigma^{t_1} U(t_1, s)v(s)ds, \tag{10.4.6}$$

where $v(s) = f(0, s)$.

Now take any point $q \in Q$ and a time $t \in [\sigma, t_1]$, and consider the piecewise constant quarry control $q_0$,

$$
q_0 = \begin{cases} 0 & \text{in } [\sigma, t], \\ q & \text{in } [t, t_1]. \end{cases}
$$

Apply (10.4.5) to obtain

$$
x(t_1, \phi, 0, 0) = \int_\sigma^{t_1} U(t_1, s)u(s) + \int_t^{t_1} U(t_1, s)(f(q, s) - q)ds. \tag{10.4.7}
$$

On subtracting (10.4.7) from (10.4.6),

$$
\int_t^{t_1} U(t_1, s)(v(s) + q - f(q, s))ds = 0
$$

for all $t \in [\sigma, t_1]$. Observe that the integrand is independent of $t$, so that by differentiation

$$
U(t_1, t)(v(t) + q - f(q, t)) = 0 \text{ almost everywhere}.
$$

Use the kernel to interpret this as

$$
v(t) + q \in f(q, t) + \ker U(t_1, t) \text{ almost everywhere } t \in [\sigma, t_1]. \tag{10.4.8}
$$

Since $f$ has values in $P$,

$$
v(t) + q \in P + \ker U(t_1, t).
$$

From the continuity of solutions of (10.1.1) with respect to initial conditions, $\ker U(t_1, t)$ is closed. Hence $(P + \ker U(t_1, t))$ is closed. Because of this, Hájek's Lemma [11, p. 59] yields

$$
v(t) + Q \subseteq P + \ker U(t_1, t) \text{ almost everywhere},
$$

or

$$
v(t) \in (P + \ker U(t_1, t)) \underset{*}{} Q \equiv V(t) \text{ almost everywhere}.
$$

Hence $v \in L_\infty([\sigma, t_1], V)$ where $V$ is as defined in (10.4.4). It now follows that $v$ is an admissible control for (10.4.2), and hence from (10.4.6)

$$
x(t_1, \phi, 0, 0) = \int_\sigma^{t_1} U(t_1, s)v(s)ds
$$

or

$$
0 = x(t_1, \phi, 0, 0) = \int_\sigma^{t_1} U(t_1, s)v(s)ds.
$$

Observe that $x_\sigma(\sigma, \phi, 0, 0) = \phi$, so that $x(\sigma, \phi, 0, 0)$ is a solution of (10.1.2). Note that (10.4.8) proves the last assertion of the theorem. Hence $v$ steers $\phi$ to 0 in time

$t_1$. Next $v$ and $0$ are pursuer and quarry controls ($0 \in Q$, $P \underline{*} Q \subset P$); for this choice the dynamical equations of the game and the control systems coincide; so do their solutions. Hence the optimality problems are the same.

Conversely, let an admissible control $v$ steer $\phi$ to $0$ at $t_1$ within the control system (10.4.2). To show that $\phi$ is in position to win at $t_1$, take quarry control $q$. Then

$$0 = x(t_1, \phi, 0, 0) - \int_\sigma^{t_1} U(t_1, s)v(s)ds \tag{10.4.9}$$

where $v(s) \in V(s)$ is such that $v(s) + q \in P + \ker U(t_1, s)$. We now construct a pursuer-control strategy as follows: by Filippov's Lemma (see the form in [11, p. 119]) there exist measurability preserving maps

$$f : Q \times [\sigma, t_1] \to P\,,$$

$$\lambda : Q \times [\sigma, t_1] \to \ker U(t_1, s)$$

for each $\sigma \le s \le t_1$ with $\lambda(q(s), s) \in \ker U(t_1, s)$ such that

$$v(s) + q = f(q, s)\,.$$

Since $f(q, s) \in P$ is a compact set, we have $f \in L_\infty([\sigma, t_1], P)$.

We now verify that to any quarry control $q$, $f$ thus constructed, will force $\phi$ to $0 \in E^n$. Indeed, if $q \in L_\infty([\sigma, t_1], Q)$, $f - q = v - \lambda$ so that the solution of (10.4.1) at time $t_1$ with the pair of $f, q$ and initial function $\phi$ is

$$x(t_1, \phi, \sigma, q) = x(t_1, \phi, 0, 0) + \int_\sigma^{t_1} U(t_1, s)(f(q(s), s) - q(s))ds$$

$$= x(t_1, \phi, 0, 0) - \int_\sigma^{t_1} U(t_1, s)(u(s) - \lambda(q(s), s))ds$$

$$= x(t_1, \phi, 0) - \int_\sigma^{t_1} U(t_1, s)v(s)ds$$

$$+ \int_\sigma^{t_1} U(t_1, s)\lambda(q(s), s)ds$$

$$= 0\,,$$

from (10.4.9) and the definition of $\lambda$.

Hence on taking $f$ and $q$ as controls we find again that the dynamical equations coincide, so that steering to $0$ in $t_1$ and the minimal times are preserved. This completes the proof. $\qquad\square$

For more general targets

$$G : [\sigma, \infty) \to r^n$$

that are continuous we do not have the duality of Theorem 10.4.1. Instead we have the following:

**Theorem 10.4.2** *Let the pursuer-constraint set be compact and let $G : [\sigma, \infty) \to r^n$ be continuous. An initial position $\phi \in C$ can be forced to a target $G$ (using the usual information pattern) at time $t_1$ within the game (10.4.1) whenever in the associated control system (10.4.2) $\phi$ can be steered to the target in time $t_1$. Furthermore,*

$$f(q, t) = v(t) + q \ \text{modulo} \ \ker U(t_1, t)$$

*determines a control strategy for (10.4.1) that counters any quarry action $q \in L_\infty([\sigma, t_1], Q)$ where $v$ is an admissible control for (10.4.2).*

**Proof** Assume that in (10.4.2) $\phi$ can be steered to $G$ in time $t_1$. Then there exists $v \in L_1([\sigma, t_1], V)$ such that the solution $x = x(\sigma, \phi, v)$ of (10.4.2) satisfies

$$x(t_1) = x(t_1, \phi, 0) - \int_\sigma^{t_1} U(t_1, s)u(s)ds \in G(t_1) \,. \tag{10.4.10}$$

Because $v : [\sigma, t_1] \to V$,

$$v(s) + q \in P + \ker U(t_1, s) \quad \text{for} \quad s \in [\sigma, t_1] \quad \text{and every} \quad q \in Q \,.$$

With $v$ fixed, we now construct a pursuer-control strategy $f$ by applying Filippov's Lemma. We obtain measurable functions

$$f : Q \times [\sigma, \infty) \to P \,,$$

$$\lambda : Q \times [\sigma, \infty) \to \ker U(t_1, s) \,, \qquad \lambda(q(s), s) \in \ker U(t_1, s)$$

such that

$$v(s) + q(s) = f(q(s), s) + \lambda(q(s)s) \,.$$

Just as in Theorem 10.4.1, $f \in L_\infty([\sigma, t_1], P)$. For any quarry action $q$, $f - q = v - \lambda$ so that the solution of (10.4.1), with this $f$ and $q$ and initial data $\phi$, satisfies

$$x(t_1, \phi, f, q) = x(t_1, \phi, 0, 0) - \int_\sigma^{t_1} U(t_1, s)(f(q(s), s) - q(s))ds$$

$$= x(t_1, \phi, 0, 0) - \int_\sigma^{t_1} U(t_1, s)v(s) + \int_\sigma^{t_1} U(t_1, s)\lambda(q(s), s)ds$$

$$= x(t_1, \phi, 0, 0) - \int_\sigma^{t_1} U(t_1, s)v(s)ds \in G(t_1)$$

by (10.4.10). This proves the Theorem. $\qquad\square$

We now examine the quantitative properties of $P$ and $Q$ from a close examination of Theorem 10.4.1.

**Definition 10.4.2**   Consider a mapping $\tau : Q \times E \to P$, where $\tau(q,t)$ is a point in $P$ whenever $q$ is a point of $Q$ and $t \in E$. Suppose $\xi$ is an induced mapping on the collection of $Q$ of functions $q(\cdot)$, which are defined by $\xi[q](t) = \tau(q(t),t)$. Then $\xi$ is called a stroboscopic function if it preserves measurability. We describe the following example: Suppose $t \to u(t)$ is given in advance. Then

$$\xi(q)(t) = q(t) + u(t)$$

is a stroboscopic strategy if $u$ is measurable and $P \supset Q + u(t)$ for all $t$.

**Theorem 10.4.3**   *In* (10.4.1)*, assume that* $0 \in Q$, $P$ *is compact,* $P, Q$ *are nonvoid, convex, and symmetric. Then the set inclusion*

$$P + \ker U(t_1,t) \supset Q, \quad \sigma \le t \le t_1 \tag{10.4.11}$$

*is necessary and sufficient for the presence of initial* $\phi$*, which can be controlled to zero in some time* $t_1$ *when the composite system* (10.4.1) *uses stroboscopic strategy. The inclusion*

$$\mathrm{Int}(P + \ker U(t_1,t)) \supset Q, \quad \sigma \le t \le t_1 \tag{10.4.12}$$

*is necessary and sufficient for the domain of null controllability* (10.4.1) *to have zero in its interior.*

**Proof**   Theorem 10.4.2 states that the set of functions $\phi$ that can be driven to zero at some time $t_1$ using the dynamics (10.4.1) is the same as the set of initial functions that can be controlled to zero when System (10.4.2) is used. Because

$$V(t) = (P + \ker\ U(t_1,t)) \underset{*}{\ast} Q$$

is the control set for (10.4.2), a set which is convex and symmetric, we deduce that $V(t)$ is nonvoid if and only if $0 \in V(t)$, i.e.,

$$P + \ker U(t_1,t) \supset Q, \quad \sigma \le t \le t_1 \,.$$

But then with

$$1R(t_1) = \int_\sigma^{t_1} U(t_1,s)V(s)\,,$$

there are initial positions that can be forced to zero stroboscopically at time $t_1$ if and only if $1R(t_1)$ is nonvoid. By Filippov's Lemma (see also Chukwu [17, pp. 437–439]), $1R(t_1)$ is nonvoid if $V(t)$ is nonvoid, for $\sigma \le t \le t_1$, so that

$$P + \ker U(t_1,t) \supset Q\,.$$

For the converse, if $P + \ker U(t_1, t) \supset Q$, then $0 \in (P + \ker U(t_1, t), 0$ is an admissible control steering zero to zero. The second assertion is valid if and only if

$$0 \in \operatorname{Int} V(t) = \operatorname{Int}((P + \ker U(t_1, t)) * Q).$$

But $0 \in \operatorname{Int} V(t)$ is a requirement for (10.4.2) to have zero in the interior of the Euclidean reachable set, and therefore in the domain of null controllability of (10.4.2), which by the duality Theorem 10.4.2 coincides with that of (10.4.1).

For ordinary differential linear systems, Hájek has obtained the following results [11, pp. 60–87]: Consider the system

$$\dot{x}(t) = Ax - p + q; \quad (p(t) \in P, \ q(t) \in Q) \tag{10.4.13}$$

in $E^n$, where $A$ is an $n \times n$ matrix. $\qquad \square$

**Proposition 10.4.1** *In (10.4.13) assume that $P, Q$ are nonvoid, convex, and symmetric. Then*

$$P \supset Q \tag{10.4.14}$$

*is a necessary and sufficient condition for presence of initial positions that can be forced to zero stroboscopically in strictly positive time. If*

$$\operatorname{Int} P \supset Q, \tag{10.4.15}$$

*then for every $t > 0$ the set of positions that can be forced to 0 stroboscopically at time $t$ is a neighborhood of 0. If*

$$G = \{x : x \in E^n; \ Mx = Mb\} \tag{10.4.16}$$

*where $M$ is an $m \times n$ matrix and $b \in E^n$, then*

$$M(P - P) \supset M(Q - Q) \tag{10.4.17}$$

*is necessary, and if $(Q$ is compact),*

$$\operatorname{Int} MP \supset MQ \tag{10.4.18}$$

*is sufficient for the presence of positions that can be forced to $G$ at strictly positive times. Suppose the pursuer's (firm's) control order $k$ is defined as that $k$ such that*

$$MA^j(P - P) = 0, \tag{10.4.19}$$

*(the set consisting of the zero vector, not the empty set) holds for $j = 0, \ldots, k - 2$ but not for $j = k - 1$. Assume $P$ is compact. A necessary condition for the presence of initial points that can be forced to $G$ at $t > 0$ is that*

> (1) *firm's control order (pursuer's) $\geq$ solidarity control order; and if $k$ is the solidarity (quarry) control order then*

(2)

$$MA^{k-1}(P - P) \supset MA^{k-1}(Q - Q). \qquad (10.4.20)$$

(i) *Suppose $P, Q$ are compact, convex, and symmetric, and*
(ii) *firm's control order $\geq$ solidarity control order;*
(iii)

$$\text{Int } MA^{k-1}P \supset MA^{k-1}Q. \qquad (10.4.21)$$

*Then for sufficiently small $t_1 > 0$, the set of initial endowments that can be forced to $G$ at time $t_1$ has a nonempty interior.*

Let (10.4.13) describe the growth of capital stock with any initial value $x(0) = x_0$ and target 0 or target $G$ in (10.4.16). If the target is 0, we can define the firm's initiative (investment, productivity wages consumption) as the control set $P$, and the set $Q$ as solidarity (e.g., money supply government taxation or subsidy). If $G$ is the target, $MA^{k-1}P$ is defined to be the firms control initiative and $MA^{k-1}Q$ is solidarity. With this terminology we deduce the following universal principle:

**Principle 10.4.1**    *No initial value of capital stock can be controlled to the target unless the firm's initiative contains solidarity as a subset. To ensure growth of any initial value of capital stock to the target that is sufficiently close, the firm's initiative must dominate solidarity. These principles provide a broad policy prescription for national economies, which will be explored in a subsequent communication.*

Fundamental principle 10.4.1 emerges from studies of linear systems. Consideration of general nonlinear dynamics in Theorem 3.2 of [15] show that the solidarity assumption is the generalization of fundamental Principle 10.4.1.


## 10.5    Applications and Economic Growth

From the discussions of Sec. 1.8 that yielded (1.8.5) and the insight of (1.8.4), it is not unreasonable to postulate that the dynamics of capital function of $n$ stocks in a region may be described by the equation

$$\dot{x}_i(t) - A_i\dot{x}_i(t - h) = L_i(t, x_t) - p_i(t) + q_i(t) \qquad (10.5.1)$$

where $p_i(t) \equiv B_i(t)u_i(t)$, $q_i(t) = C_i(t)v_i(t)$ and $q_i(t)$ is the solidarity function that is the effect of government intervention (taxation, subsidy) in economic growth. Equation (10.5.1) describes the net capital formulation of the $i$th subsystem. The effects of government acts are measured locally. We now interpret Theorem 10.4.2. Let $\phi \in C$ be any initial capital endowment function and $G$ the desired target consisting of the range of values of capital desired at time $t_1$. Let $q(t)$ represent the effects (measured locally at the firm) of government economic and other policies.

The firm can grow from $\phi$ to $G$ in time $t_1$ using its investment and other strategies $p(t)$ if and only if the control system

$$\dot{x}_i(t) - A_i\dot{x}_i(t - h) = L_i(t, x_t) - u(t) \tag{10.5.2}$$

can grow from $\phi$ to $G$ at time $t_1$. The control investment strategy is $f(q,t) = v(t) + q$ modulo $\ker U(t_1, t)$, which reacts to any government action $q$ where $v$ is an admissible control for (10.5.2). If $Q$ is the totality of "government power", and if it is permissible for government not to intervene ($0 \in Q$), and if the firm's investment capacity $P$ is limited and small (compact), then the optimal-control strategy is constrained to lie in $V$, which is defined by $V(t) = (P + \ker U(t_1, t) * Q)$. The second of the universal principles of control on the limitation of government power is valid:

$$Q \subset \text{Int}(P + \ker U(t_1, t)).$$

**Principle 10.4.2** *To guarantee economic growth from any initial endowment to a target, it is necessary that the firm's capacity for investment and its internal power for waste dominate whatever government can do. Government intervention is needed, but it should not be too big.*

We now consider System (10.4.1) very carefully. First, consider the system

$$\frac{d}{dt}D(t, x_t) = L(t, x_t) - p(t), \quad p(t) \in P. \tag{10.5.3}$$

If (10.5.3) is controllable on $[\sigma, t_1]$ with $p(t) \in P$, then

$$0 \in \text{Int}\, \mathcal{A}(t_1) \tag{10.5.4a}$$

where

$$\mathcal{A}(t_1) = \{x(p)(t_1) : p(t) \in P\}.$$

The attainable set of (10.4.1) is then

$$A(t) = \{x(p, q)(t) : p(t) \in P, q(t) \in Q\}.$$

If

$$Q \subset P \quad \text{or} \quad Q \subset \text{Int}\, P,$$

System (10.4.1) is such that

$$0 \in \text{Int}\, \mathcal{A}(t_1) \subset A(t), \tag{10.5.4b}$$

provided (10.5.3) is controllable. Thus with

$$-p(t) = B_1(t)u(t) \quad \text{and} \quad q(t) = B_2u(t),$$

for example, the system

$$\frac{d}{dt}D(t, x_t) = L(t, x_t) + (B_1(t) + B_2(t))u(t) \tag{10.5.5}$$

is made controllable. It is possible that the intervention of $q$ may make matters worse. Indeed, even if

$$\frac{d}{dt}D(t, x_t) = L(t, x_t) + B_1 u(t) \tag{10.5.6}$$

is controllable, the composite system (5.4.1) may not be controllable. A "proper amount" of $q$ is needed. We can formulate an economic interpretation with the following remarks:

To ensure growth from an initial endowment to the target we may assume (10.5.5) is Euclidean controllable. Hence matters should be so arranged that the solidarity function $q$ brought to bear on the isolated system ensures controllability.

These observations are formalized in the following principle:

**Principle 10.5.3** *If an isolated system is not proper and is not locally controllable with constraints, the composite system may be made "proper" and locally controllable provided there is an external source of power or initiative $q$ available to enforce controllability and proper behavior. It is possible that the intervention of solidarity can make matters worse. Only a "proper" amount is needed. There is no theorem at the present time that states that the interconnected system is well behaved when it is misbehaving in isolation, and there is no compensating external solidarity $q$.*

We observe that the intervention of $q$ can make matters worse. If (10.5.4a) holds and we consider (10.5.1) with $q(t) \equiv q_0 \neq 0$, then 0 need not be in $\mathcal{A}(t)$, the attainable set of (10.4.1) and controllability may fail. In this case, the solidarity function is not flexible. This seems to be the case in centralized economies.

**Principle 10.5.4** *If the solidarity function is inflexible and rigid, or if the isolated systems controlling initiatives are ignored in the construction of a nontrivial solidarity function, the interconnected system will fail to be proper and locally null controllable.*

We have isolated null controllability in $E^n$ as our objective in the analysis of the growth of capital stock. Though zero target at the final time seems artificial, the theory incorporates nontrivial targets. Indeed, if $x_1$ is a nontrivial arbitrary target and $y(t) = y(t, \phi, u^0)$ is any solution of

$$\dot{x}(t) - A_{-1}\dot{x}(t - h) = f(t, x_t, u_t)$$

with $y_\sigma = \phi$ and $u^0$ admissible such that $y(t_1) = x_1$, one can equivalently study the null controllability of the system

$$\dot{z}(t) - A_{-1}\dot{z}(t - h) = f(t, z_t + y_t, u_t) - f(t, y_t, u_t),$$

where $z_\sigma = 0$, so that $x_\sigma = y_\sigma = \phi$ and $x(t) = z(t) + y(t)$. If we can show that there is a neighborhood $\mathcal{O}$ of the origin in $z$ space such that all initial $\phi \in \mathcal{O}$ can be brought to $z(t_1, u, \phi, u^0) = 0$ by some admissible control at time $t_1$, then

$$z(t_1, u, \phi, u^0) = x(t_1, \phi, u) - y(t_1) = 0 \,,$$

so that

$$x(t_1, \phi, u) = y(t_1) = x_1 \,.$$

The remarks we have made on the consequences and insights of Theorems 10.4.1 and 10.4.2 justify a qualitative description of the solidarity function $q$ and its constraint $Q$. How big should it be? This $q$ can be viewed as a control disturbance, and we can then study System (10.4.1).

## 10.6  Optimal Control Theory of Linear Neutral Systems

In this section we study the optimal problem of the system

$$\frac{d}{dt}\left[ x(t) - \sum_{j=1}^{m} A_{-1j} x(t - h_j) \right]$$

$$= A_0 x(t) + \sum_{j=1}^{m} A_j x(t - h_j) + B(t)u(t), \quad t > 0, \qquad (10.6.1)$$

$$x(0) = g^0 \in E^n, \quad x(t) = g^1(t), \quad t \in (-h, 0], \quad g^1 \in C \,.$$

Define the fundamental matrix $W$ of 10.6.1 by

$$W(t)g^0 = \begin{cases} x(t, 0, g^0, 0), & t \geq 0, \\ 0, & t < 0, \end{cases}$$

where $x(t, u, g^0, g^1)$ is the solution of (10.6.1). Hence $W(t)$ is the unique solution of

$$W(t) = e^{A_0 t} + \int_0^t e^{A_0(t-s)} \cdot \sum_{j=1}^{m} \chi(s - h_j)(A_j + A_0 A_{-1j}) \cdot W(s - h_j)ds$$

$$+ \sum_{j=1}^{m} \chi(t - h_j) A_{-1j} W(t - h_j) \qquad (10.6.2)$$

where $\chi(s) = 0$ if $s < 0$, $\chi(s) = I$ if $s \geq 0$, $I$ identity.

**Proposition 10.6.1**  *The variation of constant formula for (10.6.1) is*

$$x(t, u, g^0, g^1) = \left[ W(t) - \sum_{j=1}^{m} W(t - h_j) A_{-1j} \right] g^0$$

$$+ \sum_{j=1}^{m} \int_{-h_j}^{0} W(t - s - h_j)[A_j g^1(s) + A_{-1j} g^1(s)]$$

$$+ \int_0^t W(t - s)B(s)u(s)ds$$

$$\equiv x(t, 0, g^0, g^1) + \int_0^t W(t - s)B(s)u(s)ds. \tag{10.6.3}$$

We study the optimal control and the adjoint system. Consider (10.6.1), where

$$u \in U_{\mathrm{ad}} \subset L_p([0, T], E^n), \quad p \in [1, \infty), \quad B \in L_\infty([0, T], E^{n \times m}).$$

Assume $U_{\mathrm{ad}}$ closed and convex set in $L_p$. We denote the solution of (10.6.1) by $x(t, u)$. Let the integral cost function be given by

$$J = J(u, x),$$

$$J = \phi_0(x(T)) + \int_0^T f_0(x(t), t) + k_0(u(t), t), \tag{10.6.1b}$$

$$\phi_0 : E^n \to R, \quad f_0 : E^n \times I \to E, \quad k_0 : L_p \times [0, T] \to E.$$

We study the following problems:

$P_1$ Find a control $u \in U_{\mathrm{ad}}$ that minimizes the cost $J$ subject to (10.6.1).

$P_2$ Find optimality conditions for the optimal pair $(u^*, x^*(u^*)) \in U_{\mathrm{ad}} \times C([0, T], E^n)$ such that

$$\mathrm{Inf}\, J(u, x) = J(u^*, x^*(u^*)), \quad u \in U_{\mathrm{ad}}.$$

We shall prove the existence of optimal controls for $P_1$, and solve $P_2$ by deriving necessary optimality conditions. Let $H$ be a compact target set in $E^n$. Suppose

$$U_0 = \{u \in U_{\mathrm{ad}} : x(u, t) \in H \text{ for some } t \in [0, T]\},$$

and suppose $U_0 \neq \emptyset$. This is the constrained-controllability assumption. We now formulate the time-optimal problem.

$P_3$ Find a control $u^* \in U_0$ such that

$$t^*(u^*) \leq t(u) \quad \text{for all} \quad u \in U_0$$

subject to (10.6.1). The number $t^*(u^*)$ is the first time $x(t^*, u^*) \in H$, i.e., $t^*$ is the optimal time.

We now consider the adjoint system. Let $q_0 \in (E^n)^*$ be a row vector,

$$q_1^* \in L_1([0, T], E^{n*}).$$

The adjoint system for (10.6.1) is

$$\frac{d}{dt}\left[y(t) - \sum_{j=1}^{m} y(t + h_j)A_{-1j}\right] + y(t)A_0(t) + \sum_{j=1}^{m} y(t + h_j)A_j - q_1(t) = 0, \quad \text{a.e. } t \in I,$$

$$y(T) = -q_0, \quad y(s) = 0, \quad \text{a.e. } s \in [T, T + h]. \tag{10.6.4a}$$

The solution of the adjoint equation is given by the adjoint state

$$y(t) = W^*(T - t)(-q_0^*) + \int_t^T W^*(s - t)(-q_1^*(s))ds,$$

where $W^*(t)$ is the adjoint of $W(t)$, $t \in [0, T]$, and is the fundamental matrix solution of (10.6.4a), which is unique. We note that

$$Y(s, t) \equiv W^*(t - s) = W(t - s)$$

a.e. in $s$, where $W(t)$ is the fundamental solution of (10.6.4a). Thus

$$Y(s, t) = I + \int_s^{t^+} d\alpha Y(\alpha, t)\mu(\alpha, s) - \int_s^t Y(\alpha, t)\eta(\alpha, s)d\alpha, \quad s \in [\sigma, t],$$

$$Y(t, t) = I, \quad Y(s, t) = 0 \quad \text{for} \quad s > t,$$

where

$$\int_{-h}^t d_\alpha \mu(t, s)x(s) \equiv \sum_{j=1}^{M} A_{-1j}x(t_1 - h_j),$$

$$\int_{-h}^t d\eta(s)x(s) \equiv \sum_{j=1}^{M} A_j x(t - h_j).$$

We next state conditions for the existence of optimal controls for problem $P_1$.

**Theorem 10.6.1** *Assume that:*

(i) *$\phi_0 : E^n \to E$ is continuous and convex.*

(ii) *$f_0 : E^n \times [0, T] \to E$ is measurable in $t$ for each $x \in E^n$, and continuous and convex in $x \in E^n$ for a.e. $t \in I$, and assume further for each bounded set $K \subset E^n$ there exists a measurable function $M_k \in L_1([0, T], E)$ such that*

$$\sup_{x \in K} |f_0(x, t)| \le M_k(t) \quad \text{a.e.} \quad t \in I.$$

(iii) *$k_0 : L_p([0, T], E^m) \times I \to E$ is such that for any $u \in U_{\text{ad}}$, $k_0(u(t), t)$ is integrable on $I$ and the functional $\Gamma_0 : U_{\text{ad}} \to E$ defined by*

$$\Gamma_0(u) = \int_0^T k_0(u(t), t)dt$$

*is continuous and convex.*

(iv) $U_{\mathrm{ad}}$ *is bounded.*

*Then there exists a control $u_0 \in U_{\mathrm{ad}}$ that minimizes the cost $J$*

$$J = \phi_0(x(T)) + \int_0^T f_0(x(t), t) + k_0(u(t), t)dt. \qquad (10.6.4b)$$

The proof is an easy adaptation of its infinite dimensional analogue developed recently by Nakagiri [27, Theorem 4.1].

**Proof**    Let $\{u_n\}$ be a minimizing sequence of controls for $J$ and $x_n$ the corresponding trajectory:

$$\operatorname*{Inf}_{u \in U_{\mathrm{ad}}} J = \lim_{n \to \infty} J(u_n, x_n) = M_0.$$

Since we assumed that $U_{\mathrm{ad}}$ is bounded and since it is weakly closed, there is a subsequence (again denoted by $\{u_n\}$) of $\{u_n\}$ and a $u_0 \in U_{\mathrm{ad}}$ such that

$$u_n \to u_0 \quad \text{weakly in } L_p(I; E^m). \qquad (10.6.5)$$

Suppose $x_0$ is the trajectory corresponding to $u_0$. Let $c$ be a row vector in $E^{n^*}$, and $t \in I = [0, T]$ be fixed. Because the fundamental matrix $W$ is such that $W(t) = 0$ if $t < 0$, we have that

$$(x_n(t), c) = (x(t, u, g^0, g^1), c) + \int_0^t (u_n(s), B^*(s)W^*(t - s)c)ds. \qquad (10.6.6)$$

Recall that $B \in L_\infty(I, E^{n \times m})$ and $W$ is by Tadmor piecewise analytic

$$B^*(\cdot)W(t - \cdot)c \in L_2(I, E^{n^*}).$$

It now follows from (10.6.5) and (10.6.6) that

$$(x_n(t), c) \to (x(t, u, g^0, g^1), c) + \int_0^t (u_0(s), B^*(s)W^*(t - s)c)ds$$

$$= (x(t, u, g^0, g^1), c) + \int_0^t (W(t - s)B(s)u_0(s), c)ds \qquad (10.6.7)$$

$$= (x_0(t), c) \quad \text{as } n \to \infty \text{ (weakly) in } E^n. \qquad (10.6.8)$$

But if a function is continuous and convex, it is weak lower semicontinuity. Therefore Assumption (i) and (10.6.8) imply that

$$\lim_{n \to \infty} \phi_0(x_n(T)) \geq \phi_0(x_0(T)). \qquad (10.6.9)$$

In the same way,

$$\lim_{n \to \infty} f_0(x_n(t), t) \geq f_0(x_0(t), t), \quad \text{a.e. } t \in I. \qquad (10.6.10)$$

By standard arguments that use Hölder inequality, the set

$$K = \bigcup \{x_n(t) : t \in I, \ n = 1, 2, \ldots\}$$

is bounded in $E^n$. Since there exists an $m_k \in L_1(I; R)$ such that

$$|f_0(x_n(t), t)| \le m_k(t), \quad \text{a.e. } t \in I. \tag{10.6.11}$$

Lebesgue–Fatou Lemma yields the following assertion, which follows from (10.6.10) and (10.6.11):

$$\lim_{n \to \infty} \int_0^T f_0(x_n(t), t) dt \ge \int_0^T \left( \lim_{n \to \infty} f_0(x_n(t), t) \right) dt,$$

$$\ge \int_0^T f_0(x_0(t), t) dt. \tag{10.6.12}$$

The following estimate is available for the terms $\int_0^T k_0(u_n(t), t) dt$:

$$\lim_{n \to \infty} \Gamma_0(u_n) \ge \Gamma_0(u_0) = \int_0^T k_0(u_0(t), t) dt, \tag{10.6.13}$$

since hypothesis (iii) is valid.

On gathering the above results in (10.6.9), (10.6.12), and (10.6.13), we have

$$M_0 = \operatorname*{Inf}_{u \in U_{\mathrm{ad}}} J \ge \lim_{n \to \infty} \phi_0(x_n(T)) + \lim_{n \to \infty} \int_0^T f_0(x_u(t), t) dt + \lim_{n \to \infty} \Gamma_0(u_n),$$

$$\ge \phi_0(x_0(T)) + \int_0^T f_0(x_0(t), t) + k_0(u_0(t), t) dt,$$

$$= J(u_0, x_0) > -\infty.$$

We have proved that $M_0 = J(u_0, x_0)$, i.e., the pair $(u_0, x_0)$ is the optimal solution for $J$. $\qquad\square$

**Remark** Note carefully that the set of admissible controls is bounded.

**Theorem 10.6.2** *For problem $P_2$, assume that:*

(i) $\phi_0 : E^n \to E$ *is continuous and Gateaux differentiable, and the Gateaux derivative* $d\phi_0(x) \in E^{n^*}$ *for each $x \in E^n$;*

(ii) $f_0 : E^n \times I \to E$ *is measurable in $t \in I$ for each $x \in E^n$ and continuous in $x \in E^n$ for a.e. $t \in I$, and the*

    (a) *value $\partial_1 f_0(x, t)$ is the Gateaux derivative of $f_0(x, t)$ in the first argument for $(x, t) \in E^n \times I$, and*

    (b) $|\partial_1 f_0(x, t)| \le \theta_1(t) + \theta_2(|x|)$ *for $(x, t) \in E^n \times I$;*

(iii) $k_0 : L_2([0,T]) \times I \to E^n$ *is measurable in $t$ for each $u \in L_p$ and continuous and convex on $L_p$ for a.e. $t \in I$ and further there exist functions $\partial_1 k_0 : L_2 \times I \to E^{n^*}$, $\theta_3 \in L_p(I,E)$ and $M_4 > 0$, such that*

(a) *$\partial_1 k_0$ is measurable in $t$ for each $u \in L_2$ and continuous in $u \in L_2$ for a.e. $t \in I$, and the value $\partial_1 k_0(u,t)$ is the Gateaux derivative of $k_0(u,t)$ in the first argument for $(u,t) \in L_2 \times I$, and*
(b) *$|\partial_1 k_0(u,t)|_{E^{n^*}} \leq \theta_3(t) + M_4\|u\|_2$ for $(u,t) \in L_2 \times I$;*

(iv) *$U_{\mathrm{ad}} = \{u \in L_2(I,E^m) : \|u\|_2 \leq \alpha\}$.*

*Let $(u,x) \in U_{\mathrm{ad}} \times C([I,E^n])$ be an optimal control solution for $J$ in (10.6.15). Then the optimal control $u$ is characterized by*

$$u = \frac{-\alpha \wedge^{-1} K(u)}{\|\wedge^{-1} K(u)\|_2},$$

*where $\wedge$ is the canonical isomorphism of $L_2(I,E^m)$ into $L_2(I,E^{m^*})$, and*

$$K(u)(t) = \partial_1 k_0(u(t),t) - B^T(t)y(t) \quad a.e.\ t \in I,$$

*and $y(t)$ satisfies the equations*

$$\frac{d}{dt}\left[y(t) - \sum_{j=1}^{M} y(t+h_j)A_{-1j}\right] + y(t)A_0$$

$$+ \sum_{j=1}^{M} y(t+h_j)A_j - \partial_1 f_0(x(t)t) = 0, \quad a.e.\ t \in I, \qquad (10.6.14)$$

$$y(T) = -d\phi_0(x(T)), \quad y(s) = 0, \quad s \in (T, T+h).$$

**Proof**   Since the cost function is Gateaux differentiable it follows from [17, p. 10] that the necessary optimality condition is given by the variational inequality

$$J'(u)(v-u) \geq 0, \quad \forall\, v \in U_{\mathrm{ad}}, \qquad (10.6.15a)$$

when $J$ is differentiable. Because of the hypothesis, and since Lebesgue's Dominated Convergence Theorem is valid, we have

$$J'(u)(v-u) = \left(\int_0^T W(T-s)B(s)(v(s)-u(s))ds, d\phi_0(x(T))\right)$$

$$+ \int_0^T \left(\int_0^s W(s-\tau)B(\tau)(v(\tau)-u(\tau))d\tau, \partial_1 f_0(x(s),s)\right) ds$$

$$+ \int_0^T (v(s)-u(s), \partial_1 k_0(u(s),s))ds. \qquad (10.6.15b)$$

All the integrands are well defined because of the hypothesis. The first term in (10.6.15) can be transformed by using Fubini's Theorem:

$$\int_0^T \int_0^s (W(s-\tau)B(\tau)(v(\tau)-u(\tau)), \partial_1 f_0(x(s), s)d\tau)ds$$

$$= \int_0^T (v(s)-u(s), B^*(s) \int_s^T W^*(\tau-s)\partial_1 f(x(\tau), \tau)d\tau)ds. \qquad (10.6.16)$$

If we let

$$y(t) = -W^*(T-t)d\phi_0(x(T)) - \int_t^T W^*(s-t)\partial_1 f_0(x(s), s)ds,$$

then from (10.6.15a)–(10.6.16) the following inequality follows:

$$\int_0^T (v(t)-u(t), \partial_1 k_0(u(t), t) - B^*(y(t)))dt \geq 0, \quad \text{for all } v \in U_{\text{ad}},$$

and this is reduced to

$$\partial_1 k_0(u(t), t) - B^*(t)y(t) = 0, \quad \text{a.e. } t \in I. \qquad (10.6.17)$$

Since $U_{\text{ad}} = \{u \in L_2(I, E^m) : \|u\| \leq \alpha\}$ and our hypothesis is valid, we easily deduce that

$$u = \frac{-\alpha \wedge^{-1} K(u)}{\|\wedge^{-1} K(u)\|_2}$$

where $\wedge$ is the canonical isomorphism of $L_2(I, E^m)$ into $L_2(I, E^m)^*$ and

$$K(u)(t) = \partial_1 k_0(u(t), t) - B^*(t)y(t), \quad \text{a.e. } t \in I.$$

$\square$

**Example 10.6.1** Let $U_{\text{ad}} = L_2([0, T], E^m)$. The cost

$$J_1 = (x(T), Nx(\tau)) + \int_0^T (x(t), M(t)x(t)) + \Gamma_Q u$$

where

$$\Gamma_Q(u) = \frac{1}{2} \int_0^T (u(t), Q(t)u(t))_{E^m} dt.$$

We assume $N$ is $n \times n$ matrix $M(\cdot) \in L_\infty([0, T], E^{n \times n})$, $Q \in L_\infty([0, T]E^{m \times m})$; and $N, M, Q$, are positive and symmetric for each $t \in [0, T]$. There is a constant $c > 0$ such that

$$(u, Q(t)u) \geq c\|u\|^2, \quad \text{a.e. } t \in [0, T].$$

Thus $\Gamma_Q$ is strongly continuous and strictly convex. Then there exists a unique optimal control for $J_1$, and the following is a consequence of Theorem 10.6.2:

**Proposition 10.6.2**   *Consider the cost function in Example* 10.6.1. *There exists a unique optimal solution*

$$(u, x) \in L_2([0, T], E^m) \times C([0, T], E^n)$$

*for $J_1$. The optimal control is given by*

$$u(t) = Q^{-1}(t)B^T(t)y(t),$$

*where*

$$\dot{x}(t) - \sum_{j=1}^{m} A_{-1j}\dot{x}(t - h_j) = A_0 x(t) + \sum_{j=1}^{m} A_j x(t - h_j)$$

$$+ B(t)Q^{-1}(t)B^T(t)y(t) + f(t),$$

$$x(0) = g^0, \quad x(s) = g^1(s), \quad \text{a.e. } s \in [-h, 0),$$

$$\frac{d}{dt}\left[ y(t) - \sum_{j=1}^{m} y(t + h_j)A_{-1j} \right] + y(t)A_0 + \sum_{j=1}^{m} y(t + h_j)A_j - M(t)x(t) = 0,$$

$$y(T) = -Nx(T), \quad y(s) = 0, \quad \text{a.e. } s \in (T, T \in h).$$

The proof follows from Theorem 10.6.2 and Condition 10.6.9.

## 10.7   The Theory of Time-Optimal Control of Linear Neutral Systems

In this section we study the time-optimal problem: Minimize

$$J(t, x_t) = t \tag{10.7.0}$$

subject to:

$$\dot{x}(t) - A_{-1}(t)\dot{x}(t - h) = A_0(t)x(t)$$

$$+ \sum_{j=1}^{N} A_j(t)x(t - h_j) + B(t)u(t), \tag{10.7.1}$$

$$x_\sigma = \phi, \quad (t, x(t, \sigma, \phi, u) \in [0, \infty) \times H, \quad H \subset E^n, \tag{10.7.2}$$

where $A_j$ are analytic $n \times n$ matrix functions and $B$ is an $n \times m$ analytic matrix function. The controls are constrained to lie in

$$U = \{u \text{ measurable}, u(t) \in E^m, |u_j(t)| \le 1, \ a.e. \ j = 1, \ldots, m\}. \tag{10.7.3}$$

For conditions for the existence of analytic solutions of

$$\dot{x}(t) - A_{-1}\dot{x}(t-h) = A_0 x(t) + \sum_{j=1}^{N} A_j x(t-h_j) \qquad (10.7.4)$$

see Tadmor [29]. If we designate the strongly continuous semigroup of linear transformation defined by solutions of (10.7.4) by $T(t,\sigma)$, $t \geq \sigma$ so that

$$T(t,\sigma)\phi = x_t(\sigma,\phi,0)\,,$$

then the solution $x(\sigma,\phi,u)$ of (10.1.1) with $x_\sigma(\sigma,\phi,u) = \phi$ satisfies the relation

$$x_t(\sigma,\phi,u) = T(t,\sigma)\phi + \int_\sigma^t X_t(\cdot,s)B(s)u(s)ds \qquad (10.7.5)$$

where $X$ is defined as follows: Let

$$X_0(\theta) = \left(\begin{array}{cc} 0, & -h \leq \theta < 0 \\ I, & \theta = 0 \end{array}\right).$$

We are justified in writing

$$T(t,\sigma)X_0 = X_t(\cdot,s)$$

where $X_t(\cdot,s)(\theta) = X(t+\theta,s)$ $\theta \in [-h,0]$, and where $X$ is the fundamental matrix solution of (10.7.4) or

$$x(t) = W(t)g^0 + \int_{-h}^0 U_t(s)g^1(s) + \int_0^t W(t-s)B(s)u(s)ds \qquad (10.7.6)$$

where

$$W(t)g^0 = \left\{\begin{array}{ll} x(t;u,g^0,0)\,, & \text{if } t \geq 0, \quad g^0 \in E^n\,, \\ 0\,, & \text{if } t < 0\,, \end{array}\right. \qquad (10.7.7)$$

and $x(t,u,g^0,g^1)$ is a solution of (10.6.1) with

$$x(0) = g^0(s) = x(t) = g^1(s), \quad \text{a.e. } s \in [-h,0], \quad g^0 \in E^n, \quad g^1 \in W_2^{(1)}\,.$$

**The solutions of the time-optimal problem.** We now solve the time-optimal problem as formulated by reinterpreting Nakagiri in Euclidean $n$-dimensional space. Thus in our case the state space is $E^n$, and the target is a fixed convex compact subset $H$ of $E^n$ with nonempty interior.

Define

$$U_{\text{ad}} = \{u \in L_2(I,E^n) : u(t) \in C^m, \text{ a.e. } t \in I\}\,,$$

where $C^m$ is the unit $m$-dimensional cube, i.e.,

$$C^m \subset E^m, \quad |u_j| \leq 1, \quad j = 1,\ldots,m\,. \qquad (10.7.9)$$

Define

$$U_0 = \{u \in U_{\mathrm{ad}} : x(t, u) \text{ is a solution of (10.7.1) and } x(t, u) \in T\}$$

$$\text{for some } t \in I\}. \tag{10.7.10}$$

**Theorem 10.7.1** *Suppose the system is controllable, i.e., $U_0 \neq \phi$. Then there exists a time-optimal control for $P_3$.*

The proof is standard and can be modified from Nakagiri [27, p. 199].

We know from our existence result that if System (10.7.1) is controllable (null controllable) with constraints, an optimal control exists. To further explore its properties, we consider the possibility of a maximum principle and a bang-bang principle. We have indicated earlier that for retarded finite-dimensional space, a bang-bang principle is false in function space [9, p. 60]. But if we restrict $J$ to be a terminal value cost

$$J = \phi_0(x(T)) \tag{10.7.0}$$

where $\phi_0$ satisfies some regularity conditions $(R_1)$ $(R_2)$ below, it can be demonstrated that under some conditions of the adjoint system a bang-bang principle is valid. We require the following assumptions:

$R_1$: The function $\phi_0 : E^n \to E$ is continuous and Gateaux differentiable, and the Gateaux derivative $\mathcal{D}\phi_0(x) \neq 0$ for each $x \in E^{n^*}$.

$R_2$: $\phi_0 : E^n \to E$ is continuous and convex.

**Theorem 10.7.2**   **(Optimality Conditions)** *For the time-optimal problem with $H$ as target where $H$ is convex, closed, and nonempty, assume that $t^*$ is the optimal time. Then there exists a nonzero $n$-row vector $q^* \in E^{n^*}$ such that*

$$\max_{v \in U_{\mathrm{ad}}} \int_0^{t^*} (v(s), B^*(s)W^*(t^* - s)q^*)ds$$

$$= \int_0^{t^*} (u(s), B^*(s)W^*(t^* - s)q^*)ds, \tag{10.7.11}$$

*where $(\cdot, \cdot)$ denotes inner product in $E^n$. If $U_{\mathrm{ad}}$ is as defined in (10.7.10) and $U$ is the unit $m$-dimensional cube, then*

$$\max_{v \in U}(v, B^*(t)W^*(t^* - t)q^*) = (u(t), B^*(t)W^*(t^* - t)q^*), \quad a.e. \ t \in [0, t^*]. \tag{10.7.12}$$

Consider the adjoint system

$$\dot{y}(t) + \sum_{j=1}^{N} \dot{y}(s + h_j)A_j(s + h_j) = -\sum_{j=1}^{N} y(s + h_j)A_j(s + h_j) - q_1^*(t), \quad a.e. \ t \in I,$$

$$\tag{10.7.13}$$

$y(T) = -q_0^*$, $y(s) = 0$, a.e. $s \in (T, T + h)$. It is said to be regular (or proper) if whenever there is a set of positive measure $I_1 \subset I$ such that $m(I_1) > 0$ and $y(t; q_0^*, 0) = 0$ for all $t \in I$, then $q_0^* = 0 \in E^{n^*}$. Systems (10.7.4) that are pointwise complete for all $t > 0$ are regular. Recall that (10.7.4) is pointwise complete if

$$x(t, \cdot, \cdot) : E^n \times W_2^{(1)} \to E^n$$

is a surjection, a.e.

**Theorem 10.7.3**  *Consider the optimal problem with cost*

$$J = \phi_0(x(T)) . \tag{10.7.0}$$

*Assume that the adjoint system is regular and the matrix $B^T(t)$ is one to one, a.e. $t \in I$. Suppose the Gateaux derivative $d\phi_0(y) \neq 0$ in $E^{n^*}$ for all $y$ in the reachable set $\mathbb{R}(t) = \{y \in E^n : y = x(t, u) : u \in U_{\text{ad}}\}$. Then the optimal control $u(t)$ is bang-bang: that means that $u(t)$ satisfies*

$$u(t) \in \partial U(t), \quad a.e. \ t \in I , \tag{10.7.14}$$

*where $\partial U(t)$ denotes the boundary of $U(t)$, the $m$-dimensional unit cube.*

**Proof**  Because of our definition of $J$, the maximum principle is stated as

$$\max_{v \in U(t)} (u, B^*(t)y(t)) = (u(t), B^*(t)y(t)), \quad \text{a.e.} t \in I , \tag{10.7.15}$$

where $y(t) = y(t; d\phi_0(x(T)), 0)$ and $x(t)$ is the trajectory corresponding to the optimal control $u(t)$. It suffices to show that

$$B^*(t)y(t) \neq 0 \text{ in } E^{m^*} , \quad \text{a.e. } t \in I . \tag{10.7.5}$$

$\square$

Suppose to the contrary that there is a set of positive measure finite $I$, $m(I_1) > 0$ with $B^*(t)y(t) = 0$, $\forall \ t \in I_1$. Since $B^*(t)$ is one-to-one and the system is regular, we have that $d\phi_0(x(T)) = 0$. But then $x(T)$ is in the attainable set, this condition $d\phi_0(x(T)) = 0$ is impossible. As a consequence of these results, we deduce the following solution of the time-optimal problem:

**Theorem 10.7.4**  *Consider the optimal problem $P_3$ with $t^*$ the optimal time. Let the target $H$ be a closed convex subset of $E^n$ with nonempty interior. Suppose the adjoint system is regular and $B^*(t)$ one-to-one. Then the time-optimal control $u(t)$ is bang-bang on $I^* = [0, t^*]$, i.e., $u(t)$ is one of the vertices of the unit $m$-dimensional cube. In addition, suppose $U(t) = \{u \in E^m : |u - y(t)|E^m \leq v(t)\} \ t \in I$, and this replaces the $m$-dimensional unit cube. Also the time $T = t^*$ is the optimal time. Then the optimal control is given by*

$$u(t) = y(t) + v(t)\frac{\wedge_{E^m}^{-1} B^*(t)z(t)}{|\wedge^{-1} E^m B^*(t)z(t)|_{E^m}}, \quad a.e. \ t \in I_0 ,$$

*where $z(t) = W^*(t^* - t)q^*$ is the solution of the adjoint equation $t \in I^*$, and $q^*$ is as given in Theorem 10.7.3. Here $\wedge_{E^m}$ is the canonical isomorphism of $E^n$ onto $E^{n^*}$.*

## 10.8   Existence Results

**Definition 10.8.1**   Let $z_t \in C([-h, 0], E^n)$ be a target point function that is time varying. System (10.1.9) is controllable to the target if for each $\phi \in C$ there exists a $t_1 \geq \sigma$ and an admissible control $u \in L_\infty([\sigma, t_1], C^m)$ such that the solution of (10.1.9) satisfies

$$x_\sigma(\sigma, \phi, u) = \phi, \quad x_{t_1}(\sigma, \phi, u) = x_{t_1}.$$

**Theorem 10.8.1**   *Assume that System (10.1.9) is controllable to the target. Then there exists an optimal control.*

**Proof**   The variation of constant formula for system (10.1.10a) is

$$x_t(\sigma, \phi, u) = T(t, \sigma)\phi + \int_\sigma^t d_s X_t(\cdot, s)[G(\sigma)(\phi) - G(s)(x_s)]$$

$$+ \int_\sigma^t X_t(\cdot, s)B(s)u(s)ds.$$

Controllability to the target is equivalent to

$$x_{t_1}(\sigma, \phi, u) = z_{t_1}, \quad \text{for some } t_1,$$

that is,

$$w_{t_1} \triangleq z_{t_1} - T(t, \sigma)\phi - \int_\sigma^{t_1} d_s X_{t_1}(\cdot, s)[G(\sigma)(\phi) - G(s)(x_s)]$$

$$= \int_\sigma^{t_1} X_{t_1}(\cdot, s)B(s)u(s)ds.$$

This is equivalent to

$$w_{t_1} \in \alpha(t_1, \sigma).$$

Let

$$t^* = \inf\{t : w_t \in \alpha(t, \sigma)\}.$$

Now $\sigma \leq t^* \leq t_1$. There is a nonincreasing sequence of times $t_n$ converging to $t^*$, and a sequence of controls $u^n \in L_\infty([\sigma, t_1], C^m)$ with

$$w_{t_n} = y(t_n, u^n) = X_{t_n}(\cdot, s)B(s)u^n(s)ds \in \alpha(t_n, \sigma).$$

Also

$$\|w_{t^*} - y(t^*, u^n)\| \leq \|w_{t^*} - w_{t_n}\| + \|w_{t_n} - y(t^*, u^n)\| \leq \|w_{t^*} - w_{t_n}\| + I,$$

where

$$I \leq \left\| \int_\sigma^{t_n} X_{t_n}(\cdot, s) B(s) u^n ds - \int_\sigma^{t^*} X_{t_n}(\cdot, s) B(s) u^n(s) ds \right\|$$

$$+ \int_\sigma^t \left\| X_{t_n}(\cdot, s) B(s) u^n(s) ds - \int_\sigma^t X_{t^*}(\cdot, s) B(s) u^n(s) ds \right\|$$

$$\leq \int_{t^*}^{t_n} \| X_{t_n}(\cdot, s) B(s) u^n(s) ds \|$$

$$+ \int_\sigma^{t^*} \| [X_{t_n}(\cdot, s) - X_{t^*}(\cdot, s)] B(s) u^n(s) \| ds.$$

Because $X_{t_n}(\cdot, s) B(s) u^n(s)$ is integrable and $[t_n, t^*] < \infty$, the first term on the r.h.s. of the inequality tends to zero as $t_n \to t^*$.

We know from Henry [25] that

$$\| X_{t_n}(\cdot, s) \| \leq \beta < \infty, \quad \text{for all } t_n, s, \quad \text{for some } \beta;$$

also $X_{t_n}(\cdot, s) \to X_{t^*}(\cdot, s)$ in the uniform topology of $C$. Hence by the bounded convergence theorem, the second summand on the l.h.s. tends to zero as $n \to \infty$.

From the continuity of solution in time and the continuity of the target, $\| w_{t^*} - w_{t_n} \| \to 0$ as $t_n \to t^*$.

Hence $w_{t^*} = \lim_{n \to \infty} y(t^*, u^n)$. Because $\alpha(t^*, \sigma)$ is closed and $y(t^*, u^n) \in \alpha(t^*, \sigma)$, $w(t^*) = y(t^*, u^*)$, for some $u^* \in L_\infty([\sigma, t_1], C_m)$, and by definition $t^*, u^*$ is optimal.

A controllability assumption was made in Theorem 3.1 for System (10.1.9). When the target is the zero function, what is required is the assumption of null controllability only. To get conditions for this, we need two preliminary results and precise definitions. We work in the space $W_\infty^{(1)}$. The argument is also valid in $C$.$\square$

**Definition 10.8.2**  System (10.1.9) is controllable on $[\sigma, t_1], t_1 > \sigma + h$ if for each $\psi, \phi \in W_\infty$ there exists a control $u \in L_\infty([\sigma, t_1], E^m)$ such that the solution of (10.1.9) satisfies $x_\sigma(\sigma, \phi, u) = \phi$ and $x_{t_1}(\sigma, \phi, u) = \psi$. If System (10.1.9) is controllable on each interval $[\sigma, t_1], t_1 > \sigma + h$, we simply say that it is controllable. It is null controllable on $[\sigma, t_1]$ if $\psi \equiv 0$ in the above definition.

**Definition 10.8.3**  System (10.1.9) is null controllable with constraints if for each $\phi \in W_\infty^{(1)}$ there exist a $t_1 \geq \sigma$ and a $u \in L_\infty([\sigma, t_1], C^m)$ such that the solution $x(\sigma, \phi, u)$ of system (10.1.9) satisfies $x_\sigma(\sigma, \phi, u) = \phi$ and $x_{t_1}(\sigma, \phi, u) = 0$.

**Proposition 10.8.1**  *Suppose System (10.1.9) is null controllable on $[\sigma, t_1]$. Then for each $\phi \in W_\infty$, there exists a bounded linear operator $H : W_\infty^{(1)} \to L_\infty([\sigma, t_1], E^m)$ such that*

$$u = H\phi$$

*has the property that the solution $x(\sigma, \phi, H\phi)$ of System (10.1.9) satisfies*

$$x_\sigma(\sigma, \phi, H\phi) = \phi, \quad x_{t_1}(\sigma, \phi, H\phi) = 0.$$

**Proof**    From the variation of constant formula (10.1.10),

$$x_t(\sigma, \phi, u) = T(t, \sigma)\phi + C(t, \sigma)\phi + S(t, \sigma)u$$

where

$$T(t, \sigma)\phi = x_t(\sigma, \sigma, 0),$$

$$C(t, \sigma)\phi = \int_\sigma^t ds\{X_s(\cdot, s)\}[G(\sigma)\phi - G(s)x_s(\sigma, \phi)],$$

$$S(t, \sigma)u = \int_\sigma^t X_s(\cdot, s)B(s)u(s)ds, \quad t \in [\sigma, t_1].$$

The null controllability of System (10.1.9) is equivalent to the following statement: For every $\phi \in W_\infty$ there exists a $t_1$ and there exists a $u \in L_\infty([\sigma, t_1], E^n)$ such that

$$T(t_1, \sigma)\phi + S(t_1, \sigma)u + C(t_1, \sigma)\phi = 0, \quad t_1 > \sigma + h.$$

This is in turn equivalent to

$$(T(t_1, \sigma) + C(t_1, \sigma))W_\infty^{(1)} \subseteq S(t_1, \sigma)(L_\infty([\sigma, t_1], E^n)). \tag{10.8.1}$$

The condition (10.8.1) is now valid by hypothesis. Denote by $N$ the null space of $S$ and by $N^\perp$ the orthogonal complement of $N$ in $L_\infty([\sigma, t_1], E^n)$. Let

$$S_0 : N^\perp \to S(t_1, \sigma)(L_\infty([\sigma, t_1], E^n))$$

be the restriction of $S(t_1, \sigma)$ to $N^\perp$. Then $S_0^{-1}$ exists and is linear though not necessarily bounded, since $S(t_1)(L_\infty([\sigma, t_1], E^n))$ is not necessarily closed. Define a mapping

$$H : W_\infty^{(1)} \to L_\infty([\sigma, t_1], E^n)$$

by $H\phi = -S_0^{-1}[T(t_1, \sigma)\phi + C(t_1, \sigma)\phi]$. Then

$$x_t(\sigma, \phi, H\phi)(\theta) = x(\sigma, \phi, H\phi)(t_1 + \theta)$$

$$= T(t_1\sigma)\phi(\theta) + C(t_1, \sigma)\phi(\theta)$$

$$+ S(t_1, \theta)[-S_0^{-1}(T(t_1, \sigma)\phi(\theta) + C(t_1, \sigma)\phi(\theta))]$$

$$= 0, -h \le \theta \le 0.$$

Since $u = H\phi \in L_\infty([\sigma, t_1], E^n)$, we deduce that $x_{t_1}(\sigma, \phi, u) = 0$.
     We now prove the boundedness of $H$ as follows:

Let $\{\phi_n\}$ be a convergent sequence in $W_\infty^{(1)}$ such that $\{H\phi_n\}$ converges in $L_\infty([\sigma, t_1], E^n)$, and let

$$\phi = \lim_{n\to\infty} \phi_n, \quad u = \lim_{n\to\infty} H\phi_n, \quad u_n = H\phi_n.$$

Since $N^\perp$ is closed in $L_\infty([\sigma, t_1], E^n)$, $u \in N^\perp$ and

$$T(t_1, \sigma)\phi + C(t_1, \sigma)\phi + S(t_1, \sigma)u = \lim_{n\to\infty} (T(t_1, \sigma)\phi_n + C(t_1, \sigma)\phi_n + S(t_1, \sigma)u_n)$$

$$= 0.$$

Thus,

$$u = -S_0^{-1}[T(t_1, \sigma)\phi + C(t_1, \sigma)\phi].$$

By the Closed Graph Theorem, $H$ is bounded. The proposition is proved. $\square$

**Definition 10.8.4** System (10.1.9) is locally null controllable with constraints if for each $\phi \in \mathcal{O}, \mathcal{O}$ an open neighborhood of zero in $W_\infty^{(1)}$, there exists a finite $t_1$ and a control $u \in L_\infty([\sigma, t_1], C^m) \equiv U$ such that the solution $x(\sigma, \phi, u)$ of (10.8.1) satisfies

$$x_\sigma(\sigma, \phi, u) = \phi, \quad x_{t_1}(\sigma, \phi, u) = 0.$$

**Proposition 10.8.2** *Suppose that System (10.1.9) is null controllable. Then System (10.1.9) is locally null controllable with constraints.*

**Proof** Because System (10.1.9) is null controllable, from Proposition (10.8.1) there exists a bounded linear operator

$$H : W_\infty^{(1)} \to L_\infty([\sigma, t_1], E^n)$$

such that for each $\phi \in W_\infty^{(1)}$ and control $u = H\phi$, the solution $x(\sigma, \phi, H\phi)$ of System (10.1.9) satisfies

$$x_\sigma(\sigma, \phi, H\phi) = \phi, \quad x_{t_1}(\sigma, \phi, H\phi) = 0.$$

Since $H$ is a bounded linear map, it is continuous at $0 \in W_\infty^{(1)}$. Therefore, for each open set containing zero in $L_\infty([\sigma, t_1], E^n)$, there is an open neighborhood $U$ of $0 \in W_\infty^{(1)}$ such that

$$H(U) \subset V.$$

Clearly $L_\infty([\sigma, t_1], C^m)$ has zero in its interior. We can choose $V$ open and contained in $L_\infty([\sigma, t_1], C^m)$. For this particular choice there exists an open set $\mathcal{O}$ around zero in $W_\infty^{(1)}$ such that

$$H(\mathcal{O}) \subset V \in L_\infty([\sigma, t_1], C^m).$$

Every $\phi \in \mathcal{O} \subset W_\infty^{(1)}$ can be driven to zero by a control $u = H\phi \in L_\infty([\sigma, t_1], C^m)$. Hence System (10.1.9) is locally null controllable with constraints. $\square$

**Theorem 10.8.2**   *Suppose in System (10.1.9), with state space $W_\infty^{(1)}$ or $C$,*

   (i) *The system (10.1.9) is null controllable.*
   (ii) *The solution $x = 0$ of*

$$\frac{d}{dt}[D(t)x_t - G(t)x_t] = L(t, x_t) \tag{10.8.2}$$

   *is exponentially stable.*

*Then System (10.1.9) is null controllable with constraints.*

**Proof**   The first assumption yields an open ball $\mathcal{O} \subset W_\infty^{(1)}$ such that every initial $\phi \in \mathcal{O} \subset W_\infty^{(1)}$ can be transferred to zero in some finite time $t$, by controls $u \in U$.

By condition (ii), every solution (10.1.9) with $u = 0$, i.e., every solution of (10.8.2) satisfies

$$\|x_t(\sigma, \phi, 0\| \le M\|\phi\|e^{-\alpha(t-\sigma)}, \quad t \ge \sigma,$$

so that

$$x_t(\sigma, \phi, 0) \to 0 \quad \text{as} \quad t \to \infty.$$

Therefore, there is a finite $t_0 < \infty$ such that $\psi = x_{t_0}(\sigma, \phi, 0) \in \mathcal{O}$. With $(t_0, \psi)$ as initial data, there exists $t_1 > t_0$ such that some control

$$u \in U = L_\infty([t_0, t_1], C^m)$$

gives a solution $x(t_0, \psi, u)$ such that $x_{t_0}(t_0, \psi, u) = \psi$, $x_{t_1}(\sigma, \psi, u) = 0$. Thus, system (10.1.9) is null controllable. The control

$$w = \begin{cases} 0, & \text{in } [\sigma, t_0], \\ u, & \text{in } [t_0, t_1], \end{cases}$$

is contained in $U$ and does the transfer of $\phi$ to 0 in time $t_1 < \infty$. $\quad\square$

**Definition 10.8.5**   System (10.1.1) is said to be proper on $[\sigma, t]$ if and only if

$$0 \in \text{Int } \alpha(t, \sigma), \quad t \ge \sigma + h.$$

It is proper if it is proper on every $[\sigma, t]$, $t > \sigma + h$.

In the above definition, $\alpha(t, \sigma)$ is assumed to be a subset of $C$ or of $W_p^{(1)}$ if the controls are $L_p$. More precisely, if we work on $C$ or $W_\infty^{(1)}$, we use controls in $L_\infty([\sigma, t], C^m) \equiv U$. However, if we are in $W_p^{(1)}$, then $U_{\text{ad}}$ is a closed and bounded convex subset of $L_p([\sigma, t_1], E^n)$ with zero in its interior.

The next result is stated for the space $W_p^{(1)}$. It is equally valid for $W_\infty^{(1)}$ or $C$.

**Theorem 10.8.3**   *System (10.1.1) is proper if and only if it is function-space controllable.*

**Proof**  Suppose system (10.1.1) is controllable on $[\sigma, t]$, $t \geq \sigma + h$. Then

$$H : L_p([\sigma, t], E^n) \to W_p^{(1)}, \quad Hu = x_t(\sigma, 0, u),$$

is onto and

$$H(U) = \alpha(t, \sigma), \quad H(\mathcal{B}) \subset H(U) = \alpha(t, \sigma) = \{x_t(\sigma, 0, u) : u \in U_{\mathrm{ad}}\},$$

where $\mathcal{B}$ is an open ball containing 0 with $\mathcal{B} \subset U$. Because $H$ is a continuous linear transformation of $L_p$ onto $W_p^{(1)}$, $H$ is an open map. $H(U)$ is open and contains zero. $0 \in \mathrm{Int}\ \alpha(t, \sigma)$. Hence System (10.1.1) is proper. Assume System (10.1.1) is proper, i.e.,

$$0 \in \mathrm{Int}\ \alpha(t, \sigma).$$

Because $0 \in \mathrm{Int}\ \alpha(t, \sigma) \subset \mathrm{Int}\ \mathcal{A}(t, \sigma)$, where $\mathcal{A}(t, \sigma) = \{x_t(\sigma, 0, u) : u \in L_p\}$ is a subspace this implies $\mathcal{A}(t, \sigma) = W_p^{(1)}$.

For the linear system (10.1.9) we have proved that uniform asymptotic stability of System (10.8.2) and null controllability of System (10.1.9) suffice for constrained null controllability of System (10.1.9). It is clear that any test for controllability, though perhaps too strong, will guarantee null controllability. Indeed, we will now show that when $D(t)x_t$ in System (10.1.1) is atomic at both 0 and $-h$, then null controllability and controllability are equivalent. $\qquad\square$

**Proposition 10.8.3**  *System (10.1.1) is null controllable if and only if it is controllable, provided $D$ is atomic at 0 and at $-h$.*

**Proof**  It is given that $D(t, \phi)$ is atomic at 0 and at $-h$, for $(t, \phi) \in E \times C$. By a theorem of Hale [9, p. 279] the operator $T(t, \sigma)$ given by the solution $x_t(\sigma, \phi) = T(t, \sigma)\phi$ of (10.1.9) is a homeomorphism for $t \geq \sigma$. As a consequence of similar arguments,

$$T(t, \sigma)W_\infty^{(1)} = W_\infty^{(1)}, \quad t \geq \sigma.$$

If system (10.1.1) is null controllable, then for each $\phi \in W_\infty^{(1)}$,

$$T(t_1, \sigma)\phi + \int_\sigma^{t_1} X_{t_1}(\cdot, s)B(s)u(s)ds = 0.$$

With $S(t_1, \sigma) : L_\infty \to W_\infty^{(1)}$ defined by

$$S(t_1, \sigma)u = \int_\sigma^{t_1} X_{t_1}(\cdot, s)B(s)u(s)ds,$$

null controllability is equivalent to

$$T(t_1, \sigma)W_\infty^{(1)} \subset S(t_1, \sigma)(L_\infty[\sigma, t_1], E^m).$$

Therefore, this implies

$$W_\infty^{(1)} \subset S(t_1, \sigma)(L_\infty[\sigma, t_1], E^m)$$

so that $S(t_1, \sigma) : L_\infty \to W_\infty^{(1)}$ is a surjection. However, $S(t_1, \sigma)$ is surjective if and only if System (10.1.1) is controllable. We have proved that null controllability implies controllability. The converse is trivially true: controllability always implies null controllability.     $\square$

**Corollary 10.8.1**    *The system*

$$\frac{d}{dt}[x(t) - A_{-1}x(t-h)] = A_0 x(t) + A_1 x(t-h) + Bu(t), \tag{10.8.3}$$

*with* $\det(A_{-1}) \neq 0$, *is null controllable if and only if*

$$\mathrm{rank}[\Delta(\lambda), B] = n, \quad \forall\, \lambda \in \mathcal{C},$$

$$\mathrm{rank}[\lambda I - A_{-1}, B] = n, \quad \forall\, \lambda \in \mathcal{C},$$

*where*

$$\Delta(\lambda) = I\lambda - \lambda A_{-1}e^{-\lambda} - A_0 - A_1 e^{-\lambda}.$$

**Proof**    $D(t)x_t - A_{-1}x(t-h)$ is atomic at 0 and at $-h$ if $\det(A_{-1}) \neq 0$. The rank conditions suffice for controllability by a result of Salamon [18, p. 157], which we now state.     $\square$

**Theorem 10.8.4**    *System* (10.8.3) *is controllable in the state space* $W_p^{(1)}$, $1 \leq p \leq \infty$, *if and only if* $\mathrm{rank}[\Delta(\lambda), B] = n$, $\forall\, \lambda \in \mathcal{C}$ *and* $\mathrm{rank}[\lambda I - A_{-1}, B] = n$, $\forall\, \lambda \in \mathcal{C}$ *where* $\Delta(\lambda) = I\lambda - \lambda A_{-1}e^{-\lambda} - A_0 - A_1 e^{-\lambda}$.

## 10.9   Necessary Conditions for Optimal Control

We now return to our original problem of hitting a continuously moving point target $w_t \in C$ in minimum time. At the time of hitting, in System (10.1.9)

$$x_t(\sigma, \phi, u) = T(t, \sigma)\phi + \int_\sigma^t [d_s X_t(\cdot, s)][G(\sigma)(\phi) - G(s)(x_t)]$$

$$+ \int_\sigma^t X_t(\cdot, s)B(s)u(s)ds$$

$$= w_t,$$

or equivalently,

$$w_t - T(t, \sigma)\phi - C(t, \sigma)\phi \equiv z(t) = \int_\sigma^t X_t(\cdot, s)B(s)u(s)ds,$$

where

$$C(t,\sigma)\phi = \int_\sigma^t [d_s X_t(\cdot,s)][G(\sigma)(\phi) - G(s)(x_s)].$$

Thus, reaching $w_t$ in time $t$ corresponds to

$$w_t - T(t,\sigma)\phi - C(t,\sigma)\phi \equiv z(t) \in \alpha(t,\sigma).$$

We shall prove shortly that if $u^*$ is the optimal control with optimal time $t^*$, i.e. if $u^*$ is the control that is used to hit $w_t$ in minimum-time $t^*$, then

$$w_{t^*} - T(t^*,\sigma)\phi - C(t^*,\sigma)\phi \equiv z(t^*) \in \partial\alpha(t^*,\sigma),$$

that is, $z(t^*)$ is on the boundary of the constrained reachable set.

**Theorem 10.9.1**  *Let $u^*$ be the optimal control with $t^*$ minimum-time. Then $z(t^*) \in \partial\alpha(t^*,\sigma)$.*

**Proof**  Suppose $u^*$ is used to hit $w_t$ in time $t^*$:

$$z(t^*) = w_{t^*} - T(t^*,\sigma)\phi - C(t^*,\sigma)\phi \in \alpha(t^*,\sigma).$$

Suppose $z(t^*)$ is not in the boundary of $\alpha(t^*,\sigma)$:

$$z(t^*) \in \text{Int } \alpha(t^*,\sigma), \quad t^* > \sigma.$$

Therefore, there is a ball $B(z(t^*),\rho)$ of radius $\rho$ about $z(t^*)$ such that

$$B(z(t^*),\rho) \subset \alpha(t^*,\sigma).$$

Because $\alpha(t,\sigma)$ is a continuous function of $t$, we can preserve the above inclusion for $t$ near $t^*$ if we reduce the size of $B(z(t^*),\rho)$, i.e., if there is a $\delta > 0$ such that

$$B(z(t^*),\rho/2) \subset \alpha(t,\sigma), \quad t^* - \delta \le t \le t^*.$$

Thus, $z(t^*) \in \alpha(t,\sigma)$, $t^* - \delta \le t$. This contradicts the optimality of $t^*$. We are led to conclude that

$$z(t^*) \in \partial\alpha(t^*,\sigma).$$

$\square$

**Theorem 10.9.2**  *Let $D$ and $L$ satisfy the basic assumptions and inequality (10.1.13). Suppose System (10.1.9) is controllable and the solution operator is a homeomorphism, and*

$$z(t) = w_t - T(t,\sigma)\phi - C(t,\sigma)\phi. \tag{10.9.1}$$

*If $u^*$ is an optimal control and $t^*$ the minimum time, then*

$$z^* = z(t^*) \in \partial\alpha(t^*,\sigma) \subset C$$

*if and only if $u^*$ is of the form*

$$u^*(t) = \text{sgn}[-y(t, t^*)B(t)], \quad \sigma \geq t \leq t^*, \tag{10.9.2}$$

*where $y(t)$ is a nontrivial solution of the adjoint equation* (10.1.6).

**Proof**   Because of Proposition 10.1.1, $\alpha(t, \sigma)$ is closed and convex. If $u^*$ is an optimal control and $t^*$ the minimum time, then by Theorem 10.9.1, $z^* \in \partial\alpha(t^*, \sigma)$. Also from controllability, and Theorem 10.8.3,

$$0 \in \text{Int } \alpha(t^*, \sigma).$$

By the Separation Theorem of Dunford and Schwartz [26, p. 148], there exists a $\psi \in B_0$ such that $\psi \neq 0$, and

$$\langle \psi, \rho \rangle \leq \langle \psi, z^* \rangle \text{ for every } \rho \in a(t^*, \sigma).$$

It follows from this that

$$\left\langle \psi, \int_\sigma^{t^*} T(t^*, s)X_0 B(s)u(s)ds \right\rangle \leq \left\langle \psi, \int_\sigma^{t^*} T(t^*, s)X_0 B(s)u^*(s)ds \right\rangle,$$

for every $u \in L_\infty([\sigma, t^*], C^m)$. On using (10.1.4), we deduce that

$$\left\langle \psi, \int_\sigma^{t^*} T(t^* s)X_0 B(s)u(s)ds \right\rangle$$

$$= \int_{-h}^0 [d\psi(\theta)] \int_\sigma^{t^*} T(t^*, s)X_0(\theta)B(s)u(s)ds$$

$$= \int_\sigma^{t^*} \left\{ \int_{-h}^0 [d\psi(\theta)]T[(t^*, s)X_0](\theta) \right\} B(s)u(s)ds$$

$$= \int_\sigma^{t^*} -y(s, t^*)B(s)u(s)ds,$$

where $s \to y(s, t^*)$ is the solution of the adjoint equation (10.1.6). (See Henry [25].) Thus

$$\int_\sigma^{t^*} [-y(s, t^*)B(s)u(s)]ds \leq \int_\sigma^{t^*} [-y(s, t^*)B(s)u^*(s)]ds \tag{10.9.3}$$

for every admissible $u$. We see from inequality (10.9.3) that $u^*$ is of the form

$$u^*(s) = \text{sgn}[-y(s, t^*)B(s)], \tag{10.9.4}$$

where $y(s, t^*)$ is the solution of the adjoint equation. We note that $u^*$ maximizes the l.h.s. of inequality (10.9.3). Thus, with $t^*$ fixed and $u^*$ of the form (10.9.4), the point $z^*$ is on the boundary of $\alpha(t^*, \sigma)$ and $\langle \psi, \rho \rangle \leq \langle \psi, z^* \rangle$ for every $\rho \in \alpha(t^*, \sigma)$, $\psi \in B_0$.   $\square$

In Theorem 10.9.2, we have the form of optimal control for the time-optimal problem in the space of continuous functions $C$. We now show that this form is also valid in $E^n$, without the controllability assumption. This is true since the Separation Theorem is true in $E^n$ for any closed and convex set, that is, through each boundary of a closed convex set in $E^n$ there is a support plane. We now state the following theorem:

**Theorem 10.9.3** *Let $D$ and $L$ satisfy the basic assumptions and Inequality* (10.1.13). *Let $w(t)$ be a point function that is continuous in $E^n$. Let*

$$z(t) = w(t) - T(t,\sigma)\phi(0) - C(t,\sigma)\phi(0). \tag{10.9.5}$$

*If $u^*$ is an optimal control and $t^*$ the minimum time, then $z(t^*) \in \partial R(t^*,\sigma) \subset E^n$ if and only if*

$$u^*(s) = \mathrm{sgn}[-B^*(s)y(s)] = \mathrm{sgn}[B^*(s)[T^*(t,s)\psi](0)]_j u_j(s),$$

*where $\psi(0) \neq 0$ and $y$ is a nontrivial solution of the adjoint equation.*

## 10.10   Normal Systems

It is important to know when the necessary condition for optimal control in Theorems 10.9.2 or 10.9.3 uniquely determines an optimal control. This occurs when System (10.1.1) is normal in the sense to be made precise below. First observe that the necessary condition for optimal control in $C$ states that

$$u^*(t) = \mathrm{sgn}[-B^*(s)y(s)], \quad \sigma \leq s \leq t^*, \tag{10.10.1}$$

where $y(s)$ is a nontrivial solution of the adjoint equation. This condition states that

$$u_j^*(s) = \mathrm{sgn}[-b^{*j}(s)y(s)] \text{ on } [0,t_j^*], \quad \mathrm{sgn}[-b^{*j}[T^*(s,t^*)\psi](0)],$$

for $j = 1,\ldots,m$, where $b^{*j}(t)$ is the $j$th row vector of $B^*(t)$. Define

$$g_j(\psi(0)) = \{s : b^{*j}(s)y(s) = 0,\ s \in [0,t^*]\}$$

$$= \{s : b^{*j}(s)[T^*(s,t^*)\psi](0) = 0,\ s \in [0,t^*]\}.$$

Define

$$y_i(\psi) = \{s : y(s,t^*,\psi)b^j(s) = 0,\ s \in [\sigma,t^*]\},$$

$$g_{ij} = \{s : -y(s,t^*,\psi)b^j(s) = 0,\ s \in [\sigma,t^*]\}.$$

**Definition 10.10.1**   We say that System (10.1.1) is $E^n$ normal on $[\sigma,t^*]$ if $y_j(\psi)$ has measure zero for each $j = 1,\ldots,m$, and for each nontrivial $\psi$ with $\psi(0) \neq 0$. The system is $E^n$ normal if it is normal on every interval $[\sigma,t^*]$.

Similar definitions can be made for $C$-normality if we replace $y_j(\psi)$ with $g_{ij}(\psi)$ in the above definition. Utilizing the ideas of Gabasov and Kirillova [12] we are led to a necessary and sufficient condition for normality that is easily computable. It is Theorem 10.3.1.

## 10.11    The Geometric Theory of Time-Optimal Control of Linear Neutral Systems

In this section we study the time-optimal problem: Minimize

$$J(t, x_t) = t \tag{10.11.0}$$

subject to:

$$\dot{x}(t) - A_{-1}(t)\dot{x}(t - h) = A_0(t)x(t) + \sum_{j=1}^{N} A_j(t)x(t - h_j) + B(t)u(t), \tag{10.11.1}$$

$$x_\sigma = \phi, \quad (t, x_t(\sigma, \phi, u)) \in [0, \infty) \times H, \quad H \subset C, \tag{10.11.2}$$

where $A_j$ are analytic $n \times n$ matrix functions, and $B$ is an $n \times m$ analytic matrix function. The controls are constrained to lie in

$$U_{\mathrm{ad}} = \{u \in L_\infty^{\mathrm{loc}}([\sigma, \infty), E^m) : u(t) \in U \quad \text{a.e. } t \in [\sigma, \infty);$$

$$U \subset E^m, \text{ compact and convex}; \ 0 \in \text{Int } U\}.$$

Here $L_\infty^{\mathrm{loc}}([\sigma, \infty), E^m)$ represents the space of measurable functions defined on each interval $[\sigma, t]$ having values in Euclidean $n$-space $E^n$ and having essentially bounded values. We work in the space $C = C([-h, 0], E^n)$ of continuous functions from $[-h, 0] \to E^n$ with the sup norm. For these two spaces, the appropriate control set is

$$U = \{u \text{ measurable}, \ u(t) \in E^m, \ |u_j(t)| \leq 1, \text{ a.e. } j = 1, \ldots, m\}. \tag{10.11.3}$$

In what follows, $x_t \in C$ is defined by $x_t(s) = x(t+s)$, $s \in [-h, 0]$. For the existence, uniqueness, and continuous dependence on initial data of the solution $x(\cdot, \sigma, \phi, u)$ of (10.11.1), see Hale [9, pp. 25, 301]. See also Henry [16]. For conditions for the existence of analytic solutions of

$$\dot{x}(t) - A_{-1}\dot{x}(t - h) = A_0 x(t) + \sum_{j=1}^{N} A_j x(t - h_j), \tag{10.11.4}$$

see Tadmor [29]. If we designate the strongly continuous semigroup of linear transformations defined by solutions of (10.11.4) by $T(t, \sigma)$, $t \geq \sigma$ so that

$$T(t, \sigma)\phi = x_t(\sigma, \phi, 0),$$

then the solution $x(\sigma, \phi, u)$ of (10.1.1) with $x_\sigma(\sigma, \phi, u) = \phi$ satisfies the relation

$$x_t(\sigma, \phi, u) = T(t, \sigma)\phi + \int_\sigma^t X_t(\cdot, s)B(s)u(s)ds, \qquad (10.11.5)$$

where $X$ is defined as follows: Let

$$X_0(\theta) = \begin{pmatrix} 0, & -h \le \theta < 0 \\ I, & \theta = 0 \end{pmatrix}.$$

We are justified in writing

$$T(t, \sigma)X_0 = X_t(\cdot, s),$$

where $X_t(\cdot, s)(\theta) = X(t + \theta, s)$ $\theta \in [-h, 0]$ and where $X$ is the fundamental matrix solution of (10.11.4).

**Definition 10.11.1** The reachable set of time $t$ is defined by

$$\alpha(t, \sigma) = \left\{ \int_0^t T(t, s)X_0 B(s)u(s)ds : u \in U_{\text{ad}} \right\},$$

and it is a subset of $C$.

The following results are easily proved:

**Theorem 10.11.1** *The following are equivalent:*

    (i) *System* (10.1.1) *is controllable on* $[\sigma, t]$, $t \ge \sigma + h$.
    (ii) $T(t, s)\alpha(s, \sigma) \subset \text{Int } \alpha(t, \sigma)$, $\sigma \le s < t$.
    (iii) $0 \in \text{Int } \alpha(t, \sigma)$, $t \ge \sigma + h$.

**Proof** The equivalence of (i) and (iii) is established in Theorem 3.3 of [4]. To see that (ii) and (iii) are equivalent, we first note that $0 \in \alpha(t, \sigma)$ for each $t \ge \sigma$, so that

$$0 \in T(t, s)\alpha(s, \sigma) \subset \text{Int } \alpha(t, \sigma), \quad 0 \le s < t,$$

if we assume (ii). Conversely, assume (iii). Suppose $T(t, s)q \in \partial\alpha(t, \sigma)$, the boundary of $\alpha(t, \sigma)$ where $q \in \alpha(s, \sigma)$, and aim at a contradiction, since indeed

$$T(t, s)\alpha(s, \sigma) \subset \alpha(t, \sigma).$$

By a separation theorem of Dunford and Schwartz [26, p. 418], there exists $\psi \ne 0$, $\psi \in B_0$ the conjugate space of $C$, such that

$$\langle \psi, \rho \rangle \le \langle \psi, T(t, s)q \rangle \qquad (10.11.6)$$

for every $\rho \in \alpha(t, \sigma)$, where $\langle \cdot, \cdot \rangle$ is the outer product in $C$. From (10.11.6),

$$\langle \psi, \rho - T(t, s)q \rangle \le 0, \quad \forall \, \rho \in A(t, \sigma).$$

But for some $u \in U$,

$$\langle \psi, T(t,s)q \rangle = \left\langle \psi, \int_{\sigma}^{t} T(t,s)T(s,\tau)X_0 B(\tau)u(\tau)d\tau \right\rangle$$

$$= \left\langle \psi, \int_{\sigma}^{t} T(t,\tau)X_0 B(\tau)u(\tau)d\tau \right\rangle .$$

If we define

$$u^*(\tau) = \begin{cases} u(\tau), & 0 \leq \tau \leq s, \\ \operatorname{sgn}[g(t,\tau,\psi)B(\tau)], & s < \tau \leq t, \end{cases}$$

where

$$g(t,\tau,\psi) = \langle \psi, T(t,\tau)X_0 \rangle = \int_{-h}^{0} [d_\theta \psi(\theta)][U(t+\theta,\tau)], \tag{10.11.7}$$

then $u^* \in U$ determines a $\rho \in \alpha(t,\sigma)$, so that

$$\langle \psi, \rho - T(t,s)q \rangle = \left\langle \psi, \int_{s}^{t} T(t,\tau)X_0 B(\tau)u(\tau)d\tau \right\rangle$$

$$= \left\langle \int_{s}^{t} \left\{ \int_{-h}^{0} d_\theta[\psi(\theta)][T(t,\tau)X_0(\theta)] \right\}, D(\tau)u^*(\tau)d\tau \right\rangle$$

$$= \int_{s}^{t} |g(t,\tau,\psi)|d\tau > 0 .$$

This contradicts our assumption that $T(t,s)q$ is in the boundary of $\alpha(t,\sigma)$ and $\langle \psi, \rho - T(t,s)q \rangle \leq 0$. Hence $T(t,s)q \in \operatorname{Int} \alpha(t,\sigma)$, so that $T(t,s)\alpha(s,\sigma) \subset \operatorname{Int} \alpha(t,\sigma)$. $\qquad \square$

**Theorem 10.11.2**    (cf. [14, Theorem 4.2]) *Suppose System* (10.1.1) *is controllable. If* $u^*$ *is an optimal control and* $t^*$ *the minimum time, then* $u^*$ *is of the form*

$$u^*(s) = \operatorname{sgn}[g(t^*,s,\psi)B(s)] \tag{10.11.8}$$

*where* $g(t^*,s,\cdot) : B_0 \to E^{n*}$ *is the row-vector function defined in* (10.11.7).

It is easy to verify that $s \to g(t^*,s,\psi)$ satisfies a linear differential equation [16]. It follows from Tadmor [29] that $s \to g(t^*,s,\psi)$ is piecewise analytic if the initial conditions are analytic.

**Remark 10.11.1**    In $E^n$ the optimal control is given by

$$u^*(s) = \operatorname{sgn}[-y(t)B(t)], \quad \sigma \leq t \leq t^* ,$$

where $y : [\sigma, t^*]$ is an $n$-row vector function of bounded variation, which is a non-trivial solution of the adjoint equation

$$\frac{d}{ds}[y(s) - y(s+h)C(s+h)] = y(s)A(s) + \sum_{j=1}^{N} y(s+h_j)B_j(s+h_j). \qquad (10.11.9)$$

The form of optimal control in (10.11.8) states that each component $u_j^*$ of $u^*$ is given by

$$u_j^*(s) = \text{sgn}[g(t^*, s, \psi)b^j(s)]$$

on $[0, t_j^*]$ $j = 1, \ldots, m$, where $b^j(s)$ is the $j$th column of $B(s)$.

**Remark 10.11.2**  It is easy to see from (10.11.8) that if (10.11.1) is normal, then optimal control is uniquely determined by (10.11.8) and it is piecewise constant and bang-bang. This implies that optimal trajectories are unique.


## 10.12   Continuity of the Minimal-Time Functions

From the definitions of Sec. 10.1.1, we deduce that if (10.11.1) is controllable with constraints, then:

$$A(\sigma) = \bigcup \{\phi : x_t(\sigma, \phi) = 0 \text{ for some } t, \text{some } u\} = C.$$

If (10.11.1) is null controllable with constraints, then

$$-T(t, \sigma)\phi \in \alpha(t, \sigma)$$

for some $t$. In this case, $\phi$ can be steered to zero at time $t$ by some admissible control.

**Definition 10.12.1**  The minimal-time function $M$ is defined by

$$M(\phi) = \inf\{t \geq \sigma : -T(t, \sigma)\phi \in \alpha(t, \sigma)\}.$$

We have $\sigma \leq M(\phi) \leq +\infty$, with $M(\phi) < \infty$ if and only if

$$-T(t, \sigma)\phi \in \alpha(t, \sigma)$$

for some $t \geq \sigma$. We observe that

$$M : A(\sigma) \to E^1.$$

When (10.11.1) is controllable so that $A(\sigma) = C$, then $M$ is continuous. We have:

**Theorem 10.12.1**  *Let (10.11.1) be controllable. Then the minimal-time function*

$$M : A(\sigma) \to E$$

*is continuous.*

The proof is similar to that of Theorem 7.2.1.

**Corollary 10.12.1**    *For $t \geq \sigma$, we have*

$$\{-T(t,\sigma)\phi : M(\phi) \leq t\} = \alpha(t,\sigma).$$

*Also*

$$\{-T(t,\sigma)\phi : M(\phi) = t\} = \partial\alpha(t,\sigma),$$

*if* $\det \mathcal{A}_{-1}(t) \neq 0$ *for each $t$.*

**Proof**    Note that

$$\alpha(t,\sigma) \subset \{-T(t,\sigma)\phi : M(\phi) \leq t\} \subset \bigcap_{\rho=1}^{\infty} \alpha\left(t + \frac{1}{\rho}, \sigma\right).$$

We now show that

$$\alpha(t,\sigma) \supset \bigcap_{\rho=1}^{\infty} \alpha\left(t + \frac{1}{\rho}, \sigma\right).$$

Indeed, let $\phi \in \bigcap_{\rho=1}^{\infty} \alpha(t + \frac{1}{\rho}, \sigma)$. Then there is an admissible control $u_\rho : [\sigma, t + \frac{1}{\rho}]$ such that

$$\phi = \int_{\sigma}^{t+\frac{1}{\rho}} T\left(t + \frac{1}{\rho}, s\right) X_0 B(s) u(s) ds$$

$$= \int_{\sigma}^{t} T\left(t + \frac{1}{\rho}, s\right) X_0 B(s) u(s) ds + \int_{\sigma}^{t+\frac{1}{\rho}} T\left(t + \frac{1}{\rho}, s\right) X_0 B(s) u(s) ds$$

$$\equiv x_\rho + y_\rho.$$

But then

$$x_\rho = \int_{\sigma}^{t} T\left(t + \frac{1}{\rho}, t\right) T(t,s) X_0 D(s) u(s) ds$$

$$= T\left(t + \frac{1}{\rho}, t\right) \int_{\sigma}^{t} T(t,s) X_0 D(s) u(s) ds$$

$$\in T\left(t + \frac{1}{\rho}, t\right) \alpha(t,\sigma).$$

Therefore,

$$x_\rho \to T(t,t)\phi = \phi \quad \text{as} \quad \rho \to \infty.$$

Because $y_\rho \to 0$ as $\rho \to \infty$, the elements $\phi$ form the closed set $\alpha(t,\sigma)$. For proof of the second formula, we observe that

$$T(t,\sigma) : C \to C$$

is an open map, since in (10.11.1) $\det A_{-1}(t) \neq 0$ for each $t$, [9, p. 279]. Indeed, $T(t,\sigma)\phi = x_t(\sigma,\phi,0)$, $t \geq \sigma$ where $x(\sigma,\phi,0)$ is a solution of (10.1.7). Theorem 2.6 of [9, p. 279] insures that

$$T(t,\sigma)C = C, \quad t \geq \sigma + h\,,$$

so that $T(t,\sigma)$ is an open map. Since $T(t,\sigma)$ is also continuous, we have $\{-T(t,\sigma)\phi : M(\phi) < t\}$ is open, so that

$$\{-T(t,\sigma)\phi : M(\phi) < t\} \subset \text{Int } \alpha(t,\sigma)\,.$$

We know that if

$$-T(t^*,\sigma)\phi \in \text{Int } \alpha(t^*,\sigma), \quad \sigma < t^* < \infty\,,$$

then

$$-T(t,\sigma)\phi \in \text{Int } \alpha(t,\sigma), \quad \text{for } t > t^*\,.$$

Therefore $M(\phi) \leq t^* < t$. This completes the proof. Indeed, suppose $x^* = -T(t^*,\sigma)\phi \in \text{Int } A(t^*,\sigma)$, $\sigma < t^* < \infty$. Then there is a ball $\mathcal{B} = \mathcal{B}(x^*,\delta)$ of radius $\delta$ about $x^*$ such that $\mathcal{B} \subset \alpha(t^*,\sigma)$. Thus

$$T(t,t^*)x^* = T(t,t^*)T(t^*,\sigma)\phi$$

$$= -T(t,\sigma)\phi \in T(t,t^*)(\mathcal{B}) \subset T(t,t^*)\alpha(t^*,\sigma) \subset \alpha(t,\sigma)\,.$$

Since $T(t,t^*)$ is open, $-T(t,\sigma)\phi \in \text{Int } \alpha(t,\sigma)$ for $t > t^*$. $\qquad\square$

Just as in Hájek [30] and Chukwu [31], we use the continuity of the minimal-time function $M$ to construct an optimal feedback control for (10.11.1). We need a special subset of $C^* = B_0$, the conjugate space of $C$, which may be described as the cone of unit outward normals to support hyperplanes to $\alpha(t,\sigma)$ at a point $\phi$ on the boundary of $\alpha(t,\sigma)$.

**Definition 10.12.2**  For each $\phi \in A(\sigma)$, let $K(\phi) = \psi \in B_0 : \|\psi\| = 1$ such that $\langle \psi, \rho \rangle \leq \langle \psi, \phi \rangle$, $\forall\, \rho \in \alpha(M,(\phi),\sigma)$ where $\langle \cdot, \cdot \rangle$ is the outer product in $C$.

**Remark 10.12.1**  Note that $\phi \in \partial\alpha(M(\phi),\sigma)$. We now outline key properties of the set $K(\phi)$. Let $S(B_0)$ denote the collection of subsets of $B_0$. Then

$$K : C \to S(B_0)$$

is the mapping defined above.

**Definition 10.12.3**  We say that $K$ is upper semicontinuous at $\phi$ if and only if

$$\limsup K(\phi_n) \subset K(\phi) \quad \text{as} \quad \phi_n \to \phi\,.$$

We have:

**Lemma 10.12.1**   *If* (10.11.1) *is controllable, then* $K(\phi)$ *is nonvoid and* $K(-\phi) = -K(\phi)$. *Also,* $K(\phi)$ *is upper semicontinuous at* $\phi$.

**Proof**   Let (10.11.1) be controllable. Then $0 \in \text{Int } \alpha(t, \sigma)$ by Theorem 10.11.1. Because $\phi \in \partial\alpha(M(\phi), \sigma)$, there exists a $\psi \in B_0$, $\psi \not\equiv 0$ [26, p. 418] such that

$$\langle \psi, \rho \rangle \leq \langle \psi, \phi \rangle \quad \text{for all} \quad \rho \in \alpha(M(\phi), \sigma).$$

The choice of $\psi$ can be made such that $\|\psi\| = 1$. Hence $K(\phi)$ is nonempty. The second assertion is true because $\alpha(M(\phi), \sigma)$ is symmetric about zero. A simple argument that uses the closeness of $\alpha(t, \sigma)$, the continuity of $t \to \alpha(t, \sigma)$ [3] and of $\phi \to M(\phi)$, proves the last statement.      $\square$

Because $K(\phi)$ is a nonvoid subset of $B_0$, we can choose $y(\phi) \in K(\phi)$ if (10.11.1) is controllable. It is a consequence of McShane and Warfield [32] that this choice can be made in a measurable way. We have:

**Lemma 10.12.2**   *Let* (10.11.1) *be controllable. There exists a measurable function*

$$y : \alpha(\sigma) \to B_0$$

*such that*

$$y(\phi) \in K(\phi), \quad \forall\, \phi \in \alpha(\sigma).$$

**Proof**   We observe that $C$ is a measure space and $B_0$ a Hausdorff space. $C$ is also separable. Let $k : C \to B_0$ be the mapping given by the collection

$$K(\phi) = \{k(\phi) = \psi : \psi \in B_0, \|\psi\| = 1, \langle \psi, \rho \rangle \leq \langle \psi, \phi \rangle, \forall\, \rho \in \alpha(M(\phi), \sigma)\}.$$

From the continuity of $t \to \alpha(t, \sigma)$ and of $\phi \to M(\phi)$, we observe that $\phi \to k(\phi)$ is continuous. If we identify the notations of McShane and Warfield and ours as

$$M = C = A(\sigma), \quad A = B_0, \quad \psi \in B_0,$$

Theorem 4 of [32] asserts that there exists a measurable function $y : A(\sigma) \to B_0$ such that $y(\phi) \in K(\phi)$.

To construct an optimal feedback control for (10.11.1), the regularity property of

$$g(t, s, \psi) = \int_{-h}^{0} d_\theta[\psi(\theta)] X(t + \theta, s) \tag{10.12.0}$$

as a function of $s$ is important. This is determined by the same property of $s \to X(t, s)$. For this we now recall that the fundamental matrix $X(t, s)$ of System (10.11.4) satisfies as a function $t$ of the equation

$$\dot{X}(t, s) - A_{-1}(t)X(t - h, s) = A_0(t)X(t, s) + \sum_{j=1}^{n} A_j(t)X(t - h_j, s)$$

for $t \neq kh$, $k = 0, 1, 2, \ldots, t \geq \sigma$. We now assume $A_j$ are real analytic functions. It is a consequence of Theorem 4.1 of Tadmor [29] that, for each $t, s \to X(t, s)$ that is piecewise analytic in $s$ on $[0,t]$, it follows that $s \to g(t,s,\psi)$ is also piecewise analytic on $[0,t]$ and more generally on any interval $[\sigma, \sigma + T]$, for $T > 0$. $\qquad\square$

**Theorem 10.12.2** *In* (10.11.1), *assume that* $A_j$ *are each analytic functions,* $\det A_{-1}(t) \neq 0$ *for each* $t$. *Let* (10.11.1) *be controllable and normal. Then there exists a measurable function* $f : A(\sigma) \to E^m$ *that is an optimal feedback control for* (10.11.4) *in the following sense: Consider the system*

$$\dot{z}(t) - A_{-1}(t)\dot{z}(t-h) = A_0(t)z(t) + \sum_{j=1}^{n} A_j(t)z(t-h_j) + B(t)f(z_t). \qquad (10.12.1)$$

*Then each optimal solution of* (10.11.1), *is a solution of* (10.12.1) *and each solution of* (10.12.1) *is a solution (possibly not optimal) of* (10.11.1).

**Proof** By Lemma 10.12.2, select a measurable function $y : A(\sigma) \to B_0$, $y(\phi) = \psi \in B_0$, and define

$$f(\phi) = \lim_{t \to s+} -\mathrm{sgn}[g(t,s,y(\phi))B(s)], \qquad (10.12.2)$$

where $g$ is defined in (10.12.0). Because $B$ is analytic, each coordinate of

$$h(s) \equiv g(t,s,y(\phi))B(s)$$

is piecewise analytic on $[\sigma, M(\phi)]$. That is, $h(s)$ is analytic on $[s_{i-1}, s_i]$, $i = 1, 2, \ldots, \nu$ for a partition

$$\sigma = s_0 < s_i < \cdots s_\nu = M(\phi).$$

Therefore

$$\mathrm{sgn}[g(t,s,y(\phi))B(s)]$$

is piecewise constant on each $[s_{i-1}, s_i]$, $i = 1, 2, \ldots, \nu$, and therefore on $[\sigma, M(\phi)]$. Because of this, and because the right- and left-hand limits of $X(t,s)$ exist at each $kh$, $k = 0, 1, 2, \ldots$, the limit (10.12.2) exists. Since $y(\phi)$ is a measurable selection, so is $f(\phi)$. Let $x : [\sigma, M(\phi)] \to C$ be an optimal solution of (10.1.1) with $x_\sigma = \phi$. It can be verified that $x$ satisfies (10.12.1). Indeed, take an arbitrary $s$, $\sigma \leq s < M(\phi)$; then $x_s \in A(\sigma)$, from controllability of (10.11.1). There exists an optimal control through $x_s$, a control that can be taken to be

$$u_s(t) = -\mathrm{sgn}[g(t,s,y)B(s)]$$

for any choice of $y \in K(x_s)$, i.e., $-y \in K(-x_s)$. Our choice is $y = y(x_s)$. This choice is possible by Theorem 10.11.2. But then, because (10.11.1) is normal, optimal control is unique and determines uniquely a response $x$. Therefore for almost all $t \geq s$ we have

$$u_\sigma(s) = u_s(t) = -\mathrm{sgn}[g(t,s,y(x_s))B(s)].$$

On taking the limit as $t \to s^+ (\sigma \leq s \leq t)$,

$$u_\sigma(s) = u_s(t) = \lim -\text{sgn}[g(t, s, y(x_s))B(s)],$$

we obtain $u_\sigma(s^+) = f(x_s)$, $s \in [\sigma, M(\phi)]$, since $u_\sigma$ is piecewise constant $u_\sigma(s) = f(x_s)$ for almost all $s$. The response $x$ to $u_\sigma$ satisfies

$$\dot{x}(s) - A_{-1}(s)\dot{x}(s-h) = A_0(s)x(s) + \sum_{j=1}^{n} A_j(s)x(s-h) + B(s)f(x_s)$$

$$\text{a.e.,} \quad x_\sigma = \phi.$$

Clearly $x$ is a solution of (10.11.1). Now let $x$ be a solution of (10.12.1), which is a response to $\nu(t) = f(x_t)$. Then $v(t) \in U$. The proof is complete. $\qquad\qquad\square$

**Remark 10.12.2**    If in Theorem 10.12.2 we work in $E^n$, then we replace $g$ in (10.12.2) by $y$ where $y$ solves the adjoint equation (10.11.9). For $E^n$ we use the usual inner product $(\cdot, \cdot)$.


## 10.13    The Index of the Control System

In our main Theorem 10.12.2 for $C$, the function

$$k(s) = g(t, s, \psi)B(s) \tag{10.13.1}$$

is fundamental in determining optimal control strategy for our time-optimal problem. It is designated as the index of the control system. To study the index, a thorough study of the fundamental matrix $X(t, s)$ of (10.11.4) is required. We initiated such a study in Proposition 2.3.1 by considering the autonomous version of (10.11.4), namely

$$\dot{x}(t) + \sum_{j=1}^{m} A_{-1j}\dot{x}(t - h_j) = Ax(t) + \sum_{j=1}^{m} A_j x(t - h_j), \tag{10.13.2}$$

if $t \geq 0$ and

$$x(t) = \phi(t), \quad \phi \in C, \quad t \in [-h, o), \quad x(o) = x_0.$$

Here $0 \leq h_1 \leq h_2 \leq \cdots \leq h_m = h$. With $X$ determined, the index of the control system is calculated using $g$ given as

$$g(t, s, \psi) = \int_{-h}^{0} d_\theta[\psi(\theta)][X(t + \theta - s)], \quad 0 \leq s \leq t.$$

The index of the control system for

$$\dot{x}(t) - A_{-1}\dot{x}(t - h) = A_0 x(t) + A_1 x(t - h) + Bu \tag{10.13.3}$$

is given by

$$k(s) = \int_{-h}^{0} d_\theta[\psi(\theta)][X(t+\theta-s)]B, \quad 0 \le s \le t.$$

**Remark 10.13.1** In the space $E^n$, $g$ is given by

$$y(t, s, x_0) = x_0^T X(t - s),$$

which is the solution of the adjoint equation. The index becomes

$$k(s) = x_0^T X(t - s)B, \quad 0 \le s \le t.$$

## 10.14 Examples

**Example 10.14.1** We consider linear models of the equation obtained from loss-less transmission line problems, namely

$$\frac{d}{dt}[x(t) + cx(t - h)] = -ax(t) + bx(t - h) + u(t), \tag{10.14.1}$$

where $a > 0, c^2 \le 1 - \delta, \delta > 0$.

Variants of (10.14.1) were obtained by Slemrod [37] and Lopes [19, p. 200]. Hale [38, p. 349] derived conditions for (global) uniform asymptotic stability of the system

$$\frac{d}{dt}[x(t) + cx)\dot{x}(t - h) = -ax(t) + bx(t - h). \tag{10.14.2}$$

Hale showed that if

(i) $a > 0, c^2 \le 1 - \delta$, for some $\delta > 0$,
(ii) and if there exists a $\beta \in [0, 1]$ such that

$$\left[\frac{2bc}{a} - 1\right]\beta + \left[\frac{b}{a} - c\right]^2 < 0,$$

then (10.14.2) is uniformly asymptotically stable.

In (10.14.1) we assume

$$|u| \le 1. \tag{10.14.3}$$

Physically the function $x$ is the voltage at the end of the line, and $u(t)$ is related either to the current or to the voltage at the source. We use $u$ to control fluctuation of the current at the end of the line to its equilibrium position (0) as fast as possible. That (10.14.1) is controllable follows from Theorem 10.8.4. When conditions (i) and (ii) of Hale hold, (10.14.2) is uniformly asymptotically stable. Therefore (10.14.1) is null controllable with constraint. It follows that a time-optimal control exists (cf. Theorem 10.8.1). It is given by

$$u^*(s) = \text{sgn}[g(t, s, \psi)B]$$

(open loop form), or by

$$f(\phi) = \lim_{t \to s^+} -\text{sgn}[g(t, s, y(\phi))B(s)]$$

in (10.12.2), where $g$ is given by (10.11.7).

With an initial choice of $y(\phi) = \psi$, $\|\psi\| = 1$, let us calculate $f(\phi)$ for $t \in [0, 2h]$. We select the (measurable) function

$$\psi = y(\phi) \quad \text{on} \quad [-h, 0].$$

Then for System (10.14.1) where $a = 2$, $c = -\frac{1}{2}$, $b = \frac{1}{2}$,

$$x(t) = e^{-2t}, \quad t \in [0, h],$$

$$x(t) = e^{-2t} + \frac{1}{2}e^{-2(t-h)}[1 - (t - h)], \quad t \in [h, 2h],$$

so that

$$k(s) = \int_{-h}^{0} d_\theta \psi(\theta)[e^{-2(t-\theta-s)}], \quad t - s \in [0, -h],$$

$$k(s) = \int_{-h}^{0} d_\theta \psi(\theta) \left[ e^{-2(t-\theta-s)} + \frac{1}{2}e^{-2(t-\theta-s-h)} \right] [1 - (t - \theta - s - h)],$$

$$t - s \in [h, 2h].$$

**Example 10.14.2**   A two-dimensional neutral system was derived by Lopes in [40], in a transmission-line problem. Its linear version is given by

$$\dot{x}(t) - A_{-1}\dot{x}(t - h) = A_0 x(t) + A_1 x(t - h), \tag{10.14.4}$$

where

$$x = \begin{pmatrix} u \\ i \end{pmatrix},$$

and $u$ is related to the voltage $v(x, t)$ as follows: $v(L, t) = u(t) - gu(t - h)$, and $i(t)$ is the current. In (10.14.4),

$$A_{-1} = \begin{pmatrix} c & 0 \\ 0 & 0 \end{pmatrix}, \quad A_0 = \begin{pmatrix} a_1 & a_2 \\ a_3 & a_4 \end{pmatrix}, \quad a_i \neq 0, \quad A_1 = \begin{pmatrix} b_1 & 0 \\ a_3 & 0 \end{pmatrix}.$$

We consider the time-optimal control of the system

$$\dot{x}(t) - A_{-1}\dot{x}(t - h) = A_0 x(t) + A_1 x(t - h) + Bu(t) \tag{10.14.5}$$

where

$$B = \begin{pmatrix} d_1 & 0 \\ 0 & d_4 \end{pmatrix}, \quad d_i \neq 0.$$

We recall from Chukwu [39, Example 5] that the conditions for uniform asymptotic stability are:

(i)  $|c| < 1$.
(ii) Let

$$D_1 = \frac{1}{2}(A_0 + A_0^T), \quad D_2 = \frac{1}{2}(A_1 + A_1^T), \quad J_i = HD_i + D_i^T H,$$

where $H$ is a positive definite symmetric matrix. Suppose the eigenvalues $\lambda_i$ of $J_i$ satisfy

$$\lambda_{1i} < -\delta_1 < 0, \quad i = 1, 2,$$

where

$$\delta_1 > r\|J_2 + J_1 A_{-1}\|, \quad r > 1.$$

For the control system (10.14.5), we assume
(iii) $|u_j| \leq 1$, $j = 1, 2$.

We examine (10.14.4) when

$$A_{-1} = \begin{pmatrix} \frac{1}{6} & 0 \\ 0 & 0 \end{pmatrix}, \quad A_0 = \begin{pmatrix} -3 & -1 \\ 1 & -1 \end{pmatrix},$$

$$A_1 = \begin{pmatrix} \frac{1}{6} & 0 \\ 1 & 0 \end{pmatrix}, \quad B = \begin{pmatrix} d_1 & 0 \\ 0 & d_4 \end{pmatrix}, \quad d_i \neq 0.$$

For stability we find that if $H = I$, the identity matrix

$$J_1 = A_0 + A_0^T = \begin{pmatrix} -6 & 0 \\ 0 & -2 \end{pmatrix}.$$

Hence the eigenvalues are $\lambda_{11} = -6$, $\lambda_{12} = -2$, and we can choose $\delta_1$ to be 2.

$$J_2 + J_1 A_{-1} = \begin{pmatrix} \frac{1}{3} & 1 \\ 1 & 0 \end{pmatrix} + \begin{pmatrix} -1 & 0 \\ 0 & 0 \end{pmatrix},$$

$$J_2 + J_1 A_{-1} = \begin{pmatrix} -\frac{2}{3} & 1 \\ 1 & 0 \end{pmatrix}.$$

Using the matrix norm

$$\|a\| = \sup_i \left\{ \sum_{j=1}^{n} |aij| \right\}$$

we deduce that

$$2 > r\|J_2 + J_1 A_{-1}\| \quad \text{where} \quad r = 1 \cdot 1 .$$

Hence conditions (i) and (ii) of uniform asymptotic stability are satisfied. We know from Salamon [6, p. 151] that (10.14.5) is exactly controllable on $W_p^{(1)}$ if and only if

    (iv) rank $[\Delta(\lambda), B] = 2 \ \forall \ \lambda$ where $\Delta(\lambda) = \lambda(I - A_{-1}e^{-\tau h}) - A_0 - A_1 e^{-\tau}$,
    (v) rank $[\lambda I - A_{-1}, B] = 2 \ \forall \ \lambda$.

Since $B$ has full rank, conditions (iv) and (v) are satisfied; and therefore the system is exactly controllable. It follows from arguments contained in Chukwu [14, Theorem 3.2] that (10.14.5) is null controllable with measurable controls that satisfy (iii). We now deduce optimal control law in the space $E^n$. To do this we need to determine the fundamental matrix $X$:

$$X(t) = e^{A_0 t} = \begin{pmatrix} e^{-2t} - te^{-2t} & -te^{-2t} \\ te^{-2t} & e^{-2t} + te^{-2t} \end{pmatrix} \quad \text{on} \quad [o, h] .$$

We select

$$(c_1, c_2)X(t - s) = (c_1(e^{-2(t-s)} - (t - s)e^{-2(t-s)}) + c_2(t - s)e - 2(t - s) ,$$

$$- c_1(t - s)e^{-2(t-s)} + c_2(e^{-2(t-s)} + (t - s)e^{-2(t-s)})) ,$$

$$y(s)B = y(s) \begin{bmatrix} d_1 & 0 \\ 0 & d_2 \end{bmatrix}$$

$$= (d_1(c_1(e^{-2(t-s)} - (t - s)e^{-2(t-s)}) + c_2(t - s)e^{-2(t-s)}) ,$$

$$d_2(-c_1(t - s)e^{-2(t-s)} + c_2(e^{-2(t-s)} + (t - s)e^{-2(t-s)})) ,$$

$$t - s \in [0, h] ,$$

$$(u_1, u_2) = \operatorname{sgn}(y(s)B), \quad u_1 = \operatorname{sgn}(d_1(c_1(1 - (t - s))) + c_2(t - s)) ,$$

$$u_2 = \operatorname{sgn}(d_2(-c_1(t - s)) + c_2(1 + (t - s)) .$$

One can obtain values of $u$ on $[h, 2h]$. The calculations become more complicated.

## References


1. M. A. Cruz and J. K. Hale, "Asymptotic Behaviour of Neutral Functional Differential Equations," *Arch. Rational Mech. Anal.* **34** (1969) 331–353.
2. J. P. LaSalle, "The Time Optimal Control Problem," in *Theory of Nonlinear Oscillations, Vol. 5*, Princeton Univ. Press, Princeton, N.J., 1959, 1–24.
3. H. T. Banks and G. A. Kent, "Control of Functional Differential Equations of Retarded and Neutral Type to Target Sets in Function Space," *SIAM J. Control Optim.* **10** (1972), 567–593.
4. E. B. Lee and L. Markus, *Foundations of Optimal Control Theory*, Wiley, New York, 1967.
5. A. Strauss, *An Introduction to Optimal Control Theory*, Springer-Verlag, New York, 1968.
6. H. Hermes and J. P. LaSalle, *Functional Analysis and Time Optimal Control*, Academic Press, New York, 1969.
7. R. B. Zmood, "The Euclidean Space Controllability of Control Systems with Delay," *SIAM J. Control Optim.* **12** (1974) 609–623.
8. J. Kloch, "A Necessary and Sufficient Condition for Normality of Linear Control Systems with Delay," *Ann. Polon. Math.* **35** (1978) 305–312.
9. J. Hale, *Theory of Functional Differential Equations*, Springer-Verlag, New York, 1977.
10. O. Hájek, "Duality for Differential Games and Optimal Control," *Math. Systems Theory* **8** (1974) 1–7.
11. O. Hájek, *Pursuit Games, An Introduction to the Theory and Applications of Differential Games of Pursuit and Evasion*, Academic Press, New York, 1975.
12. R. Gabasov and F. Kirillova, *The Qualitative Theory of Optimal Processes*, Marcel Dekker, New York, 1976.
13. E. N. Chukwu, "The Time Optimal Control Problem of Linear Neutral Functional System," *J. Niger. Math. Soc.* **1** (1982) 39–55.
14. E. N. Chukwu, "The Time Optimal Control Theory of Linear Differential Equations of Neutral Type," *Comput. Math. Applic.* **16** (1988) 851–866.
15. E. N. Chukwu, "Control in $W_2^{(1)}$ of Nonlinear Interconnected Systems of Neutral Type," Preprint.
16. D. Henry, *Theory and Boundary Value Problems for Neutral Functional Equations, J. Austral Math. Soc.* **36**(B) (1994) 286–312.
17. E. N. Chukwu, "Symmetries of Linear Control Systems," *SIAM J. Control* **12** (1974) 436–488.
18. D. Salamon, *Control and Observation in Neutral Systems*, Pitman Advanced Publishing Program, Boston (1984).
19. D. Lopes, "Forced Oscillation in Nonlinear Neutral Differential Equations," *SIAM J. Appl. Math.* **29** (1975) 196–207.
20. R. Bellman, I. Glicksberg and O. Gross, "On the Bang-Bang Control Problems," *Q. Appl. Math.* **14** (1956) 11–18.
21. E. N. Chukwu and O. Hájek, "Disconjugacy and Optimal Control," *J. Optimiz. Theory Applic.* **27** (1979) 333–356.
22. G. A. Kent, "Optimal Control of Functional Differential Equations of Neutral Type," Ph.D. Thesis, Brown University (1971).
23. H. T. Banks and M. Q. Jacobs, "An Attainable Set Approach to Optimal Control Functional Differential Equations with Function Space Terminal Conditions," *J. Diff. Equations* **13** (1973) 127–149.

24. H. R. Rodas and C. E. Langenhop, "A Sufficient Condition for Function Space Controllability of a Linear Neutral System," *SIAM J. Linear Control Optimiz.* **16** (1978) 429–435.

25. D. Henry, "FDES of Neutral Type in the Space of Continuous Functions: Representation of Solutions and Adjoint Equation," Preprint, Math. Dept. University of Kentucky, Lexington, KY.

26. N. Dunford and J. T. Schwartz, *Linear Operators Part I: General Theory*, Interscience Publications, New York (1958).

27. Shin-Ichi Nakagiri, "Optimal Control of Linear Retarded Systems in Banach Spaces," *J. Math. Analysis and Applic.* **120** (Nov. 15, 1986) 169–210.

28. J. L. Lions, *Optimal Control of Systems Governed by P.D.E.*, Springer-Verlag, Berlin, 1971.

29. G. Tadmor, "Functional Differential Equations of Retarded and Neutral Type: Analytic Solutions and Piecewise Continuous Controls," *J. Differential Equations* **51** (1984) 151–181.

30. O. Hájek, "Geometric Theory of Time Optimal Control," *SIAM J. Control* **9** (1971) 339–350.

31. E. N. Chukwu, *The Time Optimal Feedback Control of Delay Differential Systems*, Preprint.

32. E. J. McShane and R. B. Warfield, "On Filippov's Implicit Function Lemma," *Proceedings Amer. Math. Soc.* **18** (1967) 41–47.

33. S. Saks, *Theory of the Integral Monografie Matematyceque VII*, English Translation, Warzawa, 1937.

34. W. Rudin, *Real and Complex Analysis*, McGraw-Hill, New York, 1966.

35. H. O. Fattorini, "The Time Optimal Control Problem in Banach Spaces," *Appl. Math. Optimiz.* **1** (1974) 163–188.

36. R. Datko, "Linear Autonomous Neutral Differential Equations in a Banach Space," *J. Diff. Equat.* **25** (1977) 258–274.

37. M. Slemrod, "Nonexistence of Oscillations in a Nonlinear Distributed Network," *J. Math. Anal. Appl.* **36** (1971) 22–40.

38. J. K. Hale, "Stability of Functional Differential Equations of Neutral Type," *J. Diff. Equat.* (1970) 334–355.

39. E. N. Chukwu, "An Estimate for the Solution of a Certain Functional Differential Equation of Neutral Type," in *Nonlinear Phenomena in Mathematical Science*, edited by V. Lakshmikanthan, Academic Press, New York, 1982.

40. O. Lopes, "Stability and Forced Oscillations," *J. Math. Anal. Appl.* **55** (1976) 686–698.

41. E. N. Chukwu, R. G. Underwood, and L. D. Kovari, *Global Constrained Null Controllability of Nonlinear Neutral Systems*, Preprint.

Chapter 11

# The Time-Optimal Control Theory of Nonlinear Systems of Neutral Type

## 11.1 Introduction

This chapter introduces the problem of reaching a continuously moving target $z$ by a trajectory of the control system described by the nonlinear functional differential equations of neutral type,

$$\frac{d}{dt}[D(t)x_t] = f(t, x_t, u(t)),$$

$$x_\sigma = \varphi,$$

in minimum time. The state space is either $E^n$, the Euclidean $n$-dimensional vector space; the space $C = C([-h, 0], E^n)$ of continuous functions from $[-h, 0]$ into $E^n$ with the sup norm, or the Sobolev space $W_p^{(1)} = W_p^{(1)}([-h, 0], E^n)$ of functions $[-h, 0] \to E^n$ whose derivatives are $L_p$ integrable. The admissible controls are measurable functions $u : [t, \infty] \to E^m$ whose values $u(t) \in U$, a compact convex set. The existence theory in $C$ is established in Sec. 11.3; the corresponding constrained controllability questions are answered in Chap. 12. The necessary condition for time-optimal control, the maximum principle, is stated in Sec. 11.4. It is important to refer to Sec. 1.6 for the motivation from electric current.

Before the sixties, the stability of electric power systems was seldom threatened. This was because of very conservative system's design based on fairly constant and predictable load growth. The situation has now changed. Environmental concerns, economic realities and other factors have led to the development of a new generation of equipment that is prone to cause network voltage collapse and acute instability. Power systems planners and operators have become increasingly concerned with stability problems. As the region of stability of the equations of motion of such systems becomes better understood, it is predicted that the time-optimal control of voltage and current fluctuations will assume greater importance. The natural models of fluctuations of voltage and current in problems arising in transmission lines are functional differential equations of neutral type. See Cruz and Hale [10, p. 5], Salamon [5, p. 151], and Lopes [21] for derivation and other references.

413

One is interested in driving fluctuations of voltage to its stable equilibrium state as rapidly as possible. This is the time-optimal problem. It is old, very important, and continues to be interesting even for linear ordinary differential systems (see [22]). Our aim here is to present a comprehensive theory that will be needed for the construction of an optimal feedback control similar to the development in [23].

We consider the problem of reaching, in minimum time, a continuously moving target $z_t$, in function space, by a trajectory of the control system described by the nonlinear functional differential equation of neutral type,

$$\frac{d}{dt}[D(t)x_t] = f(t, x_t, u(t)), \qquad t \geq 0,$$

$$x_\sigma = \varphi.$$

$$(11.1.1)$$

In (11.1.1), $D$ is a continuous function $D(\cdot) : [\tau, \infty) \times C \to E^n$ given as

$$D(t)x_t = x(t) - g(t, x_t),$$

$$(11.1.2)$$

and $f : [\tau, \infty) \times C \times E^m \to E^n$ is a continuous function. Here $\tau$ is a real number. The target $z$ may be taken as a point of $E^n$, or of $C$, or of $W_p^{(1)}$, $1 \leq p \leq \infty$, where $W_p^{(1)}$ is the Sobolev space of all absolutely continuous functions $x : [-h, 0] \to E^n$ whose derivative $\dot{x}(t) = \frac{dx}{dt}$ belongs to the $p$-integrable space $L_p([-h, 0], E^n)$. The admissible controls are measurable functions $u : [\tau, \infty] \to E^n$ with $u(t) \in U(t, x_t)$, where $U : [\tau, \infty) \times C \to 2^{E^n}$ is a set-valued map that is nonempty and compact. We can also consider the set of admissible controls that is a closed and bounded subset of $L_p([\sigma, \infty), E^n)$ with zero in its interior, when the state space is $W_p^{(1)}$.

The time-optimal problem is to determine an admissible control $u^*$ such that the solution $x(\sigma, \varphi, u^*)$ of (11.1.1) hits a continuously moving target $z$ (in the appropriate space) in minimum time $t^* \geq \sigma$. Such a control is called time-optimal, and $t^*$ is the optimal time. This problem has three parts:

(i) Does there exist an admissible control $u$ such that

$$x_\tau(\sigma, \varphi, u) = z_\tau \text{ for some } \tau.$$

$$(11.1.3)$$

This is the problem of controllability. With the optimal time $\tau_0$ defined by

$$\tau_0 = \inf\{\tau : \tau \text{ such that } (11.1.3) \text{ holds}\}.$$

$$(11.1.4)$$

(ii) Does there exist an optimal control $u$ such that

$$x_{\tau_0}(\sigma, \varphi, u) = z_{\tau_0}.$$

$$(11.1.5)$$

Finally:

(iii) What are the form, uniqueness, and general properties of the optimal control if it exists. Part three of the problem is inferred from the necessary condition for an optimal control, the so-called maximum principle.

Though a very large body of research is available on the control of neutral systems, most are concerned with optimal control of systems with quadratic cost functions, [1–2]. The time-optimal problem was implicitly touched upon in [1], as a consequence of a result on minimizing a general cost function. Here a necessary condition was formulated. For linear systems with unrestrained controls, controllability questions were investigated by Gabasov and Kirillova [3], Rodas and Langenhop [4], and Salamon [5] and more recently in [7] for linear systems. The present treatment gives a comprehensive theory for the nonlinear system (11.1.1).

In Sec. 11.2, we establish the existence and uniqueness and some regularity properties of solutions of the differential equations (11.1.1). We introduce some conditions that will prevent finite escape times of trajectories. This will enable us to restrict our attention to a closed and bounded set of the $(t, x_t)$-space. If $x(t, \sigma, u)$ is a solution of (11.1.1) and $D_1 f(\cdot, \cdot, \cdot)$ is the Fréchet derivative of $f$ on $E \times C \times E^m$ with respect to $i$th variable, then the linear approximation of (11.1.1) about $x(t, \sigma, u)$ is

$$\frac{d}{dt}[D(t)y_t] = L(t, y_t) + B(t)v(t) \tag{11.1.6}$$

where

$$L(t, y_t) = D_2 f(t, x_t, u)y_t, \qquad B(t) = D_3 f(t, x_t, u).$$

Some of the properties of the solution map of 11.1.16 is outlined in Sec. 11.2.

In Sec. 11.3, we prove the closure of the attainable sets of (11.1.1). Under the constrained controllability assumption, we prove the existence of a time-optimal control. Sufficient condition for controllability for (11.1.1) is taken up in Chap. 12. For completeness we include the necessary condition for an optimal control, a maximum principle in Sec. 11.4. There the theory is applied to linear systems.

## 11.2  Existence, Uniqueness, and Continuity of Solutions of Neutral Systems

In (11.1.2), assume that $g(t, \cdot)$ is a bounded linear operator from $C$ into $E^n$ for each $t \in (\tau, \infty)$, $g(t, \varphi)$ is continuous for $(t, \varphi) \in (\tau, \infty) \times C$,

$$g(t, \varphi) = \int_{-h}^{0} [d_\theta \mu(t, \theta)]\varphi(\theta), \qquad |g(t, \varphi)| \le k(t)\|\varphi\|, \tag{11.2.1}$$

for some nonnegative function $k \in C((\tau, \infty), E)$, and $\mu(t, \cdot)$ is an $n \times n$ matrix function of bounded variation on $[-h, 0]$. We assume $g$ is uniformly nonatomic at zero; that is, there exists a continuous, nonnegative, nondecreasing function $L(s)$ for $s \in [0, h]$ such that

$$L(0) = 0, \left|\int_{-s}^{0} [d_\theta \mu(t, \theta)]\varphi(\theta)\right| \le L(s)\|\varphi\| \tag{11.2.2}$$

for all $t \in (\tau, \infty)$, $\varphi \in C$.

We shall have cause to require that $D$ in (11.1.2) is uniformly stable in the following sense of Definition 11.2.1:

**Definition 11.2.1**  The operator $D$ is uniformly stable if there are constants, $\alpha$, $\beta > 0$ such that for every $\sigma \in [\sigma, \infty)$, $\varphi \in C$ the solution $x(\sigma, \varphi)$ of the homogeneous difference equation

$$D(t)x_t = 0, \qquad t \geq \sigma,$$
$$x_\sigma = \varphi, \qquad D(\sigma) = 0 \tag{11.2.3}$$

satisfies

$$\|x_t(\sigma, \varphi)\| \leq \beta\|\varphi\|C^{-\alpha(t-\sigma)}, \qquad t > 0. \tag{11.2.4}$$

In (11.1.1) we assume that

$$f : [\tau, \infty) \times C \times E^m \to E^n$$

is continuously differentiable and is $C^1$. We now establish the existence of a solution of (11.1.1) for each initial data $(\sigma, \varphi) \in E \times C([-h, 0], O)$, where $O \in E^n$ is an open convex set containing the origin, and where

$$f : E \times C([-h, 0], O) \times E^m \to E^n, \qquad g : E \times C([-h, 0], O) \to E^n$$

are continuous and $g$ satisfies (11.2.1) and (11.2.2). The existence result of Cruz and Hale [8] does not quite cover the situation here. One can use the ideas of Melvin [9] to prove existence and uniqueness with initial data $(\sigma, \varphi) \in E \times W_p^{(1)}$, and this is done by Chukwu and Simpson [24].

**Theorem 11.2.1**  *Suppose in* (11.1.1),

(i) $f(\cdot, \cdot, \cdot) : E \times C([-h, 0], O) \times E^m \to E^n$ *is continuously differentiable.*

(ii) *For each compact convex set $K \subseteq O$ there exists an integrable function $M_1 : E \to [0, \infty)$ and integrable functions $M_i : E \to [0, \infty)$, $i = 2, 3$, such that*

$$\|D_2 f(t, \varphi, w)\| \leq M_1(t) + M_2(t)\|w\|, \quad \|D_3 f(t, \varphi, w)\| \leq M_3(t) \tag{11.2.5}$$

*for all $t$, $w$, and $\varphi \in C([-h, 0], K)$ where $D_i f(\cdot, \cdot, \cdot)$ is the Fréchet derivative of $f$ with respect to the $i$th variable.*

(iii) $f(t, 0, 0) = 0, \forall t$.

(iv) *If $L(t, \cdot) = D_2 f(t, 0, 0)$, $B(t) = D_3 f(t, 0, 0)$ where*

$$L(t, \varphi) = \int_{-h}^{0} [d_\theta \eta(t, \theta)]\varphi(\theta), \quad |L(t, \varphi)| = M(t)\|\varphi\|, \tag{11.2.6}$$
$$(t, \varphi) \in (\tau, \infty) \times C,$$

*for some $M \in L_1^{loc}((\tau, \infty), E)$, and some $n \times n$ matrix $\eta(\tau, \theta)$ of bounded variation in $\theta \in [-h, 0]$.*

(v) *The function $g$ in (11.1.2) and (11.2.1) satisfies (11.2.1) and (11.2.2). Then there exists an open neighborhood $O_C$, of the origin in $C$, and an open neighborhood $O_U$ of the origin in $L_\infty([\sigma, \infty), E^m)$ such that for each $\varphi \in O_C$ and each $u \in O_U$, (11.1.1) has a unique solution that is continuously differentiable.*

**Proof**   Let $C_T = C_T([\sigma - h, T], E^n)$, and define the function $G : E \times C_T \to C_T$ by

$$G(x)(t) = \begin{cases} 0, & \text{if } \sigma - h \le t < \sigma, \\ g(t, x_t) - g(\sigma, \varphi), & \text{if } \sigma \le t \le T. \end{cases} \qquad (11.2.7)$$

Define $H : C_T \times L_\infty([\sigma, \tau], E^m) \to C_T$ by

$$H(x, u)(t) = \begin{cases} 0, & \text{if } \sigma - h \le t < \sigma, \\ \displaystyle\int_\sigma^t f(s, x_s, u(s))ds, & \text{if } \sigma \le t \le T. \end{cases} \qquad (11.2.8)$$

Finally, define $I : C_T \to C_T$

$$I\varphi(t) = \begin{cases} \varphi(t - \sigma), & \text{if } \sigma - h < t < \sigma, \\ \varphi(0), & \text{if } \sigma \le t \le T. \end{cases} \qquad (11.2.9)$$

Written in integral form, (11.1.1) is equivalent to

$$x_\sigma = \varphi \in [-h, 0],$$
$$x(t) = g(t, x_t) = g(\sigma, \varphi) + \varphi(0) + \int_\sigma^t f(s, x_s, u(s))ds, \quad t \ge \sigma. \qquad (11.2.10)$$

Using this fact and the operators $G$, $H$, and $I$, we see that $x$ satisfies (11.1.1) if and only if

$$x = Gx + H(x, u) + I\varphi \qquad (11.2.11)$$

for $x \in C_T$.

It is easy to see that (11.2.10) is well defined, since $f(\cdot, \cdot, \cdot)$ is assumed continuous and since by the mean value theorem (we use the convexity of $K$),

$$\|f(t, \varphi, w)\| \le |f(t, 0, 0)|_{E^n} + M_1(t)\|\varphi\| + M_3(t)\|w\|, \qquad (11.2.12)$$

for all $t \in [\sigma, t_1]$, $\varphi \in C([-h, 0], K)$, $w \in L_\infty$.   $\square$

Note condition (ii). We need two lemmas.

**Lemma 11.2.1**   *Let $H : C([\sigma - h, T], 0) \times L_\infty \to C_T$ be as defined in (11.2.8), $G$ be as defined in (11.2.7). Then $H$ is continuously differentiable and*

$$(D_1 H(0, 0)x)(t) = \begin{cases} 0, & \sigma - h \le t < \sigma, \\ \displaystyle\int_\sigma^t D_2 f(s, 0, 0)x_s ds, & \sigma \le t \le T, \end{cases} \qquad (11.2.13)$$

$$(D_3 H(0,0)u)(t) = \begin{cases} 0, & \sigma - h \leq t < \sigma, \\ \displaystyle\int_\sigma^t D_2 f(s,0,0)x_s ds, & \sigma \leq t \leq T, \end{cases} \qquad (11.2.14)$$

*for all $x \in C_T$, $u \in L_\infty([\sigma, T], E^m)$, $t \in [\sigma - h, T]$.*

*Also $G$ is continuously differential and for any $x \in C_T([\sigma - h, t_1], E^n)$,*

$$(DG\overline{x})x = Gx. \qquad (11.2.15)$$

**Proof**    By [12, p. 177],

$$(D_1 H(\overline{x}, \overline{u})x)(t) = \int_\sigma^t D_2 f(s, \overline{x}_s, \overline{u}(s))x_s ds, \qquad (11.2.16)$$

$$(D_2 H(\overline{x}, \overline{u})u)(t) = \int_\sigma^t D_3 f(s, \overline{x}_s, \overline{u}(s))u(s)ds \qquad (11.2.17)$$

where $t \in [\sigma - h, T]$, $\overline{x} \in C([\sigma - h, T], 0)$ $x \in C_T = C([\sigma - h, T], E^n), u, \overline{u} \in L_\infty([\sigma, T], E^m)$. The results follow quickly from these. Also, because of the linearity of $g(t, \varphi)$ in (11.2.1), (11.2.15) follows at once.      $\square$

**Lemma 11.2.2**    *Define*

$$K : C([\sigma - h, T], E^n) \to C([\sigma - h, T], E^n)$$

*by*

$$Kx = DG(0)x + D_1 H(0,0)x. \qquad (11.2.18)$$

*Then $(1 - K)^{-1}$ exists as a bounded linear map.*

**Proof**    Let $q$ be any function in $C([\tau, \infty), E^n)$. Consider

$$\frac{d}{dt}[x(t) - g(t, x_t) - q(t)] = D_2 f(t, 0, 0)x_t$$

where

$$x_\sigma = \varphi = q_\sigma. \qquad (11.2.19)$$

This is the same as

$$\frac{d}{dt}[D(t)x_t - q(t)] = L(t, x_t). \qquad (11.2.20)$$

By assumptions (iv) and (v) and Theorem 2.1 of Hale and Cruz [10, p. 333], there exists a function $x(\sigma, \varphi, q)(t)$ with initial value $\varphi$ at $\sigma$ uniquely defined and continuous on $[\sigma - h, \infty)$ and satisfying (11.2.20). Furthermore, for any fixed $\sigma$ and $t \geq \sigma$, $x(\sigma, \cdot, 0)(t)$ is a continuous linear operator from $C$ into $E^n$. The integrated form of (11.2.20) is

$$x(t) = g(t, x_t) - g(\sigma, \varphi) + q(t) + \varphi(0) + \int_\sigma^t L(s, x_s)ds.$$

But this is

$$x(t) = DG(0)x(t) + \varphi(0) + q(t) + (D_1 H(0,0)x)(t) - q(\sigma), \qquad t \in [\sigma, T].$$

Since $x_\sigma = q_\sigma = \varphi$, $q(\sigma) = \varphi(0)$, we have that

$$x = DG(0)x + D_1 H(0,0)x + q(t)$$

has a unique solution for any $q \in C_T$.

Therefore $x = Kx = DG(0)x + D_1 H(0,0)x$ has the unique solution $x = 0$. Hence $I - K$ is one-to-one mapping of $C_T$ onto $C_T$. Also $I - K$ is a bounded linear operator. As a consequence of the Open Mapping Theorem ([11, p. 99]) $(I - K)^{-1}$ is a bounded linear mapping of $C_T$ onto $C_T$.

We now complete the proof of Theorem 11.2.1. Consider (11.1.1) in its integrated form (11.2.11). By Gronwall-type argument we can easily show that its solution $\xi(\varphi, u) \in C_T$ is necessarily unique for any pair of $\varphi, u$ such that the solution exists.

Define

$$M(x, u, \varphi) = x - Gx - H(x, u) - I\varphi.$$

By hypothesis (iii), $M(0,0,0) = 0$. By Lemma 11.2.1, $M$ is continuously differentiable on

$$C([\sigma - h, T], 0) \times L_\infty([\sigma, T], E^m) \times C([-h, 0], E^n), \qquad (T \le \infty)$$

with

$$D_1 M(0,0,0) = I - DG(0) - D_2 H(0,0).$$

By Lemma 11.2.2, $D_1 M(0,0,0)^{-1}$ exists. Therefore by the Implicit Function Theorem [12, p. 270] there exists a continuously differential mapping $\xi(\varphi, u)$ defined in an open neighborhood $N$ of the origin in

$$C_T([\sigma, h, T], E^n) \times L_\infty([\sigma, T], E^m)$$

such that $M(\xi(\varphi, u), u, \varphi) = 0$ for all $(\varphi, u) \in N$. Furthermore, $x = \xi(\varphi, u)$ satisfies (11.2.11) or (11.1.1). We now choose $0_{C_T}, O_U$ so that $O_C \times O_U \subset N$ to complete the proof. $\qquad\square$

It is clear from Theorem 11.2.1 that the solution $x(\sigma, \varphi, u)$ of (11.1.1) is continuously differentiable for each $(\sigma, \varphi) \in E \times C$ and each $u \in L_\infty([\sigma, \infty), E^m)$. Corresponding to the point $(t, \sigma, \varphi, u) \in E \times E \times C \times L_\infty$, the mapping $x : E \times E \times C \times L_\infty \to C$ defined by $x(\sigma, \varphi, u)(t) = x_t(\sigma, \varphi)$ represents a point in $C$. We now give some very useful properties of this mapping.

**Lemma 11.2.3**  *Let $(t, \sigma, \varphi, u) \in E \times E \times C \times L_\infty$. Assume all the conditions of Theorem 11.2.1. Let $Dx(t, \sigma, \varphi, u)$ denote the partial derivative of $x(t, \sigma, \varphi, u)$ with respect to its last two variables at the point $(t, \sigma, \varphi, u)$. Then for every $(\psi, v) \in C \times L_\infty$ we have $Dx(t, \sigma, \varphi, u)(\psi, v) = \lambda(t, \sigma, \varphi, u, \psi, v)$, where the mapping*

$t \to \lambda(t, \sigma, \varphi, u, \psi, v)$ *of* $E$ *into* $E^n$ *is the unique solution of the linear differential equation of neutral type*

$$\frac{d}{dt}[D(t)y_t] = D_2 f(t, x_t(\sigma, \varphi, u), u(t))y_t + D_3 f(t, x_t(\sigma, \varphi, u), u)v(t),$$

$$\quad (11.2.21)$$

$$y_\sigma = \psi.$$

*Also for* $v \in L_\infty([\sigma, \infty), E^m)$ *we have* $D_4 x(t, \sigma, \varphi, u)(v) = y(t, \sigma, \varphi, u, v)$ *where the mapping* $t \to y(t, \sigma, \varphi, u, v)$ *of* $E$ *into* $E^n$ *is the unique solution of* (11.2.21) *satisfying* $y_\sigma(\sigma, \varphi, u, v) = 0.$

**Proof**    For each $(\sigma, \varphi) \in E \times C$, the unique solution of (11.1.1) with initial data $(\sigma, \varphi)$ is given in (11.2.10) by

$$x_\sigma = \varphi \text{ in } [-h, 0],$$

$$x(\sigma, \varphi, u)(t) = x(t, \sigma, \varphi, u)$$

$$= g(t, x_t(\sigma, \varphi, u)) - g(\sigma, \varphi) + \varphi(0)$$

$$+ \int_\sigma^t f(s, x_s(\sigma, \varphi, u), u(s))ds, \quad t \geq \sigma.$$

Let $D_4 x(t, \sigma, \varphi, u) = D_u$ be the partial derivative of $x(t, \sigma, \varphi, u)$ with respect to the fourth argument. Then by [12, pp. 107, 173] $D_4 x_\sigma = 0$ in $[-h, 0]$,

$$D_4 x(t, \sigma, \varphi, u) = \int_\sigma^t D_2 f(s, x_s(\sigma, \varphi, u)), u(s)D_u x_s(\sigma, \varphi, u)ds$$

$$+ \int_\sigma^t D_3 f(s, x_s(\sigma, \varphi, u), u(s))ds$$

$$+ D_2 g(t, x_t(\sigma, \varphi, u))\frac{dx_t}{du}(\sigma, \varphi, u).$$

Thus, by the same reasoning in (11.2.15), we have

$$D_2 g(t, x_t(\sigma, \varphi, u))\frac{dx_t}{du}(\sigma, \varphi, u) = g\left(t, \frac{dx_t}{du}(\sigma, \varphi, u)\right).$$

Thus,

$$D_4 x(t, \sigma, \varphi, u) - D_2 g\left(t, x_t(\sigma, \varphi, u)\frac{dx_t}{du}(\sigma, \varphi, u)\right)$$

$$= \int_\sigma^t D_2 f(s, x_s(\sigma, \varphi, u), u(s))D_u x_s(\sigma, \varphi, u)ds$$

$$+ \int_\sigma^t D_3 f(s, x_s(\sigma, \varphi, u)), u(s)ds.$$

On differentiating with respect to $t$,

$$\frac{d}{dt}[D_u x(t,\sigma,\varphi,u)v - D_2 g(t, x_t(\sigma,\varphi,u))D_u x_t(\sigma,\varphi,u)v]$$

$$= D_2 f(t, x_t(\sigma,\varphi,u), u(t))D_u x_t(\sigma,\varphi,u)v + D_3 f(t, x_t(\sigma,\varphi,u), u(t))v \,,$$

or equivalently,

$$\frac{d}{dt}[D_u x(t,\sigma,\varphi,u)v - g(t, D_u x_t(\sigma,\varphi,u)v)]$$

$$= D_2 f(t, x_t(\sigma,\varphi,u), u(t))D_u x_t(\sigma,\varphi,u)v + D_4 f(t, x_t(\sigma,\varphi,u), u(t))v \,.$$

This proves the second assertion. For the first assertion, note that

$$Dx(t,\sigma,\varphi,u)(\psi,v) = D_3 x(t,\sigma,\varphi,u)\psi + D_4 x(t,\sigma,\varphi,u)v \,.$$

Now, from (11.2.10), we deduce that

$$D_3 x(t,\sigma,\varphi,u) = D_2 g(t \cdot x_t(\sigma,\varphi,u))\frac{dx_s}{d\varphi}(\sigma,\varphi,u) - D_2 g(\sigma,\varphi)$$

$$+ I + \int_\sigma^t D_2 f(s, x_s(\sigma,\varphi,u), u(s))\frac{dx_s}{d\varphi}(\sigma,\varphi,u)ds \,,$$

$$D_4 x(t,\sigma,\varphi,u) = \int_\sigma^t D_2 f(s, x_s(\sigma,\varphi,u), u(s))D_u x_s(\sigma,\varphi,u)ds$$

$$+ \int_\sigma^t D_3 f(s, x_s(\sigma,\varphi,u), u(s))ds + g(t, D_u x_t(\sigma,\psi,u)) \,.$$

Thus

$$Dx(t,\sigma,\varphi,u)(\psi,v) = g\left(t, \frac{dx_t}{d\varphi}\psi\right) - g(\sigma\psi)\,Dx(t\cdots)$$

$$+ \psi + \int_\sigma^t D_2 f(s, x_s(\sigma,\varphi,u), u(s))\frac{dx_s}{d\varphi}(\sigma,\varphi,u)\psi ds$$

$$+ \int_\sigma^t D_2 f(s, x_s(\sigma,\psi,u), u(s))D_u x_s(\sigma,\varphi,u)v ds$$

$$+ \int_\sigma^t D_2 f(s, x_s(\sigma,\varphi,u), u(s))v ds + g(t, D_u x_t(\sigma,\varphi,u)v) \,.$$

Now take the $t$ derivative to obtain

$$\frac{d}{dt}\left[Dx(t,\sigma,\varphi,u)(\psi,v) - g\left(t, \frac{dx_t}{d\varphi}\psi + D_u x_t v\right)\right]$$

$$= D_2 f(t, x_t(\sigma, \varphi, u), u(t)) \left[ \frac{dx_s}{dx} \psi + D_u x_s(\sigma, \varphi, u) v \right]$$

$$+ D_3 f(t, x_t(\sigma, \varphi, u), u(t)) v.$$

We have used the linearity of $g$. Note that

$$D x_\sigma(\sigma, \varphi, u)(\psi, v) = \psi.$$

This concludes the proof. $\qquad\qquad\qquad\square$

We require conditions that will prevent finite escape times of trajectories of (11.1.1). It is given in the next lemma.

**Lemma 11.2.4**   *In* (11.1.1), *assume:*

(i) *$D$ is uniformly stable.*
(ii)

$$(D(t)x_t,\ f(t, x_t, u) \le k(1 + |Dx_t|^2) \qquad\qquad (11.2.22)$$

*for some constant $k > 0$, for all $t \ge \sigma$, $x_t \in C$, $u(t) \in U(t, x_t)$.*

*Then there exist an $M > 0$ and an $N > 0$ such that every solution $x(\sigma, \varphi, u)$ of* (11.1.1) *satisfies*

$$\|x_t(\sigma, \varphi, u)\| \le M e^{N(t-\sigma)},$$

*so that in any compact interval $[\sigma, \tau]$,*

$$\|x_t(\sigma, \varphi, u)\| \le M e^{N(t-\sigma)}.$$

**Proof**   Let $x$ be a solution of (11.1.1). Let

$$V(x(t)) = (Dx_t, Dx_t) = |Dx_t|^2.$$

Then

$$\frac{d}{dt} V(x(t)) = 2(Dx_t, f(t, x_t, u(t))) \le 2K(1 + |Dx_t|^2),$$

by condition (ii). $\qquad\qquad\qquad\square$

Thus

$$\frac{d}{dt} |Dx_t|^2 \le 2K(1 + |Dx_t|^2).$$

It follows that

$$|Dx_t|^2 \le (1 + |D\varphi|^2) \exp 2kt \equiv g(t).$$

Since $D$ is uniformly stable [13, Lemma 3.4], there exists some constants $a$, $b$, $c$, $d$ such that

$$\|x_t(\sigma, \varphi, u)\| \leq e^{-a[t-\sigma)} \left[ b\|\varphi\| + c \sup_{\sigma \leq u \leq t} |g(u)| \right] + d \sup_{\sigma \leq u \leq t} |g(u)|\,.$$

Since $g(t) = (1 + |D\varphi|^2)\exp 2kt$, we obtain

$$\|x_t(\sigma, \varphi, u)\| \leq e^{-a(t-\sigma)}[b\|\varphi\| + c(1 + |D\varphi|^2)\exp 2kt] + d(1 + |D\varphi|^2)\exp 2kt\,.$$

If $F = \max\{b\|\varphi\|,\ c(1 + |D\varphi|^2), d(1 + |D\varphi|^2)\}$, then

$$\|x_t(\sigma, \varphi, u)\| \leq 3e^{-a(t-\sigma)}Fe^{2kt}\,.$$

With $N = 2k - a$, $M = 3Fe^{a\sigma}$, we have

$$\|x_t(\sigma, \varphi, u)\| \leq Me^{Nt}\,.$$

**Remark** Once a value $T \geq \sigma$ is chosen $T < \infty$, then $x(\sigma, \varphi, u)$ is bounded. We work in the region that is a closed and bounded subset of $E \times C([-h, 0], E^n)$:

$$M = \{(t, x_t) : \sigma \leq t \leq T \leq \infty \|x_t(\sigma, \varphi)\| \leq Me^{NT}\}\,. \tag{11.2.23}$$

**Remark 11.2.1** If $\varphi$, the initial point, lies on a certain compact initial subset of $C$ and $u(t) \in U(t, x_t)$, then the bound $Me^{NT}$ is independent of the controls. In the region $M$, when this uniform bound is in force, we shall examine the behavior of the attainable set $\mathcal{A}$ of (11.1.1). This is done in the next section.

## 11.3 Existence of Optimal Controls of Neutral Systems

Let $M$ be the closed and bounded subset of $E \times C$ that was introduced in (11.2.23). In order to study the existence of a time-optimal control of

$$\frac{d}{dt}[D(t)x_t] = f(t, x_t, u(t))\,, \quad t \geq \sigma\,,$$
$$x_\sigma = \varphi\,, \qquad (t, x_t) \in M\,, \tag{11.3.1}$$

we introduce the associated contingent equation,

$$\frac{d}{dt}[D(t)x_t] \in F(t, x_t)\,,$$
$$x_\sigma = \varphi\,, \quad (t, x_t) \in M\,, \tag{11.3.2}$$

where the set-valued map

$$F : E \times C \to 2^{E^n}$$

is given by

$$F(t, \varphi) = \{f(t, \varphi, u) : u \in U(t, \varphi)\}\,.$$

We shall always assume that $F(t, \varphi)$ is convex for each $(t, \varphi) \in M$. Because $f$ is continuous and $U(t, \varphi)$ nonempty and compact, $F(t, \varphi)$ is nonempty and compact. The symbol $2^{E^n}$ denotes the space of subsets of $E^n$.

We assume that $U(t, \varphi)$ is upper semicontinuous with respect to inclusion in $t, \varphi$ in the following sense:

**Definition 11.3.1**    The set map

$$U : E \times C \to 2^{E^n}$$

is said to be upper semicontinuous with respect to inclusion in $t, \varphi$ if for any $\epsilon > 0$ there is a $\delta > 0$ such that

$$U(t_1, \varphi_1) \subseteq U_\epsilon(t, \varphi)$$

whenever $|t_1 - t| \leq \delta$, $\|\varphi_1 - \varphi\| \leq \delta$.

Here $U_\epsilon$ denotes the $\epsilon$-neighborhood of $U$ in $E^m$.

**Definition 11.3.2**    A solution $x(t)$ of (11.3.2) is an element $x \in C([\sigma - h, T], E^n)$ such that

    (i)   $Dx_t$ is absolutely continuous and has a continuous derivative on $[\sigma, T]$,
    (ii)   $\frac{d}{dt}(Dx_t)$ satisfies (11.3.2),
    (iii)   $x_\sigma = \varphi$, $(t, x_t) \in M$.

Clearly, any solution of (11.3.1) gives rise to a solution of the orientor problem (11.3.2). By standard measurable selection arguments of [14], we know that every solution of the relation (11.3.2) can be viewed as a trajectory of the control system (11.3.1). It is summarized in the next lemma.

**Lemma 11.3.1**    *Let $F(t, \varphi)$ be convex for each $(t, \varphi) \in M$. Then a function $x \in C([\sigma - h, T], E^n)$ is a solution of (11.3.2) if and only if $x$ is a solution of (11.3.1) for some admissible control $u$, $u(t) \in U(t, x_t)$.*

**Definition 11.3.3**    The attainable set $\mathcal{A}(t)$ of (11.3.1) at time $t \geq \sigma$ is a subset of $C = C([-h, 0], E^n)$ defined by

$$\mathcal{A}(t) = \{x_t \in C, \ x \text{ is a solution of (11.3.1) for some } u(t) \in U(t, x_t),$$

$$(t, x_t) \in M\}.$$

By Lemma 11.3.1, this is the same as:

$$\mathcal{A}(t) = \left\{ x_t \in C, \ x \text{ is a solution of (11.3.2), i.e., } \frac{d}{dt}[D(t)x_t] \in F(t, x_t), \right.$$

$$\left. x_\sigma = \varphi, \ (t, x_t) \in M \right\}.$$

We assume the conditions of Theorem 11.1.1, which ensure the existence and uniqueness of solutions. Subject to these conditions, $\mathcal{A}(t)$ is nonempty. We now prove that the attainable set is closed as well.

**Theorem 11.3.1** *In (11.3.1) we assume that:*

(i) *$f : E \times C([-h, 0], E^n) \times E^m \to E^n$ is continuously differentiable.*
(ii) *$D$ is uniformly stable and satisfies (11.2.1) and (11.2.2).*
(iii) *For any measurable $y(t)$ satisfying $y(t) \in F(t, x_t)$, we have $|y(t)| \leq m(t)$, a.e., $m \in L_1([\sigma, T], E)$.*
(iv) *$F(t, \varphi)$ is compact and convex for each $t, \varphi$.*
(v) *$U(t, \varphi)$ is upper semicontinuous with respect to inclusion.*

*Then the attainable set $\mathcal{A}(t)$ of (11.3.1) is closed.*

**Proof**   To prove the closure of the attainable set, let $x_t^n \in \mathcal{A}(t)$ $n = 1, 2, \ldots$ . Let $x_t^n \to x_t$ as $n \to \infty$. We prove $x_t \in \mathcal{A}(t)$. Because $x_t^n \mathcal{A}(t)$, for each $n$,

$$\frac{d}{dt}[D(t)x_t^n] \in F(t, x_t^n), \qquad t \in I.$$

By condition (iii),

$$\left| \frac{d}{dt}D(t)x_t^n \right| \leq m(t)$$

where $m \in L_1(I, E)$. From the above inequality, we deduce that

$$|D(t, x_t^n) - D(\sigma, \varphi)| \leq \int_\sigma^t m(s)ds, \quad |D(t, x_t^n) - D(t)x_{\bar{t}}^n| \leq \int_{\bar{t}}^t m(s)ds,$$

so that

$$\int_{E_i} D(s, x_s^n)ds \to 0, \quad i \to \infty$$

uniformly with respect to $n$ for each decreasing sequence $\{E_i\}, E_i \subseteq I$ with void intersection. Therefore, by [15, p. 292] there is a sequence (we retain the same notation weakly convergent in $L_1(I, E^n)$ to a function $\xi \in L_1(I, E^n)$). Then for each $t \in I$,

$$D(t)x_t = \lim_{n \to \infty}[D(t)x_t^n] = \lim_{n \to \infty} D(\sigma)\varphi + \int_\sigma^t \dot{D}(s)(x_s^n)ds = D(\sigma)\varphi + \int_\sigma^t \xi(s)ds.$$

It now follows from [15, p. 422] that there is a sequence $\{\xi_k\}$ of convex combinations of the functions $\{\dot{D}(t, x_t^k)\dot{D}(t, x_t^{k+1})\cdots\}$ converging in $L_1(I, E^n)$ norm to $\xi$. From this sequence $(\xi_k)$ select a subsequence that converges to $\xi$ a.e.. Thus, almost everywhere on $I$

$$\xi(t) \in \bigcap_{k=1}^{\infty} \overline{\text{co}} \left( \bigcup_{n=k}^{\infty} \dot{D}(t)x_t^n \right) \subseteq \bigcap_{k=1}^{\infty} \overline{\text{co}} \left( \bigcap_{n=k}^{\infty} F(t, x_t^n) \right) \subseteq F(t, x_t), \qquad (11.3.3)$$

where $\overline{\text{co}}(M)$ is the closed convex hull of $M \in E^n$. Hence

$$\frac{d}{dt}[D(t)x_t] \in F(t, x_t).$$

Hence $\mathcal{A}(t)$ is closed, since also $(t, x_t) \in M$ from the closure of $M$. This concludes the proof. $\qquad\qquad\qquad\qquad\qquad\qquad\qquad\qquad\qquad\qquad\qquad\qquad\qquad\qquad\quad \square$

Note that we have used the upper semicontinuity of $F$ in (11.3.3). The so-called property $Q$ of Cesari, a much milder condition, could have replaced upper semicontinuity [16, p. 7]. Condition (iv) could be replaced by the closure and convexity of $F(t, \varphi)$.

The existence of an optimal control for (11.3.1) requires a controllability assumption. This concept is defined as follows:

**Definition 11.3.4**    Let $z_t \in C([-h, 0], E^n) = C$ be a target point function that is time varying. Suppose $z_T \in \mathcal{A}(T)$ for some $T > \sigma$, then the system is controllable to the target. Thus for each $\varphi \in C$ there exists a $T \geq \sigma$ and an admissible control $u(t) \in U(t, x_t)$, $t \in [\sigma, T]$, such that the solution of (11.3.1) satisfies $x_\sigma(\sigma, \varphi, u) = \varphi$, $x_T(\sigma, \varphi, u) = z_T$. In this case we also say that system (11.3.1) is function space $z$-controllable with constraints.

**Definition 11.3.5**    System (11.3.1) is null controllable with constraints if for each $\varphi \in C$, there exists a $t_1 < \infty$ an admissible control $u(t) \in U(t, x_t)$, $t \in [\sigma, t_1]$ such that the solution of (11.3.1) satisfies

$$x_\sigma(\sigma, \varphi, u) = \varphi, \qquad x_{t_1}(\sigma, \varphi, u) = 0.$$

The main result of this section is now stated.

**Theorem 11.3.2**    *In (11.3.1), we assume that:*

   (i)  $f : E \times C([-h, 0], E^n) \times E^m \to E^n$ *is continuously differentiable.*
  (ii)  $D$ *is uniformly stable and satisfies (11.2.1) and (11.2.2).*
 (iii)  *For each measurable* $y(t)$ *satisfying* $y(t) \in F(t, x_t)$, *we have* $|y(t)| \leq m(t)$       *a.e.,* $m \in L_1([\sigma, T], E]$.
 (iv)  $F(t, \varphi)$ *is closed and convex for each* $t, \varphi$.
  (v)  $U(t, \varphi)$ *is upper semicontinuous with respect to inclusion.*
 (vi)  *System (11.3.1) is function space $z$-controllable with constraints.*

*Then there exists time-optimal control for (11.3.1).*

**Proof**    By assumption, there exists some $t_1 \leq T$ such that

$$z_{t_1} = x_{t_1} \in \mathcal{A}(t_1),$$

where $\mathcal{A}(t_1)$ is the attainable set of (11.3.1) at time $t_1$. Let

$$t^* = \inf\{t : z_t \in \mathcal{A}(t)\}.$$

It follows that $\sigma \leq t^* \leq t_1 \leq T$. There is a sequence of nonincreasing times $t_n \in [\sigma, t_1]$ converging to $t^*$ and a sequence $u^n$ of admissible controls, $u^n(t) \in U(t, x_t^n)$, such that

$$z_{t_n} = x_{t_n}(\sigma, \varphi, u^n) \in \mathcal{A}(t_n).$$

Let $x(\sigma, \varphi, u^n) = x^n$. Then

$$\frac{d}{dt}[x^n(t_n) - g(t_n, x_{t_n}^n)] \in F(t_n, x_{t_n}^n)x_\sigma^n = \varphi, \quad (t_n, x_{t_n}^n) \in M.$$

Since $z_t$ is continuous and $t_n \to t_*$ as $n \to \infty$, we have

$$z_{t_n} \to z_{t_*}.$$

Also

$$x_{t_n}^n \to x_{t_*} \quad \text{as} \quad n \to \infty.$$

By condition (iii),

$$\left|\frac{d}{dt}[x^n(t_n) - g(t_n, x_{t_n}^n)]\right| \leq m(t_n) \text{ a.e. where } m \in L_1([\sigma, T], E).$$

Let

$$y_{(t)}^n \equiv x^n(t) - g(t, x_t^n).$$

The set $H = \{y^n : y^n(t) = x^n(t) - g(t, x_t^n)\}$ is an equicontinuous family. Indeed, for every measurable subset $E$ of $[\sigma, T]$ we have

$$\int_E \left|\frac{d}{dt}y^n(t)\right| dt = \int_E \left|\frac{d}{dt}[x^n(t) - g(t, x_t^n)]\right| ds \leq \int_E m(t)dt.$$

Let $\epsilon > 0$. By integrability of $m$ we conclude that there is a $\delta > 0$ such that if $\text{meas}(E) < \delta$, then $\int_E m(t)dt < \epsilon$. For this $E$ we have

$$\int_E \left|\frac{d}{dt}y^n(t)\right| dt < \epsilon,$$

so that the functions $\frac{d}{dt}y^n(t)$, $y^n \in H$ are equi-absolutely integrable. It now follows that $H$ is an equi-absolutely continuous set of functions.

Since $M$ is bounded and $g(M)$ bounded, we readily deduce that $H$ is bounded. By Ascoli's Theorem there is a subsequence, which we again call $\{y_n\}$ and which converges in the metric form of $C([\sigma, T], E^n)$ to a function $y \in C([\sigma, T], E^n)$, that is absolutely continuous on $[\sigma, T]$. By condition (iii), we deduce $|y^n(t) - y^n(\sigma)| \leq \int_\sigma^t m(s)ds$, $|y^n(t) - y(\bar{t})| \leq \int_{\bar{t}}^t m(s)ds$, so that

$$\int_{E_i} \frac{d}{dt}[y^n(s)]ds \to 0 \quad \text{as} \quad n \to \infty$$

uniformly with respect to each decreasing sequence $\{E_i\}$, $E_i \subseteq [\sigma, T] \equiv I$ with void intersection. Therefore [15, p. 292] there is a sequence $\{\dot{y}^n\}$ (we retain the same notation) weakly convergent in $L_1(I, E^n)$ to a function $\xi \in L_1(I, E^n)$. Since the functions $y^n$ converge uniformly on $[\sigma, T]$ to $y$, certainly they converge pointwise almost everywhere to $y$ and hence by absolute continuity of these functions we have $\dot{y}(t) = \xi(t)$ a.e. Thus, for $t_n \in I$, $t_n \to t_*$,

$$x(t_*) - g(t_*, x_{t_*}) = \lim_{n \to \infty} [x^n(t_n) - g(t_n, x_{t_n}^n)]$$

$$= \lim_{n \to \infty} \left\{ [x^n(\sigma) - g(\sigma, x_\sigma^n)] + \int_\sigma^{t_n} \frac{d}{dt}[y^n(s)]ds \right\}$$

$$= \varphi(\sigma) - g(\sigma, \varphi) + \int_\sigma^{t_*} \xi(s)ds.$$

We note $x_\sigma^n = \varphi$ for each $n$, so that $\lim_{n \to \infty} x_\sigma^n = \varphi$. Because $M$ is closed, $(t_n, x_{t_n}^n) \to (t_*, x_{t_*}) \in M$ as $n \to \infty$.

It now follows from [15, p. 422] that there is a sequence $\{\xi_n\}$ of convex combinations of $\{\dot{y}^n, \dot{y}^{n+1}, \ldots\}$ that converges in $L_1(I, E^n)$ norm to $\xi$. From this sequence $\{\xi_n\}$, select a subsequence which converges a.e. to $\xi$. Thus, almost everywhere on $I$,

$$\xi(t_*) \in \bigcap_{n=1}^{\infty} \overline{\text{co}} \left( \bigcup_{k=n}^{\infty} \dot{y}^k(t_*) \right)$$

$$\subseteq \bigcap_{n=1}^{\infty} \overline{\text{co}} \left( \bigcup_{k=n}^{\infty} F(t_*, x_{t_*}^k) \right)$$

$$\subseteq F(t_*, x_{t_*}) \tag{11.3.4}$$

where $\text{co}(N)$ is the closed convex hull of $N$ in $E^n$. Hence

$$\frac{d}{dt}[x(t_*) - g(t_*, x_{t_*})] \in F(t_*, x_{t_*}).$$

We have already seen that $x_\sigma = \varphi$, $(t_*, x_{t_*}) \in M$, so that

$$z_{t_*} = x_{t_*} \in \mathcal{A}(t_*).$$

This completes the proof. $\qquad\qquad\qquad\qquad\qquad\qquad\qquad\qquad\qquad\qquad\qquad$ $\square$

We observe that we used the upper semicontinuity of $F$ in (11.3.4). It could be replaced by using the weakened version of Cesari's property $Q$. See Angell [16, p. 7].

## 11.4 Optimal Control of Neutral Systems in Function Space

In this section we present a maximum principle for optimal control of nonlinear systems of neutral type with function space initial and terminal conditions. In our setting we guarantee the existence of regular nontrivial multipliers when the controls are constrained to lie on a unit $m$-dimensional cube. This problem was broached in [2] and [26], but not completely resolved. In [26] the nontriviality of the multipliers could not be guaranteed and pointwise control constraints were ruled out. This section continues the exciting recent work of Angell and Kirsch [25] on delay equations, which was reported in Sec. 7.3. We maintain the basic definitions and notation of the corresponding treatment in (7.11.2), and consider the following problem: Find a control $u$ that minimizes

$$J(x, u) = \int_0^T f^0(t, x_t, u(t))dt, \tag{11.4.1}$$

subject to the constraints

$$\dot{x}(t) - \sum_{i=1}^N A_{-1i}(t)\dot{x}(t - hi) = f(t, x_t, u(t)), \tag{11.4.2}$$

$$x_0 = \phi, \qquad x_T = \psi, \tag{11.4.3}$$

$$u(t) \in C^m \quad \text{a.e. on } [0, T], \tag{11.4.4}$$

where

$$C^m = \{u \in E^m : |u_j| \le 1, \ j = 1, \ldots, m\}, \tag{11.4.5}$$

$$f_i^1(t, x_t) \le 0, \quad i = 1, \ldots, p, \quad t \in [0, T], \tag{11.4.6}$$

and

$$x \in C([-h, T], E^n), \qquad \dot{x}(t) - \sum_{i=1}^N A_{-1i}(t)\dot{x}(t - h_i) \in L_\infty([0, T], E^n),$$

$$u \in L_\infty([0, T]), C^m). \tag{11.4.7}$$

We assume further that the initial data $\phi$ and the terminal function $\psi$ satisfy

$$\phi \in C([-h, 0], E^n), \qquad \psi \in C([-h, 0], E^n),$$

with

$$D(T)\psi \in C^1([-h, 0], E^n). \tag{11.4.8}$$

We observe that if the admissible control $u$ is such that $u_T$ is continuous, then by (11.4.2) $\dot{x}(t) - \sum_{j=1}^N A_j(t)\dot{x}(t - h_j)$ is continuous on $(T - h, T)$, since we have assumed the same properties of $f^0$, $f$, $f^1$, as are in (11.7.3). Thus the mappings

$$f^0 : [0,T] \times C \times E^m \to E \,,$$

$$f : [0,T] \times C \times E^m \to E^n \,,$$

$$f^1 : [0,T] \times C \to E^p$$

are continuous; $f$, $f^0$ are Fréchet differentiable with respect to the second argument and continuously differentiable with respect to its third argument, the derivatives being assumed continuous with respect to all arguments. Also, $f^1 : [0,T] \times C \to E^p$ is continuously Fréchet differentiable with respect to its second argument. We associate with (11.4.2) the linear equation

$$\dot{x}(t) - \sum_{j=1}^{N} A_j \dot{x}(t - h_j) = L(t, x_t) + B(t)u(t) \,, \qquad x_0 = \phi \,, \tag{11.4.9}$$

where $u \in U$, $x \in X_1$; and these subspaces are defined as follows:

$$X_1 = \{x \in C([-h,T], E^n) : D(T)x_T \in C^1([0,T], E^m)\} \,, \tag{11.4.10}$$

$$U = \{u \in L_\infty([0,T], E^m) : u_T \in C([-h,0], E^m)\} \,, \tag{11.4.11}$$

$$Y_1 = C([0,T], E^p) \,, \tag{11.4.12}$$

$$Y_2 = \{(z, u, \beta) \in L_\infty([0,T], E^n) \times C([-h,0], E^n)$$

$$\times \, E^n : z_T \in C([-h,0], E^n)\} \,. \tag{11.4.13}$$

In (11.4.9),

$$L(t, x_t) = D_2 f(t, x_t^*, u^*(t)]x_t \,,$$
$$B(t)v = D_3 f(t, x_t^*, u^*(t))v \,, \tag{11.4.14}$$

where $D_i f(t, x_t^*, u^*)$ is the Fréchet derivative with respect to the $i$th argument at $(t, x_t^*, u^*(t))$ and $x_t^*, u^*(t)$ are the optimal trajectory and optimal control respectively.

We introduce the following maps:

$$g^0 : X_1 \times U \to E \,,$$

$$g^1 : X_1 \times U \to Y_1 \,,$$

$$g^2 : X_1 \times U \to Y_2 \,,$$

which are defined as follows:

$$g^0(x, u) = \int_0^T f^0(t, x_t, u(t))dt \,, \tag{11.4.15}$$

$$g'(x, u) = f(\cdot, x_{(\cdot)}) \,, \tag{11.4.16}$$

$$g^2(x, u) = \left( \dot{x}(\cdot) - \sum_{j=1}^{N} A_j(\cdot)\dot{x}(\cdot - h_j) - f(\cdot, x_{(\cdot)}, u(\cdot)) \right.$$

$$- \left( \dot{\psi}(0) - \sum_{j=1}^{N} A_j(T)\dot{\psi}(-h_j) \right)$$

$$\left. + \dot{x}(T) - \sum_{j=1}^{N} A_j(T)\dot{x}(T - h_j), x(T) - \psi(T) \right). \qquad (11.4.17)$$

Observe that if $g^2(x, u) = 0$, then

$$\dot{x}(\cdot) - \sum_{j=1}^{N} A_j(\cdot)\dot{x}(\cdot - h_j) = f(\cdot, x_{(\cdot)}, u(\cdot)),$$

$$\dot{x}(T) - \sum_{j=1}^{N} A_j(T)\dot{x}(T - h_j) = \dot{\psi}(0) - \sum_{j=1}^{N} A_j(T)\psi(-h_j), \quad x(T) = \psi(T).$$

$$(11.4.18)$$

Because of our previous assumptions on $f$, $f^0$, $f^1$, the functions $g^0$, $g^1$, $g^2$ are continuously differentiable at each $(x^*, u^*) \in X_1 \times U$. Indeed,

$$D_1 g^0(x^*, u^*) = \int_0^T D_2 f^0(t, x_t^*, u^*(t))x_t dt, \qquad x \in X_1, \qquad (11.4.19)$$

$$D_2 g^0(x^*, u^*)u = \int_0^T D_3 f^0(t, x_t^*, u^*(t))u(t)dt, \qquad u \in U, \qquad (11.4.20)$$

$$D_1 g^1(x^*, u^*)y = D_2 f'(t, x_t^*)y, \qquad y \in C, \qquad (11.4.21)$$

$$D_2 g^1(x^*, u^*) = 0. \qquad (11.4.22)$$

Also

$$D_1 g^2(x^*, u^*)x = \dot{x}(\cdot) - \sum_{j} A_j(\cdot)\dot{x}(\cdot - h_j) - D_2 f(\cdot, x_{(\cdot)}^*, u^*(\cdot))x_{(\cdot)},$$

$$\times \dot{x}(T) - \sum_{j=1}^{N} A_j(T)\dot{x}(T - h_j), x(T)), \quad \text{where } x \in X_1, \ (11.4.23)$$

$$D_2 g(x^*, u^*)v(t) = D_3 f(t, x_t, u(t))v(t), \quad v \in U. \qquad (11.4.24)$$

Because of the validity of the Riesz Representation Theorem, there exist

$$\eta_1(t,\cdot) \in NBV_{p \times n}([-h,0]), \quad \eta_0(t,\cdot) \in NBV_{n \times n}([-h,0]),$$

$$\eta(t,\cdot) \in NBV_{n \times n}([-h,0]),$$

such that

$$L_0(t)\phi = D_2 f^0(t, x_t^*, u^*(t)) = \int_{-h}^{0} d_\theta \eta_\theta(t,\theta)\phi(\theta), \tag{11.4.25}$$

$$L_1(t)\phi = D_2 f^1(t, x_t^*, u^*(t))\phi = \int_{-h}^{0} d_\theta \eta_1(t,\theta)\phi(\theta), \tag{11.4.26}$$

$$L(t)\phi = D_2 f(t, x_t^*, u^*(t))\phi = \int_{-h}^{0} d_\theta \eta(t,\theta)\phi(\theta). \tag{11.4.27}$$

In all these representations, $\eta$, $\eta_i$, $i = 0, 1$ are measurable. We extend these functions by defining

$$\eta(t,\theta) = 0 \quad \text{for} \quad t > T.$$

To match up our notation in the statement of the multiplier rule in Theorem 7.11.2, we set

$$X = X_1 \times U, \qquad W = \{(x,u) \in X : x_0 = \phi\},$$

$$V = \{(x,u) \in W : u \in U_{\text{ad}}\}, \tag{11.4.28}$$

$$U_{\text{ad}} = \{u \in U : u(t) \in C^m \text{ a.e. on } [0,T]\},$$

where

$$C^m = \{u \in E^m : |u_j| \le 1, \, j = 1,\ldots,m\}.$$

Note that $U_{\text{ad}}$ has a nonempty interior relative to $U$.

    We now state a local maximum principle.

**Theorem 11.4.1**    *Let $u^*$ be the optimal solution of (11.4.1)–(11.4.8) and $x_t^*$ $0 \le t \le T$ the corresponding optimal solution. Assume that $Dg^2(x^*, u^*) = D_1 g^2(x^*, u^*) + D_2 g^2(x^*, u^*)$ has a closed range in $Y_2$. Then there exists*

$$(\lambda, a, \rho, v, q) \in E \times L_\infty([0,T]) \times NBV_p([0,T]) \times NBV_n([0,T]) \times E^n,$$

$$(\lambda, a, \rho, v, q) \not\equiv (0,0,0,0,0)$$

*such that*

$$\lambda \ge 0, \quad \rho = (\rho_1,\ldots,\rho_p), \; \rho_j \text{ is nondecreasing in } [0,T] \tag{11.4.29}$$

and $\rho_j$ is constant on every interval where

$$f_j'(t, x_t^*) < 0\,; \tag{11.4.30}$$

$$a(t) = \lambda \int_t^T \eta_0(\theta, t - \theta)d\theta + \int_t^T \eta_1(\theta, t - \theta)d\rho(\theta) - q$$

$$- \sum_{k=1}^N a(t + h_k)A_{-1k}(t + h_k) + \int_t^{T^+} \eta_{-1}(\theta, t - \theta)dv(\theta - T)$$

$$- \int_t^T \eta(\theta, t - \theta)a(\theta)d\theta + \int_t^T \eta(\theta, t - \theta)dv(\theta - T) \text{ for all } t \in [0, T]. \tag{11.4.31}$$

*Also*

$$-\lambda \int_0^T B_0(t)u(t)dt + \int_t^T a(t)B(t)u(t)dt - \int_{T-h}^T dv(t - h)B(t)u(t)$$

$$\leq -\lambda \int_0^T B_0(t)u^*(t)dt + \int_0^T a(t)B(t)u^*(t)dt$$

$$- \int_{T-h}^T dv(t - h)B(t)u^*(t)\,, \tag{11.4.32}$$

*for all $u \in U_{\mathrm{ad}}$.*

**Corollary 11.4.1**  *In addition to all the conditions of Theorem* 11.4.1, *assume that the mapping*

$$Dg(x^*, u^*) : (W - (x^*, u^*)) \to Y_2$$

*is a surjection. Then we can take the multiplier $\lambda = 1$.*

**Proof**  Carefully note that the functions $g^0$, $g^1$, $g^2$ of the Multiplier Rule of Theorem 7.11.2 satisfy the same assumptions of the mappings of (11.4.15)–(11.4.17) and the spaces (11.4.10)–(11.4.13). We are guaranteed the existence of the multipliers of Theorem 11.7.32. Thus if we denote the space $\mathcal{L}$ by

$$\mathcal{L} = \{x \in L_\infty([0, T], E^n) : x_T \in C([-h, 0], E^n)\}\,,$$

and its dual by $\mathcal{L}^*$, we have the existence of $\ell \in \mathcal{L}^*$, $v \in NBV_n([-h, 0])$, $q \in E^n$, $\rho \in NBV_p([0, T])$, and $\lambda \geq 0$, $(\lambda, \ell, v, q, \rho) \neq (0, 0, 0, 0, 0)$ such that $\rho$ is nondecreasing and

$$\int_0^T d\rho(t)f^1(t, x^*(t)) = 0\,, \tag{11.4.33}$$

$$\lambda \int_0^T L_0(t)x_t dt + \int_0^T d\rho(t)L_1(t)x_t + \ell\left(\dot{x}(\cdot) - \sum_{j=1}^N A_j(\cdot)\dot{x}(\cdot) - L(\cdot)x(\cdot)\right)$$

$$+ \int_{T-h}^T dv(t-T)\left[\dot{x}(t) - \sum_{j=1}^N A_j(t)x^*(t)\right] + qx(T) = 0, \qquad (11.4.34)$$

for all $x \in X_1$, $x_0 = 0$; and

$$\lambda \int_0^T B_0(t)u(t)dt + \ell[-B(\cdot)(u(\cdot) - u^*(\cdot))] \geq 0 \qquad (11.4.35)$$

for all $u \in U_{\text{ad}}$. Recall the definition of $\eta$, $\eta_i$ and their extension outside their intervals of definition, and extend the definition of $v$ by constancy and continuity outside its original interval of definition. □

We need an expression for the linear functional $\ell$. It is contained in the next lemma.

**Lemma 11.4.1**   *The linear functional $\ell$ has the form*

$$\ell(z) = \int_0^T a(t)z(t)dt - \int_{T-h}^T dv(t-T)z(t), \qquad (11.4.36)$$

*for all $z \in \mathcal{L}$ and $a \in L_\infty([0,T], E^n)$.*

**Proof**   Consider (11.4.34). Let $z \in \mathcal{L}$ and consider the equation

$$\dot{x}(t) - \sum_{j=1}^N A_{-1j}(t)x(t-h_j) - L(t)x_t = z(t) \quad \text{in} \quad [0,T], \quad x_0 = 0. \qquad (11.4.37)$$

The variation of the constant formula of Hale and Meyer [27] gives the solution as

$$x(t,z) = \int_0^T X(t,s)z(s)ds, \qquad t \in [0,T], \qquad (11.4.38)$$

where $X(t,s) \in L_\infty([0,T], E^{n2})$ is the fundamental matrix defined by

$$X(t,s) = \frac{\partial W(t,s)}{\partial s} \quad \text{a.e. in } s.$$

(a)  $W_s(\cdot, s) = 0$.
(b)  $W(t,s) = \sum_{j=1}^N A(t)W(t-s-h_j) + \int_s^t L(\lambda, W\lambda(\cdot,s))d\lambda - (t-s)I, 0 \leq s \leq t$.

With this expression, (11.4.34) takes the form

$$\lambda \int_0^T L_0(t)x_t(\cdot, z)dt + \int_0^T d\rho(t)L_1(t)x_t(\cdot, z) + \ell(z) + \int_{T-h}^T dv(t-T)z(t)$$

$$+ \int_{T-h}^T dv(t-T)L(t)x_t(\cdot, z) + qx(T; z) = 0 \qquad (11.4.39)$$

for all $z \in \mathcal{L}$. But then $L_0$, $L_1$, $L$, are as given in (11.4.25)–(11.4.27), so that (11.4.34) yields

$$\ell(z) = -\lambda \int_0^T \left( \int_{-h}^0 d_s\eta_0(t,s)x(t+s,z) \right) dt$$

$$- \int_0^T d\rho(t) \int_{-h}^0 d_s\eta_1(t,s)x(t+s,z) - \int_{T-h}^T dv(t-T)z(t)$$

$$- \int_{T-h}^T dv(t-T) \int_{-h}^0 d_s\eta(t,s)x(t+s,z) - qx(T). \qquad (11.4.40)$$

We now substitute the expression $x(t;z)$ in (11.4.38) into (11.4.40), and change the order of integration to deduce (11.4.36), where $a \in \mathcal{L}$ depends on $\lambda$, $q$, $X$, $v$, $\eta_0$, $\eta_1$, $\eta$. With (11.4.36) established, we rewrite (11.4.39) as follows:

$$\lambda \int_0^T L_0(t)x_t dt + \int_0^T d\rho(t)L_1(t)x_t$$

$$+ \int_0^T a(t) \left[ \dot{x}(t) - \sum_{j=1}^N A_{-1j}(t)\dot{x}(t-h_j) - L(t)x_t \right] dt$$

$$- \int_{T-h}^T dr(t-T) \left[ \dot{x}(t) - \sum_{j=1}^N A_{-1j}(t)\dot{x}(t-h_j) - L(t)x_t \right]$$

$$+ \int_{T-h}^T dr(t-T) \left[ \dot{x}(t) - \sum_{j=1}^N A_{-1j}(t)\dot{x}(t-h_j) \right] + qx(T) = 0,$$

for all $x_1 \in X_1$. Once more we use the expressions for $\eta_0$, $\eta_1$, $\eta$ to deduce that

$$\lambda \int_0^T \int_{t-h}^t d_\theta\eta_0(t,\theta-t)x(\theta)dt + \int_0^T \int_{t-h}^t d\rho(t)d_\theta\eta_1(t,\theta-t)x(\theta)$$

$$+ \int_0^T a(t) \left[ \dot{x}(t) - \sum_{j=1}^N A_{-1j}(t)\dot{x}(t-h_j) - \int_{t-h}^t d_\theta\eta(t,\theta-t)x(\theta) \right] dt$$

$$+ \int_{T-h}^T \int_{t-h}^t dv(t-T)d_\theta\eta(t,\theta-t)x(\theta) + qx(T) = 0,$$

for all $x \in X_1$ with $x_0 = 0$.

We consider each of the first four terms above, beginning with

$$\lambda \int_0^T \int_{t-h}^t d_\theta\eta_0(t,\theta-t)x(\theta)dt,$$

and ending with

$$\int_{T-h}^{T} \int_{t-h}^{t} dv(t-T)d_\theta \eta_0(t,\theta-t)dt\,,$$

and integrate by parts:

$$-\lambda \int_0^T \eta_0(t,-h)x(t-h)dt - \lambda \int_0^T \int_{t-h}^t \eta_0(t,s-t)\dot{x}(s)dsdt$$

$$-\int_0^T d\rho(t)\eta_1(t,-h)x(t-h) - \int_0^T d\rho(t)\eta_1(t,s-t)\dot{x}(s)ds$$

$$+\int_0^T a(t)\left[\dot{x}(t) - \sum_{j=1}^N A_{-1j}(t)\dot{x}(t-h_j)\right]dt + \int_0^T a(t)\eta(t,-h)x(t-h)dt$$

$$-\int_0^T a(t)\left[\dot{x}(t) - \sum_{j=1}^N A_{-1j}(t)\dot{x}(t-h_j)\right]dt$$

$$+\int_{T-h}^T dv(t-T)\eta(t,-h)x(t-h)$$

$$+qx(T) - \int_{T-h}^T \int_{t-h}^t dv(t-T)\eta(t,s-t)\dot{x}(s)ds = 0\,,$$

for all $x \in X_1$ with $x_0 = 0$. As in Angell and Kirsch [25], use the following abbreviation:

$$\mu(t) = \begin{cases} -\displaystyle\int_t^T a(s)ds\,, & 0 \le t \le T-h\,, \\[2ex] -v(t-T) - \displaystyle\int_t^T a(s)ds\,, & T-h \le t \le T\,. \end{cases}$$

For any $y \in \mathcal{L}$ that is extended by zero, set

$$x(t) = \int_0^t y(s)ds \quad \text{for} \quad t \in E\,.$$

Use this above and change the order of integration:

$$-\lambda \int_0^T \int_s^{T-h} \eta_0(\theta+h,-h)d\theta y(s)ds - \lambda \int_0^T \int_s^{s+h} \eta_0(\theta,s-\theta)d\theta y(s)ds$$

$$-\int_0^T \int_s^{T-s} d\rho(\theta)\eta_1(\theta+h,-h)y(s)ds - \int_0^T \int_t^{s+h} d\rho(\theta)\eta_1(\theta,s-\theta)d\theta y(s)ds$$

$$+ \int_0^T a(t)y(t)dt + \int_0^T \int_s^{s+h} d\mu(\theta)\eta(\theta, s - \theta)y(s)ds$$

$$+ \int_0^T \int_s^{T-h} d\mu(\theta + h)\eta(\theta + h, -h)y(s)ds$$

$$+ q \int_0^T y(s)ds + \int_{-h_k}^{T-h_k} a(t) \sum_{k=1}^N A_{-1k}(t)\dot{x}(t + h_k)dt = 0$$

for all $y \in \mathcal{L}$. Note that

$$\sum_{k=1}^N \int_{-h_k}^{T-h_k} a(t)A_{-1k}(t)\dot{x}(t - h_k)dt = \sum_{k=1}^N \int_{-h_k}^{T-h_k} a(s + h_k)A_k(s + h_k)y(s)ds.$$

Because the integrand above must vanish pointwise, we have for $s \in [0, T]$,

$$-\lambda \int_s^{T-s} \eta_0(\theta + h, -h)d\theta - \lambda \int_0^{s+h} \eta_0(\theta, s - \theta)d\theta$$

$$- \int_s^{T-h} d\rho(\theta)\eta_1(\theta + h, -h) - \int_0^{s+h} d\rho(\theta)\eta_1(\theta, s - \theta)$$

$$+ a(s) + \int_s^{s+h} d\mu(\theta)\eta(\theta, s - \theta) + \int_s^{T-h} d\mu(\theta + h)\eta(\theta + h, -h) + q$$

$$+ \sum_{k=1}^N a(s + h_k)A_k(s + h_k) = 0,$$

which yields

$$a(s) = \lambda \int_s^T \eta_0(\theta, s - \theta)d\theta + \int_s^T d\rho(\theta)\eta_1(\theta, s - \theta)$$

$$- \int_s^T a(\theta)\eta(\theta, s - \theta)d\theta - \int_s^T dv(\theta)\eta(\theta, s - \theta)$$

$$- q - \sum_{k=1}^N a_k(s + h_k)A_{-1k}(s + h_k).$$

We have proved that (11.4.31) holds. To prove (11.4.32), we insert $\ell(z)$ in (11.4.36) into (11.4.35):

$$\lambda \int_0^T B_0(t)[u(t) - u^*(t)]dt + \int_0^t a(t)[-B(t)(u(t) - u^*(t)]dt]$$

$$- \int_0^T dv(t - T)[-B(t)(u(t) - u^*(t))] \geq 0$$

for all $u \in U_{\mathrm{ad}}$. The result follows at once, since $\rho$ is monotonic. Note that $(\lambda, a, \rho, v, q)$ do not vanish simultaneously, since otherwise $(\lambda, \ell, v, q, \rho) \equiv (0, 0, 0, 0, 0)$. The proof of Corollary 11.4.1 and Theorem 11.4.2 follows as in [25].     $\square$

**Remark**    The properties of the range of $Dg^2$ are conjectured as follows:

*Closure*: Proposition 1.1 of Sec. 12.1.
*Surjectivity*: Controllability of the linear system

$$\dot{x}(t) - \sum_{j=1}^{N} A_j(t)\dot{x}(t - h_j) = L(t, x_t) + B(t)u(t)$$

in $W_\infty^{(1)}$. See for example Theorem 10.8.4.

    We now formulate the necessary conditions of the time-optimal control problem.

**Theorem 11.4.2**    *Consider the following problem: Minimize $T$ subject to the constraints*

$$\dot{x}(t) - \sum_{i=1}^{N} A_{-1i}(t)\dot{x}(t - h_i) = f(t, x_t, u(t)), \tag{11.4.2}$$

$$x_0 = \phi \in W_\infty^{(1)}([-h, 0], E^n), \quad x_T = \psi, \quad D^{(T)}\psi \in C^1([-h, 0], E^n), \tag{11.4.3}$$

$$u(t) \in C^m \text{ a.e. on } [0, T], \tag{11.4.4}$$

*where*

$$C^m = \{u \in E^m : |u_j| \leq 1, \ j = 1, \ldots, m\}.$$

(i) *We assume that $f$ is continuous, and is Fréchet differentiable with respect to its second argument and continuously differentiable with respect to its third argument, the derivative being continuous with respect to all arguments. Consider the linear approximation (11.4.9) of (11.4.2), and assume that:*
(ii) *This system*

$$\dot{x}(t) - \sum_{j=1}^{N} A_j\dot{x}(t - h_j) = L(t, x_t) + B(t)u(t), \qquad x_0 = 0, \tag{11.4.9}$$

*with attainable set $\mathcal{A}$ defined by*

$$\mathcal{A} = \{x_T(u) \in C([-h, 0], E^n) \ Dx_T \in C'([-h, 0], E^n) : u \in \mathcal{U}\}$$

*where*

$$\mathcal{U} = \{u \in L_\infty([0,T], E^m) : u_T \in C([-h,0], E^n)\}$$

*is such that*

$$\mathcal{A} = \{\psi \in C([-h,0]) : D\psi \in C'([-h,0], E^n)\}.$$

*Then there exists*

$$(a,v,q) \in L_\infty([0,T], E^n) \times NBV_n([0,T]) \times E^n,$$

$$(a,v,q) \neq (0,0,0),$$

*such that*

$$a(t) = -q - \sum_{k=1}^N a(t+h_k)A_{-1k}(t+h_k) - \int_t^T \eta(s,t-s)a(s)ds$$

$$+ \int_t^T \eta(s,t-s)dv(s-T)$$

*for all* $t \in [0,T]$, *and*

$$\int_0^T a(t)B(t)u(t)dt - \int_{T-h}^T dv(t-T)B(t)u(t)$$

$$\leq \int_0^T a(t)B(t)u^*(t) - \int_{T-h}^T dv(t-T)B(t)u^*(t)$$

*for all* $u \in U_{\mathrm{ad}}$.

**Proof**  We need to prove that the attainable set $\mathcal{A}$ is all of $\mathcal{D} = \{\psi \in C([-h,0]) : D\psi \in C'([-h,0], E^n)\}$ if and only if the required map

$$Dg(x^*, u^*) : (W - (x^*, u^*)) = Y_2.$$

Suppose that

$$\mathcal{A} = \{\psi \in C([-h,0]) : D\psi \in C'([-h,0])\},$$

and fix a point $(z,v,p) \in Y_2$. We shall prove that there exists a pair $(x,u) \in X_1 \times \mathcal{U}$ such that if $\hat{x} = x - x^*$ and $\hat{u} = u - u^*$, then

$$\dot{\hat{x}}(t) - \sum_{j=1}^N A_{-1j}(t)\dot{\hat{x}}(t-h_j) - L(t,\hat{x}_t) - B\hat{u}(t) = z(t),$$

$$D\dot{\hat{x}}_T = v \quad \text{and} \quad \hat{x}(T) = p.$$

Now define $\psi \in C([-h,0], E^n)$ with $D\psi \in C'([T-h,T], E^n)$ by $\psi(t) = p - \int_t^T v(s)ds$ $T - h \leq t \leq T$. By assumption, there exists a control $\overline{u}$ such that

$$x(t; 0, \overline{u}) = \psi(t) - \int_0^T U(t,s)z(s)ds, \quad T - h \leq t \leq T.$$

It follows from the variation of constant formula in Proposition 2.3.3 that

$$x(t; z, \overline{u}) = x(t, 0, \overline{u}) + \int_0^t U(t,s)z(s)dt.$$

The required choice is the triplet $\hat{x} = (x; z, \overline{u})$, $\hat{u}(t) = \overline{u}(t)$. To see that the converse is also valid, we note that if $\psi \in C([-h,0], E^n)$, with $D\psi \in C'([T-h,T], E^n)$, then the fact that $Dg$ is surjective implies there exists a pair $(x, u)$ that is the preimage of the triple $(0, \psi, \psi(T))$. We conclude that (ii) holds if and only if the required map $D_2 g$ is surjective. But this is equivalent to the controllability of the linear system (11.4.9) with controls in $\mathcal{U}$. Theorem 11.4.1 and its corollary can therefore be invoked to conclude the proof. $\qquad\qquad\square$

**Remark 11.4.1**    For autonomous systems, necessary and sufficient conditions for exact controllability on the interval $[0, T]$ are given by Salamon [5, p. 155]. There we can identify our interval to be $[0, T-h]$ with controls $u \in L_\infty([0, T-h], E^m)$ and state space $W'_\infty([-h, 0], E^n)$. On the interval $[T - h, T]$ one can use a control that is continuous. Thus it can easily be proved that in the autonomous case Salamon's conditions suffice for the surjectivity conditions.

## References


1. G. A. Kent, *Optimal Control of Functional Differential Equations of Neutral Type*, Ph.D. Thesis, Brown University, 1971.
2. H. T. Banks and G. A. Kent, "Control of Functional Differential Equations to Target Sets in Function Space," *SIAM J. Control* **10** (1972) 567–593.
3. R. Gabasov and F. Kirillova, *The Qualitative Theory of Optimal Processes*, Marcel Dekker, New York, 1976.
4. H. R. Rodas and C. E. Langenhop, "A Sufficient Condition for Function Space Controllability of a Linear Neutral System," *SIAM J. Control and Optimization* **16** (1978) 429–435.
5. D. Salamon, *Control and Observation of Neutral Systems*, Pitman Advanced Publishing Program, Boston, 1984.
6. E. N. Chukwu, "The Time Optimal Control Problem of Linear Neutral Functional Systems," *J. of Nigerian Mathematics Society* **1** (1982) 39–55.
7. E. N. Chukwu, *The Time Optimal Control Theory of Linear Differential Equations of Neutral Type*, Proceedings, Second Bellman Continuum, Georgia Institute of Technology, Atlanta, Georgia, June 24, 1986.
8. M. Cruz and J. K. Hale, "Existence, Uniqueness and Continuous Dependence for Hereditary Systems," *Annali Mathematica Pura ed Applicata* **85** (1970) 63–82.
9. W. R. Melvin, "A Class of Neutral Functional Differential Equations," *J. Differential Equations* **12** (1972) 524–534.

10. M. A. Cruz and J. K. Hale, "Asymptotic Behavior of Neutral Functional Differential Equations," *Archives of Rational Mechanics and Analysis* **34** (1969) 331–353.

11. W. Rudin, *Real and Complex Analysis*, McGraw-Hill, New York, 1974.

12. J. Dieudonne, *Foundations of Modern Analysis*, Academic Press, New York, 1969.

13. M. A. Cruz and J. H. Hale, "Stability of Functional Differential Equations of Neutral Type," *J. Differential Equations* **7** (1970) 334–355.

14. K. Kuratowski and C. Ryll-Nardzewski, "A General Theorem on Selectors," *Bull. Acad. Polon. Sci.* **12** (1965) 397–403.

15. N. Dunford and J. T. Schwartz, *Linear Operators, Part I, General Theory*, Interscience, New York, 1958.

16. T. S. Angell, "Existence Theorem for Hereditary Lagrange and Mayer Problems of Optimal Control," *SIAM J. Control and Optimization* **14** (1976) 1–18.

17. S. Lang, *Analysis II*, Addison-Wesley, Reading, MA, 1969.

18. E. N. Chukwu, "An Estimate for the Solutions of a Certain Functional Differential Equation of Neutral Type," *Proceedings of the International Conference on Nonlinear Phenomena in Mathematical Sciences*, Academic Press, 1982, edited by V. Lakshmikantham.

19. E. N. Chukwu, "Global Asymptotic Behavior of Functional Differential Equations of the Neutral Type," *J. Nonlinear Analysis Theory, Method and Applications* **5** (1981) 853–872.

20. J. Hale, *Theory of Functional Differential Equations*, Springer-Verlag, New York, 1977.

21. O. Lopes, "Forced Oscillation in Nonlinear Neutral Differential Equations," *SIAM J. of Applied Mathematics* **29** (1975) 196–207.

22. H. J. Sussmann, "Small-Time Local Controllability and Continuity of the Optimal Time Function for Linear Systems," *J. Optimization Theory and Applications* **53** (1987) 281–296.

23. E. N. Chukwu and O. Hájek, "Disconjugacy and Optimal Control," *J. Optimization Theory and Applications* **27** (1979).

24. E. N. Chukwu and H. C. Simpson, "Perturbations of Nonlinear Systems of Neutral Type," *J. Differential Equations* **82** (1989) 28–59.

25. T. S. Angell and A. Kirsch, "On the Necessary Conditions for Optimal Control of Retarded Systems," *Appl. Math. Optimization* **22** (1990) 117–145.

26. H. T. Banks and M. Q. Jacobs, "An Attainable Sets Approach to Optimal Control of Functional Differential Equations with Function Space Terminal Conditions," *J. Differential Equations* **13** (1973) 129–149.

27. J. K. Hale and R. R. Meyer, "A Class of Functional Equations of Neutral Type," *Mem. Amer. Math. Soc.*, No. 76, 1967.

Chapter 12

# Controllable Nonlinear Neutral Systems

## Introduction

Criteria for linear systems controllability and constrained controllability are contained in Secs. 10.1 and 10.2. There Euclidean and function-space targets are considered. In this chapter we treat the general nonlinear situation in $W_p^{(1)}$.

## 12.1   General Nonlinear Systems

We now consider the general system

$$\frac{d}{dt}D(t, x_t) = f(t, x_t, u(t)),  \tag{12.1.1a}$$

where $f : E \times C \times E^m \to E^n$ is continuously differentiable in the second and third arguments, and is continuous, and where (12.1.1a) satisfies the basic assumptions on $D$ and $f$ in Sec. 11.2. In particular, we assume that there exists an $m \in L_2([\sigma, \infty, E])$ such that

$$\|D_\phi f(t, \phi, u)\| + \|D_u f(t, \phi, u)\| \le m(t),  \tag{12.1.1b}$$

and

$$D(t, x_t) = x(t) - g(t, x_t)$$

where

$$|g(t, \phi)| \le k(t)\|\phi\|,  \qquad \forall\, \phi \in C, \quad t \ge \sigma  \tag{12.1.1c}$$

for some $k$ continuous, $g(t, \phi)$ is linear in $\phi$.

We need some preliminary definitions from analysis. We work in the space $W_p^{(1)}$.

**Definition 12.1.1**   Let $X$ and $Y$ be real Banach spaces and $F$ a mapping from an open set $S$ of $X$ into $Y$. If for each fixed point $x_0 \in S$ and every $h \in X$, the limit

443

$$\lim_{t \to 0}[F(x_0 + th) - F(x_0)]/t = \delta F(x_0 h)$$

exists in the topology of $Y$, then the operator $\delta F(x_0, h)$ is called the Gateaux differential of $F$ at $x_0$ in the direction of $h$. If for each fixed $x_0 \in X$ the Gateaux differential $\delta F(x_0, \cdot)$ is a bounded linear operator mapping $X$ into $Y$, we write $\delta F(x_0, h) = F'(x_0)h$, and $F'(x_0)$ is called the Gateaux derivative of $F$ at $x_0$. If $F$ has a Gateaux derivative at $x_0$, we say $F$ is $G$-differentiable at $x_0$, and $F'(x_0) \in L(X, Y)$, the space of bounded linear operators from $X$ into $Y$.

**Definition 12.1.2**    $F : X \to Y$ is weakly $G$-differentiable at $x_0$ if there exists a bounded linear map $F'(x_0) \in L(X, Y)$ such that

$$([F(x_0 + th) - F(x_0)]/t - F'(x_0)h, y^*) \to 0$$

as $t \to 0$ for all $h \in X$, $y^* \in Y^*$. As a consequence, if $F$ is $G$-differentiable, it is weakly $G$-differentiable.

Consider

$$\frac{d}{dt}[D(t, x_t)] = f(t, x_t, u(t)), \qquad x_\sigma = \phi. \tag{12.1.1a}$$

Corresponding to the point $(t, \sigma, \phi, u) \in E \times E \times W_p^{(1)} \times L_p$, the mapping

$$x : E \times E \times W_p^{(1)} \times L_p \to W_p^{(1)}$$

is defined by

$$x(\sigma, \phi, u)(t) = x_t(\sigma, \phi, u) \in W_p^{(1)}.$$

We now give some useful properties of this mapping. We assume (12.1.1b) and (12.1.1c) are valid.

Also $f : E \times C \times E^m \to E^n$ is continuous and continuously differentiable in the second and third argument.

**Lemma 12.1.1**    *Let $(t, \sigma, \phi, u) \in E \times E \times W_p^{(1)} \times L_p$. Let $v \in L_p([\sigma, \infty), E^m)$ and $F'(u)$ be the $G$-derivative of $x(t, \sigma, \phi, u)$ with respect to $u$. Assume all the conditions of existence and uniqueness of solutions of (12.1.1a) in [5] or [11]. Then we have*

$$F'(u)(v) = D_u x(t, \sigma, \phi, u)(v) = y(t, \sigma, \phi, u, v),$$

*where the mapping $t \to y(t, \sigma, \phi, u, v)$ of $E$ into $E^n$ is the unique solution of*

$$\frac{d}{dt}[D(t)y_t] = D_2 f(t, x_t(\sigma, \phi, u), u)y_t + D_3 f(t, x_t(\sigma, \phi, u), u)v(t), \quad y_\sigma = \phi, \tag{12.1.2}$$

*with $\phi \equiv 0$. Also, for every $(\psi, v) \in W_p^{(1)} \times L_p$ we have that if $F'(u, v)$ denotes the $G$-derivative of $x(t, \sigma, \phi, u)$ with respect to its last two variables at the point $(t, \sigma, \phi, u)$, then*

$$Dx(t, \sigma, \phi, u)(\psi, v) = \lambda(t, \sigma, \phi, u, v),$$

*where $t \to \lambda(t, \sigma, \phi, u, v)$ is a mapping of $E$ into $E^n$, which is the unique solution of (12.1.2) with $y_\sigma = \psi$.*

**Proof**  For each $(\sigma, \phi) \in E \times W_p^{(1)}$, the unique solution of (12.1.1a) with initial data $(\sigma, \phi)$ is given by $x_\sigma = \phi$ in $[-h, 0]$,

$$x(\sigma, \phi, u)(t) = x(t, \sigma, \phi, u)$$

$$= g(t, x_t(\sigma, \phi, u)) - g(\sigma, \phi) + \phi(0)$$

$$+ \int_\sigma^t f(s, x_s(\sigma, \phi, u), u(s))ds. \qquad (12.1.3)$$

We claim that $x(\sigma, \phi, u)$ is Fréchet differentiable with respect to $u(t) \in E^m$. Indeed, if $D_u$ denotes this partial derivative, then by Dieudonne [10],

$$D_u x_\sigma = 0 \text{ in } [-h, 0],$$

$$D_u x(t, \sigma, \phi, u) = \int_\sigma^t D_2 f(s, x_s(\sigma, \phi, u), u(s)) D_u x_s(\sigma, \phi, (u))ds$$

$$+ \int_\sigma^t D_3 f(s, x_s(\sigma, \phi, u), u(s)]ds$$

$$+ D_2 g(t, x_t(\sigma, \phi, u)) \frac{dx_t}{du}(\sigma, \phi, u).$$

Because of the linearity of $g$,

$$D_2 g(t, x_t(\sigma, \phi, u)) \frac{dx_t}{du}(\sigma, \phi, u) = g\left(t, \frac{dx_t}{du}(\sigma, \phi, u)\right).$$

Thus

$$D_u x(t, \sigma, \phi, u) - D_2 g(t, x_t(\sigma, \phi, u)) \frac{dx}{du}(\sigma, \phi, u)$$

$$= \int_\sigma^t D_2 f(s, x_s(\phi, u), u(s)) D_u x_s(\sigma, \phi, u)ds$$

$$+ \int_\sigma^t D_3 f(s, x_s(\sigma, \phi, u), u(s))ds.$$

Furthermore, on differentiating with respect to $t$ we obtain

$$\frac{d}{dt}[D_u x(t, \sigma, \phi, u)v - D_2 g(t, x_t(\sigma, \phi, u)) D_u x_t(\sigma, \phi, u)v]$$

$$= D_2 f(t, x_t(\sigma, \phi, u), u(t)) D_u x_t(\sigma, \phi, u)v + D_3 f(t, x_t(\sigma, \phi, u), u(t))v,$$

or equivalently

$$\frac{d}{dt}[D_u x(t, \sigma, \phi, u)v - g(t, D_u x_t(\sigma, \phi, u))v]$$

$$= D_2 f(t, x_t(\sigma, \phi, u), u(t))D_u x_t(\sigma, \phi, u)v$$

$$+ D_3 f(t, x_t(\sigma, \phi, u), u(t))v. \qquad (12.1.4)$$

We now prove that

$$F'(u)v = D_u x(t, \sigma, \phi, u)(v),$$

to complete the first assertion. Consider

$$u \to F(u) = x(t, \sigma, \phi, u),$$

which is a mapping

$$F : L_p \to W_p^{(1)}.$$

Let $u_0, h \in L_p$,

$$F(u_0 + \tau h) - F(u_0) = g(t, x_t(\sigma, \phi, u_0 + \tau h)) - g(t, x_t(\sigma, \phi, u_0))$$

$$+ \int_\sigma^t f(s, x_s(\sigma, \phi, u_0 + \tau h), u_0(s + \tau h(s))$$

$$- f(s, x_s(\sigma, \phi, u_0), u_0(s))ds.$$

But

$$f(s, x_s(\sigma, \phi, u_0 + \tau h), u_0(s) + \tau h(s)) - f(s, x_s(\sigma, \phi, u_0), u_0(s))/\tau$$

$$\to \delta f(u_0(s), h(s)) \quad \text{as} \quad \tau \to 0,$$

where $\delta f$ denotes the Gateaux differential of $f$. Since $f$ is continuously Fréchet differentiable with respect to $u(t) \in E^m$ by Lemma 11.4.3 of [18], we have

$$\delta f(u_0(s), h(s)) = D_u f(s, x_s(\sigma, \phi, u_0), u_0(s))h(s).$$

Also

$$g(t, x_t(\sigma, \phi, u_0 + \tau h)) - g(t, x_t(\sigma, \phi, u_0))/\tau \to \delta g(u_0, h) \quad \text{as} \quad \tau \to 0,$$

and from linearity

$$g(t, (x_t(\sigma, \phi, u_0 + \tau h) - x_t(\sigma, \phi, u_0)))/\tau) \to g(t, \delta x_t(\sigma, u_0, h)) \quad \text{as} \quad \tau \to 0,$$

so that

$$\delta g(u_0, h) = g(t, D_u x_t(\sigma, \phi, u_0)h).$$

We now have

$$\delta F(u_0, h) = \lim_{\tau \to 0} (F(u_0 + \tau h) - F(u_0))/\tau$$

$$= g(t, D_u x_t(\sigma, \phi, u_0)h) + \int_\sigma^t D_u f(s, x_s(\sigma, \phi, u_0), u_0(s))h(s)ds \,.$$

We now prove that $\delta F$ is continuous at $(u_0, 0) \in L_p^p \times L_p$, so that by Theorem 2 [9, p. 336].

$$\delta F(u_0, h) = F'(u_0)h, \ F'(u_0 \in L(L_p, W_p^{(1)}) : F'(u_0) : L_p \to W_p^{(1)}$$

is a bounded linear map. Suppose $(u_k, h_k) \to (u_0, 0)$. Then

$$\lim_{k \to \infty} \|g(t, D_u x_t(\sigma, \phi, u_k))h_k - g(t, D_u x_t(\sigma, \phi, u_0)0)\|$$

$$+ \lim_{k \to \infty} \int_\sigma^t \|D_u f(s, x_s(\sigma, \phi, u_k), u_k(s))h_k(s) - D_u f(s, x_s(\sigma, \phi, u_0), u_0(s))0\|ds$$

$$= 0$$

since $D_u f(s, x_s(\sigma, \phi, u_k), u_k(s))h_u(s) \to 0$ as $k \to \infty$ and the $L_p$ bound assumption coupled with the use of the Dominated Convergence Theorem. Since

$$D_u x_t(\sigma, \phi, u) \,, \quad D_u f(s, x_s(\sigma, \phi, u), u(s))$$

are continuous in $u$, and $k(t)$ in (12.1.1c) is continuous on $[\sigma, \infty)$, for every finite $r > 0$ and $u \in B_r(u_0) = \{u : \|u - u_0\|_p \le r\}$ there is some $L < \infty$ such that if $t \in [\sigma, t_1]$, $t_1 < 0$, then

$$\|F'(u) - F'(u_0)\| \le L < \infty \,. \tag{12.1.5}$$

We note that

$$|F'(u_0) - F'(u_0)| \le |g(t, D_u x_t(\sigma, \phi, u)) - g(t, D_u(x_t(\sigma, \phi, u_0)))|$$

$$+ \int_\sigma^t |D_u f(s, x_s(\sigma, \phi, u), u_0) - D_u f(s, x_s(\sigma, \phi, u_0), u_0)ds) \,,$$

$$\le |g(t, D_u x_t(\sigma, \phi, u)) - D_u(x_t(\sigma, \phi, u_0))|$$

$$+ \int_\sigma^t |D_u f(s, x_s(\sigma, \phi, u), u(s)) - D_u f(s, x_s(\sigma, \phi, u_0), u_0(s))|ds \,,$$

$$\le k(t)\|D_u x_t(\sigma, \phi, u) - D_u x_t(\sigma, \phi, u_0)\|$$

$$+ \int_\sigma^t |D_u f(s, x_s(\sigma, \phi, u), u(s)) - D_u f(s, x_s(\sigma, \phi, u_0), u_0(s))|ds \,.$$

If $x(\sigma, \phi, u)$ is a solution of (12.1.1a), the associated variational control system along the response $t \to x_t(\sigma, \phi, u)$ is the system (12.1.2). In particular, if

$$f(t, 0, 0) = 0, \tag{12.1.6}$$

then the trivial solution $x(\sigma, 0, 0) = 0$ is a solution of (12.1.1a) and the variational control system along the trivial solution is

$$\frac{d}{dt}[D(t)y_t] = D_2 f(t, 0, 0)y_t + D_3 f(t, 0, 0)v. \tag{12.1.7a}$$

$\square$

The following theorem is reproduced from [9, p. 155].

**Open Mapping Theorem**   *Let $X, Y$ be Banach spaces and $B_r(x_0)$ an open ball of radius $r$ about $x_0 \in X$. Suppose $F : B_r(x_0) \subset X \to Y$ is weakly $G$ differentiable such that $R(F'(x_0)) = Y$ and $|F'(x) - F'(x_0)| \le k$ in $B_r(x_0)$ for some $k > 0$. Then $B_\rho(Fx_0) \subset F(B_r(x_0))$ for some $\rho > 0$, provided $k$ is sufficiently small.*

**Theorem 12.1.1**   *Consider System (12.1.1a) whose linear variational system along $x \equiv 0$,*

$$\frac{d}{dt}D(t)z(t) = L(t, z_t) + B(t)v(t), \tag{12.1.1b}$$

*is controllable on $[\sigma, t_1]$ where $t_1 > \sigma + h$ and where*

$$D_2 f(t, 0, 0)z_t = L(t, z_t), \qquad D_3 f(t, 0, 0)v = B(t)v.$$

*Assume that*

$$f(t, 0, 0) = 0, \qquad \forall t \ge \sigma.$$

*Then*

$$0 \in \operatorname{Int} \mathcal{A}(t_1, \sigma) \tag{12.1.8a}$$

*where*

$$\mathcal{A}(t_1, \sigma) = \{x_{t_1}(\sigma, 0, u) : u \in L_p([\sigma, t_1], E^m), \ \|u\|_{L_p} \le 1, \ x(\sigma, 0, u)$$

$$\text{is a solution of (12.1.1a) with } x_\sigma = 0\}$$

*is the attainable set associated with (12.1.1a).*

**Proof**   Let $\phi \in W_p^{(1)}$, $u \in L_p([\sigma, t_1], E^m)$, $t_1 > \sigma + h$. Let $u \to x_{t_1}(\sigma, \phi, u)$ be the mapping $F : L_p([\sigma, t_1], E^m) \to W_p^{(1)}([-h, 0], E^n)$ given by $Fu = x_{t_1}(\sigma, \phi, u)$, where $x(\sigma, \phi, u)$ is a solution of (12.1.1a). Then by Lemma 12.1.1,

$$F'(u)v = D_u x_{t_1}(\sigma, \phi, u)(v) = y_t(\sigma, \phi, u, v),$$

where $y$ is the solution of the variational equation (12.1.2b) and $F'(u) : L_p([\sigma, t_1], E^m) \to W_p^{(1)}$. Evidently $F'$ is a bounded linear surjection if and only if the control system (12.1.2b) is controllable on $[\sigma, t_1]$. As a consequence, $F'(L_p[\sigma, t_1], E^m) = W_p^{(1)}$; and Lemma 12.1.1 guarantees that all the requirements of the open mapping theorem are met for the map

$$F : L_p([\sigma, t_1], E^m) \to W_p^{(1)}.$$

Thus for $u_0 \equiv 0 \in L_p([\sigma, t_1])$, $F(u_0) = F(0) = 0 \in W_p^{(1)}$, for every $r > 0$ and open ball $\mathbb{B}(0, r) \subset L_p([\sigma, t_1], E^m)$ of center $0 \in L_p$ and radius $r$ there is an open ball $\mathbb{B}(0, \rho) \subset W_p^{(1)}$ center $0$ of radius $\rho$ such that

$$\mathbb{B}(0, \rho) \subset \mathbb{B}(F(u_0), \rho) \subset F(\mathbb{B}(u_0, r)) = F(\mathbb{B}(0, r)).$$

Thus

$$0 \in \mathbb{B}(0, \rho) \subset F(\mathbb{B}(0, r)) \subset F(L_p([\sigma, t_1], E^m)).$$

It follows from this that $0 \in \operatorname{Int} \mathcal{A}(t_1, \sigma) \subset W_p^{(1)}$ where

$$\mathcal{A}(t_1, \sigma) = \{x_{t_1}(\sigma, 0, u) : u \in L_p([\sigma, t_1], E^m), \ \|u\|_{L_p} \leq r, \ x(\sigma, 0, u)$$

$$\text{is a solution of (12.1.1a)}\}. \tag{12.1.8b}$$

We have proved that for any $r > 0$,

$$0 \in \operatorname{Int} \mathcal{A}(t_1, \sigma) \tag{12.1.9}$$

where $\mathcal{A}(t_1, \sigma)$ is defined in (12.1.8). Since $r$ is arbitrary and the set of admissible controls $U_{\text{ad}}$ with constraints is

$$U_{\text{ad}} = \{u \in L_p([\sigma, t_1]) : \|u\|_p \leq 1\}, \tag{12.1.10}$$

(12.1.3) is valid. $\qquad\square$

**Corollary 12.1.1** *Assume all the conditions of Theorem 12.1.1. Then System (12.1.1) is locally null controllable with constraints. That is, there is a neighborhood $\mathcal{O}$ of zero in $W_p^{(1)}$ such that every initial point of $\mathcal{O}$ can be driven to zero in some finite time $t_1$ using some $u \in U_{\text{ad}}$.*

Condition 12.1.3 will prove not only local null controllability but constrained null controllability as well, because (12.1.9) implies that

$$0 \in \operatorname{Int} \mathcal{D} \tag{12.1.11a}$$

where $\mathcal{D}$ is the domain of null controllability, i.e., the set of initial functions $\phi$ such that the solution $x(\sigma, \phi, u)$ of (12.1.1a) with some $u \in U_{\text{ad}}$, and some $t_1 < \infty$, satisfies

$$x_\sigma(\sigma, \phi, u) = \phi, \qquad x_{t_1}(\sigma, \phi, u) = 0.$$

Indeed, suppose (12.1.11) is false. Then there is a sequence $\{\phi_n\}_1^\infty$, $\phi_n : [-h, 0] \to E^n$, $\phi_n \to 0$ as $n \to \infty$ and no $\phi_n$ is in $\mathcal{D}$. Because $x(\sigma, 0, 0) = 0$ is a solution of

(12.1.1a), $0 \in \mathcal{D}$; we can therefore assume that $\phi_n \neq 0$ for any $n$. From the definition of $\mathcal{D}$, we have $x_{t_1}(\sigma, \phi_n, u) \neq 0$ for any $u \in U_{\text{ad}}$ and any $t_1 > \sigma + h$. Thus, on setting $\xi_n \equiv x_{t_1}(\sigma, \phi_n, u)$, we note that for $u = 0$ as $n \to \infty$, $\xi_n = x_{t_1}(\sigma, \phi_n, 0) \to x_{t_1}(\sigma, 0, 0)$ (from continuity and uniqueness of solution). Because $x_{t_1}(\sigma, 0, 0) = 0$ is the trivial solution of (12.1.1a) that is contained in $\mathcal{A}(t_1, \sigma)$, $\xi_n \notin \mathcal{A}(t_1, \sigma)$ for any $t_1 \geq \sigma + h$. We have produced a sequence $\{\xi_n\} \in W_p^{(1)}$ that has the following properties: $\xi_n \to 0$ as $n \to \infty$, $\xi_n \neq 0$ for any $n$ $\xi_n \notin \mathcal{A}(t_1, \sigma)$ for any $t_1 \geq \sigma + h$. This shows that

$$0 \notin \text{Int}\, \mathcal{A}(t_1, \sigma)$$

for any $t_1 \geq \sigma + h$, a contradiction. Thus (12.1.11) is valid. Therefore the local null controllability of (12.1.1a) is proved, and, as remarked, so is the local null controllability with constraints.     $\square$

**Corollary 12.1.2**   *Consider the autonomous system*

$$\frac{d}{dt}[x(t) - A_{-1}x(t-h)] = f(x_t, u(t)).  \tag{12.1.11b}$$

*Let*

$$D_1 f(0,0)y_t = A_0 y(t) + A_1 y(t-h), \qquad D_2 f(0,0)u = Bu,$$

*and assume that:*

  (i)  $\text{rank}[\Delta(\lambda), B] = n, \ \forall\, \lambda \in \mathbb{C}$ *where*

$$\Delta(\lambda) = I\lambda - A_0 - A_1 e^{-h} - A_{-1}\lambda e^{-\lambda h}.$$

 (ii)  $\text{rank}[\lambda I - A_{-1}, B] = n, \ \forall\, \lambda \in \mathbb{C}.$

*Then (12.1.12) is locally null controllable and locally null controllable with constraints.*

**Proof**   Conditions (i) and (ii) are the necessary and sufficient conditions for the controllability of the linear variational equation

$$\frac{d}{dt}[x(t) - A_{-1}x(t-h)] = A_0 x(t) + A_1 x(t-h) + Bu(t).  \tag{12.1.11c}$$

Theorem 12.1.1 yields the required assertion.     $\square$

In Theorem 12.1.1, the controllability of the linear system (12.1.11c) is of fundamental importance. Can this be weakened? The next result asserts that the closure of the attainable set of the linear system (12.1.11c) can replace the controllability requirement. The result is stated for linear autonomous system

$$\dot{x}(t) - A_{-1}\dot{x}(t-h) = f(x_t, u)  \tag{12.1.12}$$

and the associated linear system

$$\dot{x}(t) - A_{-1}\dot{x}(t-h) = A_0 x(t) + A_1 x(t-h) + Bu(t),  \tag{12.1.13}$$

where

$$D_1 f(0,0)x_t = A_0 x(t) + A_1 x(t-h), \qquad D_2 f(0,0)v = Bu.$$

Following [1] we introduce the following notation: Let $s$ be fixed. Define the $n \times n$-matrices

$$\Delta_j^0 = \begin{cases} I, & j = 0, \\ 0, & j = 1, \ldots, s, \end{cases} \tag{12.1.14}$$

$$\Delta_j^1 = \begin{cases} A_0, & j = 0, \\ A_{-1}^{j-1}(A_{-1}A_0 + A_1), & j = 1, \ldots, s, \end{cases}$$

$$\Delta_j^i = \sum_{k=0}^{j} \Delta_k^{i-1} \Delta_{j-k}^1, \quad i = 2, \ldots, n(s+1)-1, \quad j = 0, \ldots, s,$$

$$Z_j^i = \sum_{k=0}^{j} \Delta_k^i A_{-1}^{j-k} B, \quad i = 0, \ldots, n(s+1)-1, \quad j = 0, \ldots, s, \tag{12.1.15}$$

$$Z^i = [Z_s^i, Z_{s-1}^i, \ldots, Z_0^i],$$

$$\Gamma_j^i = Z_j^i - A_0 Z_j^{i-1}, \quad i = 1, \ldots, n(s+1)-1, \quad j = 0, \ldots, s,$$

$$\Gamma^i = [\Gamma_s^i, \ldots, \Gamma_0^i]. \tag{12.1.16}$$

The matrices $K_i$ are defined as follows:

$$K_1 = \Gamma^1,$$

$$K_i = \Gamma^i - \sum_{j=1}^{i-1} \Gamma^{i-j}(Z^0)^+ + K_j, \quad i = 1, \ldots, n(s+1)-1, \tag{12.1.17}$$

where $(Z^0)^+$ is a fixed right inverse of $Z^0$, i.e., $Z^0(Z^0)^+ = I$. Bartosiewicz [1] has proved the following:

**Proposition 12.1.1**   *Let*

$$\mathcal{A}(t) = \{x_t \in W_p^{(1)}([-h,0], E^n) : x \text{ is a solution of } (12.1.13)$$

$$\text{for some } u \in L_p([0,t_1], E^m)\}.$$

*Then the set* $\mathcal{A}(t_1)$, $t_1 = (s+1)h$, *is closed in* $W_p^{(1)}$ *if and only if*

$$K_i(\ker Z^0) \subset \operatorname{Im} Z^0, \quad i = 1, \ldots, n(s+1)-1. \tag{12.1.18}$$

*If* $t_1 > nh$, *then put* $s = n$ *in* (12.1.18).

As an immediate consequence of Proposition 12.1.1, the following is valid:

**Corollary 12.1.3**   $\mathcal{A}(t_1)$ *is closed in* $W_p^{(1)}$ *whenever*

$$\operatorname{Im} \Gamma^i \subset \operatorname{Im} Z^0 , \quad i = 1, \ldots, n(s+1) - 1 , \tag{12.1.19}$$

*or*

$$\operatorname{rank} Z^0 = n , \tag{12.1.20}$$

*or*

$$\operatorname{rank} B = n . \tag{12.1.21}$$

Note that

$$\operatorname{rank} Z^0 = \operatorname{rank}[A_{-1}^s B, \ldots, B] . \tag{12.1.22}$$

Thus (12.1.22) is equivalent to closedness and finite codimensionality of $\mathcal{A}(t_1)$ in $W_p^{(1)}$ [1, p. 181].

Theorem 12.1.1 can now be weakened.

**Theorem 12.1.2**   *In* (12.1.12), *assume that all the conditions of Proposition 12.1.1 are fulfilled. Then there is a closed subspace $Y$ of $W_2^{(1)}$ and a neighborhood $\mathcal{O}$ of zero in $Y$ such that every initial point of $\mathcal{O}$ can be driven to zero in some time $t_1$ using some $u \in U_{\mathrm{ad}}$.*

**Proof**   The proof follows the same pattern as that of Theorem 12.1.1, with the new hypothesis replacing the controllability condition. The mapping

$$F \; : \; L_p([\sigma, t_1], E^m) \to W_p^{(1)} ,$$

$$Fu = x_{t_1}(\sigma, 0, u)$$

$$= g(x_{t_1}(\sigma, 0, u)) - g(\sigma, 0) + \int_\sigma^{t_1} f(s, x_s(\sigma, u), u(s))ds , \quad t \geq \sigma$$

where $x(\sigma, 0, u)$ is a solution of (12.1.12) does not have a surjective Gateaux derivative $F'$ at $0 \in L_p([\sigma, t_1], E^n)$. Instead its range is closed: $F'(0)(L_p([\sigma, t_1], E^m)) \equiv Y$ is closed. Therefore it is a closed subspace of $W_p^{(1)}([-h, 0], E^n)$. We can therefore study the mapping $F \; : \; L_p \to Y$ of the Banach space $L_p$ into the Banach space $Y$, whose Gateaux derivative $F'(0) : L_p([\sigma, t_1], E^m) \to Y$ is a surjection. It satisfies the conditions of Corollary 15.2 in [7]. The proof ends in the same way as before.         $\square$

Global controllability can be proved by imposing a stability condition on

$$\dot{x}(t) - A_{-1}x(t - h) = f(x_t, 0) .$$

Theorems 12.1.1 and 12.1.2 are local controllability results. We are now ready for the global constrained null controllability result.

**Theorem 12.1.3**   *In* (12.1.1a) *assume that*

(i) *$f$ is smooth enough for uniqueness, existence, and continuous dependence on initial data of solutions of* (12.1.1).

(ii) *$f(t, 0, 0) = 0$ for $t \geq \sigma$.*

(iii) *System* (12.1.2b) *is controllable.*

(iv) *The system*

$$\frac{d}{dt}[D(t)x_t] = f(t, x_t, 0) \tag{12.1.23}$$

*is uniformly exponentially stable, that is, every solution of* (12.1.10) *satisfies*

$$\|x_t(\sigma, \varphi)\| \leq k\|\varphi\|e^{-\alpha(t-o)}, \qquad t \geq \sigma \tag{12.1.24}$$

*for some $k > 0$, $\alpha > 0$.*

*Then* (12.1.1) *is null controllable with constraints, i.e., with controls in $U_{\text{ad}}$ of* (12.1.10) *for some $t \geq \sigma + h$.*

**Remark 12.1.1**   Conditions that guarantee the behavior (12.1.24) for System (12.1.23) are contained in Sec. 4.2.

We now consider a special case of System (12.1.1a), namely (12.1.25). The results obtained are sharp.

We now study the nonlinear in the state difference-differential control system

$$\frac{d}{dt}[x(t) - A_{-1}x(t - h)] = f(x(t), x(t - h)) + Bu(t), \tag{12.1.25}$$

where $A_{-1}$ is an $n \times n$ constant matrix, $f$ is continuously differentiable, and $B$ is an $n \times m$ constant matrix.

Let $H_1 \triangleq D_1 f(x(t), 0)$, $H_2 \triangleq D_2 f(x(t), x(t - h))$ where $D_i f$ is the $i$th partial derivative of $f$. Let

$$A_0 = D_1 f(0, 0), \qquad A_1 = D_2 f(0, 0),$$

and denote by $D_a$ the symmetric $n \times n$ matrices

$$\frac{1}{2}(H_a + H_a^T) \triangleq D_a, \qquad a = 1, 2.$$

Define $J_a$ as follows: $J_a = AD_a + D_a^T A$, where $A$ is a positive definite symmetric $n \times n$ matrix.

We have the following special result for (12.1.25):

**Theorem 12.1.4**   *In* (12.1.25) *assume:*

(i) *$f(0, 0) = 0$.*

(ii) *$\text{rank}[\Delta(\lambda), B] = n$, for all $\lambda \in \mathbb{C}$ where $\Delta(\lambda) = I - \lambda A_{-1}e^{-\lambda h} - A_0 - A_1 e^{-\lambda h}$, $\text{rank}[\lambda I - A_{-1}B] = n$ for all $\lambda \in \mathbb{C}$.*

(iii) *All the roots of the equation*

$$\det(I - A_{-1}r^{-h}] = 0$$

*have moduli less than 1; and for some positive definite $n \times n$ matrix $A$ we have*

$$J_1 \leq -\delta I$$

*where, for some $q > 1$,*

$$\delta - q\|J_2 + J_1 A_1\| \equiv \mu > 0.$$

*Then (12.1.25) is null controllable with constraints.*

**Proof** Conditions (ii) and (iii) are the necessary and sufficient conditions for the linear approximation of (12.1.25):

$$\frac{d}{dt}[z(t) - A_{-1}z(t-h)] = A_0 z(t) + A_1 z(t-h) + Bu(t)$$

to be controllable.

Condition (iii) is the required condition for the system

$$\frac{d}{dt}[x(t) - A_{-1}x(t-h)] = f(x(t), x(t-h))$$

to have all its solution $x(\sigma, \varphi)$ satisfy

$$\|x_t(\sigma, \varphi)\| \to 0 \quad \text{as} \quad t \to \infty. \qquad \square$$

## 12.2   Nonlinear Interconnected Systems

We now investigate the $i$th interconnected subsystem described by

$$\frac{d}{dt}[D_i(t)x_t^i] = f_i(t, x_t^i, u^i(t)) + k_i(t, x_t, v^i), \qquad i = 1, \ldots, \ell, \qquad (12.2.1)$$

where $k_i$ describes the action of the whole system

$$\frac{d}{dt}D(t, x_t) = f(t, x_t, u(t)) + k(t, x_t, v(t)) \qquad (12.2\ \mathrm{S})$$

on its interconnected system (12.2.1), which when "free" and "disconnected" is given by

$$\frac{d}{dt}[D_i(t)x_t^i] = f_i(t, x_t^i, u^i(t)). \qquad (12.2.2)$$

In (12.2.1), $u^i \in L_2([\sigma, t_1], E^{m_i})$, and $v^i \in L_2([\sigma, t_1], E^m)$. We shall sometimes restrain the control $u^i$ to lie in the set

$$\mathbb{P}_i = \{u^i : u^i \in L_2([\sigma, t_1], E^{m^i}), \|u^i\|_2 \leq 1\}$$

and $v^i \in Q_i$

$$Q_i = \{v^i \in L_2([\sigma, t_1], E^{m_i}) : \|v^i\|_2 \le 1\}.$$

Let

$$\mathcal{F}_i(t, \phi^i) = \{f_i(t, \phi, u^i) : u^i \in \mathbb{P}_i\},$$

$$\mathcal{K}_i(t, \phi) = \{k_i(t, \phi, v^i) : v^i \in Q_i\}.$$

We now assume that

$$\mathcal{K}_i(t, x_t) \subset \text{Int } \mathcal{F}_i(t, x_t^i).$$

The following result is valid.

**Theorem 12.2.1**  *In (12.2.1), we assume that:*

(i)  $f_i(t, 0, 0) = 0$, $k_i(t, \phi, 0) = 0$.

(ii)  *$f_i, k_i$, and $D_i$ satisfy all the smoothness conditions of existence, uniqueness, and continuous dependence on initial data.*

(iii)  *Assume that the linear variational system*

$$\frac{d}{dt}[D^i(t)z^i(t)] = L_i(t, x_t^i) + B^i(t)w(t) \tag{12.2.3}$$

*of (12.2.2), where*

$$D_2 f_i(t, 0, 0)z_t^i = L_i(t, z_t^i), \qquad D_3 f_i(t, 0, 0)w = B^i(t)w,$$

*is controllable on $[\sigma, t_1]$, $t_1 > \sigma + h$.*

(iv)  $\mathcal{K}_i(t, \phi) \subset \text{Int } \mathcal{F}_i(t, \phi^i)$.

*Then the interconnected subsystem (12.2.1) is locally null controllable on $[\sigma, t_1]$ with constraints.*

**Corollary 12.2.1**  *Assume*

(i)  *Condition (i)–(iii) of Theorem 12.2.1.*

(ii)  *The system*

$$\frac{d}{dt}[D_i(t)x_t^i] = f_i(t, x_t^i, 0), \qquad i = 1, \ldots, \ell, \tag{12.2.4}$$

*is globally exponentially stable.*

*Then the interconnected subsystem is (globally) null controllable with constraints, that is, with controls*

$$u^i \in \mathbb{P} = \{u^i \in L_2([\sigma, t_1], E^{m_i}) : \|u^i\|_2 \le 1\},$$

$$v^i \in Q_i = \{v^i v^i \in L_2([\sigma, t_1], E^m) : \|v^i\|_2 \le 1\}.$$

**Proof of Theorem 12.2.1**    Since the hypothesis of Theorem 12.2.1 holds

$$0 \in \mathrm{Int}\mathcal{A}_i(t,\sigma) \quad \text{for} \quad t > \sigma + h, \tag{12.2.5}$$

where $\mathcal{A}_i(t,\sigma)$ is the attainable set associated with (12.2.2) where $x_\sigma = 0$:

$$\mathcal{A}_i(t,\sigma) = \{x^i(\sigma,0,u^i) : x \text{ solves (12.2.2) with } x_\sigma = 0\}.$$

This is also the solution of (12.2.1) with $u^i$ as the control of the interconnected subsystem and with $v = 0$ as the control from the large-scale system. Thus $x^i(\sigma,0,u^i,0) \equiv x^i(\sigma,0,u^i)$. Suppose $x^i$ is a solution of (12.2.1) with $x_\sigma^i = 0$, then

$$x^i(\sigma,0,u^i,v^i) = g_i(t,x_t^i) - g_i(\sigma,0) + \int_\sigma^t f_i(s,x_s^i,u^i(s))ds$$

$$+ \int_\sigma^t k_i(s,x_s,v^i(s))ds$$

for $t \geq \sigma$. On defining the set

$$H_i(t_1,\sigma) = \{x^i(\sigma,0,u^i,v^i) \in W_2^{(1)}([0,t_1], E^{n_i}) : u^i \in \mathbb{P}_i, v^i \in Q_i\},$$

we deduce (on account of the hypothesis (i) and the fact that $x^i(t,0,0,0) = 0$ is a solution of (12.2.1), that $0 \in \mathcal{A}_i(t_1,\sigma) \subset H_i(t_1,\sigma)$. Since (iv) is valid and (12.2.5) holds,

$$0 \in \mathrm{Int}\, \mathcal{A}_i(t_1,\sigma) \subset H_i(t_1,\sigma) \tag{12.2.6}$$

and

$$0 \in \mathrm{Int}\, H_i(t_1,\sigma) \tag{12.2.7}$$

as well.

If $D$ denotes the domain of null controllability of (12.2.1), one proves readily that

$$0 \in \mathrm{Int}\, D,$$

proving local null controllability with constraints.      $\square$

**Proof of Corollary 12.2.1**    One uses the control $u^i = 0 \in U_i$, $i = 1,\ldots,\ell$ to glide along System (12.2.4) and approach an arbitrary neighborhood of the origin. Since (i) guarantees that all initial states in this neighborhood can be driven to zero in finite time $t_1$, we are done.      $\square$

**Remark**    Conditions for uniform asymptotic stability are available in [3].

**Remark 12.2.1**    It is very important to study the condition (12.2.6) from which (12.2.7) was deduced. Suppose

$$0 \in \mathrm{Int}\, \mathcal{A}_i(t_1,\sigma) \tag{12.2.8}$$

fails, and the isolated system is not even controllable locally and is "not well be-haved". Condition (12.2.7) can still be salvaged and the interconnected system made controllable. What may be needed is a criteria that ensures that

$$0 \in \operatorname{Int} G_i(t_1, \sigma), \tag{12.2.9}$$

where

$$G_i(t_1, \sigma) = \left\{ \int_\sigma^t k_i(s, x_s, v^j(s)) ds : v^j \in Q_j \right\}.$$

In other words we require a sufficient amount of control impact from external sources $((k_i)$, which is monitored locally at the $i$-subsystem) to be "forced" on (12.2.2). The source of power senses the control $u^i$ available to (12.2.2) and sends a sufficient controlling signal $k_i$ to the large-scale system, which is a function of the power's state variable $x$ and the control $v^j$. With this the right behavior is enforced. Policy implications of this insight will be pursued elsewhere in [7]. There we observe that the interaction does not ignore the subsystems' (initiatives) $v^j \in E^m$; it utilizes them. For example, were they to be ignored, and

$$k_i = k_i(t, x_t)$$

still be operative and nontrivial, the condition $0 \in H_i(t, \sigma)$ and subsequently (12.2.7) will fail. The interconnected system will not be null controllable. Failure to acknowl-edge individual subsystems' initiatives $(v^j)$ in highly centralized interconnected or-ganizations monitored by our dynamics leads to lack of growth from initial en-dowment $\phi$ to desired economic target $\psi$. On the other hand, failure to enforce regulations or use subsidy, bringing $k_i$ to bear when same subsystem misbehaves or is depressed, and is uncontrollable, in "individualistic" systems may make the interconnected system uncontrollable and trigger an economic depression.

In place of (12.2.1), consider the system

$$\frac{d}{dt}[D_i(t)x_t^i] = f_i(t, x_t^i, u^i(t)) + \sum_{\substack{j=1 \\ i \neq j}}^{\ell} k_{ij}(t, x_t^j, u^i(t)). \tag{12.2.10}$$

The following can be proved:

**Theorem 12.2.2** *For (12.2.10) assume (i)–(iii) of Theorem 12.2.1. But in (12.2.2), $L_i(t, z_t^i)$ and $B^i(t)$ are defined as*

$$D_2 f_i(t, 0, 0) z_t^i = L_i(t, z_t^i),$$

$$D_3(f_i(t, 0, 0) + k_i(t, x_t, 0)) = B^i(t)$$

*with $k_i(t, x_t, u^i) = \sum_{\substack{j=1 \\ i \neq j}}^{\ell} k_{ij}(t, x_t^j, u^i(t))$.*

*Then (12.2.10) is locally null controllable with constraints.*

**Proof**  We note that condition (iii) of Theorem 12.2.2 requires the controllability of

$$\frac{d}{dt}[D^i(t)z_t^i] = L_i(t, z_t^i) + (B_{1i}(t) + B_{2i}(t))u^i(t)\,,\qquad (12.2.11)$$

where

$$B_{1i}(t) = D_3 f_i(t, 0, 0)\,,$$

$$B_{2i}(t) = D_3 k_i(t, x_t, 0)\,.$$

Suppose (12.2.5) is invalid for the isolated system (12.2.3), and this can happen when (12.2.3) is not controllable with the control matrix $B_{1i}(t)$, the "solidarity function" $k_i$ can be brought to bear to force the controllability of (12.2.11) with $B^i = B_{1i}+B_{2i}$ to ensure (12.2.5). If the system is autonomous, the requirement would be (i) and (ii) of Corollary 12.1.2. Even if (i) and (ii) hold with $B_{1i}$ and (12.2.5) is assured, the composite system need not be locally null controllable unless (12.2.11) is controllable with $B^i = B_{1i} + B_{2i}$. An adequate proper amount of intervention in the form of a solidarity function $k_i$ is needed. The linear case was explored in Chap. 11. It is quite possible to extend this treatment to the situation

$$\frac{d}{dt}[D_i^i(t)x_t^i] = f_i(t, x_t^i) - p_i(t) + k_i(t)\,,\qquad (12.2.12)$$

where $f_i(t, x_t^i)$ is independent of control $p_i$, which enters the dynamics in an additive fashion. One can then deduce the four universal principles for the control of an interconnected organization as are contained in Sec. 10.4.                        $\square$

## 12.3   An Example: A Network of Flip-Flop Circuits

The basic element in a digital computer is the flip-flop circuit, a model of which is given in Fig. 12.3.1, [17]. The section between $x = 0$ and $x = 1$ is a lossless transmission line with specific inductance $L_s$ and specific capacitance $C_s$. The voltage $v$ across the line and the current flowing through it are both functions of $x$ and of time $t$. They are described by the following partial differential equations:

$$L_s\frac{\partial i}{\partial t} = -\frac{\partial v}{\partial x}\,,\quad C_s\frac{\partial v}{\partial t} = -\frac{\partial i}{\partial x}\,,\quad 0 < x < 1,\ t > 0\,.\qquad (12.3.1)$$

The boundary conditions at the ends of the line are

$$0 = E - v(0, t) - Ri(0, t)\,,$$

$$-C_1\frac{dv(1, t)}{dt} = i(1, t) + f(v(1, t))\,,\qquad (12.3.2)$$

where $f(v(1, t))$ is a nonlinear function of $v$ and gives the current in the displayed box in the direction shown.

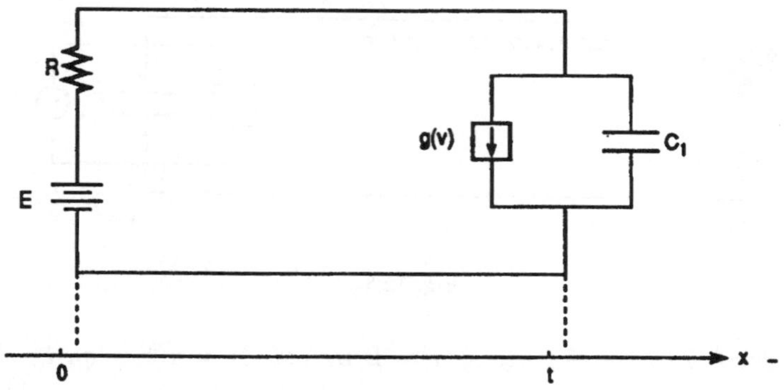


Fig. 12.3.1

If we set the left-hand side of (12.3.1) and (12.3.2) to zero and solve, we obtain a single equilibrium $(v^*, i^*)$. If it is assumed that $(v^*, i^*)$ is globally asymptotically stable, then the flip-flop circuit can be shown to operate as a memory device. To investigate conditions on the system's parameter to guarantee this stability assumption (12.3.1), (12.3.2) is converted into a functional differential equation of neutral type:

$$C_s \frac{d}{dt}[v_1(t) + kv_2(t - h)] = \frac{2E}{z + R_0} - \frac{v_1(t)}{z} + \frac{k}{z} v_1(t - h)$$

$$- f(v_1(t)) - kf(v_1(t - h)), \qquad (12.3.3)$$

where

$$z = \left[\frac{L_s}{C_s}\right]^{\frac{1}{2}}, \quad k = \frac{R_0 - z}{R_0 + z}, \quad h = 2(L_s C_s)^{\frac{1}{2}},$$

and also

$$v_0 = v(0, t), \quad v_1 = v(1, t), \quad i_0 = i(0, t), \quad i_1 = i(1, t).$$

One can pose the following question in place of the stability one: Is it possible to drive all fluctuations of $(v(x,t), i(x,t))$ to the equilibrium $(v^*, i^*)$ in finite time by introducing a control device (e.g. a stabilizer)? The same question can be posed for the circuit in Fig. 12.3.2, whose control equation is

$$C_1 \frac{d}{dt}[u(t) - qu(t - h)] = -\frac{1}{2}u(t) - \frac{q}{2}u(t - h) - g(u(t))$$

$$- qu(t - h) - i(t) + e_1(t), \qquad (12.3.4)$$

$$\frac{d}{dt}[Li(t)] = -R_1 i(t) + u(t) - qu(t - h) + e_2(t),$$

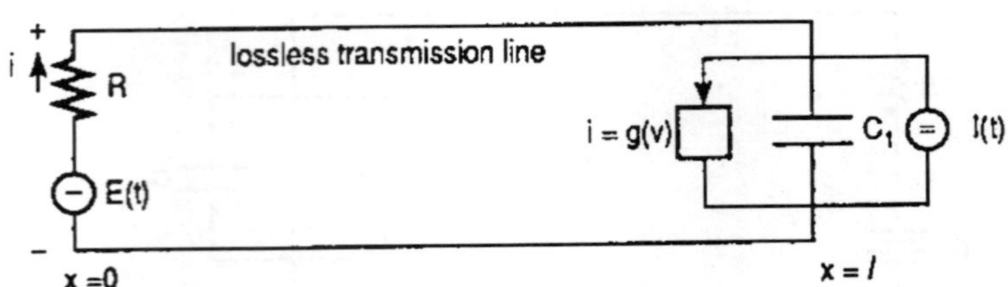


Fig. 12.3.2

where $u$ is related to the voltage at $x = L$ by $v(L,t) = u(t) - qu(t - h)$, and $L_1$, $R$, and $C_1$ are the parameters of the line. The function $e = (e_1, e_2)$ is a control function generated by the control device at $x = 0$, and it is related to $E(t)$.

Equation (12.3.4) can be put in vector form:

$$\frac{d}{dt}[x(t) - A_{-1}x(t - h)] = A_0 x(t) + A_1 x(t - h) + C_1(x(t - h)) + Be(t), \quad (12.3.5)$$

where $x(t) = (u(t), i(t))^T$, $e = (e_1, e_2)^T$,

$$A_{-1} = \begin{bmatrix} -q & 0 \\ 0 & 0 \end{bmatrix}, \quad A_0 = \begin{bmatrix} -\dfrac{1}{2C_1} & -\dfrac{1}{C_1} \\ \dfrac{1}{L} & -\dfrac{R_1}{L} \end{bmatrix}, \quad B = \begin{bmatrix} 1 & 0 \\ 0 & 1 \end{bmatrix},$$

$$A_1 = \begin{bmatrix} -\dfrac{q}{2C_1} & 0 \\ -\dfrac{q}{L} & 0 \end{bmatrix}, \quad G(x(t - h)) = \begin{bmatrix} g(u(t) - q(u(t - h)) \\ 0 \end{bmatrix}.$$

Suppose each such circuit in Fig. 12.3.1 or Fig. 12.3.2 is connected up in a complicated fashion with other such circuits to form an interconnected system. The problem of controllability can also be fruitfully posed.

For example, suppose there are $\ell = 1, \ldots, 10$ such circuits of Fig. 12.3.3. We can consider the overall system as a decentralized one in which each circuit is influenced by the interaction terms

$$y_i(t, x, e^i) = \sum_{\substack{j=1 \\ i \neq j}}^{10} g_{ij}(t, x^j, e^i)$$

as follows:

$$\frac{d}{dt}[x^i(t) - A_{-1i}x^i(t - h)] = A_{0i}x^i(t) + G_i(x^i(t - h)) + B_i e^i(t) + y_i(t, x, e^i) \quad (12.3.6)$$

is the $i$th-control system.

In (11.4.6), set

$$A_{1i} = DG_i(0),$$

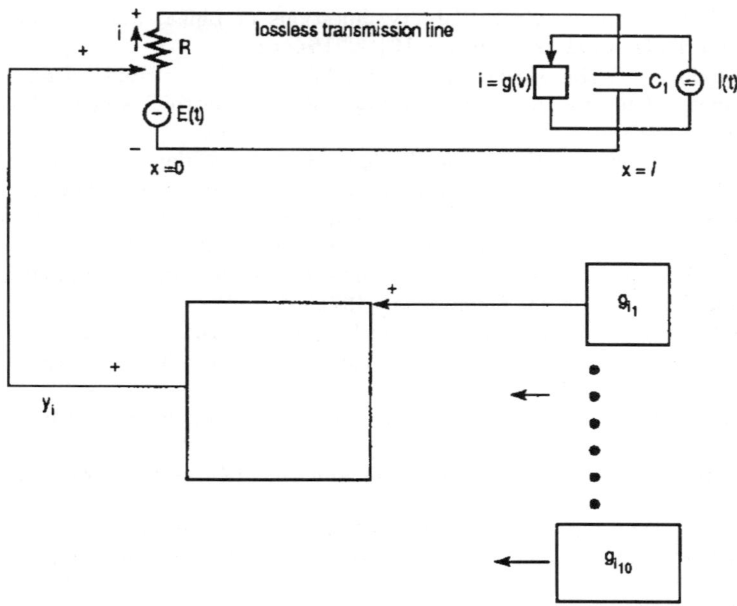


Fig. 12.3.3

$$B_{2i} = D_3 y_i(t, 0, 0),$$

and consider the companion system

$$\frac{d}{dt}[x^i - A_{-1i}x^i(t-h)] = A_{0i}x^i(t) + A_{1i}x^i(t-a) + (B_{1i} + B_{2i})e^i. \qquad (12.3.7)$$

We have the following result:

**Proposition 12.3.1** *System 12.3.6 is locally null controllable on* $[0, t_1]$ *with constraints if* $y_i(t, x, e^i)$ *is such that (12.3.7) is controllable on* $[0, t_1]$.

## References


1. Z. Bartosiewicz, "Closedness of the Attainable Det of the Linear Neutral Control System," *Control and Cybernetics* **8** (1979) 179–189.
2. E. N. Chukwu, "On the Euclidean Controllability of a Neutral System with Nonlinear Base," *Nonlinear Analysis, Theory, Methods and Applications* **11**(1) (1987) 115–123.
3. E. N. Chukwu, "Uniform Asymptotic Stability of Large Scale Systems of Neutral Type," *IEEE Proceedings of the Eighteenth Southeastern Symposium on Systems Theory*, April 7–8, Knoxville, TN, Computer Society Press, 1986.
4. E. N. Chukwu, "The Time Optimal Control Theory of Linear Differential Equations of Neutral Type," *Comput. Math. Applic.* **16**(10/11) (1988) 851–866.
5. E. N. Chukwu and H. C. Simpson, "Perturbations of Nonlinear Systems of Neutral Type," *J. Differential Equations* **82** (1989) 28–59.

6. E. N. Chukwu, "Some Controllability Questions of Linear and Nonlinear Neutral Functional Differential Equations in $W_2^{(1)}$," Preprint.

7. E. N. Chukwu, "Global Economic Growth: Will the Center Hold?" Preprint.

8. J. P. Dauer, "Nonlinear Perturbations of Quasilinear Control Systems," *J. Math. Anal. Appl.* **54**(3) (1976).

9. K. Deimling, *Nonlinear Functional Analysis*, Springer-Verlag, New York, 1985.

10. J. Dieudonne, *Foundations of Modern Analysis*, Academic Press, New York, 1960.

11. J. Hale and M. A. Cruz, "Existence, Uniqueness and Continuous Dependence for Hereditary Systems," *Ann. Math. Pura. Appl.* **85** (1970) 63–82.

12. J. Hale, "Forward and Backward Continuation of Neutral Functional Differential Equations," *J. Differential Equations* **9** (1971) 168–181.

13. J. Hale, *Theory of Functional Differential Equations*, Springer-Verlag, New York, 1977.

14. L. V. Kantorovich and G. P. Akilov, *Functional Analysis in Normed Spaces*, Macmillan Company, New York, 1964.

15. G. E. Ladas and V. Lakshmikantham, *Differential Equations in Abstract Spaces*, Academic Press, New York, 1972.

16. A. N. Michel and R. K. Miller, *Qualitative Analysis of Large Scale Dynamical Systems*, Academic Press, New York, 1977.

17. M. Slemrod, "The Flip-Flop Circuit as a Neutral Equation," in *Delay and Functional Differential Equations and Their Applications* edited by K. Schmitt, Academic Press, 1972.

18. V. N. Do, "Controllability of Semilinear Systems," *J. Optimization Theory and Applications* **65** (1990) 41–52.

19. D. Salamon, *Control and Observations of Neutral Systems*, Pitman Advanced Publishing Program, Boston, 1984.

20. H. T. Banks, M. Q. Jacobs, and C. E. Langenhop, "Characterization of the Controlled States in $W_2^{(1)}$ of Linear Hereditary Systems," *SIAM J. Control* **13** (1975) 611–649.

21. M. Kalecki, "A Macrodynamic Theory of Business Cycles," *Econometrica* **3** (1935) 327–344.

22. R. G. D. Allen, *Mathematical Economics*, Macmillan, London, 1960.

23. K. J. Arrow, *Production and Capital, Collected Papers of Kenneth J. Arrow*, Belknap Press of Harvard University Press, Cambridge, Massachusetts, 1985.

24. E. N. Chukwu, "Stability and Time-Optimal Control of Hereditary Systems," *N.C.S.U. Lecture Notes*, 1990.

## Chapter 13

# Stability Theory of Large-Scale Hereditary Systems

**Introduction**

In Secs. 2.4 to 4.4 we studied the stability properties of isolated systems and their perturbations, which are described by

$$\dot{x}(t) = f(t, x_t) + g(t, x_t), \tag{13.1.1}$$

or

$$\frac{d}{dt}[x(t) - A_{-1}x(t)] = f(t, x_t) + g(t, x_t). \tag{13.1.2}$$

But several systems of practical interest, such as the economic models of Secs. 1.8 and 9.5 or nonlinear electric circuits described in Secs. 1.6 and 12.3, may often be viewed as interconnected or composite systems. We now investigate the stability properties of such hereditary systems in terms of the same properties of the subsystems and the growth conditions of the interconnecting structure. The analysis will then be used to provide an interesting policy prescription for the growth of large-scale systems. The definitions of the various stability concepts of Sec. 2.4 will be maintained.

## 13.1 Delay Systems

To deal with the problem, we let $C^{n_i}$ denote the set of all continuous functions $\psi^i : [-h, 0] \to E^{n_i}$ with norm

$$\|\psi^i\| = \max\{|\psi^i(t)| : -h \le t \le 0\},$$

where $|\psi^i(t)|$ is the Euclidean norm of $\psi^i$. Consider the interconnected system

$$\dot{z}^i(t) = f_i(t, z_t^i) + \sum_{\substack{j=1 \\ i \ne j}}^{\ell} g_{ij}(t, z_t^j), \quad i = 1, \ldots, \ell, \tag{13.1.3}$$

where

$$f_i : E \times C^{n_i} \to E^{n_i}, \quad g_{ij} : E \times C^{n_j} \to E^{n_i} ,$$

are nonlinear functions. If we let $\sum_{i=1}^{\ell} n_i = n$,

$$x^T = [(z^1)^T, \ldots, (z^{\ell})^T] \in E^n ,$$

$$f(t, x)^T = [f_1(z_1)^T, \ldots, f_{\ell}(z_{\ell})^T] ,$$

$$g_i(t, x) = \sum_{j=1}^{\ell} {}_{i \neq j} g_{ij}(t, z^j) ,$$

and

$$[g(t, x)]^T = [(g_1(x))^T \cdots [g_{\ell}(x))^T] ,$$

(13.1.3) can be written as

$$\dot{x}(t) = f(t, x_t) + g(t, x_t) . \tag{13.1.4}$$

System (1.4) can be viewed as an interconnection of $\ell$-isolated subsystems

$$\dot{z}^i(t) = f_i(t, z_t^i) , \tag{13.1.5}$$

with interconnection described by

$$g_i(x) = \sum_{j=1}^{\ell} {}_{i \neq j} g_{ij}(z^j) .$$

We call (13.1.4) a composite system and (13.1.3) its decomposition. The basic assumptions for the existence of unique solutions are valid for (13.1.1), (13.1.3), (13.1.4), and (13.1.5).

    We recall Theorem 3.1.1 for uniform asymptotic stability of (13.1.4) in $C^n$, which is rephrased for (13.1.5) in the space $C^{n_i}$ as a definition of Property F1 for (13.1.5).

**Definition 13.1.1**    In (13.1.5), suppose $f_i : E \times C^{n_i} \to E^{n_i}$ takes $E\times$ (bounded sets of $C^{n_i}$) into bounded sets of $E^{n_i}$. System (13.1.5) is said to have *Property* F1 if there exist continuous functions $u_i, v_i, w_i : E^+ \to E^+$ that are nondecreasing with the property that $w_i(s), u_i(s), v_i(s)$ are positive for $s > 0, u_i(0) = v_i(0), w_i(0) = 0, u_i(s) \to \infty$ as $s \to \infty$; and there exists a continuous functional $V_i : E \times C^{n_i} \to E$ such that

    (i)  $u_i(|\phi^i(0)|) \leq V_i(t, \phi^i) \leq V_i(\|\phi^i\|)$,
    (ii)  $\dot{V}_i(t, \phi^i) \leq -c_i w_i(|\phi^i(0)|)$, and
    (iii)  $V_i(t, \phi_1^i) - V(t, \phi_2^i)| \leq L_i \|\phi_1^i - \phi_2^i\|$, and all these hold for all $\phi^i, \phi_1^i, \phi_2^i \in C^{n_i}$,
        where $L_i$ is a constant.

Note that Property F1 is the required criteria for uniform asymptotic stability of (13.1.5).

**Theorem 13.1.1**  *For the composite system (13.1.4) with decomposition (13.1.3), assume that the following conditions hold:*

(i)  *Each isolated subsystem (13.1.5) possesses Property* F1.

(ii)  *For each $i, j = 1, \ldots, \ell$, $i \neq j$ there are constants $k_{ij} \geq 0$ such that*

$$|g_{ij}(t, \psi^j)| \leq k_{ij} w_j(\phi^j(0)) \,.$$

(iii)  *All successive principle minors of the test matrix $S = [s_{ij}]$ ($n$ arrow) defined by*

$$s_{ij} = \begin{cases} c_i \,, & \text{if } i = j \,, \\ -L_i k_{ij} \,, & \text{if } i \neq j \,, \end{cases}$$

*are positive.*

*Then the trivial solution of the composite system (13.1.4) is uniformly asymptotically stable.*

**Proof**  Let $V_i$ be the functionals of hypothesis (i), and let $\alpha_i > 0$, $i = 1, \ldots, \ell$, be arbitrary constants. Choose $V$ as follows:

$$V(t, \phi) = \sum_{i=1}^{\ell} \alpha_i V_i(t, \phi^i) \,,$$

for $\phi = [(\phi^1)^T, \ldots, (\phi^\ell)^T]$. There exist functions $u, v, u_i, v_i$ continuous and non-decreasing with the property that $v_i(s) > 0, s > 0, u_i(s) > 0$ for $s > 0, u_i(s) \to \infty, u(s) \to \infty$, and $u_i(0) = v_i(0) = 0$, and such that

$$u(|\phi(0)|) \leq \sum_{i=1}^{\ell} \alpha_i u_i(|\phi^i(0)|) \leq V(t, \phi) \leq \sum_{i=1}^{\ell} \alpha_i V_i(\|\phi^i\|) \leq V(\|\phi\|) \,.$$

Now differentiate $V$ along the solution of (13.1.4). Then

$$\dot{V}(t, \phi) = \sum_{i=1}^{\ell} \alpha_i \dot{V}_i(t, \phi^i)$$

$$\leq \sum_{i=1}^{\ell} \alpha_i \left\{ -c_i w_i(|\phi^i(0)|) + L_i \left| \sum_{\alpha=1}^{\ell} {}_{i \neq j} g_{ij}(t, \phi^i) \right| \right\}$$

$$\leq \sum_{i=1}^{\ell} \alpha_i \left\{ -c_i w_i(|\phi^i(0)|) + \sum_{\substack{i \neq j \\ i=1}}^{\ell} L_i k_{ij} w_j(|\phi^j(0)|) \right\}$$

$$= \alpha^T S w \,,$$

where $\alpha^T = (\alpha_1, \ldots, \alpha_\ell)$, and $w_0^T = (w_1(|\phi^1(0)|), \ldots, w_\ell(|\phi^\ell(0)|))$, and $S$ is the test matrix contained in hypothesis (iii).

From the treatment of $M$-matrices in Michel and Miller [2], the constant $\alpha$ can be chosen such that $\alpha^T S > 0$. Hence we have proved that there is a function $w : E^+ \to E^+$, $w(s) > 0$, $s > 0$ such that

$$\dot{V}(t, \phi) \le -\alpha^T S u \le -w(|\phi(0)|).$$

Therefore the trivial solution of (13.1.4) is uniformly asymptotically stable.    $\square$

**Remark 13.1.1**    The interconnection may be regarded as delay-free in the sense that there are functions $G_{ij}$ such that

$$g_{ij}(\phi^i) = G_{ij}(\phi^i(0)), \quad i, j = 1, \ldots, \ell, \quad i \ne j,$$

where

$$|G_{ij}(\phi^i(0))| \le k_{ij} w_j(|\phi^i(0)|).$$

We consider Systems (13.1.3)–(13.1.5) that are time invariant, i.e., $f_i : C^{n_i} \to E^{n_i}$, $g_{ij} : C^{n_i} \to E^{n_i}$.

**Definition 13.1.2**    The isolated autonomous system (13.1.5) is said to possess Property F2 if $f_i : C^{n_i} \to E^{n_i}$ takes (bounded sets of $C^{n_i}$) into bounded sets of $E^{n_i}$, and furthermore:

(i)  $f_i(0) = 0$.

(ii) There are continuous nondecreasing functions $u_i(s), v_i(s), w_i(s)$, which are positive for $s > 0, u_i(0) = v_i(0) = 0 = w_i(0)$, $u_i(s) \to \infty$ as $s \to 0$, and constants $c_i, L_i$; and there is a Lyapunov function $V_i : E^{n_i} \to E$ such that

(iii) $V_i(0) = 0$, $u_i(|z^i|) \le V_i(z^i) \le v_i(|z^i|)$, $\forall z^i \in E^{n_i}$.

(iv) $\dot{V}_i[\phi^i] \le -c_i(|\phi^i(0)|)$ for any $\phi^i \in C^{n_i}$, such that $\max_{-h \le s \le 0} V_i[\phi^i(s)] = V_i[\phi^i(0)]$.

(v) $|V_i(z_1^i) - V_i(z_2^i)| \le L_i|z_1^i - z_2^i|$, $\forall z_j^i \in E^{n_i}$. If the isolated system

$$\dot{z}^i(t) = f_i(z_t^i) \tag{13.1.6}$$

possesses Property F2, then the trivial solution $z^i = 0$ of (13.1.6) is globally asymptotically stable and every solution $z^i(\phi^i)$ of (13.1.6) satisfies $z_t^i(\phi^i) \to 0$ as $t \to \infty$. See Theorem 3.3.2 and Corollary 3.3.2.

**Theorem 13.1.2**    *The trivial solution of the composite autonomous system*

$$\dot{x}(t) = f(x_t) + g(x_t), \tag{13.1.7}$$

*with its decomposition*

$$\dot{z}^i(t) = f_i(z_t^i) + \sum_{\substack{i \ne j \\ j=1}}^{\ell} g_{ij}(z_t^j), \tag{13.1.8}$$

*is globally uniformly asymptotically stable and every solution $x(\phi)$ of (13.1.7) satisfies $x_t(\phi) \to 0$ as $t \to \infty$ if:*

(i) *Each system (13.1.6) possesses Property F2.*

(ii) *All interconnected $g_{ij}$ satisfy the inequalities $|g_{ij}(\phi^j)| \le k_{ij}w_j(|\phi^j(0)|)$ for $\phi^j \in C^{n_i}$.*

(iii) *All the successive principle minors of the test matrix $S = [s_{ij}]$ defined by*

$$s_{ij} = \begin{cases} c_i\,, & \text{if } i = j\,, \\ -L_i k_{ij}\,, & \text{if } i \ne j\,, \end{cases}$$

*are positive.*

**Proof**   Just as before, define

$$V(x) = \sum_{i=1}^{\ell} \alpha_i V_i(z^i)$$

for $x = [(z')^T, \ldots, [z^\ell]^T]$, and observe that there are functions $u, v$ such that

$$u(|x|) \le V(x) \le v(|x|), \quad u(s) > 0, \quad s > 0, \quad u(s) > 0 \quad \text{as} \quad s > 0\,,$$

and $u(s) \to \infty$ as $s \to \infty$, $V(0) = 0$. Differentiating $V$ we deduce that

$$\dot{V} = \sum_{i=1}^{\ell} \alpha_i \dot{V}_i(z^i)$$

$$\le \sum_{i=1}^{\ell} \alpha_i \left[ -c_i|\phi^i(0)| + \sum_{j=1}^{\ell} {}_{j \ne i} g_{ij}(\phi^j) \right]$$

$$\le \sum_{i=1}^{\ell} \alpha_i \left\{ -c_i|\phi^i(0)| + \sum_{\substack{j \ne i \\ j=1}}^{\ell} L_i k_{ij} w_j(|\phi^j(0)|) \right\}$$

$$= -\alpha^T S u \equiv -w(|x|) < 0\,,$$

where $\alpha^T = (\alpha_1, \ldots, \alpha_\ell)$ $u^T = (w_1(|\phi(0)|), \ldots, w_\ell(|\phi^\ell(0)|))$, and $S$ is the test matrix given in the hypothesis. The above estimates are valid for

$$\max_{-h \le s \le 0} V[\phi(s)] = V[\phi(0)]\,.$$

It follows from Corollary 3.3.2 that the assertions of Theorem 13.1.2 are valid.   $\square$

**Remark 13.1.2**   If the interconnecting structure in $g_i(t, x)$ is replaced by

$$g_i(t, x) = \sum_{j=1}^{\ell} g_{ij}(t, z^j)\,,$$

and the estimate in condition (ii) of Theorem 13.1.1 is replaced by

$$|g_i(t, \psi)| \leq \sum_{j=1}^{\ell} k_{ij} w_j (|\phi^j(0)|),$$

then the test matrix $S$ in (iii) defined by

$$S_{ij} = \begin{cases} \alpha_i(-c_i + L_i k_{ii}), & \text{if } i = j, \\ \dfrac{1}{2}(\alpha_i L_i a_{ij} + \alpha_j m_j a_{ji}), & \text{if } i \neq j, \end{cases}$$

may be assumed to be negative definite.

If $c_i > 0$, the isolated system is stable. If $c_i < 0$, it is unstable. In this case, to ensure stability of the composite system we must have $L_i k_{ii} < 0$ and $|L_i k_{ii}| > c_i$. This has important policy implications.

## 13.2   Uniform Asymptotic Stability of Large-Scale Systems of Neutral Type

In Sec. 4.2 we reported the Lyapunov Theory of stability of the isolated functional differential equation of neutral type

$$\frac{d}{dt}[D(t)x_t] = f(t, x_t) + g(t, x_t), \tag{13.2.1}$$

where $x_t \in C = C([-h, 0], E^n)$ is defined by $x_t(s) = x(t + s)$, $s \in [-h, 0]$. Here $E^n$ is the Euclidean $n$-space with norm $|\cdot|$, and $C$ is the space of all continuous functions mapping $[-h, 0]$ into $E^n$ with the sup norm $\|\cdot\|$. The operator $D(\ )$ : $[\sigma, \infty) \times C \to E^n$ is given by

$$D(t)\phi = \phi(0) - g(t, \phi), \quad t \in E, \quad \phi \in C. \tag{13.2.2}$$

The function $g$ in (13.2.2) is given by

$$g(t, \phi) = \int_{-h}^{0} [d_\theta \mu(t, \theta)]\phi(\theta), \tag{13.2.3a}$$

where $\mu(t, \theta)$ is an $n \times n$ matrix function of $t \in [\tau, \infty)$, $\theta \in [-h, 0]$ of bounded variation in $\theta$ that satisfies the inequality

$$\left| \int_{-s}^{0} d_\theta \mu(t, \theta)\phi(\theta) \right| \leq \ell(s) \sup_{-s \leq \theta \leq 0} |\phi(\theta)|, \tag{13.2.3b}$$

for all $t \in [\tau, \infty)$, $\phi \in C$, where $\ell(s)$ is a scalar continuous nondecreasing function of $s \in [0, h]$ with $\ell(0) = 0$. We now consider equations of the form

$$\frac{d}{dt}[D^i(t)z_t^i] = f_i(t, z_t^i) + \sum_{j=1, i \neq j}^{\ell} k_{ij}(t, z_t^j), \tag{13.2.4}$$

where $z_t^i \in C^{n_i}([-h, 0], E^{n_i}) \equiv C^{n_i}$, and $f_i : E \times C^{n_i} \to E^{n_i}$, $g_{ij} : E \times C^{n_j} \to E^{n_i}$. Letting $\sum_{i=1}^{\ell} n_i = n$,

$$x^T = [(z')^T, \ldots, (z^\ell)^T] \in E^n,$$

$$[f(t, x)]^T = [(f_1(t, z')^T, \ldots, f_\ell(t, z^\ell)]^T],$$

$$k_i(t, x) = \sum_{\substack{i \neq j \\ j=1}}^{\ell} k_{ij}(t, z^j),$$

and

$$[k(t, x)]^T = [(k_1(t, x))^T, \ldots, (k_\ell(t, x))^T],$$

we may rewrite (13.2.4) as

$$\frac{d}{dt}[D(t)x_t] = f(t, x_t) + k(t, x_t), \tag{13.2.5}$$

where

$$D(t)x_t = [[D'z_t']^T, \ldots, [D^\ell z_t^\ell]^T].$$

We assume that

$$f(t, 0) = k(t, 0) = 0, \quad \forall t \in E. \tag{13.2.6}$$

The composite system (13.2.5) with decomposition (13.2.4) may be regarded as an interconnection of $\ell$-isolated subsystems

$$\frac{d}{dt}[D^i(t)z_t^i] = f_i(t, z_t^i), \tag{13.2.7}$$

with interconnecting structure described by

$$k_i(t, x) = \sum_{\substack{i \neq j \\ j=1}}^{\ell} g_{ij}(t, z^j). \tag{13.2.8}$$

We assume that conditions on $D, f$, and $k$ are sufficient for the initial-value problem to have a unique solution. We assume also that $D$ is a uniformly stable operator in the sense of Sec. 4.2. The conditions of Theorem 4.2.1 that ensure uniform asymptotic stability for (13.2.7) are now isolated and called Property A.

**Definition 13.2.1** The isolated subsystem (13.2.7) is said to possess *Property* A if there exist a continuous functional $V_i(t, \phi^i)$ on $E \times C^{n_i}$, and three continuous functions $u_i, v_i, w_i$ mapping $[0, \infty) \to [0, \infty)$, which are strictly increasing with $u_i(s) \to \infty$ as $s \to \infty$, $u_i(s), v_i(s), w_i(s) > 0$ for $s > 0$; and there exist constants $c_i > 0, M_i > 0$ such that

    (i)  $u_i(|D^i(t)\phi^i|) \leq V_i(t, \phi^i) \leq v_i(\|\phi^i\|)$,
    (ii)  $\dot{V}_i(t, \phi^i) \leq c_i w_i(|D^i(t)\phi^i|)$ or $\dot{V}_i(t, \phi^i) \leq c_i w_i(|\phi^i(0)|)$,

(iii) $|V_i(t, \phi_1^i) - V_i(t, \phi_2^i)| \leq M_i \|\phi_1^i - \phi_2^i\|$, holds for all $\phi^i, \phi_1^i, \phi_2^i \in C^{n_i}$.

**Theorem 13.2.1**   *The trivial equilibrium of the composite system* (13.2.5) *is globally uniformly asymptotically stable if the following conditions hold:*

(i)  $D^i(t)$ *is a uniformly stable operator.*

(ii)  *Each isolated subsystem* (13.2.7) *possesses Property A.*

(iii)  *For each* $i, j = 1, \ldots, \ell (i \neq j)$ *there are constants* $a_{ij} \geq 0$ *such that*

$$|k_{ij}(t, \phi^j)| \leq a_{ij} w_j(|D^j(t)\phi^j|)$$

*for all* $\phi^j \in C^{n_j}$, $j = 1, \ldots, \ell$, $t \in E$.

(iv)  *The constants* $c_i, M_i$ *in* (ii) *and* $a_{ij}$ *in* (iii) *satisfy the following conditions: There exists a vector* $\alpha^T = (\alpha_1, \ldots, \alpha_\ell)$ *with* $\alpha_j > 0$ *such that* $S = [s_{ij}]$ *specified by*

$$s_{ij} = \begin{cases} \alpha_i(c_i + M_i a_{ij}/(1 - L_i(h_0))), & i = j, \\ \dfrac{1}{2}(a_i M_i a_{ij}/(1 - L_i(h_0)) + \alpha_j M_j a_{ji}/(1 - L_j(h_0))), & i \neq j \end{cases}$$

*is a negative definite matrix. The constant* $L_i(h_0)$ *is associated with* $g_j$ *as in* (13.2.3b).

**Proof**   Let $V_i$ be given by hypothesis (ii), and let $\alpha_i$, $i = 1, \ldots, \ell$ be arbitrary constants. Define

$$V(t, x_t) = \sum_{i=1}^{\ell} \alpha_i V_i(t, z_t^i).$$

Because of Property A, there exists $u, v, w$ with the same properties as $u_i, v_i, w_i$

$$u(|D(t)x_t|) \leq \sum_{i=1}^{\ell} \alpha_i u_i(|D^i(t)z_t^i|)$$

$$\leq V(t, x_t) \leq \sum_{i=1}^{\ell} \alpha_i v_i(\|z_t^i\|)$$

$$\leq v(\|x_t\|).$$

For the second inequality in the derivative of $V$, we use Cruz and Hale Equation (7.8) of [3, p. 354] as was done by Chukwu [4, p. 354] to evaluate the derivative of $V$ along solutions of (13.2.5). Indeed,

$$\dot{V}(t, x_t) = \sum_{i=1}^{\ell} \alpha_i \dot{V}(t, z_t^i)$$

$$\leq \sum_{i=1}^{\ell} \alpha_i c_i w_i(|D^i(t)z_t^i|) + \frac{M_i \alpha_i}{1 - L_i} \sum_{\substack{i \neq j \\ j=1}}^{\ell} |k_{ij}(t, z_t^j)|$$

$$\leq \sum_{i=1}^{\ell} \alpha_i [c_i w_i(|D^i(t)z_t^i|) + \frac{M_i \alpha_i}{1 - L_i} \left[ \sum_{\substack{i \neq j \\ j-1}}^{\ell} a_{ij} w_j(|D^j(t)z_t^j|) \right].$$

Now let $R = [r_{ij}]$ be the $\ell \times \ell$ matrix specified by

$$r_{ij} = \begin{cases} \alpha_i[c_i + M_i a_{ii}/(1 - L_i)] & i = j, \\ \alpha_i M_i a_{ij}/(1 - L_i), & i \neq j. \end{cases}$$

Then

$$\dot{V}(t, x_t) = w^T R w = w^T (R + R^T)/2, \quad W = w^T S w,$$

where $S = [s_{ij}]$ is the test matrix in (iv), and where

$$w^T = [w_1(|D' z_t'|), \dots, w_\ell(|D^\ell z_t^\ell|)].$$

Because $S$ is symmetric and negative definite, all its eigenvalues are negative, so that for some $\lambda > 0$

$$\dot{V}(t, x_t) \leq -\lambda w^T w = -\lambda \sum_{i=1}^{\ell} w_i(|D^i z_t^i|).$$

Hence $\dot{V}(t, x_t)$ is negative definite. Global uniform asymptotic stability follows at once. $\qquad\square$

## 13.3 General Comments

The parameter $c_i$ in Definition 13.2.1 is the so-called degree or margin of stability. As observed by Michel and Miller [2] in ordinary differential equations, for the system to satisfy (iv) of Theorem 13.2.1, it is necessary that

$$c_i + M_i a_{ii} < 0, \quad i = 1, \dots, \ell.$$

Thus if $c_i > 0$ so that (13.2.7) is unstable, we must insist that $M_i a_{ii} < 0$ and $|M_i a_{ii}| > c_i$. Thus, to ensure stability of the large-scale system when a subsystem is unstable, we must provide a sufficient amount of stabilizing feedback from outside the subsystem. This has very great policy implications, and is now restated as a universal principle.

**Principle 13.2.1** *If a subsystem is misbehaving and unstable, the large-scale system can be made stable provided there is sufficient external feedback (external force) brought to bear on the subsystem to ensure stability. The dismantling of external feedback on a subsystem can create instability.*

This principle is analogous to those of controllability of large economic systems studied in Sec. 9.5.

## References


1. E. N. Chukwu, "Uniform asymptotic stability of large scale systems of neutral type," *IEEE Proceedings of the 18th Southeastern Symposium on System Theory*, 1986.
2. A. N. Michel and R. K. Miller, *Qualitative Analysis of Large Scale Dynamical Systems*, Academic Press, New York, 1977.
3. M. A. Cruz and J. K. Hale, "Stability of Functional Differential Equations of Neutral Type," *J. Differential Equations* **7** (1970) 334–355.
4. E. N. Chukwu, "Global Asymptotic Behaviour of Functional Differential Equations of the Neutral Type," *Nonlinear Analysis, Theory*, Method and Application **5** (1981) 853–872.

# Appendix to Section 1.10

**MATLAB Programs and Graphs**
for
**Economic Models with Delay**

**Program US2.M**

1. Definition of Economic Variables and Terms. pp. 479–481.
   In this section the economic variables extracted from the International Financial Statistic Yearbook are defined by Symbols.
2. The U.S.A. data. pp. 479–480.
   Displaced here are data extracted from the International Financial Statistic Yearbook 1994, UN Financial Statistic 1974, UN National Accounts Statistics.
3. Equations and Formulae postulated for economic variables in the body of the book are now identified, pp. 481–483.

$$
\begin{array}{lr}
ML = L - M & 481 \\
L & 481 \\
I & (1.10.5) \\
C & (1.10.3) \\
X & (1.10.11) \\
T & (1.10.2) \\
G & (1.10.7) \\
Z & (1.10.2),\ (1.10.12),\ (1.10.13) \\
\dot{R}(t) & (1.10.20),\ (1.10.21),\ (1.10.23) \\
\dot{p}(t) & (1.10.35),\ \text{inflation} \\
B = \text{Balance of payment} & (1.10.38) \\
E = \text{Cumulative Balance of Payment} & (1.10.37) \\
y = \text{National Income} & (1.10.44) \\
D = \text{deliveries of new equipment} & (1.10.48)
\end{array}
$$

$$\frac{dk}{dt} = \text{flow of capital stock} \qquad (1.10.53)$$
$$L = \text{employment} \qquad (1.10.60)$$
$$y = \text{income} \qquad (1.10.52)$$
$$Gy = \text{income government} \qquad (1.10.50)$$
$$Xy = \text{income export} \qquad (1.10.50)$$

4. MATLAB Regression Programs for Economic Variables, pp. 479–484. See "System Indentificatin Toolbox" for use with MATLAB, Lenhart Ljung, The Math Works, Inc. I.30 → I.32, 14.8, pp. 179, 2.16.
5. MATLAB Plot, Subplot programs, pp. 483–484.
6. Identification of the economic dynamics:

$$\dot{x}(t) - A_{-1}\dot{x}(t - h) = A_0 x(t) + A_1 x(t - h) + Bu(t)$$

with coefficients $A_{-1}$, $A_0$, $A_1$, $B$, $B_1$ and given in pp. 485–489 and identified in p. 490 using (Ndu.M, US3.M) program. In p. 485.

7. Diagrams and plots, Fig. U.S.1 → Fig. U.S.7.
8. We use the rank condition of Salamon and the full rank of $B$ to deduce the controllability in $W_p^{(1)}$ of the economic state variables of the dynamics

$$\dot{x}(t) - A_{-1}\dot{x}(t - h) = A_0 x(t) + A_1 x(t - h) + Bu(t).$$

If $\gamma$ is any complex number rank$[\Delta(\gamma), B] = n = 6$ and rank$[\gamma I - A_{-1}, B] = 6$, is the required condition (p. 157, D. Salamon, "Control and Observation of Neutral Systems", Pitman Advanced Publishing Program, Boston, MA). This rank condition is satisfied by our linear model, p. 491.

# U.S.A.
# Program US2.M

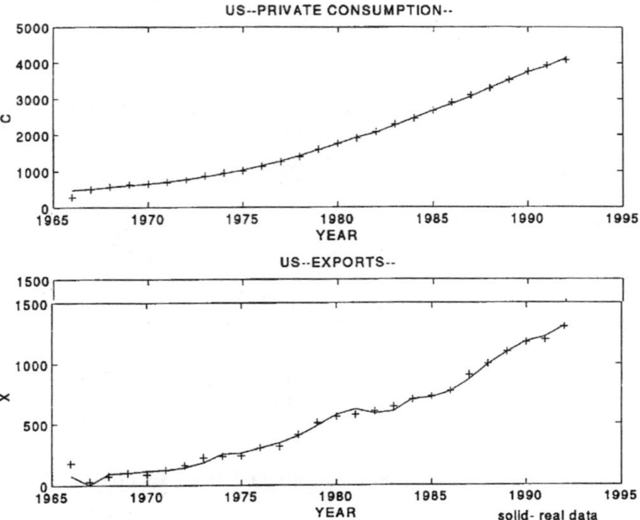



Fig. U.S.1   U.S. – Private Consumption, Exports.

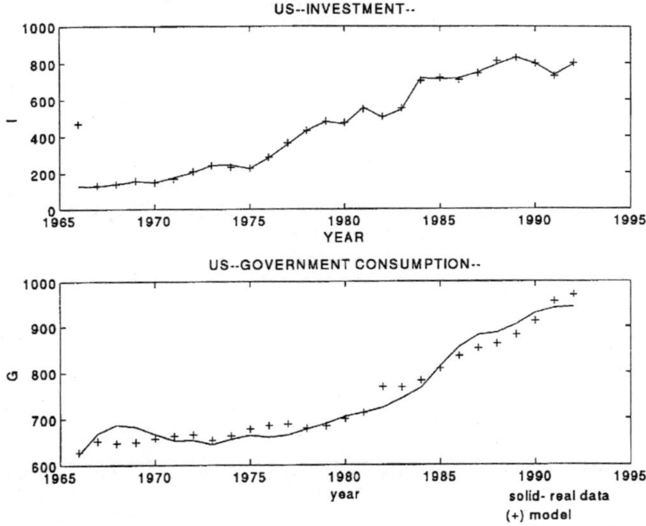



Fig. U.S.2   U.S. – Investment, Government Consumption.

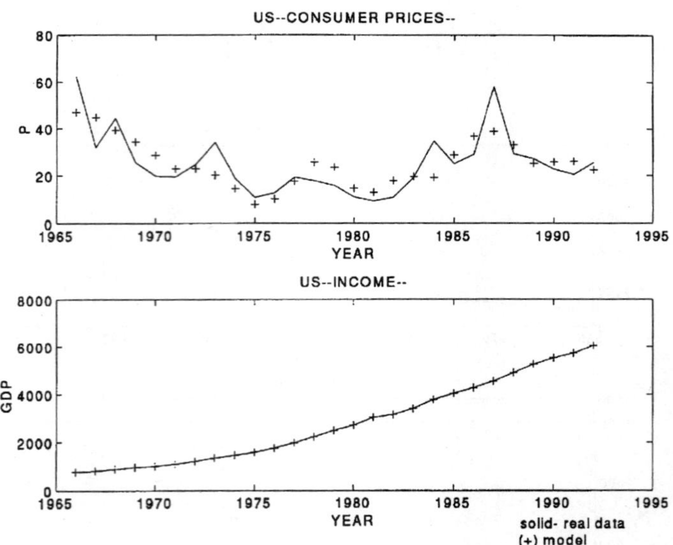


Fig. U.S.3   U.S. – Consumer Prices, Income.

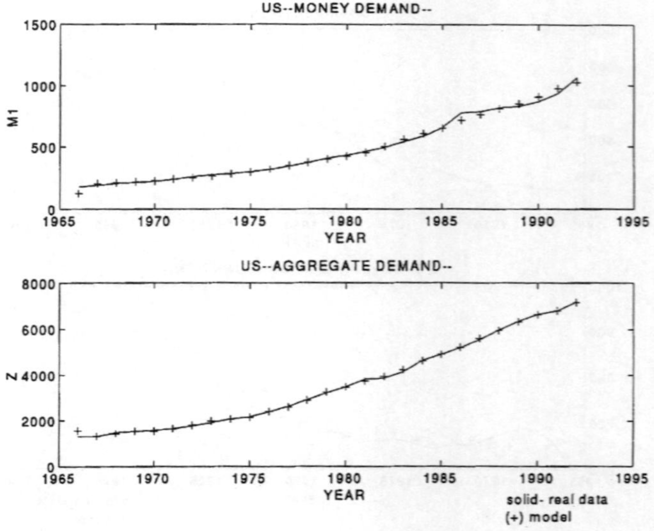


Fig. U.S.4   U.S. – Money Demand, Aggregate Demand.

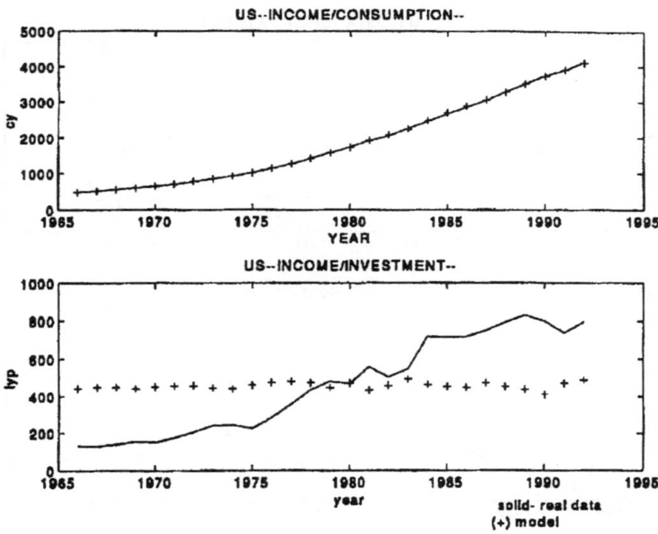

Fig. U.S.5   U.S. – Income/Consumption, Income/Investment.

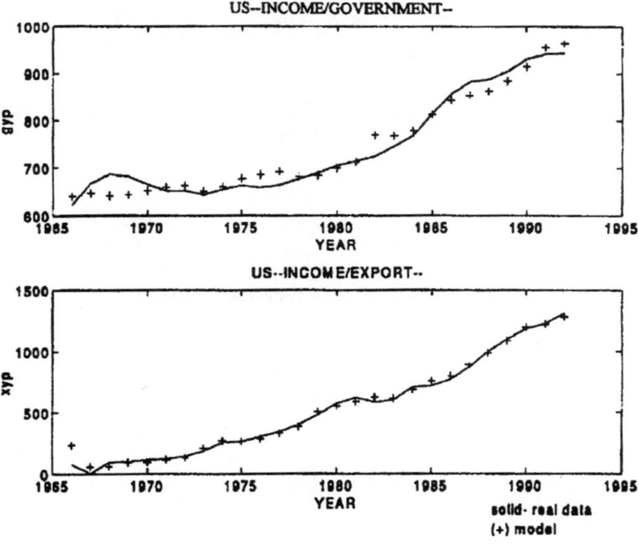

Fig. U.S.6   U.S. – Income/Government, Income/Export.

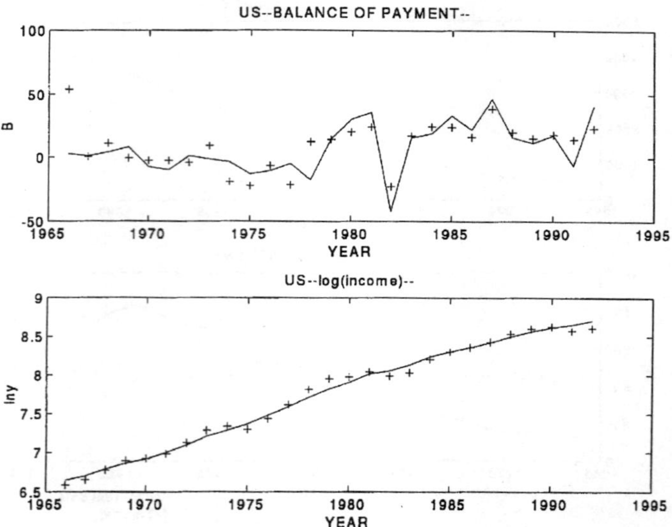



Fig. U.S.7   U.S. – Balance of Payment, Log(income).

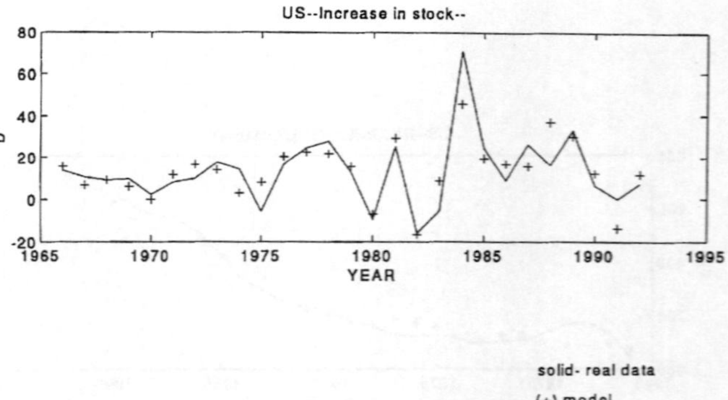



Fig. U.S.8   U.S. – Increase in Stock.

Appendix: U.S.A

% USA's data is in billions of US dollars
% adopted from International Financial Statistic Yearbook 1994, except where noted.
% Use ndu.m to obtain the matrix entries

YEAR =[1966 1967 1968 1969 1970 1971 1972 1973 1974 1975 1976 1977 1978 1979 1980 1981 1982 1983
1984 1985 1986 1987 1988 1989 1990 1991 1992]';
XX   =[39.0 41.4 45.3 49.3 57 59.3 66.2 91.8 124.3 136.3 148.9 158.8 186.2 228.9 279.2 303 282.6 276.7
302.4 302.1 319.2 364.0 444.2 508 557 601.5 640.5]';
IMP  =[-37.1 39.9 -46.6 -50.5 -55.8 -62.4 -74.2 -91.2 -127.5 -122.7 -151.2 -182.5 -212.3 -252.7 -293.9 -317.7 -
303.2 -328.1 -405.1 -417.6 -451.7 -507.1 -552.2 -587.7 -625.9 -621.1 -670.1]';
X    =XX-IMP;
%TX  =[1967-70,1972-90   233.9 269.3 305.0 258.5 292.7 291743 354182 399776 443494 458256 523873
587945 668257 757114 757114 826725 949668 970493 1029670 1142340 1239250 1311950 1424120
1513520 1630440]';
TX   =[ 190000 233900 254200 293196 299387 317571 361635 406617 453017 467002 536907 602083
668257 757114 828888 850000 974757 1036008 1148734 1248179 1322479 1435529 1526427 1657800
1749200 1829800 1877200]';
d    =[]';
%    =[112 113 117 117 100 102 106 108 104 105 108  1977-84  4.0 4.7 5.3 3.9 1.3 1.9 1.8 4.5]';
ta   =[1900 2000 2300 2400 2584 2767 3124 3620 3771 3780 4675 5485 7162 7202 7535 9993 8688 9430
12042 13067 13312 13923 15054 16096 16339 16197 17164]';
E    =[]';
F    =[-0.65 -0.88 -0.84 -0.94 -1.10 -1.11 -1.11 -1.25 -1.01 -0.92 -0.91 -0.82 -0.86 -0.91 -1.03 -4.51 -8.73 -9.06 -
9.75 -9.56 -10.12 -10.55 -11.96 -12.32 -12.39 -14.04 -14.46]';
D    =[13.8 10.5 9.1 9.7 2.3 8.1 9.9 17.7 14.3 -5.7 16.7 24.7 27.9 12.8 -9.5 25.5 -16.0 -5.5 71.1 24.6 8.7 26.3
16.3 33.3 6.3 -0.1 7.3]';
K    =[116.6 117.6 130.8 145.5 147.7 167.2 195.3 224.8 230.8 230.7 268.3 333.5 406.1 467.5 477.1 532.6 519.4
552.2 647.8 690.0 709.0 723.0 777.4 798.9 793.2 736.9 789.1]';
kk   =[116.6 117.6 130.8 145.5 147.7 167.2 195.3 224.8 230.8 230.7 268.3 333.5 406.1 467.5 477.1 532.6
519.4 552.2 647.8 690.0 709.0 723.0 777.4 798.9 793.2 736.9 789.1 876.1]';
I    =[130.4 128.1 139.9 155.2 150.0 175.3 205.2 242.5 245.1 225.0 285.0 358.2 434.0 480.3 467.6 558.1 503.4
546.7 718.9 714.6 717.7 749.3 793.7 832.2 799.5 736.8 796.4]';
B    =[2.83 1.16 4.12 8.25 -7.26 -9.88 1.25 -1.28 -3.47 -12.89 -10.39 -4.73 -17.53 14.75 30.53 35.99 -42.27
15.66 19 33.3 22.1 46.5 16.07 11.69 17.8 -6.38 41.22]';
BD   =[0.47 2.83 1.16 4.12 8.25 -7.26 -9.88 1.25 -1.28 -3.47 -12.89 -10.39 -4.73 -17.53 14.75 30.53 35.99 -
42.27 15.66 19 33.3 22.1 46.5 16.07 11.69 17.8 -6.38]';


T    = [X-B-F]';
e    =[1 1 1 1 1 1.0857 1.0857 1.2064 1.2244 1.1707 1.1618 1.2147 1.3028 1.3173 1.2754 1.1640 1.1031
1.0470 0.9802 1.0984 1.2232 1.4187 1.3457 1.3142 1.4227 1.4304 1.3750]';
w    =[25.1 26.1 27.8 29.4 31.0 33.0 35.3 37.7 40.9 44.6 48.2 52.4 57.0 61.8 67.2 73.8 78.5 81.5 84.8 88.1 89.9
91.5 94.1 96.8 100 103.3 105.8]';
G    =[622.4 667.9 686.8 682.0 665.8 652.4 653.0 644.2 655.4 663.5 659.2 664.1 677.0 689.3 704.2 713.2 723.6
743.8 766.9 813.4 855.4 881.5 886.8 904.4 929.9 941.1 943.0]';
L    =[58.4 60.1 62.1 64.3 64.8 65.1 67.3 70.2 71.5 70.3 72.5 75.4 79.2 82.1 82.6 83.3 81.8 82.4 86.3 89.0 90.8
93.2 96.2 98.6 100 98.9 99.2]';
LD   =[55.5 58.4 60.1 62.1 64.3 64.8 65.1 67.3 70.2 71.5 70.3 72.5 75.4 79.2 82.1 82.6 83.3 81.8 82.4 86.3 89.0
90.8 93.2 96.2 98.6 100 98.9]';
LP   =[1.70 2.00 2.20 0.50 0.30 2.20 2.90 1.30 -1.20 2.20 2.90 3.80 2.90 0.50 0.70 -1.50 0.60 3.90 2.70 1.80
2.40 3.00 2.40 1.40 -1.10 0.30 1.50 ]';
LPD =[ 2.90 1.70 2.00 2.20 0.50 0.30 2.20 2.90 1.30 -1.20 2.20 2.90 3.80 2.90 0.50 0.70 -1.50 0.60 3.90 2.70
1.80 2.40 3.00 2.40 1.40 -1.10 0.30]';
C    =[481.6 509.3 559.1 603.7 646.5 700.3 767.8 848.2 927.7 1024.9 1143.1 1271.5 1421.3 1583.7 1748.1
1926.3 2059.2 2257.6 2460.3 2667.4 2850.6 3052.2 3296.1 3523.1 3748.4 3906.4 4139.9]';

```
M   =[180.4 194.0 209.6 216.6 225.9 240.7 262.2 277.3 286.9 302.0 318.7 345.7 375.2 408.2 432.2 460.3
494.2 540.3 585.0 655.0 776.1 786.2 820.1 830.9 870.5 935.7 1065.8]';
MP  =[13.6 15.6 7.0 9.3 14.8 21.5 15.1 9.6 15.1 16.7 27.0 29.5 33.0 24.0 28.1 33.9 46.1 44.7 70.0 121.1 10.1
33.9 10.8 39.6 65.2 130.1 164.5]';
MPD =[4.2 13.6 15.6 7.0 9.3 14.8 21.5 15.1 9.6 15.1 16.7 27.0 29.5 33.0 24.0 28.1 33.9 46.1 44.7 70.0 121.1
10.1 33.9 10.8 39.6 65.2 130.1]';
ML  =[26.2792 23.6938 11.7627 9.7887 12.6905 14.2568 12.8253 3.9740 2.1125 14.3533 33.3342 38.6344
28.0482 7.6254 11.3427 22.2640 65.5101 64.9009 63.8482 56.3244 -13.8977 26.3147 31.4402 72.6962
91.7929 87.9106 14.1891]';
MD  =[176.2 180.4 194.0 209.6 216.6 225.9 240.7 262.2 277.3 286.9 302.0 318.7 345.7 375.2 408.2 432.2
460.3 494.2 540.3 585.0 655.0 776.1 786.2 820.1 830.9 870.5 935.7]';
P   =[24.9 25.5 26.6 28.1 29.7 31.0 32.0 34.0 37.8 41.2 43.6 46.4 49.9 55.6 63.1 69.9 73.9 76.2 79.5 82.4 83.9
87.0 90.5 94.9 100.0 104.2 107.4]';
Pf  =P.*e;
Pfe =Pf.*e;
R   =[4.5 4.5 5.5 6 5.5 4.5 4.5 7.5 7.75 6 5.25 6 9.5 12 13 12 8.5 8.5 8 7.5 5.5 6 6.5 7 6.5 3.5 3 ]';
RD  =[4.5 4.5 4.5 5.5 6 5.5 4.5 4.5 7.5 7.75 6 5.25 6 9.5 12 13 12 8.5 8.5 8 7.5 5.5 6 6.5 7 6.5 3.5]';
RP  =[0  1.0 0.5  -0.5 -1.0  0 3.0 0.2 -1.75  -0.75 0.75 -0.75 0.75  3.5 2.5 1.0 -1.0 -3.5  0 -0.5 -0.5 -2.0 0.5 0.5
0.5 -0.5  -3.0]';
RPD =[0 0  1.0 0.5  -0.5 -1.0  0 3.0 0.2 -1.75  -0.75 0.75 -0.75 0.75  3.5 2.5 1.0 -1.0 -3.5  0 -0.5 -0.5 -2.0 0.5 0.5
0.5 -0.5]';
GDP =[769.8 814.3 889.3 959.5 1010.4 1096.8 1206.5 1349.1 1458.8 1584.8 1767.1 1974.1 2232.7 2488.7
2708.1 3030.6 3149.6 3405.1 3777.2 4038.7 4268.6 4539.9 4900.4 5250.8 5522.2 5722.9 6038.5]';
y   =[769.8 814.3 889.3 959.5 1010.4 1096.8 1206.5 1349.1 1458.8 1584.8 1767.1 1974.1 2232.7 2488.7
2708.1 3030.6 3149.6 3405.1 3777.2 4038.7 4268.6 4539.9 4900.4 5250.8 5522.2 5722.9 6038.5]';
yD  =[702.7 769.8 814.3 889.3 959.5 1010.4 1096.8 1206.5 1349.1 1458.8 1584.8 1767.1 1974.1 2232.7
2488.7 2708.1 3030.6 3149.6 3405.1 3777.2 4038.7 4268.6 4539.9 4900.4 5250.8 5522.2 5722.9]';
yP  =[67.10 44.50 75.0 70.2 50.90 86.40 109.70 142.60 109.70 126.0 182.30 207.0  258.60 256.0 219.40
322.50 119.0 255.5  372.10 261.5 229.9  271.3 360.5  350.4 271.40 200.70 315.60]';
yPD =[54.60 67.10 44.50 75.0 70.2 50.90 86.40 109.70 142.60 109.70 126.0 182.30 207.0  258.60 256.0
219.40 322.50 119.0 255.5  372.10 261.5 229.9  271.3 360.5  350.4 271.40 200.7]';
YT  =y-TX;
YTD =[516760 579800 580400 635100 666304 711013 779229 844865 942483 1005783 1117798 1230193
1372017 1564443 1731586 1879212 2180600 2174843 2369092 2628466 2790521 2946121 3104371
3373973 3593000 3773000 3893100]';
YTP =[600 54700 31204 44709 68216 65636 97618 63300 112015 112395 141824 192426 167143 147626
301388 -5757 194249 259374 162055 155600 158250 269602 219027 180000 120100 268200 316600]';
YTPD=[63040 600 54700 31204 44709 68216 65636 97618 63300 112015 112395 141824 192426 167143
147626 301388 -5757 194249 259374 162055 155600 158250 269602 219027 180000 120100 268200]';
N = y.\L;
price=[23.7 24.1 24.9 25.5 26.6 28.1 29.7 31.0 32.0 34.0 37.8 41.2 43.6 46.4 49.9 55.6 63.1 69.9 73.9 76.2 79.5
82.4 83.9 87.0 90.5 94.9 100.0 104.2 107.4]';
for i=1:27
pp(i)=price(i+1)-price(i);
kprim(i) =kk(i+1)-kk(i);
end
pie=[pp]';
KP =kprim';
DP =pie.\P;
logy=log(y);
```

```
%Net export = X
%G IS GOTTEN FROM REPORT TO THE PRESIDENT ON ACTIVITIES OF THE COUNCIL
%OF ECONOMIC ADVISER DURING 1992
% TX =total current receipts of general government(millions of united states dollars)
% TX(1966,1971) comes from UN financial statstics 1974
% n =labor productivity(A/1) = GNP/employment
```

```
%    =From Yearbook of labor statistics(international labor office)
%
%  K =capital stock(Gross fixed capital formation)
%  I =investment(Gross capital formation)
%  D =increase in stock
%    The UN(National accounts statistics) defines Gross capital
%    formation as the sum of the increase in stocks and gross fixed
%    capital formation
%
%  y =GDP(Gross domestic product)
%  yD =y(t-1)
%  yP =y'(t)
%  yPD=y'(t-1)
%  X  =Net export(export-import of goods and services)
%  d =preferential trade agreement transportation
%  Pf =import price level in foreign currency
%  ta =tariffs(from: statistical abstract of the united states-1990)
%  ta for 1966-69 comes from UN financial statstics 1974
%  E =cumulative balance of payment
%  F =Net private outflow of capital
%  C =private consumption
%  M,M1  =money supply
%  Md   =money demand(liquidity), where, ML = md-M
%  P =consumer prices
%  R =interest rate(discount rate)
%  RD =R(t-1)
%  RP =R'(t)
%  RPD=R'(t-1)
%  pie=inflation
%  L =nonagri. employment
%  B =Balance of payment(deposit money banks)
%  T =Net government transfer of capital to foreigers and firms
%  G =Government consumption
%  DP =D*log(P) = p(t)'/p(t)

%format long;
%Bert1=[X,G,I,C,GDP,R,M,YEAR]
 [years,n]=size(YEAR);

X1=[-21.8;X(1:years-1)];
G1=[475.3;G(1:years-1)];
I1=[296.4;I(1:years-1)];
C1=[1178.9;C(1:years-1)];
M1=[140;M(1:years-1)];
R1=[3.36:R(1:years-1)];
GDP1=[494.2;GDP(1:years-1)];
DGDP=GDP-GDP1;
Z=X+G+I+C;
Z1=X1+G1+I1+C;

%temp=[YEAR X1 G1 I1 C1 Z1 y1 yD M11 RD1 yPD1 RP RPD T1 T P P-PF P1 P-PF1 G L LD LP LPD L1
LD1 LP1 LPD1 t t1 e e1 d d1];
%save Bert1.asc temp -ascii -double -tabs
%clear Bert1 temp;

%ML = L-M = m1-m0 - m1*y(t) - m2*y(t-h) - m3*R(t) - m4*R(t-h) - m6*P(t) - m7*R'(t-h)
```

```
% L(t) =m0 +m1*y(t)+ m2*y(t-h) +m3*R(t) +m4*R(t-h) +m5*R'(t-h) +m6*p(t);
temp=[M ones(size(y)),y,yD,R,RD,RPD,P];
thl=arx(temp,[0,1 1 1 1 1 1 1 0 0 1 0 1 1 0]);
Lp=predict([M,ones(size(y)),y,yD,R,RD,RPD,P],thl,4);
MM=thl(6,1:7);


%L(t) =17.2335 +0.8380.*y(t) -0.3061.*yD(t-h) -0.3175.*R(t) ÷3.6165.*RD(t-1)-3.3413.*P(t) -1.7714.*RPD(t-
1);
% C(t) =c0 +c1*YT(t) +c2*YT(t-h) + c3*YT(t)' +c4*YT'(t-h) +c5*R(t) + c6*R(t-h) +c7*(17.2335
+0.8380.*y(t) -0.3061.*yD(t-h) -0.3175.*R(t) +3.6165.*RD(t)-3.3413.*P(t) -1.7714.*RPD(t)) - M(t));
temp=[C ones(size(R)),YT,YTD,YTP,YTPD,R,RD,ML];
thc=arx(temp,[0,1 1 1 1 1 1 1 1 0 0 1 0 1 0 1 0]);
Cp=predict([C,ones(size(R)),YT,YTD,YTP,YTPD,R,RD,ML],thc,4);
CC=thc(7,1:8);


% I(t) =i0 -i1*y(t) + i2*y(t-h) -i3*y'(t) + i4*y'(t-h) +i5*R(t) +i6*R(t-h) +i8*L(t) +i9*L(t-h)-i11*K(t) -i13*ML(t)
temp=[I ones(size(y)),y,yD,yP,yPD,R,RD,L,LD,K,ML];
thi=arx(temp,[0,1 1 1 1 1 1 1 1 1 1 1 0 0 1 0 1 0 1 0 1 0 0]);
Ip=predict([I,ones(size(y)),y,yD,yP,yPD,R,RD,L,LD,K,ML],thi,4);
II=thi(10,1:11);


% X(t) =x0 + x1*y(t) + x2*y(t-h) +x3*y'(t) +x4*y'(t-h) +x5*R(t)+ x8*L(t) +x10*L'(t-h) +x12*P(t) +x16*ta(t)
+x15*e(t) +x17*d(t)
temp=[X ones(size(y)),y,yD,yP,yPD,R,L,LPD,P,ta,e];
thx=arx(temp,[0,1 1 1 1 1 1 1 1 1 1 1 0 0 1 0 1 0 0 1 0 0 0]);
Xp=predict([X,ones(size(y)),y,yD,yP,yPD,R,L,LPD,P,ta,e],thx,4);
XX=thx(11,1:12);


% G(t) =g0 + g1*y(t) + g2*y(t-h) +g3*y'(t) +g4*y'(t-h) +g5*R(t) +g8*L(t);
temp=[G ones(size(y)),y,yD,yP,yPD,R,L];
thg=arx(temp,[0,1 1 1 1 1 1 1 0 0 1 0 1 0 0]);
Gp=predict([G,ones(size(y)),y,yD,yP,yPD,R,L],thg,4);
GG=thg(6,1:7);


%y(t) = zs0 +zs1*y(t) +zs2*y(t-h) +zs4*y'(t) +zs5*R(t) +zs8*L(t) +zs10*L'(t) +zs13*M1(t) -zs14*TX(t)
+zs15*e(t) +zs16*ta(t) +zs17*d(t)*(0) +1/h*K'(t); h=1
temp=[y ones(size(y)),y,yD,yP,R,L,LP,M,TX,e,ta,KP];
thy=arx(temp,[0,1 1 1 1 1 1 1 1 1 1 1 1 0 0 1 0 0 0 0 0 0 0 0]);
yp=predict([y,ones(size(y)),y,yD,yP,R,L,LP,M,TX,e,ta,KP],thy,4);
yy=thy(11,1:12);


%cy(t) = y10 +y11*y(t) + y12*y'(t) - y13*R(t) + y15*ML(t) +y18*YT(t)
temp=[C ones(size(y)),y,yP,R,ML,YT];
thcy=arx(temp,[0,1 1 1 1 1 1 1 0 0 0 0 0 0]);
cyp=predict([C,ones(size(y)),y,yP,R,ML,YT],thcy,4);
ccy=thcy(5,1:6);


%Iy(t) = I0 +1/h*KP(t)
temp=[I ones(size(KP)),KP];
thIy=arx(temp,[0,1 1 0 0]);
Iyp=predict([I,ones(size(y)),KP],thIy,4);
IIy=thIy(1,1:2);


%Gy(t) = gs0 +gs1*y(t) +gs4*y'(t) +ys5*R(t) +gs8*L(t)
temp=[G ones(size(y)),y,yP,R,L];
thgy=arx(temp,[0,1 1 1 1 1 0 0 0 0 0]);
gyp=predict([G,ones(size(y)),y,yP,R,L],thgy,4);
```

```
ggy=thgy(5,1:6);

%Xy(t) = x0 +x1*y(t) +x2*y(t-h) +x5*R(t) +x8*L(t) +x10*L'(t) +x11*e(t) +x12*ta(t) +x13*d(t)*(0)
temp=[X ones(size(y)),y,yD,R,L,LP,e,ta];
thxy=arx(temp,[0,1 1 1 1 1 1 1 1 0 0 1 0 0 0 0 0]);
xyp=predict([X,ones(size(y)),y,yD,R,L,LP,e,ta],thxy,4);
xxy=thxy(7,1:8);

%B(t) = b0 +b1*y(t) +b2*y(t-h) +b3*P(t) +b4*y'(t-h) +b5*R(t) +b6*R(t-h) +b7*e(t) +b8*R'(t-h) +b9*L(t)
+b10*L(t-h) +b12*L'(t-h) +b13*ta(t) +b15*d(t) +b17*B(t-h)
temp=[B ones(size(y)),y,yD,P,yPD,R,RD,e,RPD,L,LD,LPD,ta,BD];
thb=arx(temp,[0,1 1 1 1 1 1 1 1 1 1 1 1 1 0 0 1 0 1 0 1 0 1 0 1 1 0 1]);
Bp=predict([G,ones(size(y)),y,yD,P,yPD,R,RD,e,RPD,L,LD,LPD,ta,BD],thb,4);
BB=thb(13,1:14);

% y(t) = A*K^a*L^(1-a) ==>ln(y(t)) = a*ln(K) + (1-a)*ln(L)
temp=[log(y) ones(size(y)),log(K),log(L)];
thly=arx(temp,[0,1 1 1 0 0 0]);
lyp =predict([log(y) ones(size(y)),log(K),log(L)],thly,4);
lyy =thly(2,1:3);

%D(t) = a(1-c)*y(t) -k0*K(t) +k13*K'(t) +L4*R(t) +L5*L(t) +L6*P(t) +v*y'(t)
temp=[D ones(size(y)),y,K,KP,R,L,P,yP];
thD=arx(temp,[0,1 1 1 1 1 1 1 1 0 0 0 0 0 0 0 0]);
Dp=predict([D,ones(size(y)),y,K,KP,R,L,P,yP],thD,4);
DD=thD(7,1:8);

%DP(t) = p0 +p1*Pfe(t) +p2*w(t) -p3*n(t) -p4*P(t) +p5*M'(t) +p6*ML(t)
temp=[DP ones(size(w)),Pfe,w,N,P,MP,ML];
thp =arx(temp,[0,1 1 1 1 1 1 1 1 0 0 0 0 0 0 0]);
dlp=predict([P,ones(size(w)),Pfe,w,N,P,MP,ML],thp,4);
dp=thp(6,1:7);


subplot (2,1,1),plot(YEAR,C,YEAR,Cp,'+')
title('US--PRIVATE CONSUMPTION--');
ylabel('C');
xlabel('YEAR');
subplot (2,1,2),plot(YEAR,X,YEAR,Xp,'+')
title('US--EXPORTS--');
ylabel('X');
xlabel('YEAR');
pause
subplot (2,1,1),plot(YEAR,I,YEAR,Ip,'+')
title('US--INVESTMENT--');
ylabel('I');
xlabel('YEAR');
subplot(2,1,2),plot(YEAR,G,YEAR,Gp,'+')
title('US--GOVERNMENT CONSUMPTION--');
ylabel('G');
xlabel('year');
pause
clg
subplot(2,1,1),plot(YEAR,M,YEAR,Lp,'+')
title('US--MONEY DEMAND--');
ylabel('M1');
xlabel('YEAR');
```

```
subplot(2,1,2),plot(YEAR,Z,YEAR,[Xp+Ip+Gp+Cp],'+')
title('US--AGGREGATE DEMAND--');
ylabel('Z');
xlabel('YEAR');
pause
%--------

subplot (2,1,1),plot(YEAR,DP,YEAR,dIp,'+')
title('US--CONSUMER PRICES--');
ylabel('P');
xlabel('YEAR');
subplot (2,1,2),plot(YEAR,y,YEAR,yp,'+')
title('US--INCOME--');
ylabel('GDP');
xlabel('YEAR');
pause
subplot (2,1,1),plot(YEAR,C,YEAR,cyp,'+')
title('US--INCOME/CONSUMPTION--');
ylabel('cy');
xlabel('YEAR');
subplot(2,1,2),plot(YEAR,I,YEAR,Iyp,'+')
title('US--INCOME/INVESTMENT--');
ylabel('Iyp');
xlabel('year');
pause
clg
subplot(2,1,1),plot(YEAR,G,YEAR,gyp,'+')
title('US---INCOME/GOVERNMENT--');
ylabel('gyp');
xlabel('YEAR');
subplot(2,1,2),plot(YEAR,X,YEAR,xyp,'+')
title('US--INCOME/EXPORT--');
ylabel('xyp');
xlabel('YEAR');
pause


%--------
subplot (2,1,1),plot(YEAR,B,YEAR,Bp,'+')
title('US--BALANCE OF PAYMENT--');
ylabel('B');
xlabel('YEAR');
subplot (2,1,2),plot(YEAR,logy,YEAR,lyp,'+')
title('US--log(income)--');
ylabel('lny');
xlabel('YEAR');
pause
subplot (2,1,1),plot(YEAR,D,YEAR,Dp,'+')
title('US--Increase in stock--');
ylabel('D');
xlabel('YEAR');

% The economic dynamics can be put in matrix form
% as follows: x'(t) - A-1*x'(t-h) = A0*x(t)+A1*x(t-h)+B*u(t

%x=[y R L K P E]';
```

```
%A-1 =[am11 -am14 am13 0 0 0; 0 am22 0 0 0 0; Lm03 0 -Lm01 0 0 0; a3 0 a6 -am1 0 0; 0 -M7*p6*P(t) 0 0 0
0; b4 b8 b12 0 0 0]

%A0  =[a01 a12 a14 a16 a18 0; a21 a23 0 0 a25 0; 0 0 a33 0 a35 0; 0 0 0 a44 a45 0; a51*p(t) a52*p(t) 0 0
a55*p(t) 0; a61 a62 a63 0 0 0]

%A1  =[a11 a13 a15 0 0 0; a121 a122 0 0 0 0; a131 a132 a133 0 0 0; a141 a142 a143 0 0 0; a151*p(t)
a152*p(t) 0 0 0 0; a161 a162 a163 0 0 a166]

%B1  =[-ss1 ss1 ss1z15 ss1z18 ss1z17 0 0 0; 0 0 0 0 0 -L2 0 0; -mwzs14 mw mwzs15 mwzs16 0 mwzs13 0 0;-
zs14 1 zs15 zs16 0 zs13 0 0; 0 0 p1*pf(t) 0 0 -p6 p5 0; 0 0 b7 b8 b15 0 0 -1]

%B2  =[ss1 ss1 ss1 ss1(I13+c7) 0 0 0 0 0; 0 0 0 -L2 0 0 0 0 0;0 mw 0 0 0 0 mw mw 0; 0 1 0 0 0 0 1 1 0; 0 0 0 -
p6 -p3 p2 0 0 1;0 0 0 -1 0 0 1 0 0]

%u = [q S];

%q  =[T1 g0 e ta d M1 M1' f0]';
%q  =[T1 -9.6126 e ta d M1 M1' -347.6812]';

%S =[C0 I0 x0 M0 n w x0 y10 p0]';
%S =[-0.0002 -211.6960 -291.5135 -1.5231 n w -291.5135 -0.1225 -55.5034]';

% B=[B1 B2]
% AB=A0*B
% A2B=(A0*A0)*B
% A3B=(A0*A0*A0)*B
% A4B=(A0*A0*A0*A0)*B
% A5B=(A0*A0*A0*A0*A0*A0)*B

% G=[B AB A2B A3B A4B A5B]

% rank(B,0)= 6
% rank(G,0)= 6

% For spectral controllability

% Deltaj=eye(size(A1)).*j -Am1.*(j*2.7814^(-j)) - A0
%        - A1.*(2.7814^(-j))
% a=[Deltaj B]

% for j =0.4312, 4, 10, 23, 100, ...
% rank(a)=6

% For nonzeroeigenvalues of A-1 to be controllable via matrix B,

% cc=eye(size(Am1)).*j -Am1
% c=[cc B]

% for j =0.4312, 4, 10, 23, 100, ...
% rank(c)=6

% Since rank(B) is 6, where rank(a) and rank(c) are
% also 6, the system is function space controllable

L1 =1; L2 =0.9; L5 =DD(5)
L6 =DD(6): h =1
```

```
% for y(t) = A*K^a*L^(1-a)
a= 0.9 %    a =alpha
aa=0.1  %    aa=(1-a)=beta
W=sum(w)/27
mw   =(aa/W).^(1/a)
% cobb Douglas states: aa+a=1 =>constant return to scale.
                     % aa+a>1 =>increasing return to scale.
                     % aa+a=1 =>decreasing return to scale.
% Increasing return to scale if doubling L&K more than doubles y.
% constant return to scale if doubling L&K exactly doubles y.
% decreasing return to scale if doubling L&K less y.
%
% In cobb.m program log(A) is found to be 2 to give the best
% aproximation for y(t).


z0=GG(1)+II(1)+MM(1)*(CC(8)+II(11))+CC(1)+XX(1)
z1=GG(2)+II(2)+MM(2)*(CC(8)+II(11))+CC(2)+XX(2)
z2=GG(3)+II(3)+CC(3)+XX(3)+MM(3)*(CC(8)+II(11))
z3=GG(4)-II(4)+CC(4)+XX(4)
z4=GG(5)+II(5)+CC(5)+XX(5)
z5=GG(6)+II(6)+CC(6)+XX(6)+MM(4)*(CC(8)+II(11))
z6=II(7)+CC(7)+(CC(8)+II(11))*MM(5)
z8=GG(7)+II(8)+XX(7)
z9=II(9)
z10=XX(8)
z11=-II(10)
z13=(II(11)+CC(8))*MM(6)
z14=-CC(2)
z15=XX(11)
z16=XX(10)
z17=0
z18=XX(9)+(CC(8)+z13)*MM(7)
z19=-CC(3)
z20=CC(4)
z21=CC(5)
% z22=-(CC(8)+II(11))*M1


zs0 = ccy(1)+IIy(1)+ggy(1)+xxy(1); %constant term
zs1 = ccy(2)+ggy(2)+xxy(2)     ; %y(t) term
zs2 =        xxy(3)        ; %y(t-h) term
zs4 = ccy(3)+ggy(3)            ; %y'(t) term
zs5 = ccy(4)+xxy(4)+ggy(4)          ; %R(t)  term
zs8 = xxy(5)+ggy(5)          ; %L(t) term
zs10 =xxy(6)                     ; %L'(t) term
zs13 =ccy(5)               ; %M1(t) term
zs14 =ccy(6)                    ; %T(t) term
zs15 =xxy(7)                    ; %e(t) term
zs16 =xxy(8)                    ; %ta(t) term


s1 =1-L1*z3


% assumption
yy1 = zs0;
yy2 = zs1;
yy3 = zs2;
yy4 = zs4;
```

```
yy5 = zs5;
yy6 = zs8;
yy7 = zs10;
yy8 = zs13;
yy9 = zs14;
yy10 =zs15;
yy11 =zs16;

zs0= GG(1)+XX(1)+CC(1)+II(1)
a0 = DD(1)/(h*(1-yy(2)))
a1 = -DD(1)*yy3/((1-yy2)*h)
a2 = -a1*h
a3 = a0*h*yy4
a4 = DD(1)*yy5/((1-yy2) +DD(4))
a5 = DD(1)*yy6/(1-yy2) +DD(5)
a6 = a0*h*yy7
a8 = DD(6)

%A-1 starts here; am11 = a-11

am11 =L1*z4/s1
am22 =L2*MM(6)
am13 =L1*z10/s1
am14 =L1*z10/s1
Lm03 = DD(1)*yy4*mw/(1-yy2)
Lm01 = DD(3)*DD(1)/(1-yy2) -mw*DD(1)*yy7/(1-yy2)
a3   = DD(1)*yy4/(1-yy2)
a6   = DD(1)*yy7/(1-yy2)
am1  = DD(3)*DD(1)/(1-yy2)
M7p6P = MM(7)*dp(6)*sum(P)/27
b4 = -BB(5)
b8 = -BB(9)
b12= -BB(12)

%A0 starts here!

a01 =L1*(z1-1)/s1
a11 =L1*z2/s1
a12 =L1*z5/s1
a13 =L1*z6/s1
a14 =L1*z8/s1
a15 =L1*z9/s1
a16 =L1*z11/s1
a17 =L1*z19/s1
a18 =L1*z18/s1
a21 =L2*MM(2)
a23 =L2*MM(4)
a25 =L2*MM(7)
a33 =a0;
a35 =mw*a8
a44 =a0
a45 =a8
a51=-dp(7)*MM(2)*sum(P)/27    % includes P(t)
a52=-dp(7)*MM(4)*sum(P)/27    % includes P(t)
a55 =-(MM(7)+dp(5))*sum(P)/27 % includes P(t)
a61 =BB(2)
a62 =BB(6)
```

```
a63  =BB(10)

%A1 starts here!
a11 =L1*z2/s1
a13 =L1*z6/s1
a15 =L1*z9/s1
a22 =L2*MM(3)
a24 =L2*MM(5)
a1 = -DD(1)*yy3/((1-yy2)*h)
a2 = -a1*h
l1 = -a1
l2  =-mw*DD(1)*yy3/(1-yy2)
l4  =mw*DD(1)*yy5/(1-yy2 +DD(4))
b2  =BB(3)
b6  =BB(7)
b10 =BB(11)
b17 =BB(13)
a111 =L1*(GG(3)+II(3)+XX(3)-MM(3)*(II(11)+CC(8)))/s1
a112 =a13
a113 =a15
a121=a22
a122=a24
a131=mw*DD(1)*XX(3)/(1-(XX(2)+ggy(2)+ccy(2)))
a132=mw*DD(1)*yy5/(1-yy2) -DD(4)
a133=a5-l1*h
a141 =DD(1)*yy3/(1-yy2)
a142 =DD(1)*yy5/(1-yy2) + DD(4)
a143 =DD(1)*yy6/(1-yy2) + DD(5)
a151 =-dp(7)*sum(P)/27      %includes P(t)
a152 =-MM(5)*dp(7)*sum(P)/27   %includes P(t)
a161 =BB(3)
a162 =BB(7)
a163 =BB(11)
a166 =BB(14)

% B1 starts here!

ss1 =L1/s1 % ss1=b1
ss1z13 =ss1*z13
ss1z15 =ss1*z15
ss1z16 =ss1*z16
ss1z17 =0
ss1z18 =ss1*z18
b1z14=-BB(2)*z14;
b1 =BB(1)
b1z15=BB(2)*z15
b1z16=BB(2)*z16
b1z17=BB(2)*z17
b1z13=BB(2)*z13
mwzs14 =mw*yy9
mwzs15 =mw*yy10
mwzs16 =mw*yy11
mwzs17 =0
mwzs13 =mw*yy8

p1pf =dp(2)*sum(Pfc)/27     %dp(2)*Pf(t)
p4 =dp(5)
```

```
p5     =dp(6)
b7     =BB(8)
b8     =BB(9)
b15    =0

%B2 starts here!

ss1    =L1/s1
ss1i13c7 =ss1*(II(11)+CC(8))
l2     =mw*DD(1)*yy3/(1-yy2)
p6     = dp(7)
p3     = dp(4)
p2     = dp(3)

a17    =L1*z19/s1
a45    =a8
c0     = CC(1)
I0     = II(1)
x0     = XX(1)
p0     = dp(1)
M0     = MM(1)
y10    = ccy(1)
m0     =MM(1)
g0     =GG(1)
f0     =BB(1)-XX(1)
g2     =-L2*MM(1)


%----------------------------------------------------------------------%
```

The following data were the output results for the
program us2.m

```
L1=0.1870, L2= 1.4462e-004, L5 = 0.0525,h = 1,a = 0.9000,aa = 0.1000
W = 63.1704,mw = 7.7325e-004,z0 = -512.6880,z1 = -0.7519,z2 = 0.6830,z3 = 0.9443,z4 = -0.2400,z5 = -
15.4108
z6 = 10.3291,z8 = 12.6083,z9 = -8.8065,z10 = 45.3681,z11 = 0.3089,z13 = 0.0192
z14 = -1.4798e-008,z15 = 7.8295,z16 = -0.0429,z17 = 0,z18 =15.0963,z19 = -6.9638e-009
z20 = 6.5288e-009,z21 = 6.5699e-009,s1 = 0.0557,zs0 = -512.8223,a0 = 1.5554,a1 = -0.1870
a2 = 0.1870,a3 = -0.0022,a4 = 20.9668,a5 = -16.1365,a6 = 10.1608
a8 = -0.0243,am11 = -4.3105,am22 = -0.1961,am13 = 814.8353,am14 = 814.8353
Lm03 = -1.5478e-006,Lm01 = -0.0047,a3 = -0.0020,a6 = 9.4257,am1 = 0.0026
M7p6P = 0.2685,b4 = 0.1990,b8 = -5.9898,b12 = 4.9703,a01 = -31.4655
a11 = 12.2672,a12 = -276.7857,a13 = 185.5172,a14 = 226.4519,a15 = -158.1694
a16 = 5.5489,a17 = -1.2507e-007,a18 = 271.1380,a21 = -0.0081,a23 = 0.2186
a25 = 0.1076,a35 = -1.8772e-005,a44 = 1.5554,a45 = -0.0243,a51 = 0.0081
a52 = -0.2195,a55 = 132.4471,a61 = -0.0705,a62 = -2.4137,a63 = 34.3642
a11 = 12.2672,a13 = 185.5172,a15 = -158.1694,a22 = 0.0071,a24 = -0.1522
a1 = -0.1870,a2 = 0.1870,l1 = 0.1870,l2 = -1.4462e-004,l4 = 0.0162,b2 = 0.3164
b6 = 32.1284,b10 = -20.5989 b17 = -0.0267,a111 = 12.2672,a112 = 185.5172
a113 = -158.1694,a121 = 0.0071,a122 = -0.1522,a131 = 4.1324e-004
a132 = 0.0115,a133 = -16.3235,a141 = 0.1870,a142 = 21.0641
a143 = -16.1365,a151 = -0.9037,a152 = 0.1528,a161 = 0.3164
a162 = 32.1284,a163 = -20.5989,a166 = -0.2439,ss1 = 17.9606
ss1z13 = 0.3449,ss1z15 = 140.6221,ss1z16 = -0.7698,ss1z17 = 0
ss1z18 = 271.1380,b1 = -639.1947,b1z15 = -0.5518
b1z16 = 0.0030,b1z17 = 0,b1z13 = -0.0014,mwzs14 = 5.4905e-009,mwzs15 = 0.0956
mwzs16 = 1.0145e-005,mwzs17 = 0,mwzs13 = 4.7587e-007,p1pf = 7.6410
```

p4 = -2.3281,p5 = 0.0374,b7 = -154.5053,b8 = 5.9898
b15 = 0,ss1 = 17.9606,ss1i13c7 = -1.5833,l2 = 1.4462e-004
p6 = 0.0151,p3 = 642.1701,p2 = 2.5444,a17 = -1.2507e-007,a45 = -0.0243
c0 = -1.6295e-004,I0 = -211.6960,x0 = -291.5135,p0 = -55.5034,M0 = -1.5231
y10 =-0.1225,m0 = -1.5231 g0 = -9.6126,f0 =-347.6812,g2 = 1.3708


The matrices for economic dynamics are:

Am1 = [-4.3105 -814.8353  814.8353    0    0    0
        0  -0.1961      0      0    0    0
     0.0000   0       0.0047    0    0    0
    -0.0020   0       9.4257 -0.0026  0    0
        0  -0.2685      0      0    0    0
     0.1990  5.9898     4.9703    0    0    0]


A0 =
 [-31.4655 -276.7857  226.4519  5.5489  271.1380  0
  -0.0081   0.2186     0      0    0.1076   0
    0       0       1.5554    0    0.0000   0
    0       0        0     1.5554 -0.0243   0
  0.0081  -0.2195      0      0  132.4471   0
 -0.0705  -2.4137    34.3642    0      0    0]


A1 =
 [12.2672  185.5172 -158.1694   0   0    0
  0.0071  -0.1522      0      0   0    0
  0.0004   0.0115  -16.3235     0   0    0
  0.1870  21.0641  -16.1365     0   0    0
 -0.9037   0.1528      0      0   0    0
  0.3164  32.1284  -20.5989     0   0  -0.2439]


B1 =
 [-17.9606 17.9606 140.6221 271.1380 0    0        0        0
    0      0       0       0    0  -0.9000   0        0
  0.0000 0.0008    0.0956  0.0000  0  0.0000   0        0
  0.0000 1.0000  123.6930  0.0131  0  0.0006   0        0
    0    0        7.6410    0    0 -0.0151 0.0374    0
    0    0     -154.5053  5.9898  0    0        0      -1.0]


B2 =
 [17.9606 17.9606 17.9606 -1.5833     0       0        0      0    0
    0      0      0     -0.9000     0       0        0      0    0
    0    0.0008   0      0          0       0      0.0008 0.0008  0
    0    1.0000   0      0          0       0      1.0000 1.0     0
    0      0      0     -0.0151  -642.1701 2.5444    0      0    1.0
    0      0      0     -1.0000     0    0  1.0000     0            0]


q  =[T1 -9.6126 e ta d M1 M1' -347.6812]';
S =[-0.0002 -211.6960 -291.5135 -1.5231 n w -291.5135 -0.1225 -55.5034]';


% Here we experiment with the values given by cobb
% Douglas(alpha=0.9 and beta=0.1) to obtain the best

```
% possible constant term A in the equation: y =A*K^a*L^(1-a).
%
% In the final analysis for United States data log(A) = 2.

logk =log(K)*(.9);
logL =log(L)*(.1);
logy =log(y);

 llpy= 2 + logL + logk;

subplot (2,1,1),plot(YEAR,logy,YEAR,llpy,'+')
title('cobb douglas on United States--log(income)--');
ylabel('llpy');
xlabel('validity of the use of cobb douglass alpha and beta');
```

The polynomial coefficients and their standard deviations
for US3.m are:

```
% C(t) =c0 +c1*YT(t) +c2*YT(t-h) + c3*YT(t)' +c4*YT'(t-h) +c5*R(t) + c6*R(t-h) +c7*(17.2335
+0.8380.*y(t) -0.3061.*yD(t-h) -0.3175.*R(t) +3.6165.*RD(t)-3.3413.*P(t) -1.7714.*RPD(t)) - M(t));

c0=-25.2258, c1=-0.0017, c2 =0.0002736, c3 =-0.0001802, c4 =-0.0001135, c5 =0.6036, c6= 3.9353, c7=-
0.4041
    16.5075     0.0004     0.1777          0.0808          0.1668          2.6985          4.1023     0.2254


% I(t) =i0 +i1*y(t) + i2*y(t-h) -i3*y'(t) + i4*y'(t-h) +i5*R(t) +i6*R(t-h) +i8*L(t) +i9*L(t-h)-i11*K(t) -i13*ML(t)
i0 =-53.5756, i1 =-0.0326, i2= 0.0477, i3= 0.2187, i4 =-0.1292, i5 = -2.6440, i6 =5.7933,
    95.7551          0.06310.0615     0.0482          0.0465     1.2571          3.2116


i8 = 7.5730, i9= -7.1687, i11 = 0.8737, i13 =0.0439
    3.8055          2.8628 0.1146          0.0800


% X(t) =x0 + x1*y(t) + x2*y(t-h) +x3*y'(t) +x4*y'(t-h) +x5*R(t)+ x8*L(t) +x10*L'(t-h) +x12*P(t) +x16*ta(t)
+x15*e(t) +x17*d(t)
x0 =-89.1936, x1= 0.6169, X2= -0.1565, x3= -0.3130, x4= -0.1824, x5 = 19.6237, x8= -2.3689, x10= -16.9896
    298.4460     0.2162     0.1735          0.1939          0.1311     4.4968          6.5091
6.7356


x12 =-8.7879, x16 =-0.0226, x15= 122.4798, x17 = 0
    3.3356          0.0142          98.0708


% G(t) =g0 + g1*y(t) - g2*y(t-h) +g3*y'(t) +g4*y'(t-h) +g5*R(t) +g8*L(t);
g0= 675.2828, g1 = 0.0436, g2 =0.0347, g3 =-0.1267, g4=-0.0527, g5=-3.9923, x6 =-0.9466
    181.3267          0.1076     0.1028          0.1143          0.1084     3.3706          3.2503


% L(t) =m0 +m1*y(t)+ m2*y(t-h) +m3*R(t) +m4*R(t-h) +m5*R'(t-h) +m6*p(t);
m0 =108.1063, m1 = -0.0195, m2 =0.1594, m3 =-3.3462, m4 =-7.2497, m5 =-0.7915, x6 = 1.9784
    35.1699          0.1250          0.0890     3.6606          6.7577     3.9036          4.9497
%-----------------------------------------------------------------------
% The economic dynamics can be put in matrix form
% as follows: x'(t) - A-1*x'(t-h) = A0*x(t)+A1*x(t-h)+B*u(t)

%x=[y R L K P E]';
%u = [q S];
```

# Index

# Series on Advances in Mathematics for Applied Sciences


## Editorial Board

**N. Bellomo**
*Editor-in-Charge*
Department of Mathematics
Politecnico di Torino
Corso Duca degli Abruzzi 24
10129 Torino
Italy
E-mail: bellomo@polito.it

**F. Brezzi**
*Editor-in-Charge*
Istituto di Analisi Numerica del CNR
Via Abbiategrasso 209
I-27100 Pavia
Italy
E-mail: brezzi@dragon.ian.pv.cnr.it

**M. A. J. Chaplain**
Department of Mathematics
University of Dundee
Dundee DD1 4HN
Scotland

**C. M. Dafermos**
Lefschetz Center for Dynamical Systems
Brown University
Providence, RI 02912
USA

**S. Kawashima**
Department of Applied Sciences
Engineering Faculty
Kyushu University 36
Fukuoka 812
Japan

**M. Lachowicz**
Department of Mathematics
University of Warsaw
Ul. Banacha 2
PL-02097 Warsaw
Poland

**S. Lenhart**
Mathematics Department
University of Tennessee
Knoxville, TN 37996–1300
USA

**P. L. Lions**
University Paris XI-Dauphine
Place du Marechal de Lattre de Tassigny
Paris Cedex 16
France

**B. Perthame**
Laboratoire d'Analyse Numerique
University Paris VI
tour 55–65, 5ieme etage
4, place Jussieu
75252 Paris Cedex 5
France

**K. R. Rajagopal**
Department of Mechanical Engrg.
Texas A&M University
College Station, TX 77843-3123
USA

**R. Russo**
Dipartimento di Matematica
Università degli Studi Napoli II
81100 Caserta
Italy

**V. A. Solonnikov**
Institute of Academy of Sciences
St. Petersburg Branch of V. A. Steklov
Mathematical
Fontanka 27
St. Petersburg
Russia

**J. C. Willems**
Mathematics & Physics Faculty
University of Groningen
P. O. Box 800
9700 Av. Groningen
The Netherlands

# Series on Advances in Mathematics for Applied Sciences

## Aims and Scope

This Series reports on new developments in mathematical research relating to methods, qualitative and  numerical analysis, mathematical modeling in the applied and the technological sciences. Contributions related to constitutive theories, fluid dynamics, kinetic and transport theories, solid mechanics, system theory and mathematical methods for the applications are welcomed.

This Series includes books, lecture notes, proceedings, collections of research papers. Monograph collections on specialized topics of current interest are particularly encouraged. Both the proceedings and monograph collections will generally be edited by a Guest editor.

High quality, novelty of the content and potential for the applications to modern problems in applied science will be the guidelines for the selection of the content of this series.

## Instructions for Authors

Submission of proposals should be addressed to the editors-in-charge or to any member of the editorial board. In the latter, the authors should also notify the proposal to one of the editors-in-charge. Acceptance of books and lecture notes will generally be based on the description of the general content and scope of the book or lecture notes as well as on sample of the parts judged to be more significantly by the authors.

Acceptance of proceedings will be based on relevance of the topics and of the lecturers contributing to the volume.

Acceptance of monograph collections will be based on relevance of the subject and of the authors contributing to the volume.

Authors are urged, in order to avoid re-typing, not to begin the final preparation of the text until they received the publisher's guidelines. They will receive from World Scientific the instructions for preparing camera-ready manuscript.

SERIES ON ADVANCES IN MATHEMATICS FOR APPLIED SCIENCES

SERIES ON ADVANCES IN MATHEMATICS FOR APPLIED SCIENCES